# Practical VCR Repair

David T. Ronan

# Practical VCR Repair

# David T. Ronan

**Delmar Publishers, Inc.**
I(T)P™ An International Thomson Publishing Company

New York • London • Bonn • Detroit • Madrid • Melbourne • Mexico City • Paris
Singapore • Tokyo • Toronto • Washington • Albany NY • Belmont CA • Cincinnati OH

## NOTICE TO THE READER

Publisher does not warrant or guarantee any of the products described herein or perform any independent analysis in connection with any of the product information contained herein. Publisher does not assume, and expressly disclaims, any obligation to obtain and include information other than that provided to it by the manufacturer.

The reader is expressly warned to consider and adopt all safety precautions that might be indicated by the activities herein and to avoid all potential hazards. By following the instructions contained herein, the reader willingly assumes all risks in connection with such instructions.

The publisher makes no representation or warranties of any kind, including but not limited to, the warranties of fitness for particular purpose or merchantability, nor are any such representations implied with respect to the material set forth herein, and the publisher takes no responsibility with respect to such material. The publisher shall not be liable for any special, consequential, or exemplary damages resulting, in whole or part, from the readers' use of, or reliance upon, this material.

Cover photo courtesy of Neil Moore, photographer
Cover Design: Megan K. DeSantis

**Delmar Staff**
Publisher: Michael McDermott
Administrative Editor: Wendy J. Welch
Production Editor: Barbara Riedell
Production Supervisor: Larry Main
Art/Design Coordinator: Lisa Bower

COPYRIGHT © 1995
By Delmar Publishers Inc.
an International Thomson Publishing Company

 The ITP logo is a trademark under license

Printed in the United States of America

For more information contact:

Delmar Publishers, Inc.
3 Columbia Circle, Box 15015
Albany, New York 12212-5015

International Thomson Publishing
Berkshire House 168–173
High Holborn
London, WC1V7AA
England

Thomas Nelson Australia
102 Dodds Street
South Melbourne 3205
Victoria, Australia

Nelson Canada
1120 Birchmont Road
Scarsborough, Ontario M1K5G4
Canada

International Thomson Editores
Campos Eliseos 385, Piso 7
Col Polanco
115460 Mecixo D F Mexico

International Thomson Publishing GmbH
Königswinterer Strasse 418
53227 Bonn
Germany

International Thomson Publishing Asia
221 Henderson
#05–10 Henderson Building
Singapore 0315

International Thomson Publishing Japan
Hirakawacho Kyowa Building, 3F
2-2-1 Hirakawacho
Chiyoda-ku, Tokyo 102
Japan

All rights reserved. No part of this work covered by the copyright hereon may be reproduced or used in any form or by any means—graphic, electronic, or mechanical, including photocopying, recording, taping, or information storage and retrieval systems—without the written permission of the publisher.

1 2 3 4 5 6 7 8 9 10 XXX 01 00 99 98 97 96 95 94

**Library of Congress Cataloging-in-Publication Data**

Ronan, David T.
    Practical VCR repair / David. T. Ronan.
       p.    cm.
    Includes index.
    ISBN 0-8273-6583-7
    1. Videocassette recorders--Repairing.   I. Title.
TK6655.V5R66   1994                                 94-20858
621.288'337--dc20                                            CIP

# CONTENTS

Dedication x
Preface xi
Acknowledgments xii
Introduction xiii

## CHAPTER 1  VCR Operations and Controls — 1

What VCRs Can Do 2
Operator Controls 5
What Happens after a Tape Is Put into a VCR? 8
VHF, UHF, and Cable TV Signals 10
VCR Tuners 12
Baseband Signals 15
Hookups and Connections 17
Summary 19

## CHAPTER 2  Removing Covers, Getting Started — 21

Doing it Right! 22
Safety—for You and the VCR 24
ESD Precautions 27
VCR Cover Removal 28
Shield Plates 34
Front Panel Removal 35
Releasing Circuit Boards 38
Major Components and PCBs 45
Summary 47

## CHAPTER 3  Understanding the Videotape Path — 50

Introduction to Tape Loading 50
Supply Spindle and Tape Holdback Tension Band 52
Back Tension Guide Pole 54
Full Erase Head 56
Inertia Roller 57
Drum Entrance and Exit Roller Guides P-2 and P-3 58
Video Drum 61
Audio/Control Head Assembly 64
Capstan and Pinch Roller 65
Take-up Spindle 68
Servicing the Tape Transport 71
Tape Transport Switches and Sensors 71
Dew Sensor 80
Summary 82

## CHAPTER 4  Video Cassette Examination and Repair — 84

Outside Parts and Features 84
How to Take a Videocassette Apart 88
Inside Parts 89

Videocassette Repair 92
Inspecting for Tape Damage 97
Making a Videocassette Test Jig 100
Summary 102

## CHAPTER 5  Troubleshooting Loader and Transport Malfunctions 104

Systematic Approach to Diagnosing Problems 105
Cassette Load and Unload Problems 107
Tape Load and Unload Problems 117
Tape Motion Problems 124
Summary 130

## CHAPTER 6  How to Perform VCR Maintenance and Common Repairs 133

Tools and Supplies 133
Cleaning, Lubrication, and Inspection 144
Rotary Heads 149
Stationary Tape Heads 154
Capstan and Pinch Roller 155
Belts, Idlers, and Pulleys 159
Cams and Gears 160
Switches, Connectors, and Sensors 164
Summary 170

## CHAPTER 7  How to Align Tape Path and Make Adjustments 173

Tape Tension 174
Supply-Side Guide Pole 178
Roller Guide Posts 180
V-Mounts 184
A/C Head 184
Take-up Side Guide Pole 190
Pinch Roller 191
Supply and Take-up Spindles and Brakes 192
Mode Switch 194
Summary 196

## CHAPTER 8  Understanding Basic Electronics 200

Introduction to Electronics 201
Electron Flow 201
Electrical Units of Measure 206
Ohm's Law 209
Series and Parallel Circuits 224
Alternating and Direct Current 228
Analog and Digital Circuits 234
Summary 237

## CHAPTER 9  How to Use a Multimeter 240

Introduction to Multimeters 240
How to Take Voltage Measurements 243
How to Take Current Measurements 247
How to Take Resistance Measurements 249
Practice Exercises 256
Selecting a Multimeter 259
Summary 267

## CHAPTER 10 Electronic Components — 269

Wire 270
Fuses 272
Switches 273
Relays 276
Resistors 279
Diodes 286
Transistors 295
Capacitors 304
Coils and Transformers 313
Summary 319

## CHAPTER 11 How to Solder — 323

Soldering Tools and Supplies 324
Successful Soldering Techniques 325
Insulating Connections 331
Unsoldering 331
Build a Two-Way Tester 334
Using the Two-Way Tester 340
Summary 341

## CHAPTER 12 VCR Power Supplies — 343

Power Supply Overview 344
Full-Wave Rectification 345
Rectifier and Filter Failures 349
Conventional Power Supply Schematic 363
Switching Mode Power Supply 367
Switching Mode Power Supply Safety 370
Switching Mode Power Supply Schematic 371
Summary 373

## CHAPTER 13 Checking Motors, Optical Sensors, and Remotes — 375

Independently Powering a Two-Terminal DC Motor 375
Testing Two-Terminal DC Motors 377
Checking Optical Sensors 382
Remote Control Problems 385
Summary 392

## CHAPTER 14 VCR Microprocessors and Servos — 395

Integrated Circuits and Pin Identification 395
Digital Logic Levels 397
VCR Microprocessor Integrated Circuits 398
Troubleshooting Digital Circuits 404
Digital Logic Probe 407
VCR Servo Circuits 409
Summary 414

## CHAPTER 15 How a TV Picture Is Made — 417

Producing a Black-and-White TV Image 417
Basic CRT Operation 418

Video Fields and Frames   419
Producing a Color TV Image   421
Summary   425

## CHAPTER 16  Recording on Videotape   427

Basic Videotape Recording   428
VHS Format   434
Record Circuitry Block Diagram   434
Playback Circuitry Block Diagram   437
Summary   439

## CHAPTER 17  Beyond Standard VHS   441

Hi-Fi Stereo VCRs   442
VHS High Quality (HQ)   448
Super VHS   449
Flying Erase Head   451
Other VCR Features   452
Summary   456

## CHAPTER 18  Using Manufacturers' Service Manuals   460

So Many Models!   460
Obtaining Service Manuals and Parts   462
Basic Service Manual Contents   463
Supplementary Service Manual Contents   473
Summary   477

## CHAPTER 19  Common Audio and Video Problems   478

General Diagnostic and Repair Tips   478
Linear Audio Failure Symptoms   481
Video Failure Symptoms   483
Hi-Fi Stereo Failure Symptoms   486
Summary   487

## CHAPTER 20  Service Considerations: The Business Side of VCR Repairs   489

You and Your Customers   489
Receiving a Unit for Repair   492
Unit Replacements   494
Realistic Expectations   495
Beyond Economical Repair   497
Preventing Service Returns   498
Performing an AC Current Leakage Test   501
Summary   506
Conclusion   507

# APPENDICES

## CHAPTER 1 Appendix   509

Front Panel Operator Controls   509
Tuning VCR Channel Presets   514

Rear Panel Features 514
TV and Cable Channels 517
Basic Antenna-to-VCR Connections 519
Basic VCR-to-TV Connections 521
Cable System Connections 521
Selecting VCR Input Source 524
Line Jacks Versus RF Jacks 525
Cable Channel Numbers and Names 527

# CHAPTER 3 Appendix 529

How Tape Load Shuttles Extend and Retract 529
Video Heads 531
Video Drum 531
A/C Head Assembly 543
Capstan and Pinch Roller 536

# CHAPTER 8 Appendix 537

Impedance: Supplementary Information 537

# CHAPTER 10 Appendix 539

Identification of Mica and Tubular Capacitors 539
Color Codes for Ceramic Capacitors 540
Dipped Tantalum Capacitors 541
Film-type Capacitors 541

# APPENDIX M 542

Metric System and Conversions 542

Glossary 545
Index 561

## DEDICATION

To loving parents Ruth and Jack, step parents Ann and Fred, and others with whom I have lived, expressly Rick Edenfield, all of whom encouraged me and valiantly put up with my techie side.

# PREFACE

One look in the technology section of a well-stocked bookstore or library usually reveals several books on the subject of VCR repair. Many of these are excellent resources, so why write yet one more? Actually several reasons prompted the writing of this book on maintaining and repairing VCRs.

Probably the strongest incentive for creating this book was a need for a single, comprehensive textbook that could be used in both home study and classroom programs. Existing books seem to fall into two general categories, neither of which is terribly well suited for a formal VCR study program, such as might be offered at a community college or technical school or through a home study course. Some are just too simple, others too complex.

The first category of repair books is aimed at the VCR owner who might be inclined to maintain his or her own machine and repair common problems. To many, this is an attractive alternative to taking the unit to a repair shop, and paying perhaps as much as $125 for simple procedures the do-it-yourselfer could accomplish with a $25 outlay for a few hand tools and supplies. Books of this type generally concentrate on cleaning and lubricating items in the tape transport, and also often include adequate instructions for replacing parts that wear out, like rubber belts. To be sure, this *is* effective in curing a large number of VCR problems.

Although these books certainly have their place, they usually do not include any theory about how VCRs work, nor how to repair all but the simplest electrical or electronic problems. Some don't go much beyond fuse replacement. A comprehensive VCR training program requires *some* technical discussion about *how* VCRs work. Basic electronic knowledge and troubleshooting skills are also needed for a person to be considered professionally trained for VCR maintenance and repair.

Books in the second category are just the opposite: they're way *too* technical. Most are written for the experienced radio and television bench technician. They assume the reader already knows how TVs work and how to use advanced electronic testing equipment, including oscilloscopes, video signal generators, and frequency counters. A few books on the technical side delve very deeply into the operation of VCR electronic circuits and include discussions on why a particular VCR format or circuit was designed the way it was. Though this is often interesting to the highly trained reader, it is much more complex than a student of VCR repair really needs.

This textbook, therefore, bridges the gap between overly simple VCR repair guides aimed at the average do-it-yourselfer and highly technical manuals for electronics technicians who already maintain video and other electronic equipment. Practical information on how to diagnose and repair an ailing VCR is presented. Most VCR breakdowns are mechanical in nature, and can be repaired with fairly uncomplicated cleaning, alignment, and parts replacement procedures. A significant portion of this book describes these procedures, including what tools and supplies to use.

VCR theory and basic electronics skills to go beyond simple mechanical repairs are covered within this book, including: (1) how to use a multimeter, (2) how to test electronic components, and (3) how to solder. These skills are taught assuming that students have absolutely no prior knowledge of electronics or experience with a soldering iron.

Having studied this book, a student should be able to troubleshoot VCR problems and perform the vast majority of maintenance procedures, alignments, and repairs without the need to purchase or know how to use expensive test equipment. Armed with the knowledge, skills, and abilities presented in this textbook, a person should be capable of repairing about 95% of all VCR defects. This text is also an excellent springboard for those who want to learn advanced electronic diagnostic techniques to repair the few remaining VCR problems requiring test equipment beyond a handheld multimeter.

## ACKNOWLEDGMENTS

Many thanks and much appreciation to the following for assistance and support in the production of this book, especially Mr. Freedman who has been a constant source of inspiration, encouragement, and counsel throughout.

Karl Freedman
Richard Kruger
Dr. Milton Miller
Dan Payne
Lawrence Sandress
Penny Blackman
Larry Schnabel
Sencore
JCPenney
Simpson Electric
Protek
H. W. Sams & Co.
Tentel Corp.
Mondo-Tronics Inc.
Delmar Publishers staff
Neil Moore, Photographer

# INTRODUCTION

By the 1990s, the video cassette recorder (VCR) had become a well-entrenched part of the array of electronic equipment providing entertainment and education in many households. Next to the radio and TV set, probably no other consumer electronic product has as rapid acceptance in the home as the VCR. We can watch uninterrupted full-length movies, improve our golf swing or bowling score, hold an aerobics workout, or relive a family celebration any time we wish, thanks to the VCR. The popularity of consumer camcorders—combination video camera and VCR—has seen the VCR replace, to a large extent, the family photograph album or slide collection. And with a VCR we can better manage our busy schedules through time shifting, recording TV and cable programs for playback at a later, more convenient hour.

## A Brief History

Early video tape recorders (VTRs) for the television broadcast industry first came on the scene around 1953. These used an open reel-to-reel format, rather than having videotape enclosed in a cartridge or cassette. High-quality, very expensive broadcast VTRs still use the open-reel format on 1-inch-wide tape, although 3/4-inch and 1/2-inch cassette formats are also common today in the broadcast industry.

Although many improvements in videotape technology and formats steadily occurred, it wasn't until about 1976 that the first VCR made any significant impact in the consumer market. This was the *Betamax*, developed by Sony. The first Beta format produced exceptionally good quality video, but a limiting factor was the total recording time of about 1-1/2 hours. Later Beta formats, Beta II and Beta III, extended this time to 3 hours and 4-1/2 hours, respectively.

As Sony's Beta machines became popular, despite their recording time limits and high prices compared with VCRs of today, another Japanese company, JVC, introduced the *VHS format* VCR. **VHS** stands for *video home system*. Like Beta, VHS employs a 1/2-inch-wide magnetic tape enclosed in a plastic cartridge, slightly larger than a Beta cassette. VHS became more heavily promoted than Beta, and consumers were attracted to its longer recording time of six hours.

As with many consumer items, the marketplace would determine which format ultimately prevailed. During the 1980s VHS sales gained on, and surpassed, those of Beta VCRs. One look at your neighborhood video rental store will confirm the overwhelming popularity of VHS over Beta. Most outlets now do not offer any pre-recorded movies or programs in Beta. Continued improvements in VHS machines, such as Hi-Fi Stereo and Super VHS recording/playback techniques, have no doubt helped VHS become today's home video recording standard.

## Other Formats

Although VHS has surpassed Beta in home VCR popularity, two other formats have also gained ground: compact VHS, or *VHS-C*, and 8 millimeter (8mm), also called *Video 8*. These two formats have become especially popular for consumer camcorders, due to cassettes that are significantly smaller than a standard VHS cassette. Their size permits camcorders to be made smaller and lighter—much more convenient to use than the rather bulky VHS camcorders.

VHS-C cassettes are limited to a recording time of about 20 minutes at the standard play (SP) tape speed, but one significant advantage is that they can also be used in a standard VHS tabletop machine with a cassette adapter. A VHS-C cassette slips into the adapter, and the adapter is then inserted into the VCR as if it were a standard VHS cassette.

An 8mm cassette is about the same size as a standard audio cassette, only somewhat thicker. Up to two hours can be recorded. Tabletop 8mm VCRs are also available. This is an attractive alternative for some consumers, because they don't have to use the camcorder itself to play back home videos. As with Beta, though, commercial movies and programs are generally unavailable in either VHS-C or 8mm formats.

## About This Textbook

This textbook is written for students who may or may not have any prior electronics training or experience. Knowledge of basic electricity and electronics is a benefit, but persons lacking this should in no way be discouraged from forging ahead. Instruction is given on basic electronics and how to use a multimeter to measure voltage, resistance, and electric current. How to select a suitable handheld multimeter is included in the multimeter chapter. Students can also learn about electronic components, like diodes and transistors, commonly found in VCR circuits, how they operate, and how to test them with a multimeter.

However, it *is* assumed that students have a basic familiarity with common hand tools, like screwdrivers, nutdrivers, and pliers. A chapter on soldering is included for students without skills in soldering or unsoldering electronic components.

Successful, efficient repair of any device, whether an automobile, toaster, personal computer, or VCR, requires a *systematic approach to troubleshooting*. Diagnostic tips and techniques for making quality VCR repairs are a large part of the material in this book. In short, this puts theory into practice.

**The entirety of this book deals specifically with VHS VCRs.** Because of the great similarity in theory of operation, electronics, and mechanical systems between VHS and other formats, such as Beta and 8mm, students should easily be able to apply what they learn to these latter formats. The same is true for the VCR portion of camcorders.

There are some references along the way to Beta or 8mm, where the author feels it is relevant. By focusing on only the more popular VHS format, it is hoped that basic concepts are made easier to understand than if all formats were dealt with. Unless stated otherwise, the term VCR in this text should be taken to mean a standard, tabletop, consumer VHS VCR.

For the same reason—to keep things uncomplicated, especially in earlier chapters—the text concentrates on basic or "vanilla" VHS VCRs. Super VHS, Hi-Fi Stereo, High Quality (HQ) VHS, and other features, like slow motion, are covered separately in later chapters. A glance at the table of contents should give the browser and prospective reader a good idea of material covered in this textbook.

## So Many Models!

There are so many different manufacturers and models of VHS VCRs that it would be impractical, if not impossible, to discuss each and every one within this book. Although operation of all VCRs is fundamentally the same, different models accomplish the same end result in many ways. For example, a VCR *could* have anywhere from three to six separate electric motors for recording and handling videotape.

With this in mind, descriptions, figures, and explanations in this book include some of the more common, or popular, VCR mechanical and electronic configurations. Once these are understood, students should have little or no concern about working on a VCR that operates somewhat differently than those in the text.

A good automobile mechanic can work on a Buick as well as a Volvo or Ford for most automotive problems. For specific information, she or he can refer to manufacturer's service literature for a particular. It's the same with VCRs. A well-trained and experienced VCR technician will immediately perceive how to make many repairs on different VCRs without specific instructions. For some repairs, he or she will need to consult the service manual for that model.

## You and VCR Repair Opportunities

Most likely you have some interest in VCRs, or you wouldn't have read this far. Perhaps you're just curious about how they work. You might be considering maintaining your own VCR. This primary skill level can easily save you hundreds of dollars in repair shop costs. From there, you might branch out into part-time VCR repairs for friends and relatives—for which you certainly are entitled to some monetary compensation. This textbook will get you well on your way.

Other readers might consider starting a VCR repair shop of their own, working for an employer who has a VCR repair business, or adding VCR repair to an existing line of business. I know of one man who has operated a successful vacuum cleaner and sewing machine repair shop for many years. A few years ago he decided to add VCR repair to his services. Without extensive electronic training or expensive test equipment, the shop now does a substantial business in VCR repair.

After completing your study of this book, and some practical, hands-on experience with a few VCRs, you should be fairly well qualified to work as a technician in an existing VCR repair facility or shop. Some VCR repair students locate *apprentice work* at factory-authorized VCR repair centers or independently owned service shops so they can hone their skills. In a typical situation, you work for little or no pay during an apprenticeship, as you put what you've learned into practice.

As more and more VCRs are purchased each year, and as existing VCRs get older, the field of VCR repair is expanding. This might be just the spot for you, whether full- or part-time. It has been estimated that in 1993 alone there were nearly 12.5 million VCR decks sold, as well as 1.6 million TV/VCR combination units. There is money to be made in VCR repair!

I sincerely hope you enjoy this textbook. I've tried to keep things straightforward, with clear explanations. There has been a conscious attempt to keep "techno-babble" to a minimum, although sometimes there's just no substitute for technical terms. Many students find that skimming the material in each chapter before studying it is a big help. I also suggest that you read each chapter in order, without skipping around. Visit the glossary to become familiar with new terms and for definitions.

That should be enough introduction on where we're going and how we'll get there. Let's get started!

# CHAPTER 1
# VCR OPERATIONS AND CONTROLS

VHS VCRs have become extremely popular in American households. Yet it seems there are countless jokes about how difficult it is to set a VCR clock, record a TV program, or figure out how to hook a VCR up to begin with. To the person in the joke, a VCR is a confusing array of controls and hookups that can cause all sorts of frustration. To someone like yourself, who takes the time to study VCR operations and controls, it is a device to be mastered.

This chapter discusses VCR operations, controls, and input and output connections. This should be valuable reading, even if you think you already know all there is about using VCRs and connecting them to televisions and other equipment. This chapter will reinforce what you *do* know, while explaining operational differences among various VCR models. If you want an in-depth look at how to connect VCRs in various configurations, you'll want to read this chapter first, and then refer to the further discussion in the Chapter 1 Appendix of this textbook. Somewhere along the way, you'll probably learn a few connection schemes, operating modes, and VCR terms with which you are not now familiar. In addition, in this chapter you will learn about the *cassette loading* process and the *tape loading* process, both of which are fundamental steps in proper operation of any VCR.

A solid understanding of the material in Chapter 1 is essential for you to readily grasp the contents of later chapters. In addition, many VCR "problems" are really caused by consumers who either incorrectly operate their VCRs or connect them improperly to other components. Therefore, knowing how VCRs work and how their controls and connectors should be used are the first important steps in learning VCR maintenance and repair.

## CHAPTER OBJECTIVES

**After completing this chapter, you should be able to:**

1. Describe basic VCR operating modes and functions of typical front panel controls, pushbuttons, and switches.
2. Describe what is meant by the terms *cassette loading* and *tape loading*, and the basic way each of these two separate load operations is accomplished.
3. Explain briefly what the following VCR signals are, including where each type of signal would be connected on the back panel of a typical VCR:
   - VHF Low and High Band channels
   - UHF channels
   - Cable TV and Satellite channels
   - Baseband Audio and Video (composite) signals.
4. Briefly describe how to tune in a TV station with three different types of VCR tuners:
   - Mechanical turret

- Individual VCR channel presets
- Automatic/Direct entry.

As explained in the Introduction, discussions in this text are primarily about video home system (VHS) video cassette recorders (VCRs), because this is the predominant format in the home market. Much of what applies to VHS VCRs also applies to Beta, U-Matic, and 8mm formats, as these VCRs often work similarly. U-Matic is a 3/4-inch format videotape system usually found in industrial, educational, and semi-professional markets.

A VCR is probably the single most complicated appliance in most households. It has electronic circuitry similar to that in a TV set, to process video and audio signals. Its clock/timer is somewhat like a clock radio, to turn it On and Off. Its tape handling mechanics are much more complex than any audio cassette or reel-to-reel tape recorder. Its microprocessors are similar in many respects to those in personal computers. A VCR is therefore a complex system consisting of several subsystems. This chapter focuses on overall VCR operations, which are made possible by the integration of these different subsystems and technologies. Later chapters explore VCR operations in greater detail and explain troubleshooting, maintenance, and repair techniques.

## WHAT VCRS CAN DO

Essentially, a VCR is a device that records to and plays back sound and picture information from magnetic tape housed in a plastic cassette. Figure 1.1 shows a typical VHS VCR and videocassette.

**FIGURE 1.1 VHS video cassette recorder and videocassette.**

Recorded signals can be from a conventional TV antenna, cable or satellite TV system, video camera, or other video source, such as a second VCR or laserdisc video player. VCRs contain a tuner to receive the appropriate TV or cable channel, just as a TV set does. In playback, both sound and picture from the VCR are reproduced by a TV set or video monitor. Audio signals may also be routed to a stereo system to achieve better quality sound than might be available from the TV or monitor. Many people find the improved audio reproduction—and increased stereo separation with a stereo VCR—from connecting a VCR's audio output(s) to their home sound system very desirable.

Most VHS VCRs have a programmable electronic timer that can be set to turn the VCR On and go into Record mode at a preset day and time for unattended recording of a particular TV program, which is tuned in on the VCR's tuner. A user-programmed Off time powers down the VCR at the end of a TV show or time period. Some VCRs have no tuner or unattended recording capability, and are used primarily for semi-professional video production and editing.

## Cassettes and Recording Times

Today, most VHS VCRs can record and play back at three tape speeds: Standard Play (SP), Long Play (LP), and Extended Play (EP), which in older VCR models is often called Super Long Play (SLP). Some VCRs may be capable of only one or two of these speeds. Many camcorders record only at the highest speed. By far the most popular VHS cassette is the T-120, having a record time of 120 minutes at Standard Play (SP) speed, although T-15, T-30, T-60, and even T-160 cassettes are available.

In general, T-160 videocassettes, which contain thinner tape than the others, should be avoided. Thin tape causes problems with some VCRs, especially ones not in excellent mechanical condition. Tape handling problems are more apt to occur during Rewind, Fast Forward, and Pause/Play operations with T-160s than with T-120 or shorter cassettes. A very few VCRs have a two-position slide switch labeled THIN/NORMAL, which optimizes tape transport torques and tensions for the thinner T-160 videocassettes when in the THIN position.

Figure 1.2 shows a standard VHS cassette along with other types of videocassettes. A compact VHS format, called VHS-C, is used in VHS-C camcorders, which makes them significantly smaller and lighter than a standard VHS camcorder. A *camcorder* is a VCR and video camera combined in a single handheld unit. VHS-C cassettes have 1/2-inch-wide tape, just like standard VHS. They can be used in a full-size VHS VCR by placing the VHS-C cassette in an adapter, which is then loaded into the VCR. An 8mm videocassette is about the same size as a standard audio tape cassette, but slightly thicker. (Tape in a regular audio cassette is 1/8th-inch-wide, or 3.175 millimeters; 8mm videotape is about 5/16ths-inch-wide.)

A videocassette is somewhat similar in construction to a standard audio cassette. Magnetic tape is wound on a supply reel on the left side. During Play or Record, tape comes off the supply reel, threads through the VCR, and is reeled back in by the take-up reel on the right side. One big difference: you can't flip a videocassette over and play the other side. Program material is recorded in only one direction.

Unlike standard audio cassettes, videocassettes have a spring-operated door that automatically closes across the front of the cassette, protecting the videotape from dirt and damage when the cassette is not in use. This door opens to expose the magnetic tape when a videocassette is loaded into a VCR. Chapter 4 has details on the construction of a standard VHS cassette.

Table 1.1 shows maximum recording times for different length cassettes at various VCR speeds, including popular Beta and 8mm cassettes.

4 ■ *Practical VCR Repair*

**FIGURE 1.2** Videocassettes, top to bottom: 8mm, Betamax, VHS-C, standard VHS.

TABLE 1.1 Cassette Recording Times

| VHS | SP | LP | EP |
|---|---|---|---|
|  |  |  | (SLP) |
| T-60 | 1 hr | 2 hrs | 3 hrs |
| T-120 | 2 hrs | 4 hrs | 6 hrs |
| T-160 | 2 hrs, 40 min | 5 hrs, 20 min | 8 hrs |
| **VHS-C** | **SP** |  |  |
| TC-20 | 20 min |  |  |
| **BETA** | **BETA I** | **BETA II** | **BETA III** |
| L-750 | 1.5 hrs | 3 hrs | 4.5 hrs |
| **8mm** | **SP** | **LP** |  |
| P6-60 | 1 hr | 2 hrs |  |
| P6-120 | 2 hrs | 4 hrs |  |

VHS tape speeds are sometimes referred to as 2H, 4H, and 6H, reflecting recording time with a T-120 videocassette at SP, LP, and EP/SLP, respectively. Thus, a reference to 2H, or 2-hour, mode means the same as SP tape speed.

As with so many things, there is a trade-off between tape speed and the quality of the recording. Sharper, more detailed pictures and higher audio frequencies are reproduced at the faster tape speeds. This point bears repeating: The best audio and video quality is obtained at the highest speed, which is *standard play*. Almost all commercially recorded cassettes, such as rental movies, are at standard play (SP) speed. Especially when recording a tape that will be dubbed or copied later, and when copying and dubbing tapes, VHS standard speed should always be used for best results. Slower speeds give inferior reproduction, especially when copying or dubbing.

To record and play back the high-frequency signals required for video, basic VCRs have two tape heads that are mounted 180 degrees apart on a rapidly spinning video drum. Some VCRs have additional video heads, but they all have at least two. You'll learn more about rotary video heads and how they record and play back as you progress through this book, especially in Chapter 16.

## OPERATOR CONTROLS

There is great variety in the number of VCR controls, how they are labeled, their function, and their location. Few models among the hundreds of different VCRs manufactured have *exactly* the same controls, buttons, and switches. Lower priced VCRs usually have only basic controls for recording and playback. Higher priced models add various features and additional controls to operate them.

In this chapter, we'll look at the essential controls found on nearly all VCRs and discuss what they do. The Chapter 1 Appendix has more detailed explanations of these controls and features. Later chapters cover enhanced features, such as High-Fidelity Stereo and Super VHS (S-VHS).

Figure 1.3 is the control panel layout for a representative front-loading VCR. Refer to this figure during the following discussions; you'll read what these controls and front panel features do.

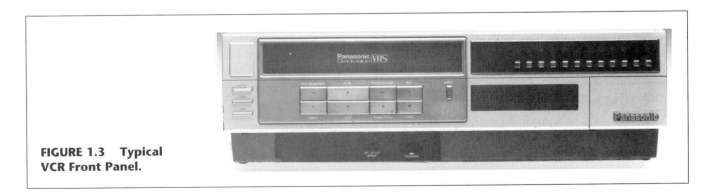

**FIGURE 1.3 Typical VCR Front Panel.**

Although there are countless variations in VCR controls, almost all have at least these basic switches and buttons:

**POWER:** Turns most of VCR's power supply on. Part of the power supply is always On to power the clock/timer and operations microprocessor, remote control sensor circuitry, and system control microprocessor.

*This button will NOT turn the VCR On if it is under control of the timer,* with the clock symbol displayed. This is a *very common* consumer mistake when they first get a VCR. They program the VCR for unattended recording, and then later press Power to turn it On. Nothing happens and they think the VCR is broken!

A VCR's AC line cord must be unplugged from a 120-volt wall receptacle to be *completely* turned Off. Be sure to unplug a VCR before removing any covers to prevent possible electrical shock and VCR damage.

**TIMER:** Turns VCR Off once unattended recording timer has been programmed. Clock face symbol appears if a cassette is loaded; clock symbol blinks if no cassette is loaded, or if loaded cassette is missing record tab.

When clock symbol is displayed, VCR is under control of timer. It is necessary to press Timer button to turn VCR On for manual operation.

On some VCRs, the Timer button is labeled Standby.

**TV/VCR:** Determines the output signal from the VCR going to the TV set. Most VCRs automatically switch to VCR mode when starting to record or play a tape, so practically the only time you need to select TV mode is to view a channel tuned by the TV while the VCR is taping another channel. When powered Off, a VCR goes into TV mode, so all antenna or cable channel signals coming into the VCR input jack go directly to the VCR's output jack, and then on to the TV.

Typically in VCR mode, **VCR** shows on the front panel display, or the words VIDEO, VCR or PLAY appears in the on-screen display. This indicates that output on the TV screen originates from the VCR:

- A tape is playing
- The channel is selected by the VCR tuner
- The VCR is receiving audio/video (A/V) input on its Line-In jacks, like from a video camera.

These signals then go to the TV on Channel 3 or 4, to which the TV tuner must also be set. Those A/V signals are also present at the audio and video Line Out jacks on the VCR.

In **TV** mode, and when VCR is powered Off, radio frequency (RF) input to VHF IN terminal goes *unaltered* to VCR VHF OUT terminal and then to TV. Channels are selected with TV tuner.

In **TV** mode, Line Out jacks have same A/V signals as in VCR mode, no switching occurs here when TV/VCR button is pressed. That is, if VCR is powered on, tuned to Channel 7, and in **TV** mode, the A/V output jacks will still have Channel 7 signals present, even though this signal is not output on VCR Channel 3 or 4. Of course if VCR is powered Off, there will be no signal at these jacks.

> NOTE: Understanding the function of the TV/VCR control and how it affects the VHF Out and Video Line Out connectors on a VCR is important. You are encouraged to read more about this control, along with the term electronic-to-electronic (E-E), in the Chapter 1 Appendix.

**REW/REV SEARCH:** Rewinds tape from Stop. Plays tape rapidly backward to review or scan tape on screen when in Play mode.

**FF/FWD SEARCH:** Fast forwards tape to take-up reel from Stop. Plays tape rapidly forward to cue or scan tape on screen when in Play mode.

**PAUSE/STILL:** Freezes video frame on screen.

**FRAME ADVANCE:** Moves video ahead one frame at a time when in Pause.

**PLAY:** Causes tape to load and play back (PB). Tape moves between supply and take-up reel under control of the capstan motor, capstan shaft, and pinch roller. Quick play models load tape even before Play button is pressed.

**REC:** Used with Play button on *some* VCRs to enter Record mode: depress and hold REC, press PLAY, release both. On other VCRs this button alone puts machine into Record mode.

**STOP:** Tape motion stops. Causes tape to unload, on some models. Quick play VCRs keep tape loaded with rotary heads spinning.

**EJECT:** Causes cassette to unload.

**RESET:** Resets tape index counter on display to "0000."

**SPEED:** Sets forward tape speed *in record mode only*. Front panel speed switch has *absolutely no effect* with VCR in Play mode. VCR automatically sets playback speed based on signal from control track on tape.

   SP: Standard Play, fastest speed, 2 hrs for T-120 videocassette, best quality recording.
   LP: Long Playing, 4 hrs on T-120.
   EP: Extended Play (also called SLP, super long play), 6 hrs on T-120.

**TRACKING:** Does not affect recording. On playback, makes minor electronic timing adjustments to compensate for recording made on other VCRs. Normally left in center position.

**OTR:** One-touch recording. Goes into Record mode for 30 minutes. Successive depressions of OTR button add 30-minute intervals to recording time, to maximum of 3 or 4 hours, depending on model.

Note: The OTR pushbutton is behind the hinge-down door with Panasonic on it, at the right of Figure 1.3. Clock and unattended recording programming pushbuttons are also behind this door.

For setting the time-of-day clock and programming unattended recording, many different methods are used. Newer VCRs have on-screen menus and prompts for setting day, channel, start, and stop times. Many VCRs have three or more pushbuttons for entering day, hour, minute, and length of unattended recording. Others use a "Select" button to go from day to hour to minute and a "Set" button to advance the day, hour, or minute. Consulting the owner's manual or experimenting will help determine how a particular VCR sets its clock and unattended recording timer. See *"Program/Clock/Timer Buttons"* in the Chapter 1 Appendix for additional information.

It is worthwhile stressing at this point that the user's manual for a particular model VCR is a valuable reference. Frequently, similar controls on two different VCRs operate quite dissimilarly, or operations vary between machines. For example, pressing Stop on some machines causes tape to unload, but on others tape remains fully loaded, and still other models cycle to the tape half-load position. The latter two types of machines go into Play faster than the first. You could go "down the tubes" trying to fix something that is normal operation for a specific model. Consulting the manual will usually clear up any doubts as to how a unit is supposed to work. Right now you may not know what "tape load" means; this is explained shortly.

## WHAT HAPPENS AFTER A TAPE IS PUT INTO A VCR?

The words *tape* and *cassette* are used very specifically in upcoming discussions. *Cassette* refers to the entire plastic box, or cartridge, with its supply and take-up reels *and* the magnetic videotape inside. *Tape* means *just the magnetic tape itself*, not its cassette housing. In day-to-day speech we often say "tape" to mean the entire cassette, as in, "Let's stop by the video store and rent a good tape for tonight." But within these pages, *tape* means *just* the magnetic videotape, whereas *cassette* is the cartridge containing videotape.

Two fundamental operations occur after a videocassette is inserted into a VCR: *cassette loading* and *tape loading*. It is important to understand these two operations as you go through this book; each is distinctly different! Here are brief overviews of cassette loading and tape loading, which are covered in greater detail in Chapters 3 and 5.

### Cassette Loading

With nearly all VHS VCRs, cassettes are inserted into a compartment opening on the front of the unit; hence these VCRs are known as *front loaders*. Most older VCRs are *top loaders*, where a videocassette is inserted into a basket that pops out of the top of the unit. Top loaders are rarely seen any more. A *very few* VCRs and some high-speed rewinders are *end loaders*: a videocassette is placed side- or end-first into these machines. Unless otherwise stated, discussions in this textbook are about front loading VCRs.

Of course before a VCR can Record or Play back audio and video, a cassette must be loaded into the machine. A front loading VCR has a door across its cassette compartment opening to keep dirt and airborne debris from entering the machine. As the first step of *cassette loading,* a videocassette is inserted with its door forward into the VCR opening. That is, the cassette door goes into the machine opening first. As the cassette is inserted, it pushes open the VCR compartment door.

A cassette should always be inserted *gently*; roughly pushing a cassette into the opening can damage the VCR loader assembly!

When a cassette is first gently inserted, it operates a *cassette-insert switch* in the front loader assembly, completing an electrical circuit. This energizes a small *cassette load motor* which draws the cassette further into the VCR and then lowers it onto the tape transport reel table. That is, a videocassette goes in, then down for *cassette loading*. As the cassette is drawn in, the door across the cassette compartment opening on the front of the VCR closes. And as the cassette goes down inside the machine, *its* door opens.

Once again, *cassette loading* starts when a videocassette is gently inserted into a VCR's compartment opening and ends when it comes to rest on the reel table inside the machine. As the cassette descends to the reel table, its door opens so magnetic tape is exposed to the tape transport. At the end of cassette loading, two *spindles* on the reel table engage the tape supply and take-up reels inside the cassette. This is essentially the end result when you load an audio cassette into a tape player.

The VCR *tape transport* is the mechanical assembly that handles and moves videotape during Play, Record, Fast Forward, Forward Scan, Rewind, Reverse Scan, Pause, and Stop operations. The *reel table* is the part of the tape transport upon which the videocassette rests at the completion of cassette loading; supply and take-up spindles on the reel table are engaged with the supply and take-up reels inside the cassette. In short, the tape transport is the place of greatest activity inside the VCR after cassette loading is complete. Except for cassette loading, which is handled by the front-loader assembly, the tape transport carries out all other VCR mechanical operations.

As you now know, there are two basic mechanisms for loading a cassette in a VCR: *top loaders* and *front loaders*. You may seldom have to service an early model top-loading VCR, but there are still many around, so it's a good idea to know something about how they work. Except for the actual cassette loading mechanisms, there are no great differences between top-loading and front-loading VCRs; nearly everything you read about VCRs in this textbook, excluding the cassette front loader mechanism itself, is also applicable to most top loaders.

Figure 1.4 shows a top loader. When the Eject button on the VCR front panel is pressed, the *cassette tray*, or *basket*, rises up from the top cover so a cassette can be inserted or removed. A videocassette is inserted, door first, into the cassette tray, just like with a front loader. Then the cassette lid is pushed down into the VCR until it latches, at which time the cassette is loaded on the reel table.

**FIGURE 1.4** VHS top loader in raised position.

Virtually all present day VCRs are front loaders. Basic cassette loading with a front loader was covered earlier. To review, a cassette is drawn in, then down to the reel table, in an inverted L-shaped path, by a motorized loading mechanism. The cassette load motor runs when the cassette-insert switch transfers as the cassette is first gently pushed into the opening. Top loader or front loader, the end result is the same: a cassette is seated with its door open on the reel table, and its supply and take-up reel hubs are engaged with spindles on the tape transport reel table at the end of *cassette loading*.

### Tape Loading

Once a cassette has been loaded on the reel table, the next operation to occur before a videotape can play or record is *tape loading*. Pressing Play or going into Record mode initiates tape loading; tape loading does not take place automatically after cassette loading on most machines. However, *some* machines *do* load tape right after cassette loading, before the VCR is put into Play or Record. These "Quick Play" VCRs go into playback mode very quickly when the Play button is pressed, because the tape load cycle has already occurred.

Other models half-load tape following cassette load, which means less time to go into Play mode than if the entire tape load operation had to be completed first.

During tape loading, magnetic tape is pulled from the cassette and threaded through the tape path. The following basic operations take place during tape loading:

- Upper portion of video drum starts to rotate counterclockwise. Video drum is also called *cylinder* and sometimes *scanner*. Its diameter is 2.44 inches (6.2 cm).
  — Upper cylinder has two or more record/playback magnetic tape heads for video; it spins at approximately 1800 revolutions per minute (rpm), or 30 times per second.
  — Lower cylinder is stationary. Entire video drum is tilted approximately 5 degrees from vertical.
- Two moveable vertical posts, called *tape guides* or *P-guides*, pull a loop of videotape from inside the front of the now-opened cassette. Tape is then threaded through the tape path; positioned against several transport parts, including stationary magnetic tape heads; and wrapped partway around the video drum.
  — Tape loop is approximately 13 inches long.
  — Tape is wrapped halfway around video drum.
  — Tape passes between vertical capstan shaft and rubber pinch roller. The rotating capstan and pinch roller together move tape during Play, Record, Forward Scan, and Reverse Scan.

You'll want to learn the details of tape loading, discussed in Chapter 3, so you can repair the many problems that can occur during this process. At this point, the important thing to know is that during tape loading a loop of videotape is pulled from the cassette shell and threaded through the tape path. Magnetic tape is positioned so it can move precisely past components that record and play back audio and video.

Most VCRs unload tape when the Stop button is pressed. That is, the two movable posts, called P-guides, retract to just inside the cassette, and tape is reeled back into the cassette so none of it is around the video drum or anywhere else in the tape path. VCRs do not automatically unload the cassette at this point!

All VCRs unload tape when Eject is pressed—if the tape was not previously unloaded by pressing Stop—and then unload the cassette itself. Remember, tape must be fully unloaded from the tape path and reeled back into the cassette before a VCR unloads or ejects a cassette.

During tape unload, the supply reel on the left side of the cassette turns counterclockwise to pull in the tape loop, while the two moveable P-guides return toward the open cassette door. Video drum rotation stops when tape is unloaded.

## VHF, UHF, AND CABLE TV SIGNALS

In addition to playing and recording tapes, VCRs contain electronic circuitry, similar to that in TV sets, that enables them to receive TV signals. Broadcast and cable signals are the primary ones you need to know about. It is also important to know generally how VCRs tune in these signals.

Before we describe how VCRs select TV channels, this is a good place to discuss the different types of channels available in the United States. Consumers can become confused by how a broadcast or cable signal reaches them. Their confusion can lead them to believe that a VCR is not working properly even when it is. The following discussion explains how a signal may reach a consumer's VCR and TV/monitor in any of several ways. This discussion will also help you if a VCR is not properly tuned to meet a consumer's needs.

Electromagnetic waves known as radio frequency (RF) signals carry sound, video, and other information either through the air or through wires. Each TV channel, whether VHF, UHF, or cable TV, occupies a slot that is 6 megahertz (MHz) wide in the RF spectrum. This slot or band of frequencies is called *bandwidth*, which is the range of radio frequencies between the lowest and highest frequency in each channel.

Each 6 MHz-wide RF channel carries both video and audio information. Just as your AM radio can tune to an RF carrier at 980 kilohertz (kHz) and your FM receiver can tune to a 98.5 megahertz (MHz) carrier, TV and VCR tuners can tune to audio and video carriers within each 6 MHz-wide TV channel. These carriers may be on a broadcast channel or on a cable channel. A carrier is an RF signal that is changed, or *modulated*, by an audio or video signal. These signals change, or vary, the strength *or* frequency of the electromagnetic carrier waves.

Don't be too concerned if you don't know how these signals are developed or how they are selected by a TV or VCR tuner. You won't need any detailed knowledge of this subject to understand material presented in this textbook—but if you want it, later chapters and the Chapter 1 Appendix have some additional information. As long as you eventually grasp the *general* concept, you're doing fine.

> *Hertz* is the international unit of frequency, or oscillations, of these waves in one second of time. (Cycles per second is another term for Hertz.) The metric prefix *mega* means one million, or 1,000,000 ($10^6$ in scientific notation). Thus, for example, a 54 MHz RF signal oscillates, or swings regularly back and forth, 54 million times a second. (By contrast, 120-volt household current *alternates*, or changes back and forth, at only a 60 Hz rate, or 60 times a second.) Alternating current (AC) and direct current (DC) are covered in Chapter 8. Additional material about RF signals is contained in the Chapter 1 Appendix to this textbook.

TV channels are located in two portions of the RF spectrum:
- VHF (very high frequency), Channels 2–13 on 54–216 MHz
- UHF (ultra-high frequency), Channels 14–83 on 470–890 MHz.

TV Channels 2 through 6 are called VHF Low band, while Channels 7 through 13 are VHF High band. This is important to know because with many VCRs, you have to set a small *band select switch* to either $V_L$ or $V_H$ to pre-tune a VHF channel, depending whether it is a VHF Low or High band channel. These same VCRs require that the band switch be set to U when pre-tuning a UHF channel. Tables A1.1 and A1.2 in the Chapter 1 Appendix list VHF and UHF TV channels and frequencies, respectively.

There is a tremendous 254 MHz gap between TV Channels 13 and 14. Because of this large frequency separation between the top of the VHF band and the bottom of the UHF band, two separate antennas are required to receive these signals, although the two antenna arrays may be combined in a single antenna system. Many rooftop TV antennas are really three antennas in one: VHF Low, VHF Hi, and UHF.

Most newer VCRs can tune in all broadcast or cable signals with an autotune feature. Some cable-ready VCRs with small band select switches for pre-tuning VCR channels have one position of each switch labeled CABLE or CATV. (CATV today is synonymous with cable TV. Originally, CATV stood for

Community Antenna TV, which was the birth of commercial cable TV systems.) Only older VCRs require a separate cable converter, or "box," to tune in cable TV channels.

Refer to Table A1.3 in the Chapter 1 Appendix for standard cable TV bands, frequencies, and cable channel numbers contained in each band. Knowing the frequency of cable channels enables you to tune some of them on non-cable-ready VCRs. Appendix Table A1.4 lists *alternate* names for cable channels.

A source of some confusion to many consumers, and some VCR technicians, is that broadcast TV channel numbers and cable channel numbers often don't logically relate to one another—*and* different cable companies can number channels any way they wish. Broadcast Channel 17 might be cable Channel 17 with one cable company, yet be cable Channel 12 on a different cable company's system. Be aware of the possibility of channel number confusion, especially when talking with someone who may not know what you now do. Local newspapers and editions of *TV Guide* cross-reference broadcast and cable channels.

Furthermore, a particular cable company may assign its channels differently than shown in Table A1.4. There's really nothing to prevent a cable company from putting *its* Channel 36 on any carrier frequency it wants! A call to the local cable company *can* help determine how it has allocated channels.

## VCR TUNERS

Now that VHF, UHF, and cable TV channels have been described, we'll move on to see how VCRs tune them in. There are essentially three different types of VCR tuners, all of which are also incorporated in various makes of TV sets. Tuner types can be categorized as follows:

- Mechanical rotary turrets
- Individual VCR channel presets
- Automatic/Direct entry (digital).

### Turret Tuners

Almost everyone is familiar with the rather clunky rotary turret tuners on TV sets. Two large rotary knobs (one VHF, the other UHF) tune in one desired channel at a time. The VHF control selects Channels 2 through 13, while the UHF knob tunes Channels 14 through 83. Usually the VHF control must be set to its UHF position before tuning with the UHF control.

Turret tuners are really a form of rotary switch, with several contact surfaces that must make solid, low electrical resistance connections for each channel. Aging turret tuners are prone to flaky contacts due to worn and corroded contact surfaces. Sometimes you have to wiggle them back and forth to get a good signal.

Cleaning the contacts with spray tuner cleaner often helps. TV tuner cleaners in spray cans usually have a lubricant in addition to contact cleaner solution to reduce wear on the mechanical contact surfaces of turret tuners. (TV tuner cleaners are fine for other electrical switches, but should not be used for general cleaning where the added lubricant could be detrimental, such as on rubber belts. You'll learn much more about VCR cleaning in Chapter 6.)

Early VCRs had turret tuners. Such machines are now pretty much antiques; mechanical tuners haven't been used for several years on VCRs.

### Individual VCR Channel Presets

Many VCRs have separate tuning controls for each VCR channel. Usually 12, or perhaps 14, individual VCR channels are pre-tuned by the user to the desired TV or cable channel. This is accomplished with a small band selector switch and tuning wheel for *each channel* on the VCR. Thus, on a VCR with 12 channels, there are 12 band selection switches and 12 small tuning wheels. These controls are either on the front panel or on the top of the VCR, behind a small access door. Figure 1.5 shows a typical VCR channel tuning panel.

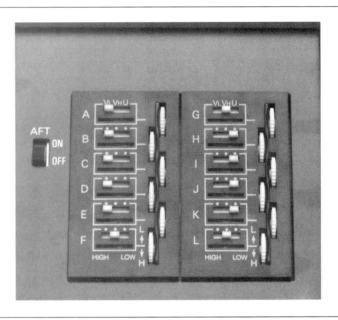

**FIGURE 1.5** Individual VCR Channel Tuning Controls.

Notice the small On/Off slide switch labeled **AFT** at the left in Figure 1.5. This turns the tuner's automatic fine tuning circuit On and Off. In most cases, this switch should be Off when pre-tuning a VCR channel. For weak stations, it may be helpful to leave this switch On with some VCR tuners.

On this style of tuner, a small mechanical indicator adjacent to each tuning wheel moves as the wheel is turned. It usually takes from 20 to 30 complete revolutions of a tuning wheel to go from the low end of the selected band to the top end. The indicator gives you a *rough* idea of where you are tuning within the band selected by the small band selector switch.

The procedure for pre-tuning a VCR channel with this style of tuner is described in the Chapter 1 Appendix.

### Automatic/Direct Entry (Digital)

Later model VCRs do away with the *individual* VCR channel preset tuning controls just described. Instead, the VCR automatically tunes TV and cable channels, or you can directly enter the desired TV channel number from a keypad, usually from the remote control unit only. The user doesn't need to be concerned whether the channel is VHF Low, VHF High, UHF, or cable (on cable-ready VCRs). Electronic tuning of this type is often called *digital tuning*, and the channel number actually displays on the florescent front panel of the VCR or on the TV screen itself.

Desired TV/cable channels are entered into VCR memory locations called *electronic presets*, one channel per preset. Models often have 16 to 20 presets. Presets are usually designated P-01, P-02, P-03, and so on, and function similarly to preset VCR channels. Think of a VCR's channel presets as storage locations for different channel numbers available in the area or from a cable company. The user *programs* one TV or cable channel number to each VCR channel preset. In this way, broadcast Channel 14 might be stored in VCR channel preset P-06.

There are many specific ways in which automatic electronic tuners work; each model seems to do things just a little differently. The following describes a typical arrangement, somewhat similar to many car radios and stereo receivers or tuners with digital tuning. User controls associated with automatic tuning and storing channel numbers into memory presets on a *representative* VCR include:

### Normal/Preset switch

In the Preset position, TV/cable channel numbers on the display/indicator panel can be entered into memory.

Once channels have been entered into memory presets, the switch is placed in its Normal position for all other operations.

### Channel Up and Channel Down buttons

These two buttons function somewhat differently depending on whether the VCR is in Normal operation or Setup operation.

*Setup Operation* moves the channel selection one channel at a time, through *all* channels that the VCR is capable of receiving. The channel number appears on the display/indicator panel, and also on-screen with some models.

*Normal Operation* moves to the next higher or next lower channel that has previously been preset. For example, suppose the following channels have been stored in preset memory:

| Preset | TV Channel |
|--------|------------|
| P-01   | 2          |
| P-02   | 5          |
| P-03   | 8          |
| P-04   | 11         |
| P-05   | 17         |
| P-06   | 30         |
| P-07   | 46         |

Pressing the Channel Up and Channel Down buttons moves to higher or lower presets, and thus the channel programmed into memory for that preset. Channels that have not been preset are skipped. The actual channel number appears on the display/indicator panel.

### Scan button

This button is used when storing TV channel numbers into VCR memory presets. Pushing Scan increments to the next higher preset number. For example, if Channel 30 has been stored in memory preset P-06, depressing Scan goes to the next preset, P-07, where the next available TV channel, 46, can be stored.

**P-07** appears in the display/indicator panel. The Channel Up and Channel Down buttons then select the desired TV channel to store in preset P-07.

*Channel Memory button*

Pressing this button stores the displayed TV channel number into the displayed preset memory location (for example TV Channel 8 stored in preset P-03).

This is a description of how just one manufacturer incorporates electronic digital tuning in a particular model. Some models can scan all possible TV/cable channels and automatically store channel numbers with a good, strong signal into memory presets.

## BASEBAND SIGNALS

In the previous two sections of this chapter, we briefly described (1) radio frequency (RF) signals, (2) how the RF spectrum is organized into frequency bands allocated to broadcast and cable TV, and (3) VCR tuners that select the desired TV or cable channel. Recall that a modulated RF TV channel signal received from either antenna or cable contains within a 6 MHz-wide band all necessary A/V information. This includes:

- Video: picture detail, brightness, and color information
- Synchronization signals: coordinate video at transmitter and receiver to produce stable picture frames
- Audio.

We now turn our attention to TV signals that do *not* modulate an RF carrier. These signals are called *baseband*; a signal is either modulated or baseband. Baseband signals can come from the following sources: video camera, camcorder, laserdisc player, computer video interface, another VCR, some TV/monitors, or audio from a tape deck, preamplifier, or stereo receiver. Notice that broadcast and cable (RF) signals are not on the list of baseband signals.

VCRs and video monitors, or combination TV/monitors, have input and output connections for *two separate* baseband signals: video and audio. Older TV sets and new, smaller, or inexpensive TVs usually do not have these connections.

Baseband video signals, called *composite video*, contain all picture brightness, color, and synchronization information. Composite video signals between a VCR and a video monitor travel on a single-conductor shielded cable called coaxial cable. Often called just "coax" for short, this cable consists of one wire surrounded by a conductive metallic shield. The inner wire and the outer shield allow a complete circuit to be made between the two pieces of equipment.

Baseband audio signals contain only the sound portion of a program. As with composite video, a single-conductor shielded cable between VCR and video monitor carries this baseband signal (two cables for stereo).

Baseband audio is described first, since this type signal is more likely familiar to you.

### Baseband Audio

If you've ever connected a CD player or cassette deck to a stereo receiver, you have had your hands on baseband audio cables. Audio on stereo component interconnect cables *is not specifically referred to as baseband*, but that's really what it is; there is no RF carrier and no modulation has occurred. Audio signals of this type are normally referred to as *line level*, as opposed to weaker signals such as microphone level, or stronger signals like speaker level. From now on, we'll refer to baseband audio signals as line level.

Line level audio signals between a VCR and a video monitor (or TV/monitor) are totally compatible with line level stereo system signals. As a result, a VCR AUDIO OUT jack can be connected to TAPE IN, CD, or AUX IN jacks on a stereo receiver, preamplifier, or audio mixer. This allows you to play audio from the VCR through speakers connected to that stereo system. Similarly, you can connect line level audio outputs, like TAPE OUT on a stereo receiver, to the AUDIO IN jacks on a VCR. You could do that to record the stereo simulcast from an FM radio station during a televised concert.

> On the technical side: Audio line level signals cover the frequency range from 20 Hertz to 20,000 Hz for high-fidelity sound. Non-hi-fi VCRs typically have an audio frequency response from 100 Hz to 8,000 Hz at SP speed. High-frequency audio response rolls off (deteriorates) somewhat at slower tape speeds, typically 6,000 Hz at LP and 5,000 Hz at EP/SLP speeds, respectively. Line level audio signals typically have an amplitude around 1 volt peak-to-peak (p-p), compatible with audio tape decks. (*Peak-to-peak* means the alternating current (AC) amplitude in volts between the most negative-going signal and the most positive-going signal.)

## Composite Video

All components to produce an image are contained in the composite video signal (see Figure 1.6):

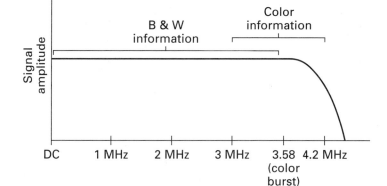

**FIGURE 1.6 Composite video frequency bandwidth.**

- Picture detail and brightness information (luminance)
- Color information (chroma)
- Synchronization (sync) pulses that keep receiver and transmitter in step with each other. In this context, a playback VCR is the transmitter, while the TV/monitor is the receiver.

A composite video signal has an amplitude of approximately 1 volt peak-to-peak. Other consumer video electronics, such as camcorders and video laserdisc players, have composite video connections compatible with VCRs. Composite video input and output jacks are also called video line level or just video line jacks. A/V inputs and outputs refer to baseband or line level audio and video signals.

## HOOKUPS AND CONNECTIONS

Now that you know the basics of RF and baseband signals, it's time to describe the various inputs and outputs on the rear panel of a typical VCR. This is where connections are made between a VCR and:

- VHF antenna
- UHF antenna
- CATV cable or cable converter
- TV receiver
- Video monitor and stereo system

Figure 1.7 shows a typical VCR rear panel layout. We'll explain what each feature or connector is used for, followed by some typical hookups between VCR, antenna, cable box, and TV. The terms *connector*, *jack*, and *terminal* are used somewhat interchangeably.

**FIGURE 1.7  Rear panel of typical VCR.**

### AC Line Cord

Primary VCR power is 120 volts at 60 Hertz. It is important that a VCR be plugged into a wall outlet that is On or "hot" all the time, to keep the clock running even when the VCR is powered Off. Many homes have one outlet in a room (usually the one closest to the door) controlled by a wall switch near the door. Often, only one of the two AC outlets in a duplex wall receptacle is switched, while the other is hot all the time.

### AC Convenience Outlet

Some VCRs have an AC outlet on the rear panel, usually adjacent to the line cord entrance. If present, this outlet is *unswitched*, that is, it is *not* controlled by the VCR in any way. It's just an extension 120-volt outlet, which is often handy. For example, a TV set line cord could be plugged into the back of the VCR. Total power consumption of any device(s) plugged into this convenience outlet *must not exceed* the maximum wattage rating that is printed adjacent to the outlet, usually 300 to 400 watts.

Rear panel connectors and components include:

**Ch 3/Ch 4:** Switch on back (or bottom) panel. When in **VCR** mode, determines whether RF modulator in VCR outputs on TV Channel 3 or 4. Ch 3/4 output can be tape playback, or channel selected by VCR tuner, or program input to A/V Line In jacks.

> NOTE: Additional information about this important VCR switch is located in the Chapter 1 Appendix.

**VHF IN:** 75-ohm unbalanced RF input F connector from VHF antenna, cable, or cable converter.

**VHF OUT:** 75-ohm output F connector from RF modulator on TV Ch 3 or 4 when in **VCR** mode. VHF OUT is same as VHF input when in **TV** mode, or when VCR is powered Off.

*It is very important to realize that the signal at the VHF Out terminal can come from two distinctly different sources, depending on the mode of the front panel TV/VCR switch when the VCR is powered On.* VHF Out is:

- Same as VHF IN when VCR is powered Off, or when powered On and in **TV** mode. Antenna switcher in VCR connects VHF IN connector directly to VHF OUT.
- On VHF Channel 3 or 4 when in **VCR** mode. Signal is tape playback *or* TV channel tuned by VCR tuner, like when monitoring a program being recorded.

VHF output on Channel 3 or 4 of the program selected by the VCR tuner is called *electronic-to-electronic*, or *E-E, operation.* That is, in E-E, VCR output is essentially the same signal received by the VCR tuner (or line input jacks).

**UHF IN/OUT:** 300-ohm balanced RF input from UHF antenna and output to UHF input of TV. UHF IN and OUT terminals are essentially in parallel, with *no antenna switching in VCR mode* as with VHF IN and OUT. A UHF antenna is normally the only device that would be connected to a VCR's UHF IN terminals.

**VIDEO IN:** Baseband (not RF modulated) composite Video In jack. (Jack is RCA style, like on an audio cassette deck or CD player.) Possible input sources: video camera, laserdisc player, computer video interface, other VCR. On many VCRs, Video In jack automatically disconnects video signal received by tuner when RCA plug is inserted. Other VCRs have front panel switch to select TV or Line.

> A TV/monitor is a fully functional TV, with tuner, that also has separate line input jacks for composite video and line level audio. It's a combination TV and video monitor. A switch on the set or remote control unit determines whether picture and sound are from the set's tuner or from an external source connected to audio and video line input jacks.
>
> A video monitor has no tuner to receive broadcast or cable TV signals, but has a Video In jack to receive composite video signals from a videotape player, laserdisc player, video camera, or computer video interface. A video monitor may or may not contain audio amplifiers and speakers.

**VIDEO OUT:** Composite output signal RCA jack. Unaffected by VCR/TV mode button. Output is from (a) tape playing, (b) VCR tuner video, or (c) Video In jack. Output to video monitor, other VCR, or computer video digitizer.

**AUDIO IN:** Line level signal fully compatible with consumer audio high-level outputs on tape deck, preamplifier, or stereo receiver. Left (L) and right (R) jacks on stereo VCRs. On many VCRs, RCA line jack disconnects audio received by tuner when RCA plug is inserted. Others have front panel switch to select audio from tuner or line jack. On some stereo VCRs, if audio cable is plugged into only R Ch Audio In jack, R Ch audio also goes to L Ch record circuits. (This saves having to use an RCA Y adapter to put mono audio on both stereo channels.)

**AUDIO OUT:** Line level signal compatible with consumer audio gear: High level Aux, CD, or Tape In on stereo receiver or preamplifier.

### Basic VCR Connections

There are so many possible configurations for connecting a VCR to antennas, cable systems, TVs, and related home entertainment equipment that it is nearly impossible to cover them all. Things can get especially complicated when connecting to a cable TV system. Perhaps the VCR is cable-ready, but the TV is not. This calls for a different hookup than if both were cable-compatible. Some people may want to watch one cable channel while recording another, requiring yet a different connection scheme.

Understanding the previous material in this chapter should assist you in connecting a wide variety of equipment to produce the desired results. In the most basic VCR hookup, one or more antennas connect to the RF input terminals on the VCR; then the RF output from the VCR connects to the antenna terminals of the TV set. Various adaptations of this basic VCR-to-TV hookup configuration are described in the Chapter 1 Appendix.

Higher quality audio and video signals are derived from separate A/V Line Output jacks than from modulated RF. For the best picture, a TV/Video Monitor should be connected to a VCR's Line Out jacks when playing tape. Line inputs/outputs likewise provide better quality than an RF interconnect when copying, or *dubbing*, tapes with two VCRs.

Quality cables and connectors, cable of the correct impedance, and clean, tight electrical connections are required for best results when interconnecting video equipment. Excessive length and coiling of extra cable should be avoided.

## SUMMARY

Knowing how a VCR is *supposed* to work, understanding what the various operator controls *do*, and recognizing the proper methods for connecting a VCR to other components are the first steps in performing VCR maintenance and repair. Sometimes customers bring VCRs in for repair when the problem is not with the VCR, but in how it is operated or hooked up.

The Chapter 1 Appendix contains additional information about subjects covered in this chapter.

## SELF-CHECK QUESTIONS

1. What is the total length of time that may be recorded on a T-120 VHS cassette at the standard play (SP) speed?
2. What is the first thing that should take place when a cassette is first gently inserted into the cassette compartment opening of a front loader?
3. Briefly describe how video record-and-playback magnetic tape heads are mounted in a VCR.
4. Describe what must be done so that a VCR power supply is completely powered Off.
5. What does it mean if there is a blinking clock face symbol on the front panel display of a VCR?

6. When a VCR is turned On and placed in TV mode, what signal is available at the **VHF OUT** connection, or terminal, on the rear panel of the VCR?
7. Briefly describe what affect the manual tracking control on the front panel of a VCR has when recording a cable TV program.
8. In which RF band is broadcast TV Channel 7 located?
    a. UHF band
    b. VHF Low band
    c. VHF High band
    d. MHz band
9. Another term frequently used for *baseband video* signals is
10. Briefly describe what the Ch 3–Ch 4 slide switch, found on the rear panel of most VCRs, is for.

# CHAPTER 2
# REMOVING COVERS, GETTING STARTED

Most VCR repair and maintenance work requires removing one or more of the unit's covers and getting inside. True, some problems may be corrected without removing a VCR's cabinet, such as improper connections to TV and other gear, or operator error, but in most cases access to the inner mechanics and electronics is needed. Various VCR hookups and modes of operation were discussed in Chapter 1. Be sure that a VCR "problem" is not really user error or a fault in attached equipment or cabling before taking screwdriver in hand to remove VCR covers.

Within this chapter, you'll learn the *right* way to proceed when getting under the covers, along with precautions to take to ensure your safety and to prevent damage to VCR components. A little care and taking a few safeguards beforehand can save you from frustration and wasted time. What's more, this also keeps the VCR from becoming worse off than it was to begin with due to parts being accidentally broken during disassembly and repair. A VCR has many delicate parts and sensitive adjustments; loosening the wrong screw or handling a component incorrectly can easily complicate what otherwise might have been a straightforward repair.

Proper procedures for accessing different VCR areas and a survey of major components found in the vast majority of VCRs are covered in this chapter. Typical interior layouts and subassemblies are described, along with some general tips and techniques for smooth sailing during repair and maintenance operations.

## CHAPTER OBJECTIVES

Upon completing this chapter, you should be able to:
1. List preliminary steps to take *before* removing VCR covers.
2. Describe personal safety precautions when working under the covers of a VCR.
3. Describe *electrostatic discharge* (ESD), its possible effect on electronic circuitry and components, and safeguards to avoid ESD when working on a VCR.
4. Describe how to remove external covers and panels on front loader and top loader VCRs.
5. Describe how to release and swing out electronic printed circuit boards (PCBs).
6. Describe major components in a typical VCR and their functions.

Information in this chapter is fundamental to all VCR repair and maintenance activities. An overview of several internal VCR components and subassemblies provides a foundation for later chapters. Knowing how to get started, what to look for, and the correct steps to take for trouble-free, efficient repair can certainly make your time spent troubleshooting easier. And sometimes, knowing what *not* to do can be just as important as knowing what *to* do!

# DOING IT RIGHT!

A saying that's been around for a long time in the world of technicians and repair persons goes, "If it ain't broke, don't fix it!" The idea behind this rather cryptic remark is that sometimes it's better to leave well enough alone than to try to get just a little better performance by opening a machine up and tweaking its innards. Often in the attempt to urge better performance from a machine, a part gets broken or lost or an adjustment gets disturbed; the end result is worse than if nothing had been done in the first place. This is something to keep in mind when it comes to VCRs. Before opening up a VCR and crawling inside, it's best to first make sure there's a *need* to go under the covers.

Let's face it, most VCRs are *not* built like tanks. In most, the main chassis or frame is plastic, with many plastic tabs, catches, and latches that hold things together. Opening up a VCR unnecessarily can put unwarranted strain on fragile parts. In reality, it just gives us one additional opportunity to break a part or make some other mistake that might not have happened if we weren't so eager to see what's going on inside. It's especially true that many people new to VCR or personal computer repair can't wait to poke around under the hood. But wait! Read on a little bit before you dig in; it could save you some grief.

### Try Things Out; Take Some Notes

Prior to opening a VCR cabinet, take the time to try various machine operations. Give it a thorough checkout! Is it just one function that's failing, or several? For example, does it fail to fully rewind a tape, yet Fast Forward and Play work fine? Try as many operations as you can, and be sure to take notes as you go. It's all too easy to take apart a VCR and then wonder to yourself, "Did the Cassette-in indicator come on when I loaded a cassette or not?" "Was playback OK on LP and EP speed, even though there was a barely viewable picture in SP?" Trying as many operating modes as possible and jotting down what the unit does is an important first step in problem diagnosis.

Don't overlook the fact that what appears at first glance to be a problem may be perfectly normal operation for a particular machine. For example, it's normal for most VCRs *NOT* to power On fully if there is excessive moisture inside the cabinet. Is the message **DEW** displayed on the indicator panel? Leaving the machine plugged in at room temperature for a few hours could correct this so-called problem. Heat from portions of the power supply and circuits that are powered whenever the VCR is plugged in can help dry out excessive moisture in the unit.

Loading tape against a rotating video drum that has a slight film of moisture—as might occur when a VCR is brought in from the cold—could cause damage. Tape would probably stick to the drum instead of gliding over its surface, ruining the tape and perhaps damaging a rotary head. The system control microprocessor monitors a dew sensor in the transport and prevents the VCR from fully powering up and loading tape if moisture is present.

Reading the owner's manual, observing the front panel indicators, and operating the controls correctly may disclose that what at first was thought to be incorrect operation is really quite normal for that particular VCR. For example, some VCRs prompt for the *actual* Off time, by hour and minute, when you program unattended recording, whereas others prompt for the *length of time* the machine is to record. Confusing these two operations could lead to the mistaken conclusion that the timer circuit was defective.

VCR interconnections to TVs, antennas, cable systems, and other audio/video (A/V) gear are another source of mistaken VCR problems. Case in point: One customer had for several years owned a VCR with a two-position slide switch on the front panel to select TV or LINE as the input source. In TV, input was from the VCR tuner; LINE selected the A/V Line In RCA jacks as a signal source. When this customer purchased a newer VCR, he connected it just like the old one had been, with RCA cables plugged into the rear panel Audio and Video IN jacks, so he could conveniently hook up his camcorder to the VCR to copy a cassette whenever he wanted. But the new VCR wouldn't record anything from its tuner; just a blank screen and silence would result. Well, the new VCR *didn't have* an input selector switch. Instead, audio and video from the tuner was automatically disconnected whenever an RCA cable was inserted into an A/V Line In jack. This was proper operation for that particular VCR, but it appeared as a problem to the customer. A little time with the new VCR user's manual would have saved a service call! (This is the type of inside knowledge you will gain over time as you work on numerous VCRs.)

As shown in the Chapter 1 Appendix, some VCR hookups can get fairly complex, such as when you connect a cable box, VCR, and TV with a signal splitter and A/B switch so a scrambled channel can be recorded while you watch another nonscrambled cable channel. All connections and cable ends should be checked before opening a VCR when signals don't appear to be going where they should. Cables can be tested for shorts or opens with a multimeter (which is covered in Chapter 9).

Did the setup ever work? Perhaps things were connected inappropriately in the first place. For example, if the VHF OUT terminal on a VCR is connected to a TV set with combined VHF/UHF input on a single F connector, the TV won't be able to tune UHF channels, but the VCR will! This, of course, assumes that a UHF antenna is connected to the VCR UHF IN terminals. Similarly, if a customer has connected a signal splitter/combiner so that its UHF input or output is connected to what should be VHF, things just won't work right.

Some VCR problems may not be caused by the VCR itself but by the videotape. Perhaps a cassette or the tape inside it was damaged on another machine. It's a good idea to check operation with *known good* tapes before determining that the VCR is at fault. Chapter 4 explains how to check for videotape and cassette damage.

Customers can also supply information that is helpful to jot down. Take time to question them about what they observe the VCR doing, or *not* doing. They may not always explain a problem in technical terms, but usually there's an element of truth in what they say that could just lead you to the problem area. For example, a customer might say that a sound like crinkling cellophane sometimes comes from the VCR before it shuts Off. Well, it could be this sound is caused by tape bunching up about the capstan because the take-up spindle has stopped turning. You've got a good place to start your diagnosis.

Chapter 20 explores in greater depth ways to interact with your customers, especially when they bring you an ailing VCR.

Once you determine that there *is* a problem in a VCR, rather than with how it's being used or how it's connected, there are still several things you should do before diving under the covers. Again, any notes you take on how a VCR behaves *before* you open it up become very helpful in tracking down a problem. Following are examples of some items that might be noted, although these symptoms wouldn't necessarily all be present in one VCR at the same time:

- What functions work 100 percent OK?
  — Cassette load/unload OK.
  — No problems with clock/timer.
  — Tracking control *does* change position of horizontal noise bars in picture.

- What functions fail completely?
  — Won't rewind tape.
  — Cassettes eject about three seconds after loading.
- What operations are marginal or intermittent?
  — Unloads tape sometimes near end of two-hour program.
  — Cassette-in icon sometimes flickers.
  — Have to fuss with Stop button for it to work.
  — VCR sometimes shuts itself Off right after a cassette loads.
- Description of audio and video problems during playback:
  — Severe tearing at bottom of picture, LP and EP speeds only.
  — Color comes and goes at regular two-second interval, with some jitter in picture.
  — Picture very snowy uniformly on playback, all speeds.
  — Constant waver or flutter in audio (sounds sick!).
- Any unusual or missing display panel indicator or icon?
  — **VCR** doesn't display when VCR/TV button is operated.
  — Index counter hesitates during Fast Forward.
- Any unusual mechanical sounds coming from VCR?
  — Occasional high-pitched squeal in Play, no problems in picture or sound.
  — Rumbling or grinding noise during tape load cycle.
  — Crinkling sound from machine, like crumpling cellophane candy wrapper; comes and goes when playing a tape.
- Any odor coming from VCR?
  — Slight burnt plastic or enamel smell preceded failure to load cassette.
- Any obvious physical cassette or tape damage caused by VCR?
  — Front-to-rear scratch along bottom right of cassettes.
  — Cassette ejects with tape hanging out cassette door.
  — Crinkled bottom edge of videotape.
- Customer reports:
  — "It worked fine until that lightning storm we had last Friday."
  — "I think my son pushed a toy or something in where the tape goes."

Most of this probably seems pretty obvious, but, especially if you're working on several VCRs, writing a few notes before taking one apart can often help locate problems. In fact, taking notes throughout disassembly and repair, and working in a systematic manner, saves time and prevents costly mistakes when working on a unit.

Scribing a line with a sharp pointed tool around the edge of a bracket or other mechanical part *before* it is removed is really helpful in realigning the part during reassembly. Learning to be well organized in your repair activities definitely pays off in the long run. Take time to do it right ... and make it a habit.

## SAFETY—FOR YOU AND THE VCR

VCRs are fairly safe devices on which to work, but some safety precautions must be taken if you want to prevent personal injury *and* avoid damaging delicate VCR components. Understanding what to do and what *not* to do when opening up and repairing a VCR will keep both you and the machine healthy.

### Avoiding Electrical Shock

First off, you would almost have to work at it to receive an electrical shock from a VCR. Almost all areas you can *easily* touch with your fingers have only low voltages, with 35 volts being about the highest. The majority of VCR circuitry operates on voltages in the 5- to 12-volt range. These low voltages are generally safe to work around. *However*, all VCRs are powered by 120-volt alternating current (VAC) from the wall outlet, and *this* is *a dangerous voltage!* Power supplied to a machine from a 120-VAC outlet is referred to as *line voltage*.

Line voltage comes into a VCR on a two- or three-conductor *line cord*, also called a *power cord*. On nearly all VCRs, line cord wires attach to terminals on the power supply, right inside the cabinet where the line cord enters the back panel. On some VCRs, these terminals are exposed, and you can easily contact them if you aren't careful.

Some VCRs have a 120-volt convenience outlet located close to where the AC line cord enters. This is an area you want to be very careful not to touch when the VCR is plugged in! Line voltage is exposed in most VCRs at metal clips that hold a small glass main line fuse. Dangerous line voltage is present on some power supply printed circuit board wires and conductors, at input terminals to the power transformer, and in some power supplies at a line voltage rectifier.

Figure 2.1 illustrates points on a representative VCR where line voltage is present. **Use extreme caution so you do not touch anything in this area when a VCR is plugged in.** Remember, the VCR doesn't have to be powered On, just plugged in to have dangerous line voltage present!

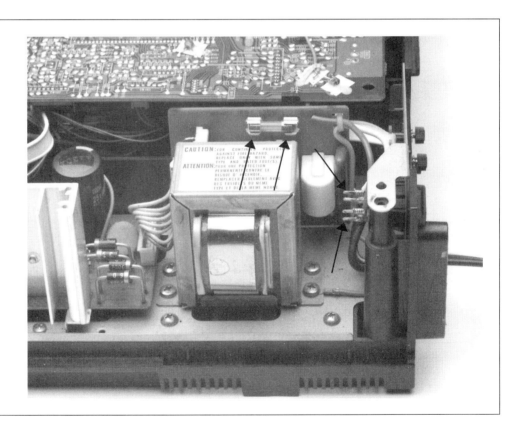

**FIGURE 2.1** Areas on typical VCR with dangerous line voltage exposed. Arrows point to exposed 120-volt AC terminals.

Always unplug a machine from the AC wall receptacle before removing any covers. Once the covers are off, locate where the line cord enters the cabinet and connects to the power supply. Locate any exposed line voltage terminals, and stay clear of this area of the machine. It's a good idea to place some electrical tape around exposed line voltage terminals, or form a piece of cardboard around exposed terminals and line fuse. Secure the cardboard with a piece of electrical tape.

Be especially careful around the power supply in a VCR; this is generally the only place where you could receive a dangerous electrical shock. Some VCRs have a plastic shield around exposed line voltage and high-voltage power supply terminals, which offers some protection from electrical shocks while working on the unit. (A plastic safety shield of this type was removed before taking the photo in Figure 2.1.)

Following are electrical safety precautions you should *always* observe when working on electrical and electronic equipment:

- Work on a clean, well-lit, nonconductive surface.
- Never remove a cover or cabinet panel when equipment is powered up.
- Do not work on equipment that is powered up except when absolutely necessary.
- Remove all metallic jewelry, including rings, watches, bracelets, chains, necklaces, and pendants.
- Never stand on a damp floor or in water.
- Never work on equipment when your feet are bare.
- Never touch electrical components with *both hands* at the same time; an electrical current could flow through your body. Keep one hand behind your back! This also means do not touch a metal chassis ground with one hand when there is a possibility of touching an exposed electrical contact with the other hand.
- Don't assume that just because a machine is unplugged it is totally safe electrically. Some VCR power supplies (which are discussed in Chapter 12) can retain over 200 volts of direct current stored in a capacitor. It usually takes a few minutes for this charge to fully drain off. It *could* take several hours or even longer!

Some of these safety practices are to protect the VCR as well as yourself. For example, if a metal watchband or ring comes in contact with the wiring side of a printed circuit board, it could easily short two or more circuits together. A "fried" integrated circuit, diode, or transistor might be the result. You probably wouldn't feel a thing!, but you just added another problem to the one you were originally trying to locate and fix.

Electrical circuits should always be respected, even those with low voltage. I recall working under the hood of my first automobile while adding an ammeter to the dashboard. Somehow the metal band on my wristwatch got between an exposed 12-volt headlight relay terminal and the metal car body while I was looking elsewhere, working with my right hand. Although I didn't get an electrical shock, I did receive a burn as the 12-volt short tried to melt a link in the watchband.

Often it's not the electric shock itself that causes injury but a resulting secondary reaction. A *secondary reaction* is involuntary muscle movement, for example suddenly jerking your arm aside when you receive a shock, which could cause your hand to hit against another part of the machine, cutting you, breaking a machine part, or both. The worst electric shock I've experienced occurred while repairing a tube-type radio receiver (years ago!). I accidentally got zapped as my fingers touched 350V DC, with my bare forearm resting on the metal radio chassis. The shock itself didn't hurt much, but involuntary muscle contractions literally "threw" me across the room! I got a bruise on my head from hitting the wall and a gash on my arm where it scraped the radio chassis during my hasty exit.

The intent here is not to scare you, but to let you know the importance of working safely, especially around live circuitry. Following the preceding safety procedures and using some common sense will keep both you and the unit on which you're working accident-free. Understanding what *could* happen and respecting electricity are themselves good safety measures.

Many pieces of electrical equipment have an outside sticker or label picturing a lightning bolt or exclamation point (!) inside a triangle to warn of the possibility of electrical shock when going under the covers.

### Avoiding Eye Injury

Be sure to wear eyeglasses or safety glasses when removing springs and clips inside a VCR. Many portions of a VCR's mechanical assembly have small coiled springs that provide tension on arms, latches, and other components. It's very easy for these springs to fly off, possibly into your face, when you are removing or installing them. The same is true for the tiny C-clips and E-clips that typically attach a pulley or pivot arm to a shaft.

Good practice is to take a small cloth and cup the cloth around a spring or clip when removing or installing it. Then, should the spring or clip fly off, it will be caught in the piece of cloth. This not only protects your eyes, but also helps prevent losing a small part.

## ESD PRECAUTIONS

ESD stands for electrostatic discharge. You know what an electrostatic discharge is, especially if you live in an area of the country where the relative humidity gets low: you walk across a carpet, touch a door knob, and *Zap!* Well, you need to take antistatic precautions into consideration whenever you work on a VCR—or nearly any sophisticated electronic equipment, for that matter. ESD can easily destroy or weaken several kinds of electronic components.

ESD dangers and the importance of preventing it when working on electronics cannot be stressed enough. Many people think that unless they experience an electric shock, no static discharge has taken place. This, however, is far from the truth! To feel an ESD shock, a relatively high electrical charge must be built up on your body—in fact, close to 1,000 volts or more. This can occur when you walk across a carpet when the relative humidity is low. Smaller amounts of ESD, roughly 400 volts and lower, won't be felt in the least. But these smaller voltage discharges, of which you are totally unaware, can spell doom for delicate electronic components, like a microprocessor integrated circuit chip in a VCR. Many newer VCRs have CMOS memory chips that are also very susceptible to destruction from ESD. (*CMOS* stands for complementary metallic oxide semiconductor, very-low-power-consumption integrated circuits (ICs), often used to store channel presets and timer programming information.)

Even normal body movements in clothing, like brushing a shirt sleeve across a trouser leg, or shrugging your shoulders, can generate *several hundred volts* of electrostatic charge. This charge stays on your body until it can be discharged in some way, such as by touching a water pipe or electrical appliance. Again, the discharge may be so slight that you don't feel a thing.

Now suppose you're working on a VCR and touch a printed circuit board with your hand. As you do, any static charge on your body transfers, or discharges, to the circuit board. The static electricity can flow along a printed circuit board (PCB) wire to an integrated circuit. To an IC that has been designed to work on only 5 volts, this ESD can be damaging or even lethal! Specifically, the discharge can weaken or destroy microscopic transistor junctions within the chip.

Please don't underestimate the importance of taking ESD precautions. In electron microscope pictures, the tiny silicon junctions within an IC chip that were exposed to electrostatic discharges of only 200 volts (which would not be felt by a person delivering the static charge) look like they were bombed, with gaping craters. Sometimes a crater doesn't completely destroy a transistor junction, but only blasts part of it away. The chip may still work, but it is severely weakened. In time, if the chip becomes overheated, the junction will fail. ESD is the silent, unseen enemy of much of today's sophisticated microprocessor-controlled equipment, including VCRs.

What can be done to control ESD? The best thing, before ever working on a VCR, is to purchase an antistatic kit from a computer or electronic supply store. Kits vary, but the main component is an ESD wrist strap, with a wire several feet long having an alligator clip at the other end. You put the adjustable strap around your wrist (usually the left wrist if you're right-handed). The clip at the other end of the wire is attached to an electrical ground connection or metal framework on the VCR.

ESD wrist straps are also called *conductive wrist straps or bands*. They are usually made of elastic cloth, but contain conductive strands that touch your skin, and so bleed off any static electricity on your body. Some straps are made of conductive rubber. The wire attached to the strap contains a large value resistor, usually one megohm (1,000,000 ohms), that allows any charge on your body to be slowly transferred to a VCR ground point. This built-in resistor also limits the amount of current to a safe level in case the other end of the wire touches a power supply voltage.

Some ESD kits also contain a conductive pad on which you can place components, like circuit boards and ICs, when they're out of the VCR. Often there's an additional clip on the wire leading from the wrist strap that you can attach to this conductive pad, so that any charge on the pad will also flow to the VCR ground point. Read and follow the instructions that accompany the ESD kit you purchase.

So start off right when troubleshooting VCRs; get an ESD wrist strap and wear it *whenever* you go under the covers or handle ICs and circuit boards. Most electronic parts suppliers have a conductive wrist strap for under fifteen dollars; this is money well spent. Zapping an integrated circuit could put an additional problem in a VCR, or weaken it so that it fails prematurely in the future. One replacement IC could easily cost more than a conductive wrist strap. When handling ICs and circuit boards that are not installed, place them on antistatic pads or in the conductive plastic bags in which they are usually packed.

If you must work on a VCR without a conductive wrist strap, touch an electrical appliance, such as a lamp or the screw in a wall outlet cover plate, before touching inside the cabinet. Then when you do reach inside the VCR, touch a ground point, like the metal tape transport baseplate, before touching anything else. This should ground out any static charge on your body. Also, avoid wearing clothes that easily produce a static charge, such as knits and wools, and limit your movements as much as possible, especially on carpets, while working on electronic circuits.

Taking a few precautions when working on electrical/electronic equipment is truly worthwhile, for your safety *and* the protection of delicate circuit components.

## VCR COVER REMOVAL

Most VCRs have two covers. A metal top cover wraps around to also enclose the two sides, and a separate metal cover is fastened across the bottom of the unit. All these covers are fairly easy to take off, requiring only the removal of a few screws. But before we get into cover removal, first be aware of some general comments about disassembling a VCR.

## Keeping Track

It is very easy to start taking a VCR apart and soon have a pile of different-size screws and other hardware. Some screws are for the top cover; some are for the front loader assembly; others are for the bottom cover; still others attach a shield plate over the video drum on the tape transport. (These components are described further in Chapter 3.) You work on the VCR, and find a stretched belt, hard rubber idler wheel tire, and a deformed rubber pinch roller. You order replacements from a mailorder supplier of VCR parts.

A few days later you install the replacement parts, check all VCR functions (it works great!), and start to put everything else back together. You *thought* you'd remember which screw went where, but now you're not too sure. You pick out two screws and start to reinstall the metal shield plate over the video drum. "Oops!" You realize they're the wrong ones as you almost strip out the threads in the thin metal. You pick two other screws and they work OK, but when you go to put the bottom cover back on, you discover the screws that remain are too long; they bottom out in the plastic VCR frame before tightening against the bottom cover.

This little scenario is unfortunately all too common. Similar-looking screws with different threads can strip out plastic threads. Screws that are too long can crack a printed circuit board, puncture a cable when the unit is reassembled, or prevent some mechanical part from moving freely. Sorting through a variety of hardware wastes time and is downright frustrating. Fortunately, these problems can be easily avoided with a systematic approach when taking a VCR apart in the first place.

A muffin tin, plastic ice cube tray, or three-ounce disposable bathroom cups are ideal for keeping track of mounting hardware and other small parts when working on a VCR. Simply number the containers, put like parts in a container as you take things apart, and make a list identifying what parts each container holds. For example, put top cover screws in cup #1, then stack cup #2 inside cup #1. Jot down on paper what's in each cup, something like this:

| Cup | Contents | Notes |
|---|---|---|
| #1 | 6 top cvr scrws | 2 Lng ones from *bottom, sides* |
| #2 | 4 loader scrws | |
| #3 | loader gnd wire scrw | |
| #4 | idler C-clip & wshr | Wshr goes *under* idler |
| #5 | pinch roller cap | sml dia end of roller goes *down* |

Now when you put things back together, there will be no confusion, stripped threads, or similar problems. Sure, it takes a little extra time in the beginning, but later you'll be glad you did!

It's also helpful to mark where you remove screws with an indelible felt tip pen. For example, there might be ten screw holes in a bottom cover, but only six are actually used to mount the panel in that particular VCR. Marking the holes on the cover as you remove screws helps later when you reattach the cover.

## Which Screws to Remove

Take time to look things over before you remove any screws. Often, screws that hold on covers or other components that are frequently removed during service are identified in some way. Screw heads may be tinted pinkish/red or blue, or there may be an arrow pointing to the screw. Cover mounting screws may be identified with a stamped circle around each screw head.

> CAUTION: Some screws or locking nuts on the tape transport assembly may have a blob of red paint on them. These screws and nuts have been paint-locked at the factory to prevent them from turning. *Do NOT loosen paint-locked screws or nuts unless you are absolutely sure what you are doing!* They usually lock in a critical factory adjustment that is extremely difficult to make in the field. In most cases, you will never need to disturb paint-locked hardware during VCR maintenance and repair, unless an associated part is broken.

Most covers are attached with four, five, or six Phillips or flathead screws. Covers should come off easily, without forcing. Many top covers must be slid back, usually less than an inch, to clear L-shaped latches molded into the plastic sides of the cabinet. If the cover doesn't lift off easily, look for additional screws that might be holding it on.

### Top Loader Top Cover Removal

There are two common ways to remove the main top cover on VCRs with a top loader, depending on whether there are two visible screws attaching the cassette lid, as shown in Figure 2.2.

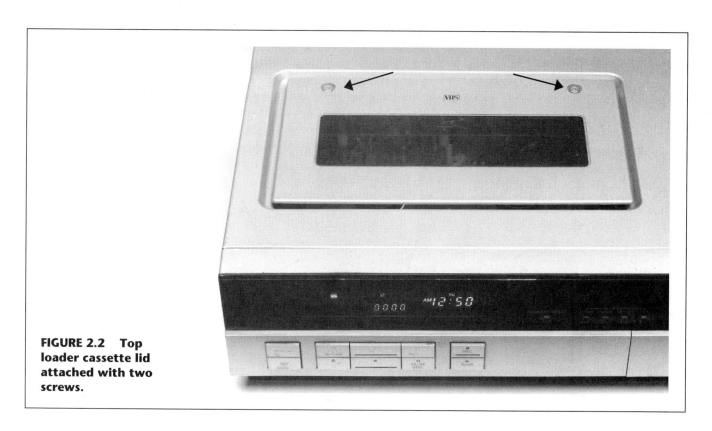

**FIGURE 2.2** Top loader cassette lid attached with two screws.

For this type of top-loading VCR, perform the following steps to remove the top cover:

1. Remove the two cassette lid screws.
2. Power On the VCR and press Eject to pop up the cassette tray or basket.
3. Unplug the VCR.

4. Push forward on the rear of cassette lid, to release the latches on the front of the lid that attach it to the loader assembly.
5. Remove the cassette lid.
6. Push down on the cassette tray so it latches back down inside the VCR.
7. Remove all screws on the rear edges or along the top edge of the main top cover. (VCRs with channel tuning controls beneath a door in the top cover may have a cover screw hidden in this compartment.)
8. Lift the rear of the cover up and pull the cover straight back to release the front latches, as shown in Figure 2.3.

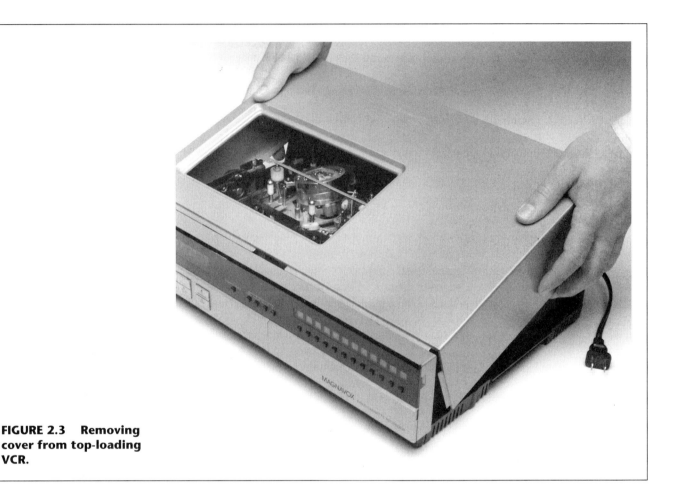

**FIGURE 2.3** Removing cover from top-loading VCR.

On top loading VCRs *without* visible screws attaching the cassette lid, it is usually not necessary to first remove the lid. Proceed as follows:

1. Eject the cassette tray.
2. Unplug the VCR.
3. Remove all top cover mounting screws.
4. Lift up the rear of the cover and pull straight back a short distance to release the front of the cover.
5. Move the top cover forward over the cassette tray to remove the cover.

To repeat, don't use excessive force when removing a VCR's top cover. Most need to be lifted slightly at the rear and pulled back a small amount to clear latches on the two sides of the machine. Also, the front edge of many top covers sits about an eighth of an inch under a lip in the front panel. After the cover is pulled back to clear the side latches or the front panel lip, or both, the cover can be lifted completely off. Figure 2.4 illustrates typical latches on the side of a VCR that engage slots along the bottom edge of the top cover.

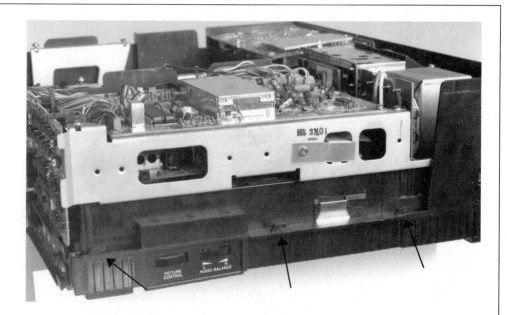

**FIGURE 2.4** L-shaped latches that engage VCR top cover.

### Front Loader Top Cover Removal

The top covers on VCRs with front loaders are usually removed in a very similar manner to VCRs with top loaders. Covers are held on with four to six screws. Frequent locations for these screws are along the top edge at the rear of the cover, on the sides of the cover, or both. On some VCRs, the top cover is also held on with two screws *through the bottom cover!* These top cover screws are often indicated with arrows marked on the bottom cover. As with top loader covers, it is usually necessary to lift the rear of the top cover slightly and pull back to clear the front panel lip, as illustrated in Figure 2.5, before the cover can be completely removed.

When replacing most VCR top covers, be sure that the front edge fits under the lip along the top edge of the front panel, and that any latches in the VCR side frame are properly engaged with the sides of the cover. *All* cover screws should be replaced. Metal covers act as RF shields and require that screws be in place for good contact with metal assemblies within the VCR. Shielding prevents outside radio frequency energy from interfering with signals in the VCR, as well as containing VCR-generated signals within the cabinet so they don't interfere with nearby A/V equipment.

Cover screws should be tightened until snug, not overtightened. Many cover screws, as well as numerous others in a VCR, screw into the plastic frame assembly. Too much force when replacing these screws can easily strip out threads in the plastic.

### Bottom Cover Removal

When it comes to the bottom cover of a VCR, there's really not much to say. In most cases, a single flat metal plate is held on with six screws, as shown in Figure 2.6. Removal of this panel is required to get at the underside of the tape transport mechanism, also called the undercarriage. Most VCRs have one or more printed circuit boards (PCBs) behind the bottom cover. (A PCB may also be called a PWB, for printed wiring board.)

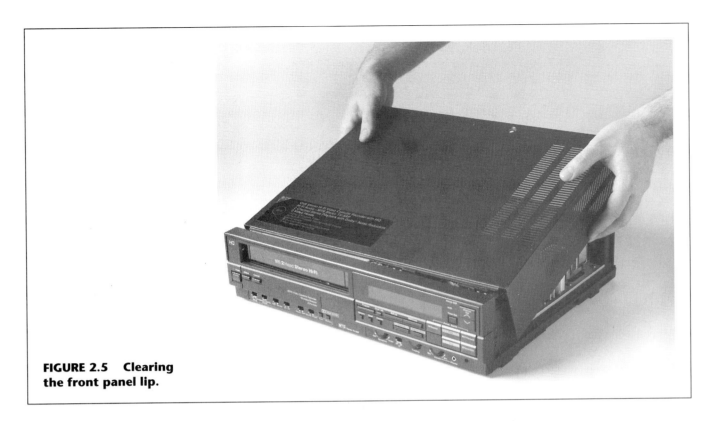

**FIGURE 2.5 Clearing the front panel lip.**

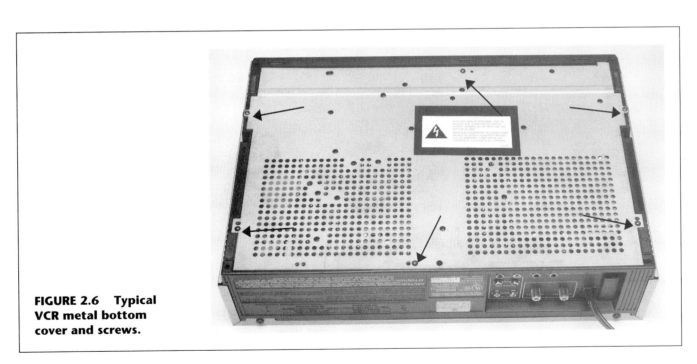

**FIGURE 2.6 Typical VCR metal bottom cover and screws.**

Unplug the VCR before removing the bottom cover, then place the VCR on its side or completely upside down, as in the previous Figure. Be careful when placing a unit upside down, if the top cover is already off, so that exposed components don't get damaged. Some units may be unstable when on one side; use one hand to steady the unit during cover removal. Do not roll a VCR onto its side; instead, pick it up, turn it, and then set it down on its side. With some models, rolling the unit on a workbench after the covers have been removed puts stress on a circuit board or other component, and may cause the part to break. Some bottom covers may have to be slid a short ways forward or backward to clear plastic tabs before they can be lifted off.

If a VCR has rubber or plastic feet with a screw in the center of each foot, see if the panel will come off after removing other screws first. If there are no other screws, or the panel stays on after you remove other screws in the bottom cover, then remove the screws in the center of the feet. Don't use excessive force when reinstalling screws in feet; this could warp or crack the feet. Be sure that any washers between screw heads and feet are reinstalled.

On some VCRs, you must first remove the plastic front panel before the bottom cover will come off. This procedure is covered shortly in this chapter.

Just like the metal top cover, the bottom panel serves as an RF shield, so it is important to replace *all* screws when it is reinstalled. Usually only one or perhaps two screws actually make an electrical connection between the bottom panel and a chassis ground strap or grounding bracket. As mentioned earlier, it is helpful to mark the screw locations on the cover with an indelible felt-tip pen; bottom covers frequently have several more screw holes than are actually used on any specific model. Some units have screws of different lengths, thread size, or thread pitch, so note which screw goes where for hassle-free re-assembly.

Don't overtighten screws when putting the cover back on; you could easily strip out or break the plastic frame.

## SHIELD PLATES

As you've just read, both top and bottom metal covers serve an additional function besides enclosing VCR mechanics and electronics. They also keep stray radio frequency energy, like that from a nearby radio or TV station antenna, from getting inside the unit, where it could interfere with VCR signals. Cover shielding works the other way around too, keeping VCR signals from leaking out and possibly interfering with a nearby FM radio or television receiver.

Within the VCR, there is usually at least one, and perhaps more, metal shield plates. They prevent radiated signals from one part of the VCR from interfering with low-level or weak signals in other areas of the unit.

Weak signals within a VCR include the RF inputs from the VHF and UHF IN terminals, and playback signals from magnetic tape heads, especially video heads in the rotating upper cylinder of the video drum. Radiated electromagnetic "noise" from the VCR power supply or from a motor could possibly be picked up by magnetic tape heads or by RF amplifiers in the VCR tuner section. Shield plates act as barriers to electromagnetic noise, absorbing and routing unwanted signals to ground.

*Noise* is a general term meaning *any unwanted signal*, whether audio, video, or RF, that rides along with the desired signal. For example, hum or crackling in the sound that is not part of the program is audio noise. Black-and-white horizontal snowy bands that appear at the bottom of the screen while a tape is playing are a form of video noise. Ghosts (offset double images) on a screen that is receiving a TV picture are caused by one or more reflected RF signals reaching the antenna a short time after the primary signal transmitted by the TV station. Said another way, ghosts are a form of RF noise, caused by reflections off large buildings or metal structures; the reflected signal arrives at the receiving antenna after the primary signal and causes a weak image to appear to the side of the primary image.

Most VCRs have a metal shield plate covering the top, rear portion of the transport mechanism. This is where the rotary video heads and stationary audio heads are. Usually two or three screws attach the shield plate. After the

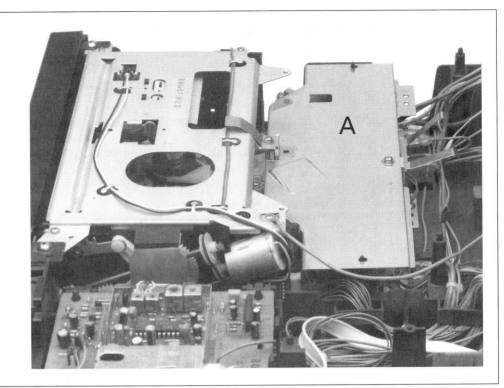

**FIGURE 2.7 Shield plate (A) over top, rear portion of tape transport.**

VCR top cover is removed, you will see it. A typical shield plate over the video drum and rear portion of the tape transport assembly is shown in Figure 2.7.

Before removing a shield plate, take an indelible felt marker and label the shield, along with an arrow indicating which is the front or back of the plate. This helps greatly when reassembling; you won't have to turn the plate around and around to see which way it goes—and on a VCR having more than one shield plate, you won't have to guess and experiment to see which goes where. You might even mark the container number for the mounting screws on the shield plate. This may seem to you like a lot of unnecessary part-marking and documentation. But when you have to remove many covers and parts to make a repair—and especially when it may be several days before reassembly—you'll really appreciate how easily things go back together when you've marked, labeled, and documented during disassembly. You owe it to yourself to make things as easy as possible while accomplishing the best result. Figure 2.8 shows the same VCR pictured in Figure 2.7, but with the tape transport shield plate removed.

*All* mounting screws must be used when reinstalling a shield plate. Shield effectiveness decreases if one or more screws are left off. A missing screw could also cause the plate to vibrate in one or more VCR operating modes, producing an undesirable mechanical sound.

## FRONT PANEL REMOVAL

For many VCR repairs you won't need to remove the front panel. Don't remove it unless you have to; on most VCRs it's held on with plastic latches, which are easily broken. Front panel removal is required if a pushbutton, switch, or indicator behind the panel is not working. Sometimes, though, you'll need to remove the front panel to service or remove the front loader assembly. With some VCRs, the front panel has to come off before the bottom cover can be removed. In general, here is what to do:

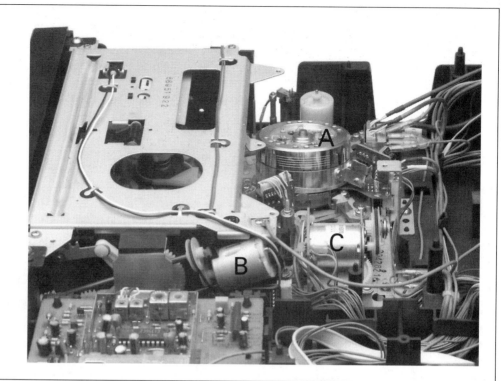

**FIGURE 2.8** Shield plate over top, rear portion of tape transport removed. A: Video drum B: Cassette load motor C: Mode motor

First, unplug the unit and remove the top cover. On some VCRs you will need to take the bottom cover off before removing the front panel. Most front panels mount with three or four plastic latches, or a few screws, or both latches and screws across the top edge of the panel. These panels usually hinge out and unlatch along the bottom edge for complete removal. Figure 2.9 shows typical front panel mounting with screws and plastic locking latches.

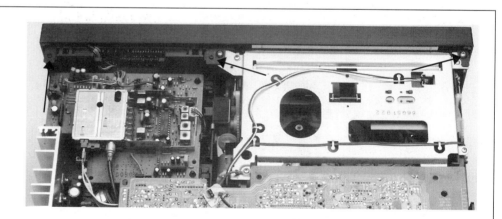

**FIGURE 2.9** Mounting screws and plastic latches along top edge of front panel.

After removing any front panel mounting screws, *carefully* lift up on the plastic tabs to release them from the round pins or rectangular studs along the top edge of the VCR frame. Then pull outward on the top of the front panel to unlatch its bottom edge. Typically, three or four plastic tabs along the bottom edge of the panel slip into slots in the plastic chassis frame. Sometimes the slots are in the front panel with matching tabs along the bottom of the plastic chassis. In either design, pivoting the top of the front panel downward disengages

**FIGURE 2.10** Plastic tab-in-slot latches along bottom edge of front panel.

the tabs-in-slots so the panel can be lifted off. Figure 2.10 shows tab-in-slot latches along the bottom edge of a front panel.

Although this is how many front panels are mounted, there are several variations. Carefully examine all around the front panel for additional screws, plastic locking latches, or both. Front panels should come off *very easily*, without forcing, *once you have released all latches and removed all screws*.

Following are variations in front panel mounting:

- Screw(s) on bottom edge
- Screw(s) along each side
- Screw(s) behind flip-down door(s)
- Plastic locking latch on each side
- Locking latches along bottom.

Plastic latches can break *very easily*, especially on older units where the plastic has become brittle. Do not lift or bend plastic latches any more than absolutely necessary to unlock them from the pin or tab with which they are mated. Carefully align pins in latches and tabs in slots when reinstalling a front panel. Tighten all screws, if used, so they are just snug; plastic chassis threads strip easily.

With the front panel removed, you can get at one or more control/indicator printed circuit boards (PCBs). Figure 2.11 shows the control panel PCBs on a two-head, Hi-Fi Stereo VCR with digital tuning. Two PCBs contain all VCR controls and indicators (except for the Ch 3/4 modulator select switch on the rear panel).

Note the following control panel components in Figure 2.11:

A. **Microswitches:** Operated by front panel pushbuttons: Power, Play, Pause, FF, etc.
B. **Slide switches:** Record speed: SP, LP, EP
   Input selector: Tuner, Simul, Line
   Dolby NR: On/Off

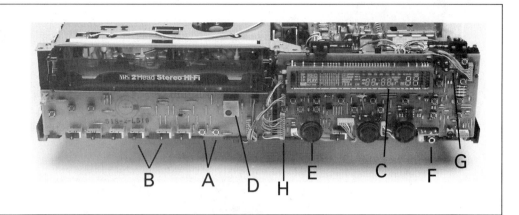

**FIGURE 2.11** Control panel printed circuit boards, buttons, and display.

C. **Integrated display indicator panel:** Fluorescent panel for day, time, index counter, channel #; mode: Play, Rec; Timer and Cassette-in icons, etc.
D. **IR photodetector:** Receives infrared pulses from remote control unit.
E. **Tracking control potentiometer:** Variable resistor, like radio volume control.
F. **Headphone jack:** 1/8-inch phone jack for stereo headset (volume control to its left).
G. **Individual LED indicators:** Stereo; Hi-Fi; SAP (second audio program).
H. **Multi-pin connectors:** Wiring interface between printed circuit boards and VCR components.

Figure 2.11 shows only one particular VCR model, of course, but most units have similar control panel components. Nearly all incorporate the same small microswitches and a single display/indicator panel for showing day, time, and various operating modes, like >> for Fast Forward. You are looking at the component side of the PCBs; the back side has printed circuit wires connecting the various components and multi-pin connectors.

Control panel circuit boards are usually very dependable. Occasionally a slide switch or microswitch may need replacement, but most often this is because of some physical abuse the VCR has suffered. A plastic shaft or nub on the back of a front panel pushbutton that presses against a microswitch may break with abuse, or if something bangs into the VCR. This can often be repaired with epoxy cement or cyanoacrylate glue.

The entire florescent display/indicator panel is a single, reliable unit. If it gets broken or fails, it *may* be less expensive and easier in the long run to replace the entire PCB, including the display (if the board is available), rather than unsoldering and soldering display assemblies, which can have over thirty connections to the printed wiring board. (Soldering and unsoldering techniques are described in Chapter 11.)

## RELEASING CIRCUIT BOARDS

By far the most frequent VCR repairs require work on the tape transport assembly, either on the top, bottom, or both. Some of the more common maintenance and repair procedures on the top of the tape transport include:

- Cleaning rotary heads and video drum surfaces
- Cleaning stationary audio/control (A/C) head
- Cleaning tape guides
- Cleaning capstan shaft and lubricating bearing
- Cleaning or replacing rubber pinch roller

- Cleaning or replacing rubber idler wheel
- Adjusting supply spindle back tension.

On the underside of the tape transport, called the *undercarriage*, the following are fairly regular procedures:

- Replacing rubber drive belts
- Cleaning pulley grooves
- Lubricating cam assembly
- Cleaning and adjusting mode switch
- Replacing slip clutch assembly.

On nearly all VCRs, to get at the top of the transport assembly, or access the undercarriage, or both, you must move one or more printed circuit boards out of the way. Frequently a circuit board covers part of the tape transport on the top of the VCR, while one or perhaps two PCBs are beneath the undercarriage. On some VCRs, you can get at everything in the undercarriage without repositioning any circuit board; on other models, you can't even get a glimpse of the undercarriage mechanics without first moving a PCB.

This section describes some of the many ways to release and pivot circuit boards out of the way so you can get at the top and bottom of the tape transport assembly. You'll need to release circuit boards for most electronic testing, adjustments, and PCB part replacement. Usually circuit boards can pivot or hinge open without unplugging any multi-conductor connectors or unsoldering wires going to them.

> CAUTION: Be sure to take ESD precautions, such as wearing a grounded wrist strap, when working with circuit boards.

The majority of printed circuit boards are mounted to the plastic VCR frame with screws, plastic latches, or both. Many have a hinge at one edge of the board so they can swing open like a door, out of the way, after mounting screws have been removed or after plastic latches have been released, or a combination of both. *Circuit boards are delicate and are easily cracked with excessive force.* Careful observation of circuit board mounting schemes and *gentle handling* are musts to keep PCBs alive and well.

All screws on a PCB are *not* necessarily for board mounting; some may attach a subassembly, driver transistor, regulator, or some other device to the opposite side of the board. On some boards, screws attach a board edge to a metal hinge piece; do not remove these screws. PCB mounting screws are usually Phillips head. The ones you need to remove to release a board are often colored red, pink, or blue, whereas screws that should be left alone are not colored. Board mounting screws sometimes have circles printed around them or arrows that point to them.

Some PCB mounting screws may have an insulating fiber washer underneath the screw head. Be sure to make a note of this! Replace these insulators during reassembly. Some areas of a circuit board may have a piece of insulating material, called *fish paper*, protecting them. Fish paper is usually gray, and is similar to very stiff shirt cardboard. Make a sketch or otherwise note the location of any insulating material so everything gets put back together correctly.

Figure 2.12 shows a large circuit board that must be pivoted back to get at the top of the tape transport. This particular board has two screws attaching its front edge to the front loader assembly, and two plastic latches holding down the right side. Plastic locating pins protrude through holes in the board to position it. Removing the two screws and releasing the latches allows the board to pivot back.

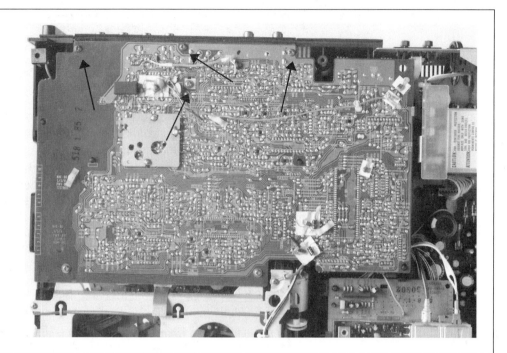

**FIGURE 2.12** PCB over top rear of tape transport.

Notice the three screws along the rear edge of the PCB in Figure 2.12; these mount a subassembly to the opposite side of the board. The subassembly contains the RF modulator/antenna switcher, with Ch-3/4 switch, and all rear panel input/output jacks. Also note the single screw toward the rear of the board, left of center; this holds a bracket for a sub-PCB mounted on the component side of the large board. These four screws *do not* have to be removed to pivot the board back to service the tape transport.

Figure 2.13 shows the board in the previous figure pivoted back. (A shield plate covering the rear portion of the transport has been removed.) The large circuit board is screwed to a section of the rear panel containing the RF modulator/antenna switcher, VHF and A/V input and output jacks, and the Ch-3/4 selector switch. The lower edge of this rear panel section pivots open on a plastic hinge at each end.

Figure 2.14 is a VCR with two printed circuit boards that open on hinges. Each board is held closed with two screws; removing these allows you to pivot the boards up to get at the other side. On this particular VCR, neither board needs to be moved to service the top of the tape transport mechanism. In Figure 2.14, the board at left rear of the VCR has two white plastic hinges along the front edge. The board at front right has a single metal bracket along its left edge that pivots on two plastic pins, one at each end of the bracket.

Figure 2.15 shows the two printed circuit boards in the previous figure released and pivoted up, exposing the underside of each. (The front loader and a metal shield plate over the rear portion of the tape transport has been removed in Figures 2.14 and 2.15.)

Figure 2.16 is the underside of a VCR having two large circuit boards. The undercarriage is under the lower board in the photo. To get at the undercarriage, remove a screw in each corner of the board, release three plastic latches along the lower edge and one latch at the front left, and pivot the board open on the three white flexible plastic hinges.

After mounting screws are removed and plastic latches released, the board pivots open on the three flexible plastic hinges that connect the two boards. Figure 2.17 shows the board opened up, revealing the undercarriage.

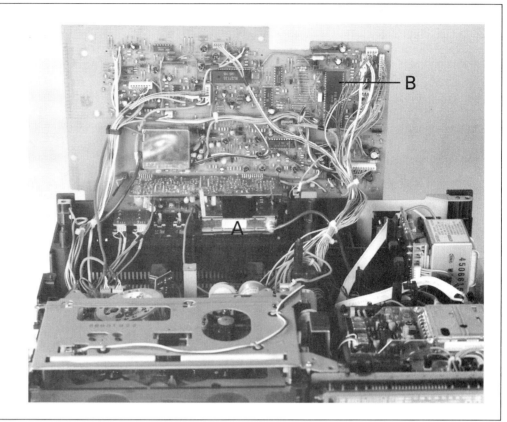

**FIGURE 2.13** PCB pivoted back to expose top rear of tape transport.
A: RF Modulator and VHF antenna switcher on rear panel section
B: System control microprocessor IC

**FIGURE 2.14** Two hinged printed circuit boards.

**FIGURE 2.15** Two printed circuit boards hinged open.

There are several variations for circuit board mounting, but most are similar to those just described. Some PCBs are held on one or more edges by plastic grooves in the VCR frame. Slight bowing of the plastic allows the board edge to pop out of a groove. Other PCBs have front panel controls, such as the tracking potentiometer or the speed select slide switch, directly mounted to the board. On these, the board may have to be slid toward the rear slightly before pivoting so these controls clear the front panel.

On most VCRs, releasing circuit boards is fairly straightforward. However, there are some where it isn't immediately obvious how to proceed. Look on these as a challenging puzzle! With some models, you must remove the front panel, unclip or unscrew one or more PCBs at the front of the unit, and then pivot the whole works toward the rear of the machine to get at the undercarriage. Look for those red- or blue-tinted screws and remove them first. Most often they need to come out, and there may be others. Screws aren't always along the edge of a board; there may be one or more near the center.

Figure 2.18 shows one VCR model where almost everything hinges open at the bottom rear to expose the undercarriage. Front panel control PCBs, top of cabinet tuning controls, rear panel, and even the power supply pivot as a unit to reveal the underside of the transport. On this particular machine, 16 screws mount 2 large circuit boards to a metal frame, but removing only 4 screws and releasing 6 plastic clips or latches allows the works to pivot open.

Notice in Figure 2.18 that line voltage terminals on the power supply are exposed, and could easily be touched during service. Take the time to cover these terminals with electrical tape or other insulating material before plugging

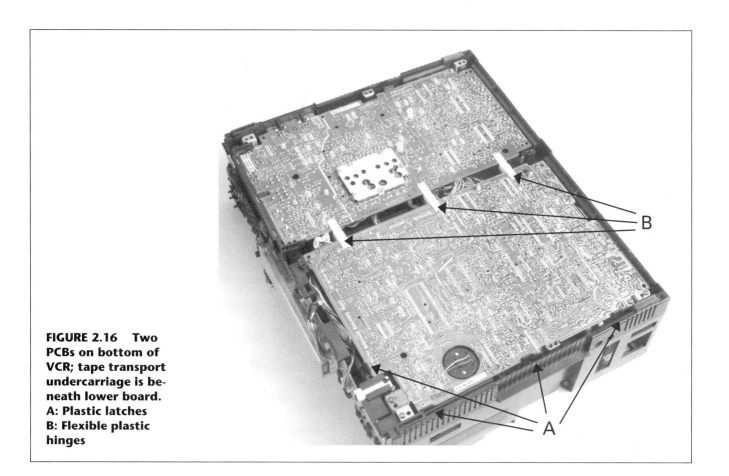

**FIGURE 2.16** Two PCBs on bottom of VCR; tape transport undercarriage is beneath lower board.
A: Plastic latches
B: Flexible plastic hinges

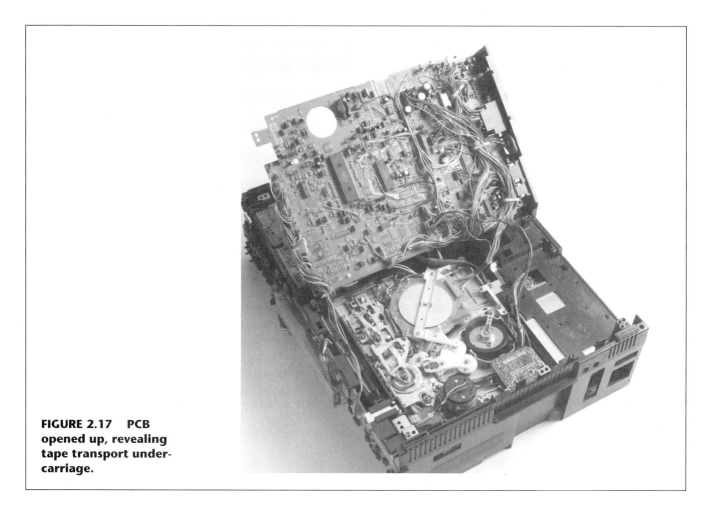

**FIGURE 2.17** PCB opened up, revealing tape transport undercarriage.

FIGURE 2.18 Bottom view of VCR model where almost everything pivots open to expose tape transport undercarriage.

the unit in. When service work is completed, unplug the VCR and remove the tape before closing things up. Leaving the tape on could restrict air flow and cause overheating.

The main thing when releasing circuit boards is to not get frustrated, nor hurry, nor use force. Cracked circuit boards are an almost guaranteed outcome of brute-force attempts to release a board. Look things over thoroughly before you start removing every screw in sight. Usually only a half-dozen must be removed. It's all right if you take out additional screws. Once you see that they don't secure a board, you can always put them back. As you proceed, it's very important to take notes and keep screws separated and labeled. It's not unusual to have to remove three or more different types of screws to release a circuit board. Some screws may thread into plastic posts in the VCR frame, whereas others screw into metal parts having different threads.

Start to remove screws very slowly. Sometimes screws threaded into plastic posts seize up after they've been installed for awhile. Rapid removal can snap the plastic part. "Nice 'n' easy" usually backs the screw out without breaking the plastic.

Once you've released a PCB, you may need to prop it open with a wooden stick or cardboard tube while you work on the VCR. Before plugging a VCR into line voltage while boards are hinged open, be sure no board wiring or components are touching metal machine parts or shorting out in any way. Folded newspaper or pieces of cloth make convenient insulators between circuit boards and other parts with which they may come in contact when you work on a unit with power applied.

## MAJOR COMPONENTS AND PCBS

Now that top and bottom covers have been removed, we'll take a look inside a typical VCR and identify some of the major components. Again, there is wide variation among manufacturers, and among different models by the same manufacturer, as to where components are located and in the overall VCR packaging scheme. In some VCRs, two large circuit boards contain nearly all electronic circuitry except for the power supply. Other models may have six, eight, or more separate PCBs, interconnected with many multi-conductor cables.

In this final section of Chapter 2, *some* of the major components inside a VCR are identified. You will read further descriptions of these components and how they operate in later chapters. For now, this is just a big picture of what you'll typically see inside the top and bottom of a VCR with its covers removed. Figure 2.19 is a view of the top of one unit, with the circuit board over the rear portion of the tape transport still in place. The following are major components visible on the topside of this VCR:

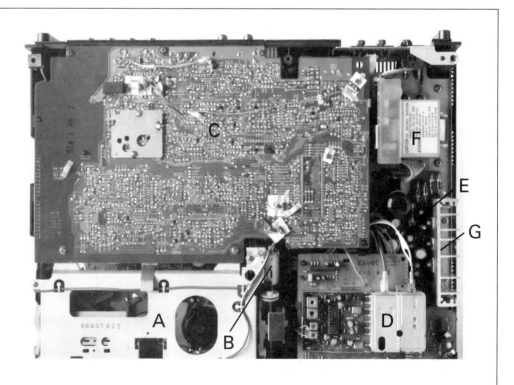

**FIGURE 2.19** Top view of representative VCR model.

A: **Front loader:** Draws in video cassette and then lowers it onto reel table. Cassette takes an inverted L-shaped path as it loads: in and down. When Eject is pressed, loader assembly reverses operation, lifting cassette up off transport and expelling it about one-half inch out compartment door.

B: **Cassette load motor:** Reversible direct current (DC) motor turns gears on each side of loader assembly to load and to eject a cassette. Twelve-volt motor in most VCRs.

C: **System control, servo, and video board:** PCB has circuits that (a) manage VCR operations with *system control microprocessor*, (b) precisely monitor and regulate video drum and capstan rotation with *servo* circuits, and (c) process video signals during Record and Playback.

D: **RF tuner section:** Receives RF input from rear panel terminals; has 300-ohm twin-lead from UHF IN and 75-ohm coax lead with RCA plug from VHF antenna switcher. Selects TV/Cable channel to be recorded.

E: **Power supply:** Produces several low voltage DC outputs, such as 5, 9, 12, and 35 volts. Some outputs are present whenever VCR is plugged in, to power timer/operations microprocessor, system control microprocessor, and remote control circuits.

F: **Power transformer:** Changes line voltage to lower AC voltages that are then rectified with diodes to produce low DC voltages.

G: **Voltage regulator heatsink:** Husky aluminum piece with fins draws off excessive heat produced by solid state voltage regulators. Two, three, or more integrated circuit regulators, similar in appearance to some power transistors, regulate a few output voltages, like +5 VDC. (Motor driver transistors or ICs may also be mounted to power supply heatsink.)

The transport, video drum, and tape load motor are beneath the large circuit board (C) in Figure 2.19.

- **Tape transport** Mechanical assembly that loads, positions, and moves videotape. Chief transport components include: (a) supply and take-up spindles that engage cassette tape reels; (b) several tape guides; (c) video drum; (d) stationary tape heads; and (e) capstan and pinch roller, which pull tape through tape path.
- **Video drum** Has direct drive motor to spin upper cylinder and rotary heads. Lower part of drum is stationary. Videotape wraps half-way around drum when tape is loaded.
- **Tape load motor** Reversible DC motor turns one or more cams and gears to thread tape through tape path and around video drum. Cam mechanism loads pinch roller against rotating capstan shaft and performs other tape transport mechanical operations, like ejecting cassette tray on some top loaders. Usually a 12-volt motor, also called *mode motor* or *cam motor*.

Figure 2.20 is a view of the underside of the same VCR. The bottom side of the tape transport assembly is called the *undercarriage*. This particular unit has no printed circuit board covering the undercarriage, so it can be easily serviced.

The following are some principal components visible on the underside of this VCR:

A: **Audio PCB:** Contains all circuitry for linear audio reproduced by stationary head *and* Hi-Fi Stereo, which uses two separate rotary heads on video drum upper cylinder. Also has circuits for second audio program (SAP).

B: **Capstan motor and flywheel:** This model has a direct drive (DD) capstan motor, where the motor shaft *is* the capstan shaft. Large diameter flywheel smooths out capstan speed. Rubber belt from capstan motor provides drive power for turning cassette supply and take-up reels.

C: **Mode cam gear and sensor:** Turned by mode or cam motor. Supplies motion to load and unload tape, and brake capstan flywheel. A cam on the transport topside turns with this cam to either move the pinch roller up against the capstan or to retract the roller. Sensor reports mechanical position of cams to system control microprocessor.

D: **Videodrum motor:** Part of videodrum assembly; rotates upper cylinder with rotary heads one complete revolution 30 times each second (1800 RPM).

E: **Tape load ring gears:** Driven by load or mode motor. Cause guides P-2 and P-3 to pull loop of tape from cassette and thread it through tape path during tape loading.

There are many differences in undercarriages, depending on the VCR model. Some are complex, whereas others use fewer parts. For example, the

**FIGURE 2.20** Bottom view of representative VCR model.

12-volt tape load motor, also called the *cam* or *mode* motor by some manufacturers, may be on the top or bottom of the tape transport; it may be mounted horizontally or vertically. The mode motor could be connected to the cam and load mechanism by a belt and pulley arrangement or with a worm gear.

This is just a brief look at the basic functions of parts on the underside of a typical tape transport. Chapter 3 provides details on these VCR operations and individual transport components.

## SUMMARY

It's important to be well organized when disassembling a VCR and while performing repairs. Keeping track of parts in numbered containers, taking notes identifying what goes where, marking parts with an indelible marker, scribing the outline of a part on the chassis or baseplate before removing it, and making a paper sketch are tremendous aids to reassembling a unit *correctly*. A clean, well-lit, nonconductive work surface is a must for performing a safe, quality repair.

Save repair time by documenting what *is* working and making notes about incorrect operations and failure symptoms before digging under the covers. Having these facts on paper helps focus on the area of the machine most likely causing trouble and can make troubleshooting easier.

Safety is extremely important when working on a VCR. Always unplug the power cord before removing covers, and at all times when it isn't necessary to have the unit powered during repair activities. Always remove metallic rings, watches, bracelets, and pendants. Dangerous line voltage is frequently present on exposed terminals near where the AC line cord enters the back panel and connects to the power supply. These areas should be taped over if the VCR is plugged in during service.

Always use extreme care so there is no chance of current flow through your body, such as from right hand to left hand. Respect all electrical circuits; a secondary reaction (for example, an arm muscle involuntarily retracting as a result of even a small electrical shock) can cause accidents, personal injury, and broken parts. Some VCRs have switching mode or transistor switching power supplies, which can have voltages even higher than AC line voltage.

Eyeglasses or safety glasses should be worn whenever springs, C-clips, or E-clips are being removed or installed. These parts can easily fly into your face. Cupping a cloth around a spring or similar part as it is taken off or replaced can help prevent injury and part loss.

Electrostatic discharge (ESD) from your body can destroy or weaken electronic components. Microprocessor and CMOS memory chips are especially vulnerable to ESD damage. Wearing an ESD wrist strap connected to frame or chassis ground drains static charges off your body, preventing damaging ESD. Without a wrist strap, you should first touch a ground point with your fingers before touching VCR circuitry or electronic parts. An electrostatic discharge of just a few hundred volts is too small to feel in most cases, but can be deadly to ESD-sensitive devices.

Most VCR top covers are held on with four to six screws along the top and rear edges, or along the sides of the cabinet. A few VCRs have two screws through the bottom panel that also must be removed to take off the top cover. After cover screws are removed, it is often necessary to slide the top cover back about an inch, and then lift it straight off to clear L-shaped latches molded into the sides of the VCR frame.

A metal shield plate over the top, rear portion of the tape transport keeps interfering signals from being picked up by magnetic tape heads. All shield plates should be replaced after completion of service work.

Bottom covers are normally held on with six or more screws. Covers may have to be slid out from under plastic latches to be removed. *All* screws should be reinstalled to retain the electromagnetic shielding provided by VCR top and bottom covers and shield plates.

Many VCR front panels are held on with plastic locking latches or clips, which break easily if panel removal is not gentle. Look thoroughly for additional mounting screws to remove or plastic latches to release rather than using force to remove VCR panels and printed circuit boards; they usually release quite easily with little force when done correctly.

Printed circuit boards often must be released from the VCR frame and pivoted open like a door to get at either the top or bottom of the tape transport, or both. Mounting screws that have to be removed to swing boards open are frequently colored red or blue. A combination of screws and plastic locking latches holds many boards in place. Care must be taken that no PCB wiring or component can short out to metal parts or other boards while they are swung open. Folded newspaper or pieces of cloth may be needed to insulate boards when power is applied.

Chapter 2 concluded by showing top and bottom views of a representative VCR with its covers removed. Although diverse VCR models will look different, they all have similar major components that perform the same types of operations, like loading a cassette, loading tape, spinning the capstan shaft, and providing several DC voltages to the various VCR circuits.

## SELF-CHECK QUESTIONS

1. What does your textbook recommend that you do *before* taking the cover off a VCR?
2. Explain why a VCR that has just been brought inside from the cold may not power up right away.
3. What area(s) on a typical VCR should you be particularly careful working around in order to avoid a dangerous electrical shock?
4. What is the principal method for protecting electronic components inside a VCR from ESD damage while you are working on the unit?
5. What should you do when removing the cover mounting, screws, and other parts during VCR disassembly?
6. Explain the purpose of screws or nuts inside a VCR that have a small dab of paint on them. What should you do with this type of hardware?
7. Why will you know which screws to remove in order to swing out a printed circuit board?
8. Describe the function of the *tape load motor* in a typical VCR. This motor also goes by two other names; what are they?
9. Describe the purpose of the *transformer* and *power supply* in a VCR.
10. On the particular VCR in Figure 2.20, what are the functions of the two large ring gears in the undercarriage?

# CHAPTER 3 UNDERSTANDING THE VIDEOTAPE PATH

This chapter examines the path taken by videotape within the transport when tape is loaded in a VHS VCR. There are many items in the tape path between where videotape leaves the cassette supply reel and where it returns to the take-up reel. The function and operation of each part in this tape path are discussed.

## CHAPTER OBJECTIVES

**Upon completing this chapter of study, you should be able to:**

1. Describe the basic operation of loading a VHS tape along the tape path.
2. Describe the principal function and operation of the following items in the tape path:
   - Supply and take-up spindles
   - Tape holdback tension band
   - Back tension guide pole
   - Full erase head
   - Inertia roller
   - Drum entrance and exit roller guides (P-2 & P-3)
   - V-blocks (V-mounts or V-stops)
   - Video drum
   - Audio/Control/audio erase (A/C) head
   - Capstan and pinch roller.
3. Describe the purpose of each of the following switches or sensors associated with the tape path:
   - Cassette-in switch
   - Record interlock switch
   - Supply and take-up reel sensors
   - Tape end sensors
   - Mode or cam switch
   - Dew sensor.

## INTRODUCTION TO TAPE LOADING

After a video cassette is initially put into a VCR, it is placed on what is called the *reel table* portion of the tape transport. This is called *loading a cassette*. Once the cassette itself is loaded, videotape is pulled from the open front door of the cassette and threaded through the tape path in order to play or record. The tape path consists of many components that handle videotape during Play, Forward Scan (search forward or cue), Reverse Scan (search reverse or review), Record, Fast Forward, and Rewind modes.

The path taken by videotape between the cassette supply and take-up reels is without a doubt the most critical operation of a video cassette recorder. Many components along the tape path ensure proper tape positioning, tension, and movement. Other components, namely magnetic record/playback and erase heads, write and read audio and video information on the magnetic tape.

Many VCR maintenance procedures and repairs are concerned with videotape movement along the tape path. In fact, the majority of VCR repairs deal with tape path parts. You'll learn about cleaning, lubricating, aligning, and replacing worn tape path components in later chapters. Right now, it's important to know what the tape path is and what the various parts do. Knowing how things are *supposed* to work is half the battle in diagnosing and repairing any mechanical or electronic machine!

As briefly discussed in Chapter 1, *cassette loading* consists of the proper placement of a video cassette onto the reel table by either a top loader or front loader carriage mechanism. Once the cassette is loaded, the next step before recording or playback can occur is *tape loading*. During tape loading, a loop of videotape is drawn out of its cassette and threaded through the tape path. As part of this process, videotape wraps half-way around a spinning video drum. Be certain to make the distinction between cassette loading and tape loading!

During cassette loading, the door on the front of the cassette opens as the cassette descends onto the reel table. When the VCR is put into Play or Record mode, the tape load motor runs to pull a loop of tape out of the cassette and thread it through the tape path. Looking down at a loaded cassette, videotape comes off the supply reel on the left, threads through the tape path, and returns to the take-up reel on the right. Several tape guides correctly position the tape while other components move it smoothly, which is absolutely necessary for stable picture and sound.

Figure 3.1 illustrates the VHS videotape path with tape loaded, along with the major components. Notice that the tape itself forms the shape of the letter **M**. For this reason, VHS tape loading is sometimes called M-loading. You'll find it helpful to refer to Figure 3.1 often during the following descriptions of tape path components and functions.

Supply and take-up reels inside the cassette rest on transport *spindles* that drive the reels in one direction or another during Play, Rewind, Fast Forward, and Forward or Reverse Scan modes. When a cassette is loaded, the supply and take-up roller P-guides actually poke into the underside of the cassette in the

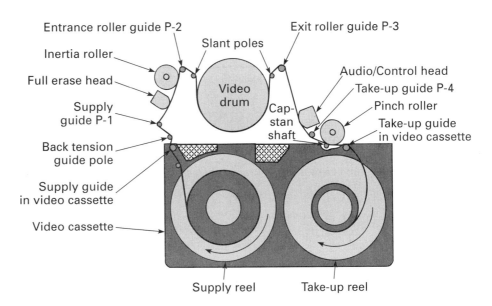

**FIGURE 3.1 VHS videotape path and major components, with tape loaded.**

cross-hatched areas in Figure 3.1. These two P-guides move forward during tape loading to thread tape through the tape path and position it around the video drum.

Once loaded, tape is pulled from left to right through the tape path by a round rubber pinch roller that presses tape against a rotating capstan shaft.

Various manufacturers use different methods to load videotape. For this reason, there are variations in tape path components, the names of the components, their locations on the reel table, and the particular ways they work. Figure 3.1 is representative of typical VCRs. Nevertheless, the basic principles of tape handling remain the same among VCRs.

Following are explanations of tape transport components and what they do. The terms *tape transport* and *reel table* both refer to the mechanical assembly that controls tape movement during all operating modes, such as Play, Fast Forward, Rewind, and Pause. Several different names are frequently used for the same part. As you've already learned, the small motor that loads tape and performs other transport functions may be called the *tape load motor, cam motor, mode motor*, or simply the *load motor*, depending on the particular VCR manufacturer and model. This is why you will come across such a variety of terms and names for similar items throughout this textbook.

## SUPPLY SPINDLE AND TAPE HOLDBACK TENSION BAND

The supply *spindle* on the tape transport reel table engages with and controls the supply *reel* on the left side of a videocassette. When a cassette is loaded, the spindle is directly beneath the supply reel. Grooves or slots in the cassette reel hub, called *splines*, engage three small projections on the transport spindle, which then controls rotation of the cassette reel. (See Figure 3.2.)

Be sure to make the distinction between *spindle* and *reel*. Supply and take-up *spindles* are part of the reel table portion of the tape transport. Supply and take-up *reels*, upon which videotape is wound, are inside the cassette housing or shell.

The supply spindle has these primary functions:

- Provides proper tape holdback tension during record, play, and fast-forward. Tape moves left to right through the tape path; supply reel turns clockwise (CW).

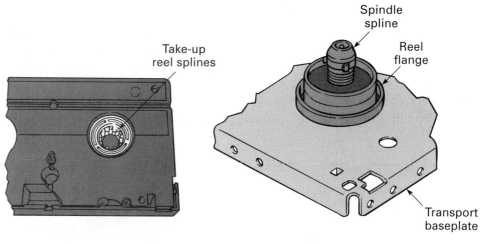

**FIGURE 3.2** Cassette tape reel and transport spindle.

- Stops or brakes the supply reel when the machine is stopped, to prevent tape spillage.
- Winds tape back onto the supply reel during rewind. Tape moves from right to left; supply reel turns counterclockwise (CCW).
- Reels in loop of tape when tape is unloading.

Tape *holdback tension* is often supplied by a felt-covered band that wraps around approximately 180 degrees of the lower portion, or flange, of the supply spindle. This provides a controlled amount of drag on the tape as it exits the supply reel. Proper tape tension is critical during play and record to keep tape taut throughout the tape path, between supply reel and capstan.

Without adequate holdback tension, tape is not held firmly against the erase head, audio head, and video drum. This can cause video and audio *dropout* during playback, which is momentary loss of signal transfer between videotape and magnetic tape heads. The image on screen may just flicker, have horizontal noise lines running through it occasionally, or be totally absent for several seconds, depending on how long the dropout occurs. Some VCRs output a solid blue screen during dropout, just as if there were no material recorded on tape. Audio dropout usually results in a large decrease in volume and a very muddy, bassy sound, with the higher frequencies totally absent.

When the VCR is in Stop mode, a brake pad comes into contact with the supply spindle, preventing rotation.

During rewind, or when reviewing a tape in Reverse Scan mode, the supply spindle rotates counterclockwise to reel in tape. In many VCRs, the supply spindle is driven by a rubber *idler wheel*.

An *idler* is a gear or wheel placed between two other wheels or gears that transfers motion from one to the other without changing speed or direction. An idler wheel is essentially a pulley with a rubber tire around its circumference, which transfers rotary motion from a motor or drive pulley to the supply spindle lower flange. Figure 3.3 illustrates an idler wheel powering a reel table spindle.

**FIGURE 3.3 How an idler wheel operates in the tape transport.**

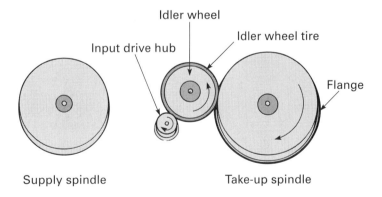

Some VCRs have an *idler gear* that powers the spindle instead of a rubber idler wheel. The idler gear engages teeth around the lower circumference of the spindle and teeth on a drive gear or drive pulley.

In Play or Record modes, the supply spindle must place a small amount of holdback tension on the tape so it runs smoothly through the tape path. Videotape has to be held taut against the various tape guides, heads, and video drum as it is pulled through the transport by the capstan and pinch roller, located near the take-up reel. Without holdback tension—with the supply reel just free-wheeling—tape would be loose in the tape path. To prevent this, holdback tension is provided with a tension band around the lower portion of the supply spindle.

**FIGURE 3.4 Supply spindle with tape hold-back tension band.
A: Supply spindle
B: Felt-lined tension band C: Back tension guide pole**

Figure 3.4 shows a typical arrangement of supply spindle with tape holdback tension band. The tension band consists of a thin, flexible metal strip that is lined with felt. It wraps around approximately 180 degrees of the supply spindle, so the felt presses against the rotating spindle, providing supply reel holdback tension. The tighter the band is about the spindle, the greater the tape holdback tension. On some models, the tension band may also brake the spindle when the VCR is put into Stop mode, and so may be called a *brake shoe band*.

Some machines have a completely separate motor directly connected to the supply spindle. In these machines, back tension, braking, rewind, and reverse scan functions are performed by electronically controlling voltages to the direct drive (DD) motor. VCRs incorporating direct drive have far fewer mechanical components, such as belts, pulleys, idlers, and gears. This generally makes them more reliable. Of course, the other side of the coin is that more sophisticated electronic circuitry is required to drive and control the various motors. Direct drive is generally found in more expensive VCRs.

Supply *spindle* height is adjusted so that the supply *reel* is elevated slightly inside the cassette housing when a cassette is loaded onto the reel table. In this raised position, the supply reel is positioned *between* the floor and ceiling of the cassette housing, so that the lower and upper reel flanges don't scrape against the inside surfaces of the cassette.

## BACK TENSION GUIDE POLE

As tape exits the supply side of the cassette it rides against a stationary guide pole. Some manufacturers refer to this as the *P-0 guide*. From there, tape crosses a movable tape guide, called the *back tension guide pole*. The back tension guide pole is situated inside the videocassette shell to the right of the P-2 slant pole when tape is unloaded; refer to Figure 3.4, item C. During tape load, the guide pole swings out to the left as P-2 and its slant pole move toward the video drum.

The back tension guide assembly consists of a spring-loaded pivot plate or bracket, to which is mounted the back tension guide pole that presses against videotape. The tension band around the supply spindle connects to this same pivot bracket. As the tension bracket pivots left or right, it tightens or loosens the band around the supply spindle, respectively. All this results in the back tension guide assembly regulating the amount of tape holdback tension provided by the back tension band.

Don't be concerned if you don't understand it all just yet; the following explanation will hopefully clarify how tape holdback tension is controlled.

When tape is loaded, the back tension guide pole is spring-loaded to the left side of the transport assembly. As tape exits the cassette, the guide pole is on the right side of the tape, and so pulls tape towards the left side of the transport.

Now if the tape tends to go slack, the guide pole moves further to the left, away from the center of the transport. This causes its bracket to pivot, which tightens the tension band around the supply spindle. The resulting increase in drag on the supply spindle puts additional holdback tension on the tape, keeping it taut.

Conversely, if tape is *too* taut, it forces the back tension guide pole to the right. Now the bracket pivots in the opposite direction, relaxing the tension band around the supply spindle. This reduces drag on the supply spindle and thus decreases tape holdback tension.

You may want to re-read the previous three paragraphs while looking back at Figure 3.4.

The back tension guide pole and supply spindle tension band thus work together as a self-regulating mechanism to maintain the proper tape holdback tension, so important to quality record and playback. Without such a compensation system, tape holdback tension would change greatly between the beginning of a tape, with a full supply reel, and further along in the tape, when less and less tape is on the reel. Feedback between back tension guide pole and spindle tension band may also be described as a mechanical *servo system* or a *governor*.

Figure 3.5 shows a typical back tension guide bracket configuration. Videotape is positioned where it would be in the tape path during normal Play mode, but the videocassette and front loader are gone for better visibility. Note the back tension guide pole, back tension bracket pivot point, and where the back tension band attaches to the pivot bracket. Once again, proper tape holdback tension provided by the supply reel is *extremely important!* Tape should pass through the transport neither too loosely nor too tightly. There are different arrangements for this mechanical servo system between back tension guide pole and supply spindle tension band in VCRs, but the results are the same.

*Supply guide P-1* is a fixed or stationary tape guide between the back tension guide pole and the full erase head.

If tape is too loose as it passes the full erase head, previously recorded material on tape may not be fully erased. If tape is too loose as it wraps around the video drum, video quality *will* suffer, causing problems such as noise lines and signal dropout, where the picture may totally disappear for a short time. Intermittent audio and tape damage are also possible with low tape tension.

On the other hand should tape tension be too high, heads and guides along the videotape path experience excessive wear, especially video head tips. Tape wear also increases. High holdback tension can cause the tape itself to be stretched or deformed, as it wraps too tightly around tape guides. Excessive tape holdback tension can even trigger a VCR's automatic shutdown circuitry; for example, the system control microcomputer can initiate tape unload and power down the VCR if it detects that the video drum or capstan is spinning too slowly, which could occur with very high tape tension within the tape path.

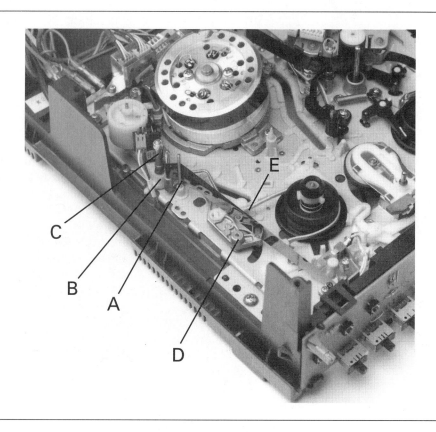

**FIGURE 3.5** Tape holdback tension components. A: Stationary guide pin P-0  B: Back tension guide pole  C: Supply side tape guide P-1  D: Holdback tension bracket pivot  E: Tension band

Obviously, both of these conditions are undesirable. Tape holdback tension adjustment is covered in Chapter 7.

## FULL ERASE HEAD

The next major tape transport component is the full erase head. Note that there may be one or two stationary tape guides between the back tension guide pole and the full erase head. These keep tape on its proper path. Tape guide height may be adjustable to position tape correctly over the full erase head.

During recording, the full erase head magnetically wipes the tape clean, removing *all* previously recorded information: audio, video, and control track signals. Of course, the full erase head is also energized whenever a recording is being made on a blank, or virgin, tape. The vertical magnetic gap in the full erase head covers the entire half-inch videotape width.

Figure 3.6 shows the full erase head in the tape path. For this photo the cam mechanism has been cycled to the tape-loaded position, without videotape being present: note the P-2 guide and slant pole on the left side of the video drum. Also notice that without any videotape to hold it back, the back tension guide pole has swung to the extreme left side of the transport,.

During recording, an alternating current (AC) signal is applied to the full erase head coil. This is a sine wave with a fixed frequency, generally somewhere between 65 kilohertz (kHz) and 125 kHz, depending on the particular machine. Erase frequency is not critical. Its job is to scramble the magnetic domains on the tape into a random pattern. Imagine spelling out your name on a piece of cardboard, forming block letters with lots of match sticks or toothpicks. Now shake the cardboard so the letter pattern is destroyed. That's essentially what the full erase head does with magnetic particles on the tape! Any information previously on the tape is wiped out.

Chapter 3 *Understanding the Videotape Path* ■ 57

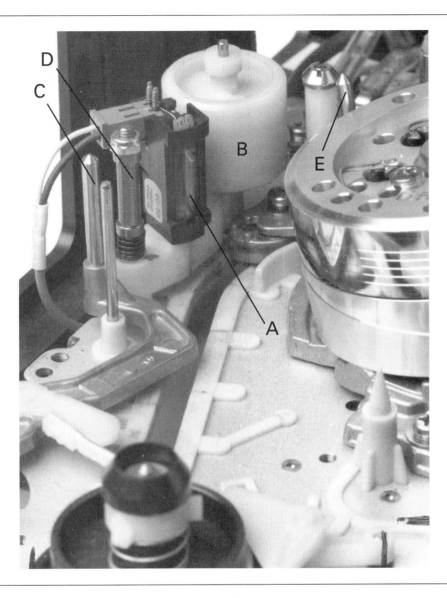

**FIGURE 3.6** Full erase head (A) and inertia roller (B) C: Back tension guide pole D: P-1 guide E: P-2 guide and slant pole

There *are* times when the full erase head is *not* energized during recording, specifically to enable either the audio or video dubbing capabilities of some VCRs (usually the more expensive ones with lots of features). In audio dubbing, a new linear audio track is laid down on the tape without altering the existing video. Some machines can also preserve existing linear audio, but record new video. Videotape editing is discussed further throughout this textbook.

When compared with video and audio tape heads, full erase head alignment with the tape path is not nearly as critical. Its job requires only brute force, rather than the precision needed for audio and video reproduction. In most VCR models, the full erase head is mounted on a small, spring-loaded pivot plate or bracket, along with the inertia, or impedance, roller.

## INERTIA ROLLER

The next item in the tape path between cassette supply and take-up reels is an *inertia roller*, also called a *tension roller* or *impedance roller*. Immediately after the full erase head, the tape contacts a spool approximately 3/4-inch in diameter. In fact, it looks rather like a small spool of thread (without the thread, of course). This inertia roller is mounted on a spring-loaded pivot plate or bracket. Usually the full erase head is mounted to the same pivot plate.

The inertia roller places a *slight* tension on the tape at this point. This helps keep the tape in intimate contact with the full erase head, and smooths out small irregularities in tape tension that may still exist after the action of the back tension guide pole and supply spindle tension band. Another way of describing the inertia roller's function is to say that it prevents the tape from undergoing small, rapid changes in tape tension, sometimes referred to as *shuttering*. A VCR designer's goal is to have tape motion as smooth and constant as possible when it wraps around the video head drum; the inertia or impedance roller helps accomplish this.

How VCR designers reach this goal varies. Figure 3.6 shows a typical inertia roller next to the full erase head in the tape path. As just mentioned, the roller may be mounted on a common spring-loaded bracket along with the full erase head. In some designs, the roller is *before* the full erase head in the tape path. Yet another variation couples the impedance roller with the back tension guide bracket, so the roller also becomes part of the mechanism regulating the all-important tape tension.

Some models, especially those with direct drive spindle motors that monitor and adjust tape tension electronically, may not have an impedance roller at all. Other models may have an inertia roller only on the *exit* side of the video drum. Still other models have *two* inertia rollers, one on the entrance side and one on the exit side of the drum. Whatever the design, the function is essentially the same for all: to help maintain extremely uniform tape tension around the video drum.

## DRUM ENTRANCE AND EXIT ROLLER GUIDES P-2 AND P-3

Tape guides, also called *P-guides*, *position* tape correctly in its path around the video drum and throughout the entire tape path. Both the drum entrance roller guide (P-2) and exit roller guide (P-3), and associated parts, are described in the following paragraphs. These components do essentially the same thing on each side of the video drum during tape loading, and during Play when tape moves over the drum surface:

- Pull loop of tape out of cassette and into tape path.
- Thread tape through tape path.
- Wrap tape 180 degrees around spinning video drum.
- Keep tape at precise height as it enters and exits video drum, positioning tape with respect to video heads.

A roller guide has stationary upper and lower flanges, also called shoulders or ridges, spaced 1/2-inch apart (tape width). A central sleeve freely turns, providing a low-friction surface on which tape rides. Guide posts themselves are usually stainless steel, whereas the roller is a plastic, nylon, glass, or ceramic sleeve that rotates about the guide post. Roller guide height is adjusted by turning the guide clockwise or counterclockwise on a threaded mounting stud, as shown in Figure 3.7.

Roller guide P-2 on the left side of the video drum is called the *drum entrance guide*. On the right side of the drum, roller guide P-3 is also referred to as the *drum exit guide*.

Each roller guide (P-2 and P-3) is mounted on a *shuttle assembly*, which also has a stationary, non-rotating guide pole. These stationary guides, called

**FIGURE 3.7 Roller guides control tape height, correctly positioning tape vertically in tape path.**

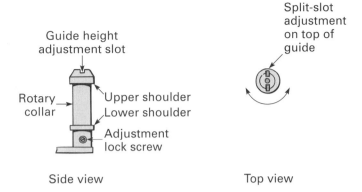

the *drum entrance and exit slant poles*, respectively, are at a slight angle from vertical to align videotape with the tilted video drum. Slant poles are sometimes called *thrust poles*.

Height adjustment of the two roller guides is critical to correctly position videotape around the drum. P-guide adjustment is covered in Chapter 7. Normally you should not adjust the slant poles, which sit between their respective drum entrance and exit roller guides and the video drum when tape is loaded. These are factory adjustments, requiring a special alignment tool or jig. Figure 3.8 shows the drum exit shuttle assembly with roller guide P-3 and slant pole. In this photo the cam assembly has been cycled so that the shuttle is part-way back in its track, between tape unloaded and loaded positions.

**FIGURE 3.8 Shuttle assembly with roller guide P-3 and slant pole.**

Consider how the two shuttle assemblies, with their roller guides and slant poles, load videotape. When a cassette is loaded into a VCR, by either top loader or front loader, drum entrance and exit shuttle assemblies are in their tape-unloaded position, all the way toward the front of the machine in their tracks. In the final part of cassette loading, the bottom of the cassette settles down on the reel table against four locating projections: Two flat-topped studs support the rear of the cassette (where the long label is), while two cone-shaped pins engage holes in the cassette bottom near the front (where the door is). These studs and pins precisely position the cassette on the reel table. In this position the cassette door is held open, exposing tape to the tape path. At this time, both drum entrance guide P-2 with slant pole and exit guide P-3 with slant pole protrude through cutouts in the bottom of the cassette. In this position, they are *just inside* the cassette shell, behind the videotape.

When the machine enters Play or Record mode, the upper portion of the video drum starts spinning and the shuttle assemblies move out on their tracks, away from the cassette, toward the drum. The two roller guides and slant poles pull a loop of videotape through the tape path. During this operation, tape comes off the supply reel; the take-up reel in most cases remains stationary during tape loading and unloading. Figure 3.9 shows roller guides P-2 and P-3 with associated slant poles about halfway through the tape load operation.

**FIGURE 3.9** Approximate tape load halfway point. A: Guide P-2 B: Guide P-3 C: V-stops or V-mounts

The two shuttle assemblies continue in their tracks toward the back of the machine, where the roller guides come to a precise stopping point, determined by *V-stops*. Another name these roller guide stop blocks go by is *V-mounts*. V-stops are precisely positioned at time of manufacture and normally do not require any adjustment for the life of the machine. V-stop mounting screws are normally dabbed with red lock paint.

CAUTION: In general, do not loosen lock-painted transport screws without being *absolutely certain* of what you are doing. These points are factory adjustments that can be expensive and time-consuming to readjust in the field, requiring special alignment jigs and gauges.

**FIGURE 3.10** Roller guides P-2 and P-3 fully seated in V-mounts at completion of tape loading.

Figure 3.10 shows roller guides P-2 and P-3 fully seated into their associated V-mount, or V-stop, at the completion of tape loading. A fully-seated P-guide is said to be *locked* in its V-mount. This is a critical point along the tape path, for it defines the precise positioning of videotape around the drum. If either P-guide is not fully seated in its V-mount at the end of shuttle travel, video problems will result. The cause of that condition should be investigated and corrected.

As elsewhere in VCRs, there are differences in manufacturers' V-mounts. Some engage only the lower portion of the P-guides, whereas others have the V-shaped stop engage the top of the roller guide. Again, don't adjust V-mounts under normal conditions.

A description of how the roller guide shuttles move to load and unload tape is in the Chapter 3 Appendix.

## VIDEO DRUM

The video drum assembly is the most complicated and delicate mechanical component in a VCR. *Scanner* and *cylinder* are other names for the video drum, which contains at least two combination record/playback heads for writing and reading video information on magnetic videotape.

Four-head VCRs have two additional video heads. Hi-Fi Stereo VHS machines have two separate heads in the video drum for recording and playing back signals for high-fidelity audio. Some models have a flying erase head in the video drum, which enables clean video editing. The functions of these additional heads are described in later chapters, especially 17. For right now, we'll stick with just the two essential video heads on a basic, no frills VCR.

In VHS machines, the video drum consists of a lower, stationary cylinder and an upper cylinder that rotates counterclockwise at 1,800 rpm. The upper cylinder rotates at 1,800 rpm regardless of the tape speed (SP, LP, or EP/SLP).

**FIGURE 3.11** Entire VHS video drum assembly removed from VCR. A: Tape head slot.

Videohead tips are exposed to the tape through slots at the bottom edge of the spinning upper cylinder, as shown in Figure 3.11. Also notice the five small grooves encircling the highly polished upper cylinder surface. They reduce what is called *stiction*, a tendency of the tape to adhere to the cylinder. (Think of stiction as a combination of the words "sticking" and "friction.")

To record or play videotape in a VHS VCR, two video heads, Head A and Head B, are required. Heads A and B are 180 degrees apart, and so are exactly opposite each other on the upper cylinder. With a 180-degree tape wrap around the cylinder, only one head contacts the tape at a time. One full revolution of the drum equals one complete video picture, or TV *frame*. Each video frame is composed of two *fields*. Head A records or plays back Field #1, then Head B does the same for Field #2. Each of these two video heads must play back the same tracks it originally laid down—even though Heads A and B that originally recorded the video may have been on a different machine. (Chapter 14 describes the servo circuits that ensure this happens.)

Figure 3.12 shows the underside of the top portion of a video drum that has been removed from the drum assembly. Here you can easily see the two rotary video heads at 10 o'clock and 4 o'clock. Two wires from each head go to the top of the assembly, where they are soldered to rotary transformer terminals.

CAUTION: Rotary head tips are *extremely fragile!* Take extraordinary care that no tool or removed part ever touches them.

In Chapter 6 you'll learn proper procedures to demagnetize and clean rotary heads. The Chapter 3 Appendix has additional information about video heads.

Within the tape path, the video cylinder is tilted slightly more than 5 degrees from vertical; thus each rotating head is offset from the horizontal tape by this amount. With this arrangement, each head actually traces a diagonal track on tape as both tape *and* heads move in relationship to each other. Video tracks on tape are illustrated in Figure 3.13. This method of laying down video tracks with heads mounted on a rotating cylinder is called *two-head helical scan format*. Diagonal video tracks are also called *RF*, for radio frequency, *tracks*.

Chapter 3 Understanding the Videotape Path ■ 63

**FIGURE 3.12** Bottom view: Upper cylinder and rotary video heads.

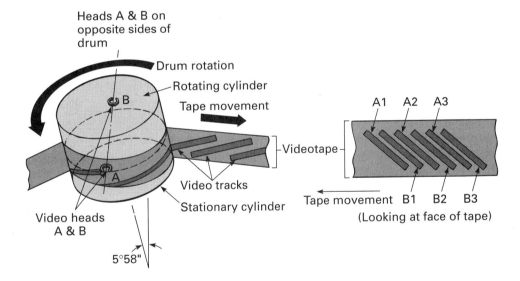

**FIGURE 3.13** Video tracks produced by Heads A and B.

Because the drum is rotating at 1,800 rpm, and it takes a full revolution of the drum for a complete video frame, we can determine that 30 complete video frames are handled by Heads A and B each second (1,800 rpm divided by 60 seconds per minute equals 30 revolutions each second). Thirty frames-per-second is the standard TV frame rate in the United States, as defined by the National Television Systems Committee (NTSC) (pronounced "NIT-see").

Additional information on the drum and video recording is in the Chapter 3 Appendix. You'll also learn much more about how information is recorded on videotape in Chapter 16. At this point, just realize that Head A is responsible for the first half of a video frame and Head B for the second half. One complete revolution of the upper cylinder therefore equals one video frame.

Since we've already described the exit, or P-3, roller guide on the right side of the drum, we'll move on to the next major tape path component. You may wish to look back at Figure 3.1.

## AUDIO/CONTROL HEAD ASSEMBLY

The Audio/Control (A/C) tape head records and plays back two signals on the videotape: linear audio and control pulses.

Audio signals occupy a 1mm-wide track at the top edge of the videotape. Sound recorded by the audio/control head is called *linear audio*, because the track is along the tape length, like a conventional audio tape recorder. The same audio head both records and plays back.

A linear audio erase head is part of the A/C head assembly. The erase head gap is positioned just before the audio head and erases only previously recorded linear audio. This is important during audio editing, when new audio is added to an existing tape without altering any video tracks. Figure 3.14 shows the Audio/Control/audio Erase head assembly.

**FIGURE 3.14** Audio/Control and audio Erase head assembly, usually called the A/C head.

Because the linear **A**udio record/playback head, **C**ontrol track record/playback head, and linear audio **E**rase head are all contained in a single assembly, this part is sometimes called the *ACE head*.

The linear audio erase head works exactly like the full erase head described earlier. A high-frequency sine wave, usually between 65 kHz and 125 khz, scrambles any magnetic patterns on the tape, wiping out previously recorded audio.

It's imperative that rotating video Heads A and B be in precise alignment with their recorded tracks on tape during playback to obtain a steady, noise-free picture. This is where the control track at the bottom of the tape comes in. During playback, a recorded signal on the control track enables the VCR servo circuit to govern tape and drum movement so the spinning video heads are precisely positioned over the video tracks on tape.

Another way of stating this is that control track pulses identify the location of video tracks on tape originally recorded by Heads A and B. Then, in playback, Heads A and B have to trace the exact same tracks originally recorded by Heads A and B, respectively. This is what *tracking* is all about! It just won't work if Head A plays back tracks laid down by Head B, or vice versa, or if either spinning head doesn't align perfectly with its respective video tracks.

The control head gap in the A/C head assembly records and plays back a 0.75mm-wide track at the bottom edge of the tape, as illustrated in Figure 3.15. A VCR capable of stereo linear audio records two separate tracks, for left and right channels, within the same 1mm-wide track at the top of the videotape, as shown in Figure 3.15.

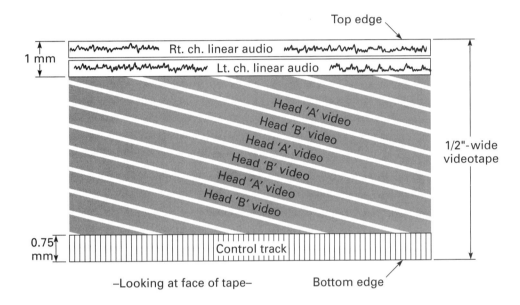

**FIGURE 3.15** VHS linear audio, video, and control tracks. For monophonic sound, the audio head gap covers the entire 1mm track at the top of the tape.

## CAPSTAN AND PINCH ROLLER

Moving along the tape path toward the take-up reel, tape next passes across stationary take-up guide P-4 and then between the capstan shaft and rubber pinch roller. In some VCRs P-4 is a roller guide. Guide P-4 may also be called the take-up tape guide.

During play and record, the free-wheeling pinch roller presses videotape against the rotating capstan to draw tape through the transport at a constant speed. As you can see back in Figure 3.1, the capstan shaft sits just inside the video cassette housing. Figure 3.16 shows take-up guide P-4, capstan shaft, and rubber pinch roller. The cam mechanism is in the tape unloaded or *home* position, so the pinch roller is retracted from the capstan.

The slanted plastic cap atop P-4 holds the tape door open after a cassette is loaded onto the reel table.

Usually the cam or mode mechanism moves the pinch roller up against the capstan, where it is firmly held under spring tension. You can see the spring on the right side of the pinch roller pivot plate in Figure 3.16. Other arrangements, such as a solenoid-operated pinch roller, are possible.

There are two basic methods for rotating the capstan shaft: (1) belt drive and (2) direct drive (DD). Both have a large-diameter flywheel attached to the bottom of the capstan shaft in the undercarriage. The flywheel supplies rotational inertia, smoothing out short-term speed variations of the capstan shaft. In the first type, a rubber belt goes between a small-diameter pulley on the capstan motor to the outside of the relatively large-diameter flywheel, as shown in Figure 3.17.

66 ■ *Practical VCR Repair*

**FIGURE 3.16** Take-up guide P-4 (A), Capstan (B), and Pinch Roller (C).

**FIGURE 3.17** Belt-driven capstan. A: Capstan motor pulley B: Capstan flywheel

The second way of turning the capstan shaft is with a direct drive capstan motor, where the motor shaft itself *is* the capstan. A direct drive capstan also has a flywheel, generally incorporated as part of the capstan motor assembly. Figure 3.18 shows a direct drive capstan motor.

In many VCRs, the capstan motor, whether belt or direct drive, also powers the supply and take-up spindles for all modes of operation. This is done with various combinations of pulleys, belts, gears, slip clutch, and idler wheel. Most of these drive components are in the undercarriage, but the idler wheel is usually on top of the reel table, between the supply and take-up spindles.

For these VCRs, the capstan motor is energized even during rewind or fast forward to power the supply and take-up reels, respectively, when the pinch roller is actually retracted from the capstan shaft. Some high-quality VCRs have two separate direct drive spindle motors that eliminate the need for the capstan motor to do anything but spin the capstan.

Once tape passes between the capstan and pinch roller, it goes back into the cassette. There may be an additional non-rotating tape guide in the transport along the way on some VCR models. If present, this guide is usually on a pivot plate, and may be called a *reverse arm*. During tape loading, it moves forward a small amount to position the tape for a straighter path back into the cassette. This puts less dependence on the take-up guide within the cassette itself. That is, tape entering the cassette doesn't wrap as many degrees around the take-up guide inside the cassette. When tape is unloaded, the guide moves

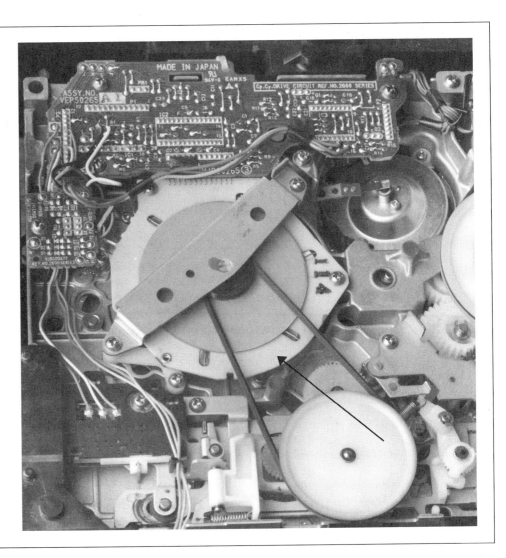

**FIGURE 3.18 Direct drive capstan motor.**

back, just inside the cassette next to the capstan. It holds tape slightly away from the rotating capstan shaft during fast forward and rewind. Of course, during fast forward or rewind tape is unloaded.

You'll find additional information about the capstan, capstan motor, and pinch roller in the Chapter 3 Appendix.

## TAKE-UP SPINDLE

Just like the cassette supply reel, cassette take-up reel splines engage the take-up spindle when a cassette comes to rest on the reel table at the end of cassette loading. Spindle height is adjusted so the take-up reel within the cassette is lifted off the bottom of the cassette, but not pushed too close to the top. Take-up and supply reels inside the cassette should be positioned midway between the top and bottom of the cassette shell.

Take-up spindle functions include:

- Smoothly reeling in tape during record, play, forward scan, and fast forward. (The take-up reel turns clockwise.)
- Braking the cassette take-up reel when the machine is stopped, to prevent tape spillage.
- Providing holdback tension, or drag, on tape during rewind or reverse scan. (The take-up reel turns counterclockwise)

Typically, in normal forward play or record, power is delivered to the take-up spindle like this: a belt around the capstan flywheel pulley connects to a slip clutch assembly; an idler wheel then transfers power from the output of the slip clutch to the take-up spindle. There may be gears or an intermediate belt/pulley combination somewhere along the way. Many variations exist, but the basic principles of operation are the same. Generally, all except the final idler that transfers power to the spindle are in the undercarriage.

Probably the most critical component in this drive train is the slip clutch assembly. Its purpose is to allow power to be transferred, *with controlled slippage*, between the constant rpm input from the capstan-powered pulley and the take-up spindle, by way of an idler wheel. Slippage is necessary because the take-up spindle must operate over a somewhat wide rpm range: it turns much faster at the beginning of a tape, when the take-up reel diameter is small, than toward the end of the tape, when take-up reel diameter is much larger.

Slip clutch assemblies are fairly simple. In most designs, two round plastic disks on a common shaft are spring-loaded toward each other, with a felt washer between the two disks. The "input" disk, either gear- or belt-driven, turns at a constant rpm. The "output" disk transfers power to an idler wheel, which then turns the take-up spindle. The felt washer between the disks allows slippage and transfers rotary motion from the input disk to the output disk. The

**FIGURE 3.19  Typical VCR clutch assemblies.**

more spring force there is clamping the two disks against the felt washer, the less the clutch slips. Figure 3.19 illustrates representative VCR clutch assembles.

The clutch on the right in Figure 3.19 is quite different from the others. It is an electrical spindle clutch; a reel table spindle fits over the stationary upright shaft. When no voltage goes to the coil, the spindle is completely free-wheeling on its shaft. With different amounts of voltage applied to the coil, this clutch supplies both holdback tension, or soft braking, and full braking to the spindle.

In several VCR designs the slip clutch and idler are incorporated into a single, unitized assembly. In other VCRs, the idler and slip clutch can be replaced separately. If either of these parts is worn sufficiently to require replacement, it's a good idea to replace or rebuild the other one at the same time. Some clutches can be rebuilt by simply replacing the worn felt washer(s) between the clutch disks. And just the rubber tire can be replaced on some idler assemblies rather than putting in an entire new part.

Slippage of one style of clutch can be adjusted by changing the spring tension that forces the clutch disks against each other. The position of a three-legged, bowed spider spring can be altered. Refer to Figure 3.20. The center of the spider spring is around the clutch shaft and acts to force the two disks together. The three spider legs contact the outside perimeter of the disk, which has different height notches into which the legs seat. Clutch adjustment is made by rotating the spider spring so that the legs fit in different height

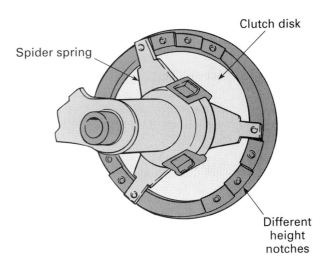

**FIGURE 3.20** Adjustable clutch. Spring legs are rotated to different height notches on perimeter of disk to change clutch torque.

notches. For example, rotating the legs to a higher notch increases spring tension on the disks, boosting clutch output torque.

Power is transmitted from the output clutch disk to an idler wheel. A common arrangement is to have the rubber idler wheel tire contact a small grooved or serrated plastic collar, or hub, on the top side of the clutch output disk. The idler wheel assembly itself then pivots left or right so its tire contacts the lower flange of either the supply or the take-up spindle. In most VCRs, the direction of rotation of the clutch output disk pivots the idler. That is, the spinning output disk collar or hub against the idler wheel causes the idler to pivot in the direction of rotation. Figure 3.21 shows a typical idler wheel sitting between the two transport spindles.

For forward scan during play, with tape loaded, the capstan motor speeds up and the clutch/idler operates as in normal play.

In fast forward or rewind, with tape *unloaded*, clutch slippage is not desired. In some designs, a mechanical linkage to the cam motor locks the two clutch disks together for positive drive.

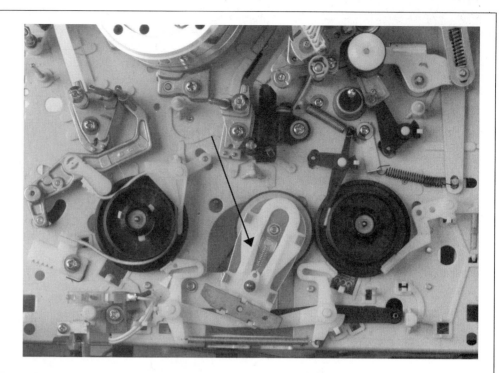

**FIGURE 3.21** Idler wheel and pivot arm.

It is necessary, therefore, to place *some* drag on the take-up reel during rewind. Take-up spindle back tension is often provided by a "soft" brake, a felt brake pad that is spring loaded against the lower spindle flange. The soft brake allows the spindle to turn, but adds a bit of drag. In like manner, soft braking may be applied to the supply spindle during fast forward.

Take-up spindle hard braking is accomplished with a felt brake pad against the spindle's lower flange when the machine is in stop. Often a common mechanical linkage applies braking to supply and take-up spindles at the same time. Momentary braking is also applied between fast forward and rewind. That is, if the VCR is in rewind and then fast forward is selected without first hitting the Stop key, the system control microprocessor applies the brake for about a second before changing capstan motor direction. Figure 3.22 shows a take-up spindle with both *soft* and *hard* brake pads and operating arms.

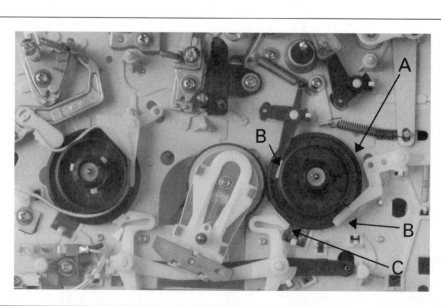

**FIGURE 3.22** Take-up spindle (A), soft brake pads (B), hard brake pad (C).

*Chapter 3 Understanding the Videotape Path* ■ 71

The previous description on how the take-up spindle is driven in various operating modes is representative; there are many differences in mechanical assemblies and the ways in which they connect and operate. If you understand the basic concepts of how the capstan motor delivers power to take-up and supply spindles through slip clutch and idler, you should have no trouble figuring out the details on any VCR.

## SERVICING THE TAPE TRANSPORT

Common transport problems are due to slipping belts and idlers and worn clutch assemblies. Weak springs, bent brackets, lack of lubrication, dirt, broken parts, and foreign objects are also frequent causes of tape transport malfunction. Diagnosing and repairing these areas are covered in Chapter 6.

Troubleshooting transport problems is much simpler on models with separate direct drive (DD) spindle motors. Motors themselves, under electronic control, perform spindle functions without any belts, clutches, idlers, or linkages from the cam. A spindle may be directly attached to its motor shaft, or there may be a small plastic intermediate gear between an output gear on the motor shaft and the underside of the spindle.

As explained in the Chapter 3 Appendix, *hand cycling* is often a good way to determine how things are supposed to work. Operating the VCR under power with a "scratch" cassette or a videocassette test jig, while carefully observing the action on both sides of the transport, is also very helpful in figuring out transport operations. You can often see where drive train parts are slipping or are maladjusted. We'll fully describe what a scratch cassette and a test jig are in Chapter 4, as well as how to make a videocassette test jig. Essentially, a *scratch cassette* does not contain a recording of any value and a *videocassette test jig* is a cassette with no magnetic tape, with which you can exercise some transport operations without actually having videotape in the machine.

One little detail about working on the tape transport: Certain transport mechanics may not work properly with a VCR on its side. For example, a pivoting idler assembly typically will *not* swing against the force of gravity to contact a drive spindle unless the transport is in a horizontal position.

This completes the *basic* discussion of a typical VCR tape path and the function and operation of the major components along the way. Additional detail is included in the Chapter 3 Appendix and in later chapters.

Another important tape transport function is to send information to electronic circuitry about what is actually going on during various modes, such as tape loading, Play, Rewind, and so on. The next discussion is about how transport status is accomplished.

## TAPE TRANSPORT SWITCHES AND SENSORS

As you now know, the tape transport is a busy place, with several different modes of operation. A variety of switches and sensors report mechanical status to the system control microprocessor. The following are primary feedback methods employed in the majority of VCRs.

### Cassette-in Switch

Either a small microswitch or leaf switch on the upper surface of the reel table is operated when a cassette is fully loaded and resting on the four locating studs. This is usually a *normally open* (N.O.) switch. When a cassette loads, switch contacts close, making, or completing, an electrical circuit. This sends a signal to the system control microcomputer so it "knows" there is a cassette loaded on the reel table.

With most VCRs, system control then turns On a cassette-loaded icon on the front panel display, like that shown in Figure 3.23. (An *icon* is a simple graphical representation, like the outline of a lit cigarette on an automobile cigarette lighter.) Another name for the cassette-in switch is *cassette-down switch*.

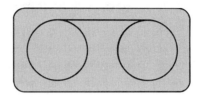

**FIGURE 3.23  Front panel cassette loaded icon.**

Location of the cassette-in switch can vary greatly. As a rule, the switch operating arm is on the top surface of the reel table. In most designs, the bottom of a loaded cassette presses down on the switch lever arm. In others, the switch is operated by the left or right side of the cassette as it comes to rest on the reel table. Just look for some switch that transfers as a cassette completes the load cycle.

In some circuit designs, if the cassette-in switch transfers even momentarily, which could occur with faulty switch contacts, system control shuts down the VCR if in other than Stop mode. Other VCR models consider the cassette-in switch only at the completion of the cassette load operation; once the switch transfers, system control assumes that a cassette is loaded until a cassette eject operation takes place.

It is easy to confuse the reel table cassette-in switch with the cassette-*insert* switch on the front loader assembly. These names can vary among models and manufacturers. The *cassette-in* switch transfers at the completion of cassette loading, with the cassette seated on the reel table. This causes the cassette-loaded icon to display on the front panel.

The *cassette-insert* switch transfers when a cassette is gently pushed into the front loader opening. This signals system control to energize the cassette load motor, which then moves the cassette in the inverted L-shaped path—in and down to the reel table.

And now another one of those VCR exceptions, which by now you should realize are prevalent. It is amazing in how many different ways various models accomplish the same end result. Some VCRs do not have a separate cassette-in switch. Instead, if either or both tape-end sensors is dark, this indicates to system control that a cassette is loaded on the tape transport.

### Record Interlock Switch

To prevent accidental recording over previously recorded material, cassettes have a plastic record safety tab on the back left side. When this tab is pried out or broken off, the machine is prevented from going into Record mode. This is the same system used with audio cassettes.

With a cassette loaded, a microswitch or leaf switch senses the presence or absence of the record safety tab. This is generally a *normally closed* (N.C.) switch. When a record safety tab is present, the switch lever arm moves, opening switch contacts. With the record safety tab broken out, the switch arm is in its N.C. position, in the cassette notch. This signals system control, which prohibits the VCR from going into Record mode.

Figure 3.24 shows typical cassette-in and record interlock switches. Other names may be used, such as record proof, record inhibit, or record safety

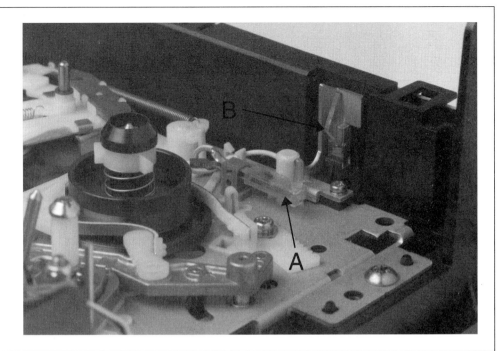

**FIGURE 3.24** Cassette-in (A) and Record Interlock (B) switches. The front loader has been removed so the two switches can be seen.

switch. In some designs, if the record interlock switch bounces—that is, if the contacts momentarily open during recording—system control goes out of Record mode, retracts the pinch roller, unloads tape, and shuts Off the VCR. Other designs only consider the switch position when the VCR is initially placed in Record mode.

In some VCR models, the record safety switch is part of the front loader assembly. As the cassette drops down on the reel table, the record safety tab, if present, transfers a leaf switch in the lower, front part of the loader.

## Supply and Take-up Reel Sensors

The purpose of take-up and supply reel sensors is to signal whether the take-up and supply spindles are actually turning. Most VCRs have a take-up sensor, and some models sense rotation of both spindles. System control uses the output of these sensors to determine if the transport is operating properly.

For example, suppose a tape is playing. Everything works fine until three-quarters of the way through the tape, when the worn slip clutch becomes unable to supply enough torque to the take-up reel. (It takes more torque to reel tape in with more tape on the take-up reel.) A slipping idler wheel could also cause the take-up reel to stop turning.

Although the capstan and pressure roller keep turning, the take-up reel no longer reels in enough tape. Tape spills into the transport, and will probably wind around the capstan shaft. This describes a VCR "eating" a tape. To prevent this unfortunate occurrence, system control monitors a take-up reel sensor. If the take-up spindle stops turning, system control releases the pinch roller, unloads tape, and shuts down the VCR, just as if you had turned the VCR Off. Some models indicate that there's been some abnormal condition with a front panel error display, like a blinking block or even the word ERROR. Other models have no indicator.

The same basic operation takes place in other modes. For example, should the take-up reel stop turning in Reverse, Forward Scan, Fast Forward, or Rewind modes, system control shuts down the VCR. Of course, in the latter two modes, tape should already be unloaded, with the pinch roller retracted from the capstan.

The most common type of reel motion sensor consists of an infrared light-emitting diode (IR LED) and phototransistor directly beneath the take-up spindle in the undercarriage. The underside of the spindle has alternating reflective and nonreflective areas or strips. When the spindle turns, the phototransistor sends pulses to system control when IR energy is reflected back onto it. Figure 3.25 shows typical reel sensor components.

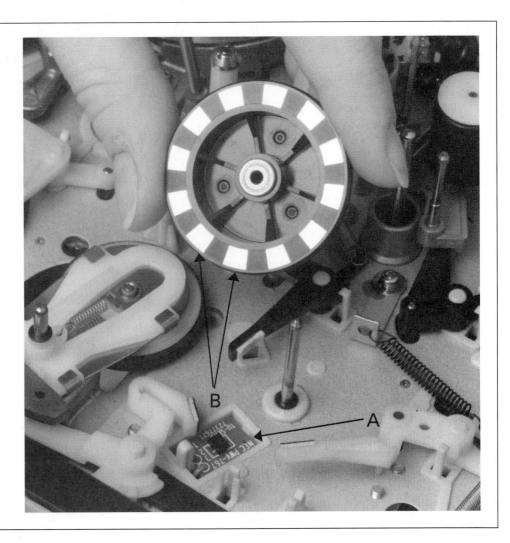

**FIGURE 3.25** Reel sensor. IR LED/phototransistor circuit board (A) and reflective areas on underside of take-up reel spindle (B).

An optical *encoder disk* with slits or holes and IR LED/phototransistor pair, described in the Chapter 3 appendix, is another possible reel sensor configuration. Direct drive spindle motors may include a Hall effect device to produce reel sensor pulses.

The main point is that if system control does not receive reel sensor pulses when it should, the VCR unloads tape, and usually powers down, to prevent tape and machine damage. This is a fairly common VCR problem. Low spindle drive torque, caused by worn or dirty belts, idler, or slip clutch, causes the take-up reel to stop turning at some point during play or rewind.

When you're working on a VCR either without any cassette or with a videocassette test jig, you will probably need to turn a spindle by hand to prevent auto shutdown. For example, suppose you've manually operated or shorted the cassette-in switch with no cassette loaded. You've also blocked the two tape-end sensors (described next). System control now thinks a cassette is loaded *and* that the tape is neither fully wound nor fully rewound. You then push the Rewind button to observe what happens. The supply spindle starts turning to reel in tape, but a second or so later the VCR goes into auto shutdown.

Why? With no tape to turn the take-up spindle, system control receives no take-up reel pulses, assumes something is amiss, and goes into Stop mode and shuts down the VCR. By manually turning the take-up spindle, system control receives take-up reel pulses and remains in Rewind mode. Dirt accumulation, such as cigarette smoke film on the IR LED, phototransistor, or on reflective encoder strips, can cause a reel sensor to malfunction.

With all that said, there is one additional function of the take-up reel sensor in most newer VCR models: output pulses are used to increment or decrement an electronic index counter display on the front panel. Early VCRs have belt-driven mechanical counters.

### Tape End Sensors

VHS VCRs use an optical system to detect when the tape comes to an end, at both the beginning and the end of a tape. Within a cassette, the two magnetic videotape ends are attached to the supply and take-up reels with transparent leaders, which are roughly six inches long. These leaders are sensed at both the supply reel side and the take-up reel side of the cassette. Here's how.

When a cassette is loaded onto the reel table, a post containing a light source protrudes through a round hole in the bottom of the cassette. This opening is approximately 1/2-inch in diameter and is midway between the two sides and about one-inch back from the front of the cassette. Light shines through channels inside the cassette toward the left (supply) and right (take-up) sides.

Small openings in the left and right sides of the cassette line up with the channels and light post. When the end of the tape is reached at either the supply or the take-up reel, light shines through the transparent leader, out the cassette side openings, and onto a phototransistor mounted in the transport at each side of the cassette. Figure 3.26 illustrates VHS end-of-tape sensing.

When light shines on either phototransistor, a signal is sent to system control, which acts upon this information. For example, when during play the end of the tape is detected at the supply reel, system control (syscon) causes the VCR to go out of Play and into Stop mode, unloading tape. In some models, syscon also puts the machine into rewind. Similarly, when the end of the tape

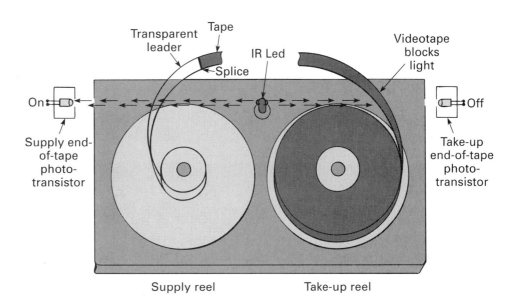

**FIGURE 3.26** End-of-tape sensing in VHS VCRs.

is detected at the take-up reel during rewind (which means the tape is fully rewound), syscon puts the machine into Stop mode. These operations prevent snapping the tape or pulling the leader out of either cassette reel hub.

On earlier top loader VCRs, the light source within the cassette is often a small incandescent lamp. In some models, this lamp does not turn On until the cassette-in switch transfers when a cassette seats on the reel table. This lamp, of course, can burn out in time, and is a common replacement item on these older machines.

Later models, and virtually all front loaders, use an infrared LED as the light source that protrudes up into the cassette. Usually this LED is powered whenever the machine is turned On. On some models, if there is no output from either phototransistor when the VCR is turned On, *but without a cassette loaded*, system control powers down the VCR after a few seconds. This bears repeating: with no cassette loaded, there *should* be output from both phototransistors, since there is nothing to block the light from the light post. If system control doesn't receive output from both phototransistors, it knows something must be wrong with the light source or one of the phototransistors, and so it shuts the VCR Off.

You cannot directly see infrared radiation. However, you can purchase an infrared sensor card, about half the size of a credit card, at electronic supply houses for approximately $6.00. An infrared-sensitive area on the card glows reddish-orange in the presence of near-infrared light, such as that emitted by an IR LED or by a diode laser in a CD player. Figure 3.27 illustrates checking tape-end sensor IR LED light post output with an infrared sensor card. In Chapter 11

**FIGURE 3.27** Checking IR LED output with an infrared sensor card.

you will learn how to build your own low-cost IR detector with a handful of electronic components.

Some VCRs have the two tape-end phototransistors mounted on the front loader assembly; others mount these components to the reel table. There is usually no problem operating a VCR with these sensors on the loader when the assembly is removed and unplugged, as is necessary for many reel table maintenance and repair procedures. System control (syscon) "sees" the two phototransistors as dark all the time, because they are unplugged. With the cassette-in switch transferred, either manually or with a cassette test jig, it's as if a cassette *were* loaded, with tape blocking both end sensors.

Just be careful when using a real tape under these conditions, that you stop before the end of the tape in Fast Forward and Rewind modes. If not, the splice between tape and leader may separate, or the leader may break away from a reel hub, because system control cannot detect tape end and will not stop the VCR.

Depending on the model, a VCR may refuse to load a cassette, or may load and then immediately eject the cassette, if the end-of-tape light source is not working. Also, an incandescent shop light has a large amount of infrared output. If this light strikes the IR phototransistors while a VCR is on the service bench, it can create misleading results. The system control microprocessor may lock up. Florescent lighting is low in IR output and will not cause this problem. Once you build the IR detector described in Chapter 11, you can see for yourself the relative amount of IR output from different light sources.

Eight-millimeter VCRs and camcorders use a similar optical system for detecting tape ends. Beta machines employ a different system: metallic leaders at the tape ends are sensed with electromagnetic coils.

## Mode or Cam Switch

Within the undercarriage, a mode switch tells system control what is going on in the tape transport. It senses mechanical positions, and so is also called the *mode* or *cam sensor*. As you've read, the cam/tape-load/mode motor is responsible for several mechanical operations, most of which take place in the undercarriage. To review, these are typical operations it performs:

- Load tape by moving two shuttle assemblies with roller guides P-2 and P-3 along guide tracks. Roller guides fully seat in their respective V-stops at each side of the video drum when movement is complete
- Unload tape by moving shuttles and P-guides back into cassette (supply spindle turns CCW during tape unload to reel in tape)
- Position the pinch roller against the capstan
- Apply braking to capstan flywheel during Pause
- Operate linkage to either lock or disengage the slip clutch during fast forward or rewind.

There are many variations on what the cam accomplishes, but these operations are fairly typical. Here's an example of what the cam mechanism might do when Reverse Scan is pressed to back up tape and view it during play:

1. Cam turns a little in one direction to retract pinch roller from the capstan and brakes capstan flywheel for about a second.
2. Cam then turns in opposite direction to release capstan brake and move pinch roller back against capstan. The capstan motor now turns faster than play speed in the opposite direction.

In these and other operations, it is important for syscon to know the transport's mechanical status. This is the function of the *mode* or *cam switch*. It signals syscon when tape is fully loaded or fully unloaded so direct current to the cam motor can be cut off. It also indicates other positions of the cam after tape is loaded so syscon knows in what direction to energize the mode motor for subsequent operations, and when to stop the motor. In short, the mode or cam switch tells system control the cam's position, and therefore what's going on in the tape transport.

There are two basic types of mode switches: mechanical and optical. Either type is generally located near the cam in the undercarriage, with four or more wires connected to the switch or sensor assembly.

Here's how a multi-position mechanical mode switch functions. As the cam rotates through its various positions, an operating arm moves the switch lever to different positions. Different electrical contacts are made within the switch depending on switch position.

Suppose on a particular VCR the switch has four wires; one is a common wire. Each of the other three wires goes to a switch contact labeled either A, B, or C. Any or all of contacts A, B, C can be connected to the common wire, depending on the switch position. For different cam positions, switch contacts A, B, and C are made or connected to the common in different ways to tell system control the cam's position, and thus the position of components in the transport. For example:

- No contacts made: Indicates tape fully unloaded and Stop mode (Home position)
- Common to A: Indicates tape unloaded, but in Rewind or Fast Forward mode
- Common to B: Indicates tape loaded; pinch roller retracted
- Common to A & B: Indicates Play mode; pinch roller against capstan
- Common to C: Indicates Pause; pinch roller against capstan and brake applied to capstan flywheel
- Common to C & B: Indicates play, but in Forward or Reverse Scan mode.

This example is just meant to give you an idea of how a mode switch operates; it doesn't necessarily correspond with an actual VCR's operation.

Figure 3.28 shows a typical mechanical mode switch. Note how the switch slider is operated by the cam arm, at point B in the photo.

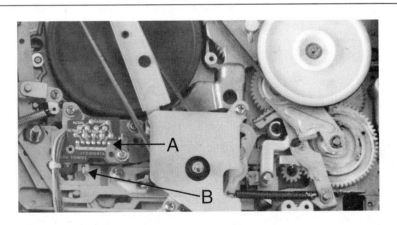

**FIGURE 3.28** Mechanical mode switch. A: Switch B: Switch slider

Many VCR problems are caused by faulty switch contacts or a maladjusted mode switch. For example, if the cam switch fails to indicate that the cam has arrived at the position where tape is fully loaded, system control may unload tape and shut down, or perhaps unload and retry the load operation.

Poor switch contacts may cause system control to keep the cam motor energized when it should be stopped. This may stall the motor as it exceeds its travel limitations. Generally, system control will shut down the VCR or retry an operation if the expected status is not returned within a predetermined number of seconds programmed into the microprocessor. System control microproces-

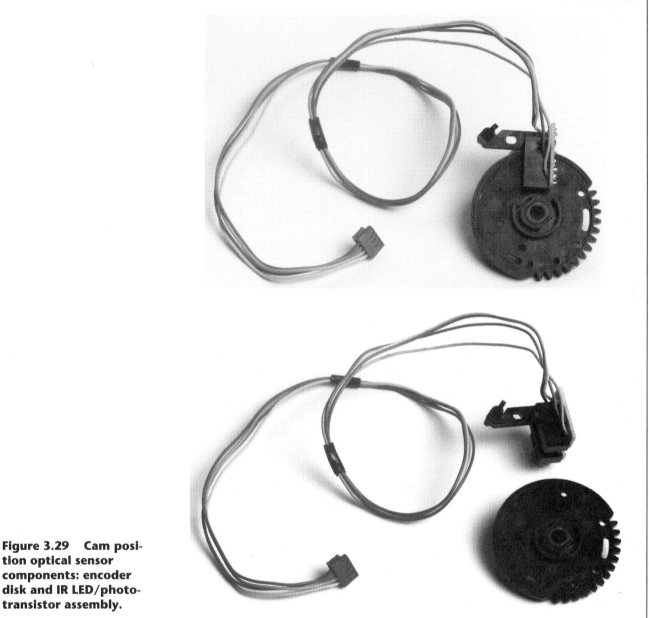

**Figure 3.29** Cam position optical sensor components: encoder disk and IR LED/phototransistor assembly.

sors in diverse models are programmed differently, and so handle incorrect status from the mode switch in various ways. But the basic operation is the same, often with a forced tape unload operation and VCR shutdown in the absence of correct mode switch signals.

With mechanical mode switches, a high resistance contact—caused by cigarette smoke film, for instance—is a fairly common problem. Cleaning the switch with contact cleaner often restores correct operation. Direct canned spray cleaner with a nozzle extension tube into switch openings. On some mode switches, you can carefully pry up a plastic switch cover to get cleaner to the contacts. Operate the switch through all its positions several times to clean dirty or corroded contacts and respray. If bad switch contacts cannot be cleaned, or mode switch problems persist, replace the switch.

The second type of mode switch is less prone to problems. It isn't really a switch at all, but an optical sensor like the encoder disk with slits or holes described in the Chapter 3 appendix under *"Capstan and Pinch Roller."* Typically,

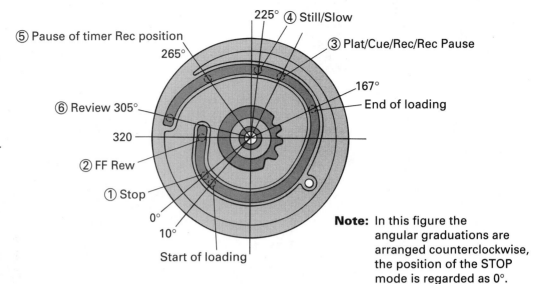

**FIGURE 3.30** Typical cam positions and operating modes.

in this design, two pairs of IR LEDs and phototransistors are mounted on opposite sides of an opaque disk. There are several holes or apertures in the disk, which turns along with the cam.

As IR light shines through disk apertures, either of the phototransistors, or both, conduct, informing syscon of the cam position. For example, suppose we call the two phototransistors A and B. When IR light from the LEDs strikes them, they conduct, or turn On. This is essentially the same as closing a mechanical switch contact: electric current flows through the phototransistor back to system control. Mode switch and syscon might work together something like this: A On indicates one cam position, B On a second position, and A and B On together indicate a third cam position. Figure 3.29 shows an IR LED/phototransistor sensor assembly and cam encoder disk. The top photo shows the disk positioned within the optical sensor assembly, as it would be in the machine, while the bottom photo is of the two parts separated. Notice the slot in the sensor assembly for the encoder disk.

As the cam turns, perhaps A turns On and then shuts Off; then B turns On, then Off; followed by both A and B turning On together. With these signals, system control knows in what direction the cam is turning, and specific cam positions. In other words, by keeping count of the On/Off cycles of A and B and the sequence, system control knows where the cam is positioned at all times. Figure 3.30 illustrates cam, or mode, positions for a typical VCR. An encoder disk with two pairs of IR LED/phototransistors geared to the cam informs system control of the operating mode.

If a VCR exhibits erratic, intermittent, or bizarre behavior of cam-operated functions, yet the cam-operated components such as gears and linkages appear in good order, suspect problems with the mode switch or encoder IR LED/phototransistors. Many variations exist from model to model in the details of cam sensor or mode switch operation. Essentially they all perform the same task of letting the system control microprocessor know the position of transport components operated by the cam.

## DEW SENSOR

Many, but not all, VCRs have a dew sensor. Its purpose is to inform system control when humidity within the VCR is high. This could be the result of bringing a VCR in from the cold, when condensation could develop inside the unit, or from a very high humidity environment. Moisture could cause tape to cling to the spinning video drum and possibly to other components in the tape path. Tape sticking or adhering to the video drum, called *stiction*, is not good! Severe tape and upper cylinder damage could result if this were to happen.

To prevent this, system control prohibits tape loading, and may even prevent powering up the VCR, if the dew sensor detects moisture or dew within the transport area. Leaving the machine plugged into AC power will normally dry out any moisture within the VCR cabinet. Portions of the power supply and timer/operations and system control microprocessors are active as long as the VCR is plugged in. This produces a small amount of heat that helps dry out the inside of the VCR, even when the main power is Off.

A dew sensor is a component whose electrical resistance changes *in direct proportion* to changes in levels of humidity. If humidity increases, so does the resistance value of the dew sensor. For example, a particular dew sensor may be around 25,000 ohms in very dry air, but over 1,000,000 ohms (one megohm)

**FIGURE 3.31** Dew sensor on top side of tape transport.

when the air is extremely moist. If it senses dew sensor resistance that is higher than a certain value, system control won't allow the VCR to operate.

The dew sensor, if present, is generally located on the top of the tape transport, somewhere near the video drum. It is usually a flat component about one-fourth the size of a regular postage stamp with two wires leading from it. Figure 3.31 shows a dew sensor. In this model, the sensor is mounted to the right of the video drum, on a bracket just above the cam motor gears. (The belt on the pulley in the lower right goes around a smaller pulley on the cam motor shaft.)

If you feel that high humidity, or dew, is *not* present, but the word **DEW** appears on the front panel, you can try placing a 10,000 ohm resistor across the dew sensor to see if this allows normal operation. The value of the resistor you use is *not* critical; a 1,000-ohm resistor will probably work just as well. The sensor could be defective, falsely indicating a dew condition. If the sensor is *unplugged*, this is a *very high* resistance, and so syscon will prevent power On.

## SUMMARY

This chapter has described the major tape transport components and their operation. Tape loading and unloading are controlled by the tape load motor, also called the cam or mode motor. During tape loading, roller guides P-2 and P-3 pull a loop of tape from the cassette, thread it through the tape path, wrapping it halfway around the spinning video drum.

A back tension guide pole works with a tension band around the lower flange of the supply spindle to regulate holdback tension as tape comes off the supply reel. Several different tape guides and an impedance roller position tape and ensure its smooth flow throughout the tape path.

As tape travels through the tape path, it passes a full erase head and then the spinning video drum with rotary heads. After exiting the video drum, tape passes an assembly containing: (1) linear audio erase head, (2) linear audio Rec/PB head, and (3) control track head. This assembly is called the A/C (audio/control) head assembly. After passing the A/C head, tape runs between a rubber pinch roller pressed against a rotating capstan shaft, which together pull tape through the tape path.

Rotary head tips protruding from slots in the upper video cylinder are delicate; extreme care must be taken when working around them.

Major components involved with transport operations include a motor-driven cam, tape load assembly gears and movable shuttles, slip clutch, and idler wheel. Most VCRs have a system of belts, pulleys, gears, slip clutch, and idler to deliver power to the take-up and supply spindles from the capstan motor. Some VCRs eliminate these components by using separate direct drive spindle motors.

Principal trouble areas in the transport area include:

- Incorrect tape holdback tension at supply spindle
- Slipping belts, pulleys, and rubber idler tire
- Worn slip clutch
- Flaky mode switch contacts
- Dirty photo sensor components.

Most audio and video problems are caused by a problem in the tape path, such as a clogged video head, a tape guide that is out of alignment, dirty capstan shaft, or deformed pinch roller.

A common cause of unexpected tape unload and auto shutdown during play or rewind is insufficient torque on the take-up spindle or on the supply spindle. When system control doesn't receive reel pulses from a sensor, it unloads tape and powers Off the VCR.

Many transport operational problems, such as failure to start the tape load cycle, unloading tape unexpectedly, and automatic shutdown, are caused by a problem with a mechanical switch or optical sensor. Dirty mode switch contacts are a frequent reason why transport operations don't work at all, or work incorrectly, operate intermittently, or function in a bizarre fashion. Transport malfunctions occur when the system control microprocessor gets conflicting or inappropriate information from one or more of the many switches and sensors in the VCR.

## SELF-CHECK QUESTIONS

1. What term sometimes is used to describe tape loading in a VHS VCR because of the path taken by videotape through the transport?
2. In a sentence or two, explain why it is important for videotape to have proper tension as it goes through the tape path.
3. Briefly describe how the back tension guide pole and supply spindle tension band work together to control tape tension.
4. Briefly describe what the full erase head in a VCR does.
5. Give another name for the P-2 guide. Give the other name for the P-3 guide.
6. In addition to a roller guide, each of the two shuttle assemblies carries a non-rotating, or stationary, tape guide. What is the tape guide called?
7. Briefly describe what each rotary video head, A and B, does when recording or playing back videotape.
8. Briefly describe the purpose of the *control track* on videotape.
9. In many VCRs, what is responsible for moving the pinch roller up against the rotating capstan shaft?
10. Briefly describe the purpose of the *slip clutch* in a VCR.
11. Describe the purpose of the *cassette-in* switch.
12. Give at least two names for the switch that transfers when a cassette having its *record safety tab* present is loaded.
13. What is the purpose the the *take-up reel sensor?*

# CHAPTER 4  VIDEOCASSETTE EXAMINATION AND REPAIR

A major video home system component is the videocassette itself. Quality record and playback depend not only on the VCR machine, but also on videotape and the cassette shell in which it is housed. This chapter focuses on the features, component parts, and operation of a VHS videocassette. Steps to take apart a cassette and repair a broken leader, plus instructions for making and using a videocassette test jig, are also explained.

One can learn a lot about how a particular VCR is working from cassettes themselves. For example, examining tape damage can help identify several tape transport problems. Therefore, knowing how a cassette interacts with the tape transport is essential to diagnosing many VCR mechanical problems. These topics are covered in this chapter, too.

## CHAPTER OBJECTIVES

**Upon completing this chapter, you should be able to:**

1. Describe the basic structure, features, mechanical parts, and operation of a VHS videocassette.
2. Describe steps to disassemble a videocassette and repair a broken leader.
3. Match various types of videotape damage with probable causes and the appropriate corrective steps.
4. Explain how to make a test jig from a videocassette and how to use it for VCR troubleshooting.

From Chapter 3 you have become familiar with many tape transport components—such as roller guides, capstan and pinch roller, supply and take-up spindles—that position tape and control tape motion. This chapter closely examines the VHS videocassette itself to learn how it works in a VCR.

## OUTSIDE PARTS AND FEATURES

If at all possible, have a VHS videocassette with you as you read this chapter. It will make things easier to understand. Keep in mind that the *front* of the cassette is where the door is; the *rear* of the cassette is where the title label is placed. Also, remember that supply and take-up *spindles* are parts on the tape transport, whereas supply and take-up *reels* are inside a cassette.

Figure 4.1 is a view from the right side of a VHS videocassette. The front of the cassette is away from us, as when it is inserted into a VCR. In Figure 4.1, the spring-loaded door is closed, protecting the videotape. This door closes and locks whenever a cassette is *not* loaded onto the VCR reel table. During *cassette*

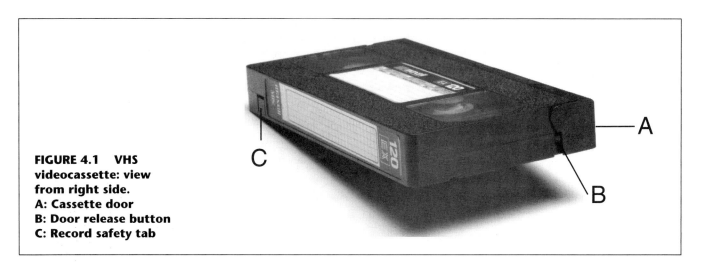

**FIGURE 4.1 VHS videocassette: view from right side.**
A: Cassette door
B: Door release button
C: Record safety tab

*loading,* the door opens as the cassette descends to the reel table. Then, during *tape loading,* a loop of videotape is drawn from the opening.

The cassette door is kept closed with a mechanical latch; a release button on the right side of the cassette unlocks the door. During cassette loading, a finger in the loader carriage presses against the release button to unlock the door. As the cassette lowers onto the reel table, a stationary pin or arm contacts the bottom edge of the door, opening it. This happens in approximately the final inch of cassette downward travel. The door is held open as long as the cassette remains loaded in the VCR.

The very top edge of the cassette door is beveled at a 45-degree angle. On some VCRs, this ensures that a cassette is inserted correctly: a spring-loaded finger or catch just inside the loader opening rides up over the bevel and allows the cassette to be inserted. If the cassette is put in backward, the leaf spring or spring-loaded catch hits against the straight rear edge of the cassette, preventing it from seating in the loader.

At bottom center of the cassette door, there is a small cutout or notch. A small pin or flange on many loader assemblies slides into this notch when the cassette is inserted. If the cassette is inserted backward, the absence of a notch on the backside of the cassette prevents it from going into the loader.

Referring again to Figure 4.1, the supply reel is beneath the left cassette window; the take-up reel is under the right window. There's an opaque label area between the two reel windows.

A square plastic *record safety tab* at the left rear of a cassette allows a recording to be made. The record safety tab covers a small square cavity in the cassette shell. With the tab in place, a new recording can be made over existing material on tape. When the record safety tab is broken off, a switch lever in the tape transport falls into the cavity. Syscon senses this position of the record interlock switch and prevents the VCR from going into Record mode.

You can record on a cassette with a missing record safety tab by forming a small square pellet from a strip of paper. Fit the pellet into the record interlock cavity and secure it with adhesive tape. If possible, use splicing tape for this purpose. It has less tendency to become gooey and loose with the warm temperatures inside a VCR than masking tape, transparent tape, or electrical tape.

## Keeping the Door Open

There will be times when you need to prop a cassette door open. For example, to service reel table components, you will generally remove the cassette loader assembly (also called the *cassette basket assembly*) so you can get at the tape transport. Most front loaders detach easily by removing four screws and unplugging a multi-pin electrical connector from the assembly.

With the loader removed, there is nothing to release or open the cassette door. Simply prop it open! Prepare a 4-1/2 inch length of some stiff insulating material, such as a wooden popsicle stick. Manually operate the door release button, open the door, and slip the prop beneath the front lip of the door. Position the prop just to the right of the supply reel window, as shown in Figure 4.2. Tape the prop to the cassette top to hold it in position. In most cases, positioning the door prop as shown lets you manually put the cassette onto the reel table and perform further operations. Position the prop so that its front end does not extend more than necessary to hold the door open.

**FIGURE 4.2 Propping open VHS videocassette door.**

You can also take a scratch cassette and totally remove the door. This is a handy tool to have: with the cassette loaded onto the reel table, you can look straight down and see where roller guides P-2 and P-3 come to rest inside the cassette shell at the completion of tape unloading.

NOTE: Do not try to put a cassette *without* a door into a loader assembly cassette tray. In many cases it will not go in; without the 45-degree bevel at the top of the door, the cassette is prevented from sliding beneath the spring-loaded catch, which is at the top, center of many cassette trays.

### Cassette Underside

Now flip your VHS cassette over to locate features on its underside. Figure 4.3 is the view with the door propped open as just explained. Note the vertical splines in the white supply reel hub, near the top of the photo, and in the take-up reel hub. Three projections or vanes near the top of both supply and take-up spindles in the VCR engage with these splines when the cassette lowers onto the reel table to drive the cassette reels.

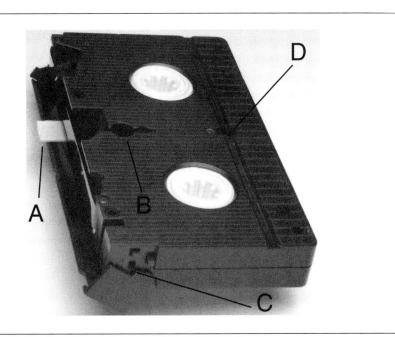

**FIGURE 4.3** Underside of VHS cassette. A: Wooden stick door prop B: Light source hole C: End-of-tape light exit (take-up side) D: Reel lock release hole

## Cassette Reel Locks

Whenever a cassette is *not* in a VCR, the supply and take-up reels are prevented from turning in *one* direction. This prevents tape from loosening or spilling within the cassette. It is important that tape be taut across the front of the cassette—no loose tape or loops—when the cassette loads in a VCR. When not in a machine, either cassette reel *can* be turned manually in the direction that *reels in* loose tape, but not the other way. Here's how this is done.

Inside a cassette, each tape reel has many notches forming a *ratchet* around the circumference of its lower flange. Spring-loaded *pawls* press against these ratchet wheels, keeping the tape reels from rotating in one direction, but allowing them to turn the other way. The supply reel ratchet and pawl prevents clockwise rotation, while permitting the reel to go counterclockwise. Likewise, the take-up reel is prevented from turning counterclockwise, but it can go clockwise.

When a cassette is loaded, a metal post on the reel table protrudes through the reel lock release hole on the bottom of the cassette. The reel table post presses a lever in the cassette, which pivots the pawls away from the reel ratchets, allowing them to turn in *either* direction.

Recall from Chapter 3 that the end of a tape is sensed by light shining through a transparent leader at each end of the videotape. As the cassette settles down onto the reel table during cassette loading, a light post protrudes into the light source hole. Inside the cassette, light is channeled toward two small openings, one on each side of the cassette. You can see the take-up side light exit in Figure 4.3. Directly above this opening is the latch that keeps the cassette door closed when the cassette is not loaded. A similar light opening is on the supply reel side of the cassette.

Precise cassette positioning on the reel table is important. Two flat-topped locating studs on the reel table contact round support pads on the bottom rear of the cassette, while two cone-shaped studs at the front of the reel table engage two locating holes on the cassette bottom. Together, the four reel table studs position the cassette in three dimensions: left to right, front to rear, and height.

Referring again to Figure 4.3, notice the four openings, wells, or cavities, along the bottom front of the cassette. Starting near the top of the photo, the first cavity is where drum entrance roller guide P-2 and slant pole protrude when a cassette is loaded, *between* the videotape and the back of the cavity. The tape holdback tension guide pole also resides in this cassette well when tape is unloaded. Moving downward in the photo, the cavity to the left of the light source hole is home for drum exit roller guide P-3 and slant pole when tape is unloaded. During tape loading, guides P-2 and P-3 and their associated slant poles pull tape from the cassette at these points.

Moving downward in the Figure 4.3 photo once again, the next cassette well is for the rotating capstan shaft. The capstan is *positioned between* tape and cassette body. This is why it's so important that tape be taut across the front of a cassette during cassette loading; loose tape can tangle with roller guide P-2 or P-3, or get on the wrong side of the capstan.

Finally, the small rectangular opening at the bottom of the cassette photo is where an arm or post in the VCR opens the cassette door. This arm may be part of the front loader assembly or it may be mounted on the reel table.

Notice in Figure 4.3 the narrow groove or channel running from side to side along the cassette bottom, about 7/8-inch in from the rear. The reel lock release hole is mid-way along the groove, which is used by a very few machines that are *end loaders*, including separate high-speed rewinders. A projection in the machine opening fits into the channel when a cassette is inserted correctly. This prevents you from inserting a cassette backward.

Those are the principal features that can be seen on the outside of a cassette. Next we'll take a cassette apart and explore its internal components.

## HOW TO TAKE A CASSETTE APART

You will seldom need to disassemble a videocassette, but this is the only way to salvage a priceless tape with a broken leader, or to transfer videotape to a new, empty shell if the cassette shell is damaged. Read each step thoroughly before actually performing the operation.

1. Select a clean, well-lit surface on which to work.
2. Peel off any label along the back edge of the cassette, and remove any adhesive tape that may be substituting for a broken-out record safety tab. If the label doesn't peel off easily, use a thin, sharp knife or single-edge razor blade, as illustrated in Figure 4.4, to cut through the center of the label,

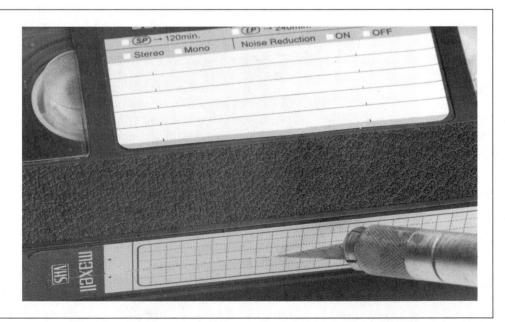

**FIGURE 4.4** Cutting through label to separate cassette halves.

where the cassette halves meet. The idea here is that we don't want any adhesive tape or label holding the two halves of the cassette shell together.
3. Place the cassette upside down on your work surface so the cassette door is away from you. Remove the five small Phillips head screws indicated in Figure 4.5. Remove the center screw last, while pushing down on the bottom half of the cassette to keep the two halves together. Place the screws in a small container to prevent loss.

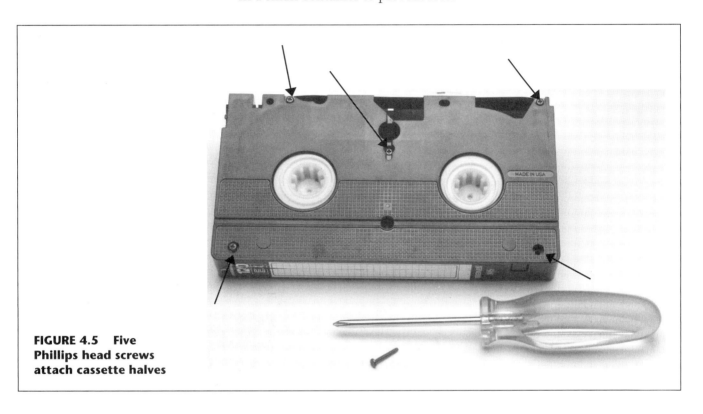

**FIGURE 4.5  Five Phillips head screws attach cassette halves**

4. Holding the two halves together, carefully turn the cassette right-side up on the work surface with the door facing you.

> NOTICE!: If you take the cassette apart by lifting up on the bottom half before turning it right-side up, all kinds of parts *will fall out*, and you'll have a puzzle on your hands to put back together!

5. Grasp the top half of the cassette with both thumbs near the top of the door and your other fingers along the rear edge. Lift straight up to separate the halves. *Be careful* as you lift the top half, not to snag tape on the bottom edge of the cassette door. Pivot the door open slightly with both thumbs as you lift off the cassette top. Propping the cassette door open first, as previously described, makes cassette disassembly easier. Set the top half with the door aside.

## INSIDE PARTS

With the cassette apart, observe the tape path and location of components, such as reel lock release lever, pawls and springs, and door latch mechanism.

Jot down notes about the tape path and part positions, or make a sketch. It's *easy* to misthread tape or misposition a reel lock part. A sketch helps get everything back together correctly. If you dislodge a reel pawl, spring, lock release lever, or tape guide within the cassette, a pair of needlenose pliers or tweezers is handy for placing parts back where they belong.

> CAUTION: Wear eye protection whenever you are working with springs and spring-loaded parts. A small spring or part could easily "fly" and cause eye damage.

Figure 4.6 shows interior cassette parts and tape path. Notice the reel lock release lever. When a cassette is loaded, a post on the reel table protrudes into the cassette and presses this lever; the two spring-loaded pawls then pivot away from the reel ratchets. When the cassette is not loaded, pawl tips are held against the reel ratchets by small springs, preventing reel rotation in one direction.

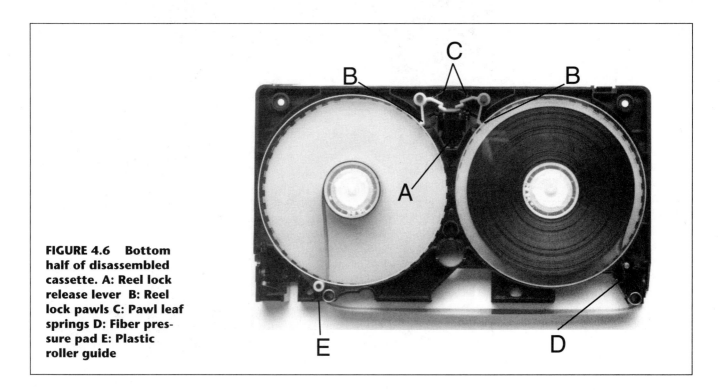

**FIGURE 4.6** Bottom half of disassembled cassette. A: Reel lock release lever  B: Reel lock pawls  C: Pawl leaf springs  D: Fiber pressure pad  E: Plastic roller guide

With the disassembled cassette positioned as in Figure 4.6, the supply reel is on the right. Observe how tape threads from the supply reel, *between* a fiber pressure pad and stationary metal guide pin, and then around a stationary metal supply guide pole. Tape crosses the front of the cassette and then between another stationary metal guide pole and a plastic or nylon roller guide before entering the take-up reel on the left.

The two metal guide poles and the roller guide sit on plastic pins in the lower half of the cassette. Be sure these parts are properly positioned and that tape is in the correct path before reassembling a videocassette.

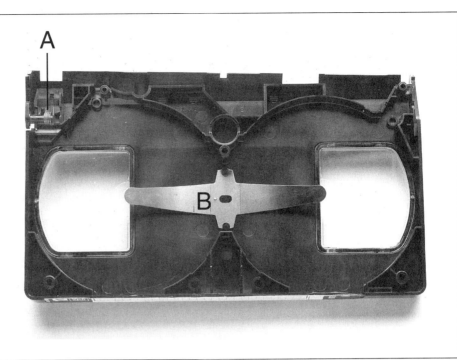

**FIGURE 4.7** Top half of disassembled cassette. A: Door spring B: Reel hold-down leaf spring

Now take a look at the top half of the cassette shell you previously set aside. It looks like Figure 4.7, with the cassette door still attached. Note the small spring on the left side of the door; it closes the door during cassette unloading.

Within a cassette, the thickness of tape reels, from top of upper flange to bottom of lower flange, is slightly less than the inside height of the cassette shell. When a cassette is loaded, reel table spindles should elevate the two tape reels slightly within the cassette, so they are positioned midway between the floor and ceiling inside the cassette shell.

A large leaf spring in the top half of the cassette presses down on the supply and take-up reel hubs. This downward force *tends* to lift the entire cassette up off the reel table. That is, downward spring pressure on cassette reels against transport spindles attempts to lift the entire cassette vertically—if the cassette itself is not held down against the four transport locating studs. When a cassette is held down firmly, the supply and take-up spindles lift tape reels slightly so that they "float" halfway between the floor and ceiling inside the cassette.

This is very important, and worth repeating!: In a properly adjusted VCR, the loader assembly holds a cassette firmly down on the four reel table locating studs. With proper spindle height, cassette supply and take-up reels ride midway between the bottom and top of the cassette shell. If the loader *doesn't* press the cassette down on the reel table, the entire cassette rises up. This causes problems, such as tape bottom edge damage and uneven tape motion.

You can easily check this: You should not detect *any* movement when you press down anywhere on the top of a loaded cassette. If you do, chances are the loader is stopping too early in its downward travel or a bent or broken part in the loader is failing to press the cassette down on the reel table.

When operating a VCR with the loader removed, a cassette will ride up, rather than resting against the four transport locating studs. You'll have to place a weight on top of the cassette to hold it firmly down. A double D-cell flashlight placed atop the center label area of the cassette does this nicely. (The flashlight should be nonmetallic, *without* a magnet on the side.) Any nonmetallic, nonmagnetic weight of approximately one pound should work fine. Just make sure there is no downward movement when you press anywhere on the top of a loaded cassette.

## VIDEOCASSETTE REPAIR

Once a cassette is apart, you can make cassette repairs. Generally, the only time you should repair a cassette is if:

- The splice between the tape and transparent leader separates
- The transparent leader separates at the reel hub
- The cassette shell is damaged, in which case tape reels and tape can be transferred to a new shell.

A properly operating VCR should not cause tape splice or leader breakage: the transparent leader is sensed and the cassette reels stop turning before there is any pull on the tape ends. Breakage can occur if the supply or take-up spindle doesn't stop in time. This can occur during repair activities, when operating a VCR with the front loader removed on models that have the end sensors mounted on the loader. With the sensors unplugged, it's the same as if they were dark all the time; system control cannot determine tape end.

Another cause of leader and tape end breakage is not the VCR but a separate high-speed rewinder. This accessory rewinds tape faster than many VCRs, freeing up the VCR and saving wear on its moving parts. However, some inexpensive rewinders do not have an optical system to detect transparent tape leaders. Instead, they stop by detecting when the rewinder spindle stops turning. This tugs on the take-up leader, splice, and videotape at the end of rewind, sometimes causing one of the first two conditions previously listed.

As an example of the third reason for repairing a cassette, suppose a portion of a cassette is damaged, but the two tape reels and videotape are still in good shape. Perhaps the door has broken off, or there's a large crack in one of the transparent cassette windows. If this is a priceless video, you may want to transfer the good tape and tape reels to a new videocassette shell.

Following are procedures to handle two types of repairs. It is *not* advisable to splice videotape other than at the leaders, which don't get as far as the rotating video drum. A splice elsewhere could snag a delicate video head tip or cause tape to stick to the drum. These are not desirable circumstances!

### Repairing a Broken Splice or Leader

Assume that videotape is broken near the leader, or a splice connecting tape and leader has come apart, or a leader has broken away from its reel hub. No doubt one cassette reel will be full and the other empty, possibly with just the leader still attached to the reel hub. Here's how to start the repairs:

1. Disassemble the cassette as described earlier. Take notes or draw a sketch of internal cassette parts, in case you accidentally dislodge them during the procedure.
2. One at a time, push in on the two reel pawls to disengage the pawl tip from the reel ratchets. Then lift straight up on the supply and take-up reels and remove them from the cassette.
3. Place the two tape reels side by side on a clean, well-lighted work surface in the same relative position as in the cassette.

These three steps will be required for all videocassette repairs.

Avoid touching videotape, as much as possible, with your fingers. Body oil can get on the tape surface, where it can attract dust and contaminate rotary heads and other parts in the tape path.

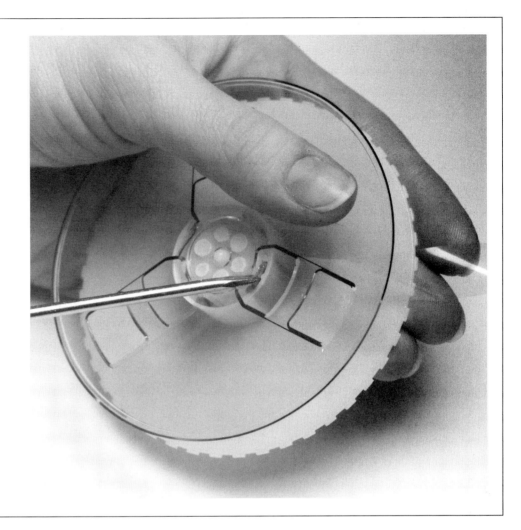

**FIGURE 4.8** Removable wedge clamps transparent leader to reel hub.

### Repairing Broken Leader at Reel Hub

To repair a leader that has broken away from a reel hub, refer to Figure 4.8. In most cassettes, the leader is clamped to the hub by a small plastic wedge. You can usually pull straight up on the wedge with longnose pliers through an opening in the upper reel flange. On most cassettes, you'll find a small hole on the underside of the reel hub, right under the plastic wedge. Simply push a scribe or straightened paper clip through this hole to pop the wedge out. Remove and discard the small length of leader, usually about 3/4-inch or so, that was left clamped to the reel hub.

On some cassettes, the upper reel flange is easily removable by holding the lower part of the reel steady and rotating the top flange counterclockwise. But don't remove the upper flange unless you have to, to get at the reel hub wedge!

Once you have removed the reel hub wedge, cut off the very end of the broken leader tape so it has a fresh end, without any creases. Place about 1/2- to 3/4-inch of the leader tape across the reel hub gap. Now press the plastic wedge back in laterally toward the center of the reel hub until it locks into place, clamping the leader to the hub. It often requires a fair amount of pressure on the wedge for it to pop into place. Check that the leader tape edges are parallel to the reel flanges. Some plastic reel hub wedges have a small metal insert or stiffener. Pull the insert out with a pair of longnose pliers before clamping the leader tape to the hub. Once the wedge is back in place, press the stiffener into the wedge with a screwdriver, as shown in Figure 4.8. Then follow the directions under *"Cassette Reassembly"* later in this chapter.

## Repairing Broken Tape Splice

As stated earlier, splicing videotape except where it has broken close to or at the leader is *not* recommended; rotary head damage could result. Repairs can be made when the splice joining videotape and leader has failed, or when videotape has broken and the tape between the break and the leader can be discarded. For either of these situations, first take the cassette apart.

If videotape has broken, but not at the leader splice, discard any tape left attached to the leader by cutting the leader just before the splice. Do the same when the splice itself has failed. Now trim the end of the videotape to be spliced to the leader so it is a clean cut, with no jagged edges or wrinkles. You may have to discard a foot or so of tape, as it may have stretched close to the break.

With clean ends of both leader and tape, overlap the ends slightly and make a diagonal cut through both at the same spot. A commercial 1/2-inch tape splicing block is ideal for making an easy, quality splice; but with great care you can make a satisfactory splice without one. Here's how.

Place the two tape reels on your work surface about a foot or so apart. Bring the leader and tape ends together so that they overlap about one inch on a smooth, hard surface, such as a ceramic tile or piece of formica. Make sure the two pieces are parallel to each other and that there is no twist in either.

Use two small pieces of splicing tape to *lightly* hold both leader and videotape in place on the hard surface, about an inch from where the ends overlap, as shown in Figure 4.9.

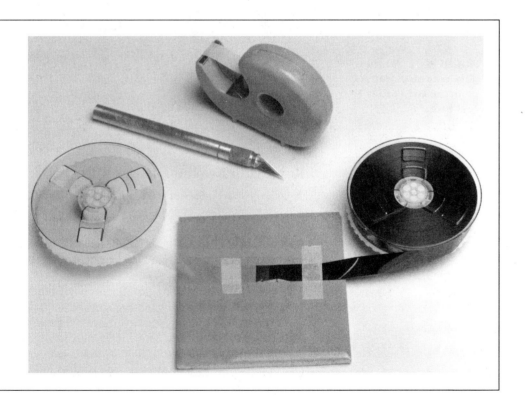

**FIGURE 4.9  Aligning leader and tape for splicing**

With a thin, sharp knife or single-edge razor blade, make a diagonal cut where tape and leader overlap. It's essential that the blade be sharp! Hold tape and leader in place as you make the cut so their ends don't move. Lay a metal ruler or other straightedge down as a cutting guide. When done correctly, videotape and leader will precisely butt each other at the diagonal cut, with no overlap or gap.

Spliced surfaces must be clean and oil-free for good adhesion. Moisten a cotton swab or chamois stick with alcohol or tape head cleaner. Lightly swab tape and leader surfaces for about 1/2-inch on each side of the diagonal cut. This is where you'll be applying splicing tape.

Be sure the cleaner has completely evaporated. Then take a piece of splicing tape and carefully place it over the diagonal cut so that the leader and tape ends remain butted against each other. With the splicing tape in place, press it firmly into both videotape and leader using a smooth, hard instrument, such as the bottom of a teaspoon or plastic pen cap. Rub lightly with the instrument until all air bubbles between splicing tape, videotape, and leader are gone. This is illustrated in Figure 4.10.

**FIGURE 4.10 Splicing transparent leader and videotape. Rub splicing tape with smooth object to remove air bubbles.**

The next step is to trim the splice so that *no* splicing tape remains beyond the bottom or top edges of the videotape/leader. Trim spliced edges with a knife or single-edge razor blade. It is actually desirable to cut away a tiny sliver of videotape and leader at the splice. A slight decrease in tape width at the splice is much better than even the *smallest* increase in tape width, even if only a .001-inch increase! Increased tape width can snag at tape guides and other transport components. Figure 4.11 illustrates proper splice trimming. Note that the trim is somewhat inside the full 1/2-inch tape width.

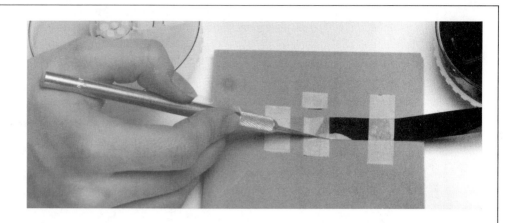

**FIGURE 4.11** Trimming a splice.

*Lightly* clean around the splice with an alcohol swab, especially at the edges of the splice. The swab should be *barely* moist, just so it can pick up any finger oils or splicing tape adhesive. Be certain no alcohol or cleaner actually contacts the edges of the splice, as this could loosen the adhesive.

Inspect the quality of the splice. You should observe that:

- Tape and leader are parallel so they form a straight horizontal path, with no angle at the splice
- Tape ends butt each other at the splice, with no overlap and no gap
- Splicing tape should adhere well at both sides of the splice. No air bubbles beneath splicing tape, no loose corners or edges of splicing tape
- No oozing of splicing tape adhesive. Splicing tape adhesive not gummy; splicing tape adheres firmly to tape surfaces. If problems of this type, probably old splicing tape; do not use
- No part of splice sticking beyond top or bottom tape edge. Clean splice trim, no ragged edges or barbs
- Clean surfaces, both sides of tape/leader.

Once you have a quality splice, carefully remove the pieces of splicing tape holding the tape to the hard surface on each side of the splice. Clean any stickiness off the videotape/leader with a moistened swab. This basic splicing procedure works well with Beta and reel-to-reel audiotape, too.

### Transferring Videotape Between Cassettes

Disassemble the damaged and new cassette shells. Remove supply and take-up reels with tape to be salvaged from damaged cassette and place into new cassette shell. You can use another cassette as the "new" cassette; just discard its tape and reels. Reassemble the good cassette as described in the following section.

### Videocassette Reassembly

Once the leader or tape has been repaired, or tape reels removed from a damaged cassette, place the supply and take-up reels into the bottom of the cassette shell while holding back the pawl for each reel. Thread tape through the cassette. Check that tape is between the fiber pressure pad and steel guide post at the supply reel, and then goes around the stationary supply guide. At the take-up side, tape must travel between the stationary take-up guide and roller guide. Turn one of the reels so tape is taut across the front of the cassette.

Check that all parts in the cassette bottom are properly positioned:

- Supply and take-up tape guides

- Reel pawls and springs
- Pawl operating lever, or pivot plate
- Cassette door latch plate and spring.

Then take the top half of the cassette shell, check that the door is properly positioned, and lower the top straight down onto the lower half of the cassette. Check that supply and take-up tape guides seat properly on their locating studs in the top half.

It is helpful, when placing the top cassette half on the lower half, to pivot the door open slightly so it doesn't catch on the tape across the front of the cassette. (You may want to prop the door open with a 4-1/2-inch length of popsicle stick, as previously described.)

When the cassette halves are aligned properly, and the cassette's internal parts are where they should be, the halves should come together easily. You should not have to exert any force! If the halves don't mate, lift the top off and check for any parts that may have shifted.

Once both halves are joined, hold them together and turn the cassette over so the bottom faces up. Replace the five screws. Turn the screws just until they are snug; excessive force can strip out the plastic threads or crack the cassette.

Check the operation of the door and latch mechanism. Also check that reel pawls operate properly. Push a small screwdriver into the hole to release them; both tape reels should turn easily in either direction. Remove the screwdriver and verify that both reels are prevented from moving in the direction that would spill tape. Manually turn one of the tape reels so tape is taut across the front of the cassette.

This completes videocassette reassembly and checkout.

## INSPECTING FOR TAPE DAMAGE

One of your most powerful diagnostic tools for VCR repair is the videotape and cassettes that have been used in a particular machine. Observing both tape and cassette can often tell you where to look for problems in the VCR, or give you some clue as to what areas may require attention.

### Outside of Videocassette

What information can be gathered by looking at the cassette? Carefully look for scratches, digs, and abrasions on cassettes that have been in a VCR. This may help you identify an area on the loader assembly that is catching on the cassette. For intermittent cassette load or eject failures, including jamming, telltale marks on the cassette could point you in the right direction.

For example, suppose a VCR fails to load a cassette. The cassette is drawn in by the loader, but a few seconds later the cassette ejects. You also observe that the cassette-loaded icon on the front panel never comes On. Close examination of the cassette may reveal a slight scratch mark just to the right of the light post hole on the underside. Perhaps the light post mount on the reel table has come loose; the post doesn't line up with the descending cassette, but hits on the edge of the hole.

Looking through the transparent windows over supply and take-up reels can tell you how consistently tape is wound during play, fast forward, and rewind. Top edges of layers of tape on tape reels should be straight across, with possibly one or two small ridges for a full reel. Many ridges indicate uneven tape tension during the operation that placed tape on the reel.

For example, after fully rewinding an entire tape, you may observe many tape ridges, or unevenness of tape wrap, around the supply reel. This could indicate a slipping idler wheel driving the supply spindle during rewind or insufficient soft braking at the take-up reel.

Similarly, after playing an entire tape, and before rewinding, you might see that tape is unevenly wound on the take-up reel, with many ridges. This could be caused by excessive slippage of the slip clutch, or a slipping belt or idler supplying power to the take-up spindle during play.

If a slight bulge in the tape wrap around a reel is observed, this most likely indicates that the tape has been badly crinkled at this point. Tape may even be folded over on itself in severe cases, perhaps caused by a sticky spot on the tape. This tape should not be used again.

## Physical Tape Damage

There are several ways in which a VCR can physically damage tape as it passes through the transport. Sometimes the damage is *very evident*. Suppose you eject a cassette and a loop of tape is still hanging outside the cassette door. This might be caused by the supply spindle failing to reel in tape during tape unloading.

Severely mangled tape probably means that tape has wrapped around the capstan. Often this is due to a worn, dirty, or gummy pinch roller, a dirty capstan, a pinch roller that is out of alignment with the capstan, or low take-up reel torque. A cassette may eject with a loop of tape still caught inside the machine. Tape wrapped many times around the capstan is a likely cause. You'll need to get at the capstan to remove the ball of tape around it. These are examples of a VCR "eating" tape. (We'll discuss what to do to correct causes of these problems in Chapters 6 and 7.) Figure 4.12 is an example of what tape looks like after it has been "eaten" by the transport.

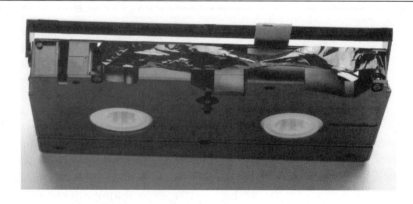

**FIGURE 4.12** Tape that has been "eaten" by the tape transport.

Other tape damage may not be as immediately evident. For example, you might notice a slight scalloped effect, or crinkling, along one edge of the tape, usually the bottom. This can be caused by a broken or out-of-adjustment tape guide or by a pinch roller that is not aligned correctly with the capstan shaft. If the pinch roller bracket is bent, the roller exerts unequal pressure against the top and bottom of the tape, causing the tape to ride up or down on the capstan shaft. Another term for this is *tape skew*. Tape hitting the base of the capstan shaft or curling over the edge of an adjustable tape guide crinkles or scallops the tape, as shown in Figure 4.13.

Videotape inspection may reveal a horizontal scratch or crease along the tape length. This most likely indicates a broken or damaged part in the tape path or a burr on some part that contacts the tape. A foreign object within the

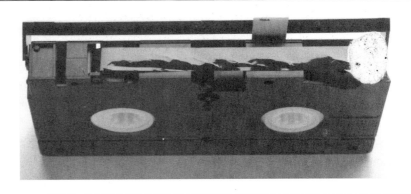

**FIGURE 4.13** Crinkled or scalloped tape edge.

tape transport could also create this type of tape damage. Tape should be absolutely smooth, with no scratches, wrinkles, scallops, or crinkles.

Once you have determined that there are no visible defects on the tape itself, running videotape through the VCR in various modes, such as Play, Forward Scan, and Rewind, can reveal problem areas. With tape moving, carefully inspect along the entire tape path with a strong light. Tape should always run smoothly, with no jerkiness or flutter. As tape goes across guides and stationary heads, it should not wander up or down, but maintain a constant height. There should be no ripples or tape skewing, especially at capstan and pinch roller during play.

A known, good-quality tape should be used for visual inspection of the tape path in motion. You don't want any defects already on the tape, such as a scalloped edge, to mislead you in diagnosing a transport problem. Detailed observation of transport operations and tape motion throughout the tape path is a principal method for determining the cause of many VCR mechanical problems.

Having a few high-quality, commercially recorded cassettes on hand is *absolutely necessary* to perform some VCR problem diagnosis and alignments, without using special (and somewhat expensive) alignment tapes. However, when you first begin to work on a VCR—when you don't know the true condition of the tape transport mechanism—it is best to start with a *scratch cassette*. This can be any tape in fairly good condition with which you can check basic VCR functions. Should the VCR damage this tape, there is no great loss.

Once you've determined that the VCR operates without damaging a scratch tape, you can proceed with a known, good-quality commercial cassette, such as a movie, for further testing and alignments. We'll cover this in Chapter 7.

You'll want several fairly good-quality scratch tapes for checking overall record, playback, and all other VCR tape-related functions. They should be T-120 cassettes, so the transport has to handle full reels of tape during all operating modes. A T-30 or T-60 cassette may not reveal a slipping belt or idler, because the smaller amount of tape on a full reel requires less spindle torque than a T-120.

You are probably better off never having anything to do with T-160 cassettes, which have a recording time of 2.7 hours at SP and 8 hours at EP speeds. The thinner tape is easily damaged and may not be handled well by some VCRs. Discourage others from using T-160s when one or more T-120s can do the job. The extra recording time may not be worth the inherent problems in using the thinner tape.

Inspect any scratch tape for damage, such as a scalloped or crinkled edge, when you are through using it on a given VCR. You don't want existing scratch tape damage to mislead you when working on another VCR in the future.

Keep all your work tapes in a clean environment and away from magnetic fields. For example, color TVs and monitors have an electromagnetic degaussing coil around the front circumference of the cathode ray tube (CRT). This coil is energized with 60 Hertz (Hz) current when the monitor is first turned On to demagnetize any stray magnetic field surrounding the CRT. Storing magnetic tapes close to a TV or monitor can cause partial erasure of recorded information. Keep tapes at least 18 inches away.

Tapes should be shelved vertically for long-term storage, not laid flat. Layers of tape stored horizontally can slide down to the bottom reel flanges over time, causing curling or bending of the lower tape edge. This is a critical area on tape, where the control track is recorded. Even slight bottom-edge tape irregularities can cause picture problems during playback.

Cassettes that have been exposed to high temperatures, like sitting on the seat of a car in the hot sun, can have warped plastic parts, which can cause cassette load and other problems. Long term exposure to high temperatures can dry out videotape, making it brittle and curling the edges.

## MAKING A VIDEOCASSETTE TEST JIG

Diagnosis of some VCR problems is easier using a *videocassette test jig*, which is a cassette without any tape, tape reels, or other inside parts. Some people refer to a videocassette test jig as a *"dummy"* cassette. At times, it's better to see what's going on without tape in the way. Suppose you want to check some mechanical function of the tape transport, like the pinch roller moving over to contact the capstan or the operation of the spindle brakes. Loading the VCR with a test jig lets you do this without tape actually being loaded in the machine. As we go along, you'll learn other ways to use this testing tool. Here's how to make a VHS test jig.

Simply follow the procedure described earlier in this chapter for taking a cassette apart. Once you've separated the two halves of the cassette, remove all interior parts that simply lift out from the bottom half: tape and reels; pawls, springs, and pivot plate; supply and take-up guides; door latch assembly and spring. Remove the cassette door and spring, and the reel hold-down leaf spring from the top half of the cassette. You may have to break the spring away with pliers, but usually you can simply pull it off by hand, breaking the plastic mounts. Wear eye protection when doing this!

Next, break out the transparent reel windows with a small hammer and pliers. Sometimes they'll push right out very nicely with just your fingers! Take care not to crack the cassette shell itself. Use a hobby saw, diagonal pliers, or nibbling tool, as shown in Figure 4.14, to enlarge the window openings. Remove the black plastic toward the center of the cassette so each opening is one inch or so wider than the original window. Smooth any sharp plastic edges with a file or sandpaper.

Reassemble the two halves of the cassette shell, leaving the door Off. At the front of each side, normally covered by the closed door, block the two tape-end sensor light exits with electrical tape. Your videocassette test jig will now operate the cassette-in switch and keep the tape-end phototransistors dark, as though tape were in the cassette.

### Using a Videocassette Test Jig

You can now use the homemade cassette test jig to run the VCR in various operating modes. For example, observe tape loading and unloading sequences. You can see if P-2 and P-3 roller guides and shuttle assemblies move in their tracks, out from just inside the front of the cassette, and come to rest in V-stops at each side of the video drum. (Tape loading and unloading operations are described in Chapter 3 and in the Chapter 3 Appendix.)

Chapter 4 Video Cassette Examination and Repair ■ 101

**FIGURE 4.14** Enlarging reel windows in top half of cassette. A metal nibbler cuts away a small bit of plastic each time the lever is squeezed in towards the tool body.

An alternative to the custom-made videocassette test jig is a commercially manufactured VHS test jig, available at many electronic supply outlets. A commercially-made test jig essentially performs as the homemade one, but has the advantage of being fabricated of clear plastic so you have a completely see-through tool. End-of-tape sensor areas at the supply and take-up sides of the test jig are blocked, so system control senses that tape is present when the jig is loaded on the reel table.

As explained in Chapter 3, for many operating modes you may have to manually turn a spindle so reel sensor pulses are produced. If system control doesn't receive take-up reel pulses during play, for example, tape is unloaded and the VCR will usually go into shutdown. You can reach through the enlarged window openings of the test jig to turn the take-up spindle with your fingers, thus preventing shutdown. Removal of the loader assembly is required to easily reach the spindles.

You can get a *rough idea* of the torque applied to supply and take-up spindles by grasping them through the cassette test jig openings while they are being driven. With experience, you'll be able to determine when there is insufficient rewind torque on the supply spindle, or low take-up spindle torque during play.

Those enlarged openings in the top of the cassette test jig may cause problems with a few front loaders. For instance, a leaf switch or tension spring may catch in an opening. You may want a second test jig with the windows intact, although the inability to reach through and turn the transport spindles by hand will limit the usefulness of this cassette in most cases.

## SUMMARY

In this chapter, you've read that a videocassette is somewhat of a machine all by itself. When a cassette is not loaded in a VCR, a reel ratchet-and-pawl mechanism prevents the supply and take-up reels from turning in one direction, and a latch keeps the front door shut. Both tape reels are spring-loaded downward by leaf springs on the top half of the cassette shell.

When a cassette is loaded, the following tape transport parts protrude into the cassette:

- Two cone-shaped locating studs
- Reel pawl release pin
- End-of-tape sensor light post
- Drum entrance roller guide P-2 and slant pole
- Tape holdback tension guide pole
- Drum exit roller guide P-3 and slant pole
- Capstan.

The last four items are *between* the videotape and the body of the cassette.

When the plastic record safety tab on the left rear of a cassette is broken off, the VCR will not go into Record mode.

Videocassette disassembly and repair procedures for broken tape and transparent leader were explained. Any label on the rear edge must first be removed. After removing five screws on the cassette bottom, hold the two halves together and turn the cassette right-side up on the work surface, then lift the top half straight off. Note location of tape path and all parts before proceeding, to help you reassemble correctly.

Splices should be made only at the leader, using magnetic tape-splicing tape. Avoid splices that will pass rotary heads. A quality splice has tape ends butted, with no gap or overlap. Splicing tape must adhere completely and be trimmed so that it doesn't extend beyond tape edges.

Leader that has broken close to a reel hub can be reattached. First remove the wedge that clamps the leader to the hub: push a straightened paper clip or scribe through the small hole on the underside of the reel hub. Then insert the leader tape end into the hub cutout and pop the wedge back into place.

Inspecting videotape for different types of damage can often tell a great deal about what transport components need attention. Faulty pinch roller and capstan alignment can cause the tape's bottom edge to be scalloped or crinkled. Badly mangled or "eaten" tape is often caused by a defective or dirty pinch roller, dirty capstan, misaligned pinch roller, or insufficient take-up reel torque.

Good-quality scratch tapes and commercially recorded movies are good tools for VCR troubleshooting. Always use a scratch tape before loading a commercially recorded tape or special alignment tape when beginning to work on a VCR. Should the VCR cause tape damage, the scratch tape is more expendable than a pre-recorded tape. Test cassettes should be checked for physical tape damage *before* being loaded into a VCR to be certain tape damage isn't present due to a previous encounter with an ailing tape transport.

VCR troubleshooting can sometimes be made easier with a cassette test jig, which can be easily made by disassembling a cassette, removing all interior parts and the front door, and blocking the light path to the two tape-end sensors. Breaking out and enlarging the reel windows in the cassette top lets you manually rotate the transport spindles to prevent auto shutdown when using the videocassette test jig.

## SELF-CHECK QUESTIONS

1. Describe the difference between the take-up *spindle* and the take-up *reel*.
2. Describe how a VHS videocassette door is unlocked and opened during cassette loading.
3. What should you normally do with a cassette before using it in a VCR that has its loader assembly removed?
4. Describe why videotape cannot unwind when a cassette is not loaded into a VCR.
5. Why is it important that tape be taut across the front of a videocassette when the cassette is loaded into a VCR?
6. When taking a videocassette apart, why should you turn the cassette right-side up after removing screws on the bottom, *before* separating the two cassette halves?
7. Describe how to check that a cassette is *fully seated* on the reel table.
8. After fully rewinding a T-120 cassette in a particular VCR you observe numerous ridges along the layer of tape on the supply reel. what is this a likely indication of?
9. What causes a slight scalloped effect or crinkling along the bottom edge of a videotape?
10. Why might you want to Grasp the supply spindle with your fingers while using a videocassette test job in Rewind mode?

# CHAPTER 5: TROUBLESHOOTING LOADER AND TRANSPORT MALFUNCTIONS

There's a problem with the VCR sitting on the bench, but how do you go about determining what should be done? What is the correct way to diagnose it? This chapter describes a systematic approach to troubleshooting three areas of the VCR. Troubleshooting narrows down possible sources of failures to the one or two that need repair. Frequent VCR malfunctions occur in these three areas of operation:

- Cassette loading
- Tape loading
- Videotape traveling through the tape path.

*Cassette* loading has to complete successfully before *tape* loading can even begin. Then tape loading must occur without problems before videotape can move through the tape path. Finally, videotape motion through the entire tape path must be precise to reproduce quality video and audio during record and playback. Common problems that can occur in these three operations are described in this chapter, along with what generally has to be done to correct them. Subsequent chapters explain loader and transport repairs in greater detail, and later chapters describe common problems that arise in other VCR areas and how to correct them.

In other words, here is what to expect in Chapters 5, 6, and 7. In this chapter, you'll learn how to observe what problems may be contributing to the symptoms exhibited in three areas of a VCR. In Chapter 6, you'll learn to clean these areas of a VCR. In Chapter 7, you'll learn to adjust parts in these important areas. This will allow you to follow these simple principles of troubleshooting a VCR:

1. Observe.
2. If cleaning might help, clean, then observe again.
3. If cleaning does not help, but adjusting a part might help, adjust, then observe again.
4. If neither cleaning nor adjusting works, then perhaps there is a broken or bent part, or there is an electronic problem.

Finally, before studying this chapter, review the material presented in Chapter 3.

## CHAPTER OBJECTIVES

**Upon completing this chapter of study, you should be able to:**

1. Describe in general terms what is meant by a systematic approach to troubleshooting and problem resolution.

2. Describe common *cassette loading* failures and basic repair procedures to correct them.
3. Describe common *tape loading* problems and what to do to repair them.
4. Describe typical VCR tape motion problems and how the most common ones can be corrected. (Additional problem areas are addressed in Chapters 6 and 7.)

The intent of this chapter is to determine likely causes of VCR tape transport malfunctions, and the overall approach to take to correct the problem, rather than describing *specific* steps to fix a particular model.

VCR repair and maintenance activities usually require performing one or more of the following activities:

- Cleaning
- Lubrication
- Mechanical part replacement
- Mechanical adjustment
- Electronic part replacement
- Electronic adjustment.

Later chapters in this text explain how to make repairs by completing these tasks. Because most VCR failures are mechanical, not electronic, this chapter deals primarily with these types of problems.

## SYSTEMATIC APPROACH TO DIAGNOSING PROBLEMS

Troubleshooting a VCR requires clear thinking and a relaxed, yet methodical approach. A systematic approach is just the opposite of helter-skelter. As stated at the beginning of Chapter 2, don't dive right in and start taking covers off before observing what the VCR *is* doing, what it *isn't* doing, and jotting down a few notes.

Accurate, thorough note taking during diagnosis and repair are extremely helpful in getting things right the first time, without wasting time and making mistakes. Notes help keep you organized and prevent you from forgetting what went where when you're putting parts back together. Your notes can also serve as excellent references for fixing other VCRs later on.

A systematic approach to troubleshooting, whether VCRs or almost anything else for that matter, involves a step-by-step, logical approach to the problem. The first step is information gathering. What symptoms and errors do you observe? Under what conditions do they occur? Is the problem consistent, or solid, or is it intermittent, where a VCR function works some times and fails at others? For example, does the picture tear along the top when playing back *all* tapes, or just tapes that were recorded on *this* VCR? Is the picture bad only at LP and EP/SLP speeds, or at SP speed as well? Does the VCR shutdown at random, or only towards the end of a T-120 cassette? Is there any strange mechanical sound coming from the VCR when the picture breaks up?

### Look for the Simple Things First

Sometimes it's easy to get sidetracked and go after the "reported" trouble or the first difficulty you notice rather than a simpler problem. Take care of the little things, and often a larger problem will disappear! Here are a few examples:

### Example 1

Suppose a VCR will not correctly play back tapes that have been recorded on the same machine. First, check tapes that have run through the machine for damage, as described in Chapter 4. Run a good scratch tape through. If there's no damage to it, then play a known good commercially recorded tape. There's no sense trying to fix a record problem if the VCR doesn't play back a pre-recorded tape well. Fix the playback problem first.

### Example 2

Suppose off-the-air recordings made on this VCR play back with noise in both picture and sound. Check that the VHF output from the VCR to the TV looks OK in E-E mode first, before going after a record problem. This will show what kind of audio and video are being tuned by the VCR's tuner, *before* a recording is made.

If the picture on the TV remains bad, perhaps there's a weak signal from the antenna, or the VCR channel needs fine tuning for the particular TV station, or the VCR tuner might be defective. Do other TV stations come in poorly? Investigating this shows that all channels tuned by the VCR look and sound poor on the TV.

Play a good commercially recorded tape. It is noisy as well! Looks like there might be some kind of playback or overall output problem with the VCR.

Try a composite video input signal to the VCR from a second, known good VCR, for instance. Is the E-E picture on the TV OK? No, the picture and sound are still bad.

Since VCR tuner, tape playback, and video line in signals all produce a bad TV picture, how about the composite video output on the VCR? Is the picture OK there?

Connect the Video Out RCA jack to a video monitor. Aha! The picture here is beautiful, from (a) stations tuned by the VCR's tuner, (b) videotape playback, and (c) from line input signals. This indicates that the RF modulator in the VCR is probably defective, because program material from VCR tuner, tape playback, and video line in sources all look poor, but line out signals are fine.

### Example 3

Sound seems to occasionally vary in pitch, or warble, when playing back a pre-recorded tape, as if the tape speed were sometimes jerky. Try just operating the VCR in fast forward and rewind. Does the tape wind smoothly and consistently? Does the tape index counter increment or decrement without hesitation?

If problems are encountered during fast forward or rewind, these may be easier to locate than problems in play. There are a lot fewer tape transport components involved. Fix the fast forward or rewind failure, and there's a good chance the playback problem will be taken care of at the same time.

### Example 4

Sound gets a bit muddy with a severe reduction in volume every once in a while. Picture usually drops color and has horizontal noise lines or other imperfections at the same time.

Observe tape as it passes over the stationary Audio/Control (A/C) head. Mistracking here is easy to detect. Notice if the tape rides up or down as it passes the audio head instead of staying at a constant height. Correcting this problem will most likely cure both audio and video problems. Here's why.

Remember, the control track recorded at the bottom of the videotape has to play back correctly for the two videoheads to keep in step with their corresponding diagonal RF tracks on tape. Also recall that the linear audio head and control head are combined in a single A/C head assembly. Varying tape height past the A/C heads thus causes signal reduction or loss from *both* the linear audio and control tracks.

### Example 5

VCR sometimes shuts itself off while loading or ejecting a cassette. Unit sometimes shuts down while playing or rewinding a tape. Hum heard in audio when playing a tape. Front panel time-of-day clock display sometimes dims, then comes back to normal, even when a tape *isn't* playing.

In this scenario, the simplest symptom to track down is the changing brilliance of the display; it occurs even when no tape transport motors are energized. Changing display brilliance could very well be caused by some defect in the power supply. Perhaps a rectifier diode or filter capacitor is intermittently breaking down. This could account for *all* the problems, as the power supply is common to everything else in the VCR. Monitoring power supply DC output voltages with a multimeter could help locate the section of the supply that is acting up. (Electronic components are covered in Chapter 10, power supplies in Chapter 12.)

Taking time to carefully observe and experiment with all VCR modes often provides the clue that leads to quick identification of a problem source. There is probably no single set of steps to follow in a systematic approach to troubleshooting, but it is a *methodical* technique for determining what is working, what isn't, and then tracking down logically possible causes first. That's why effective troubleshooting begins with a thorough understanding of how *all* VCR systems relate to one another.

One of the most valuable activities when troubleshooting and repairing is to take accurate notes, from start to finish. Documenting as you go along takes a little time, but in the long run you'll find it's well worth it.

One note of caution may be worthwhile at this point in the textbook. Some readers may feel they now know enough to actually tackle a VCR repair. And that may be so. But without the guidance of a classroom instructor, it might be prudent to wait until you have read more of the textbook. This could possibly prevent you from making a costly mistake. Of course, if you've got an old VCR on which you can experiment, that's another story.

## CASSETTE LOAD AND UNLOAD PROBLEMS

Recall from Chapter 2 that there are two types of mechanisms for loading a cassette: top loader and front loader. For both loaders, a cassette is inserted into the loader basket, or tray, with the cassette door facing forward (that is, toward the rear of the VCR).

Prior to about 1980, all VCRs had top loaders. On top loaders, a cassette is placed into a basket that rises vertically out of the top of the machine. Then the user pushes down on the cassette basket until it latches into place with the cassette positioned on the reel table. There is no separate cassette load motor or geared mechanism for moving the cassette basket, as with a front loader.

Once front loaders became popular, manufacturers stopped making top loading VCRs; very few are still around. For this reason, top loader assemblies and their repair are discussed only briefly.

### Top Loaders

Top loader mechanisms are fairly simple. One or more leaf springs hold the cassette against the bottom of the loader basket when a cassette is inserted.

A projection or "finger" in the basket releases the door latch on the right front side of the cassette. As the assembly is pushed down into the VCR, a stationary pin or arm on the tape transport opens the cassette door. When the basket is all the way down, a latch on the transport engages a pin or catch on the loader basket, locking it down.

When the Eject button is pressed, the tape transport latch holding the cassette basket down unlocks, releasing the basket. The basket then rises up through the top of the VCR under spring power. Of course, the mechanism is interlocked so that Eject won't function unless tape is first unloaded. Eject on many top loader VCRs causes the mode motor to turn beyond the tape unload position where the cassette basket latch is released by the undercarriage mechanism. A top loader can also be released with an electric current flowing through a *solenoid*. The moving solenoid plunger opens a latch, releasing the cassette basket catch or pin. (Early VCRs with piano-style pushbuttons often have a mechanical linkage between the Eject button and cassette basket latch.)

A *solenoid* is an electromagnetically operated plunger, as illustrated in Figure 5.1. It converts electrical energy into straight line, or linear, motion. When electricity flows through the electric coil, the resulting magnetism pulls an iron plunger into the core of the solenoid, which opens the cassette basket latch.

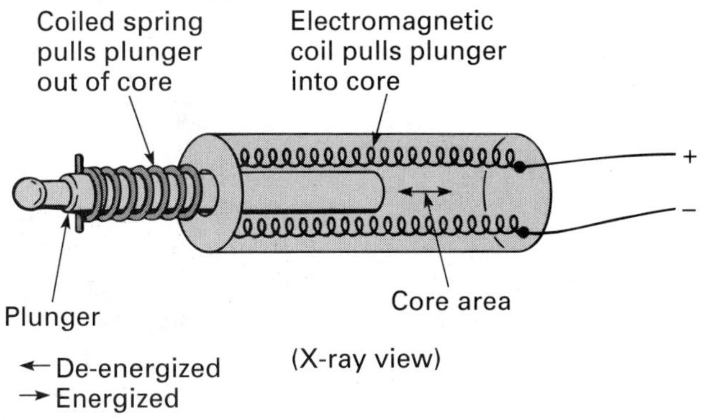

**FIGURE 5.1 Electric solenoid**

Following are common top loader problems and their causes:

- Cassette won't seat properly when inserted into basket in raised position.
  — This problem is normally easy to locate and fix. Usually a bent or broken leaf spring inside the cassette basket is the cause. If one of the screws holding the cassette lid to the top of the loader has been replaced with one that is too long, the screw can protrude into the basket area, preventing the cassette from sliding all the way in.

- Basket won't push down all the way.
  — A foreign object in the machine is often the cause of cassette loader problems. Children like to put toys, coins, and even food into the open cassette compartment opening! (Time to feed the VCR?)
  — A bent or broken catch on the cassette basket, or a bent or broken latch on the transport floor, can prevent the catch from seating properly in the open latch when the basket is pushed down.
  — A binding solenoid plunger or mechanical linkage can keep the latch closed when it should be open.

- Basket won't eject.
  — Bent, broken, or binding parts in cassette basket catch, transport latch, or solenoid and linkage.
  — Tape not fully unloaded.

— Solenoid not being energized: open solenoid coil, defective Eject button, broken wire.
— Foreign object in VCR.

There are a variety of cassette basket latches, and they are in different locations from model to model. But they are all fairly straightforward; almost all top loader malfunctions are caused by broken, bent, or binding parts.

Don't overlook bent parts in the cassette basket where a cassette is inserted. It is not unknown for both top loaders and front loaders to become bent or otherwise damaged from someone trying to force a cassette in backward, crooked, or even upside down! Use care when re-forming, or bending, sheet-metal; excessive bending will weaken and even break the part.

Most top loader assemblies mount to the tape transport with four, (sometimes six) screws. Usually these are tinted red or blue for identification. There may be a separate ground wire connected to the loader assembly that will have to be taken off before the loader can be removed from the VCR. If there is a ground wire connected to the loader, it's a good idea to put a piece of electrical tape over the terminal at the end of the wire after it's removed. This prevents the wire from possibly shorting to something if it flops around while the machine is powered up.

## Front Loaders

Nearly all of today's VCRs are front loaders. A cassette is gently inserted, door first, partway into the compartment opening on the VCR. At this point, the cassette-insert switch transfers, the DC-operated cassette load motor is energized, and loader mechanics pull the cassette the rest of the way in, and then lower it onto the tape transport reel table, in an inverted L-shaped path. Also during cassette load, the cassette door latch is released by a spring-loaded projection or "finger" on the right side of the cassette tray, and the door is opened by a stationary arm as the cassette seats on the transport spindles.

Usually power has to be turned On for the loader to operate, but some VCRs automatically power up when a cassette is inserted. With a cassette loaded on the reel table, the cassette-in switch is transferred. This switch tells the system control microprocessor that a cassette is loaded, and the cassette-loaded icon on the display/indicator panel lights up.

When Eject is pressed, system control first checks whether tape is unloaded, which it senses from the position of the mode switch or cam position sensor. If tape is not unloaded, system control initiates a tape unload operation. For example, on most VCRs it is *not* necessary to hit Stop when playing a cassette to stop and eject. Just hitting Eject stops tape motion, withdraws the pinch roller from the capstan, and unloads tape. When system control receives the signal from the mode switch or cam position sensor that tape is unloaded, then the eject operation begins.

To eject a cassette, direct current energizes the cassette load motor. This voltage is now reverse polarity compared with DC to the motor when a cassette is loaded, so the motor turns in the opposite direction. The loader tray lifts the cassette up off the reel table, and expels it about 3/4-inch out the compartment opening.

Many maintenance and repair procedures require that the front loader assembly be removed. For example, with most VCR models in order to clean or replace the rubber idler wheel between supply and take-up spindles on the reel table, the loader assembly must first be taken off.

To remove the loader, the cassette tray should first be at the Eject, or home, position if at all possible. On some VCRs, you will have to take off the front panel in order to take out the front loader assembly. Most front loaders have a mounting screw in each of the four corners which must be taken out. On some models, the two mounting screws at the back of the loader are somewhat hidden: you have to sort of snake a screwdriver through openings in the top of the loader to get at them. Look for Phillips head screws that are tinted red or blue; they are likely the ones you need to remove.

Depending on the model, a circuit board or shield plate may be attached to the front loader, requiring removal of a few more screws, as shown in Figure 5.2. There may also be a ground wire to disconnect. Finally, a multi-conductor cable that plugs into the loader assembly will generally have to be unplugged. On some models the front loader cable is long enough so that you don't have to disconnect it; simply place the loader behind the machine on the workbench while you work on the transport.

FIGURE 5.2 Typical VCR front loader assembly. A: Four loader mounting screws (left rear hidden from view) B: Two circuit board mounting screws C: Shield plate mounting screw, accessible after pivoting large circuit board up D: Ground wire and screw

One model front loader is mounted with just two red screws at the rear, and metal L-shaped hooks at the front. After the two screws are removed and the multi-conductor cable is unplugged, the loader assembly is slid toward the front of the VCR about 1/4-inch to unhook the loader from the transport. It then lifts straight up and out.

Use care when disconnecting the front loader cable; wires and connector halves are delicate and easily broken with too much force. Go slowly! Use a small, flat-bladed screwdriver to pry apart small, multi-pin connectors. Some have a plastic latch on the side that locks the connector halves together. Depending on the make, you either pull up or push down on the plastic latch to release it as you separate the connector halves.

Carefully lift the rear of the loader up several inches, and work the assembly out of the machine. Don't force it if it doesn't come out readily. Find out what is holding it in. It should come out easily if you've removed all hold-down screws, ground wire (if any), and multi-pin connector.

Front loaders are more complex than top loaders, but are not terribly complicated. Usually there is a small DC motor mounted to the right side of the assembly. A worm gear, or belt and pulleys, rotate a gear that moves a drive arm, which then moves the cassette basket to the rear and down in an inverted L-shaped slot. A transverse shaft from the right to the left side of the assembly turns a similar gear and drive arm on the left side of the loader.

Figure 5.3 is a representative loader assembly. On this particular unit a worm gear on the motor shaft meshes with a worm wheel. Other units have a pulley on the motor shaft with a small rubber belt that drives a pulley connected to gears on the loader.

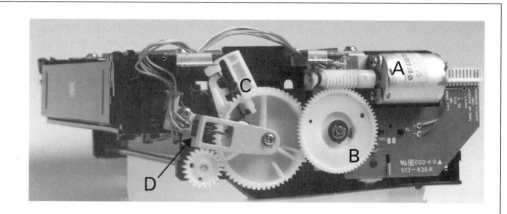

**FIGURE 5.3** Typical front loader assembly, right side. A: Motor with worm gear  B: Worm wheel  C: Lifter arm and cam gear  D: Leaf switch

On many loaders, both supply and take-up end-of-tape phototransistors are mounted at the bottom left and right, respectively, on small circuit boards. Some VCRs also have the record safety, or record interlock, switch incorporated in the loader assembly. Refer back to Figure 5.3 as you read a description of how this loader works.

- When a cassette is inserted into the compartment, the front of the cassette comes up against two small upright projections, or tabs, on the floor of the movable sheetmetal cassette tray. The two projections contact the cassette body through small cutouts at each end of the cassette door.

    As the cassette is pushed gently forward, it pushes against the tray projections and moves the tray further into the machine a short distance. This movement turns the large lifter arm/cam gear clockwise just a few degrees.

    A spring-loaded "finger" on the right side of the tray presses against the cassette door release button.

- As the lifter arm/cam gear turns, a plastic projection on the cam gear moves away from the leaf switch lever. This causes Contact #1 of the leaf switch, the cassette-insert switch, to open. That is, Contact #1 is kept *closed* by the gear projection with the cassette tray all the way to the front of the VCR in the full eject position. It *opens* when a cassette is gently pushed into the tray.

- System control senses the transfer of the cassette-insert switch (Contact #1 opening) and energizes the cassette load motor, driving the lifter arm/cam gear clockwise.

- The lifter arm moves toward the rear, pulling the cassette tray along an inverted L-shaped slot in the side frame. A transverse shaft across the bottom front of the loader operates a similar lifter arm/cam gear on the left side of the assembly.

- As the cassette lowers toward the reel table, an arm or post opens the already unlatched cassette door. This arm may be part of the loader assembly or attached to the tape transport.
- When the cassette tray is all the way down on the reel table, another projection on the lifter arm/cam gear closes Contact #2 on the leaf switch. This signals system control (syscon) to stop the motor.
- Pushing the Eject button signals syscon to apply DC voltage in reverse polarity to the cassette load motor. The motor now turns the lifter arm/cam gear on the right side of the loader counterclockwise, lifting the cassette tray from the reel table and pushing it out the cassette compartment opening. A small spring inside the cassette closes its door as the cassette raises off the reel table.
- The lifter arm/cam gear on the loader's left side operates a lever that opens the cassette compartment door at the front of the loader.
- When the first projection on the lifter arm/cam gear closes Contact #1 on the leaf switch, syscon stops the motor.

Figure 5.4 shows the bottom and left side of the same front loader pictured in Figure 5.3. Notice that this loader incorporates the record safety switch, in this case a normally closed (N.C.) leaf switch. The back of the cassette transfers this switch if the record safety tab is intact. (Quite often the record safety switch is mounted on the tape transport itself.)

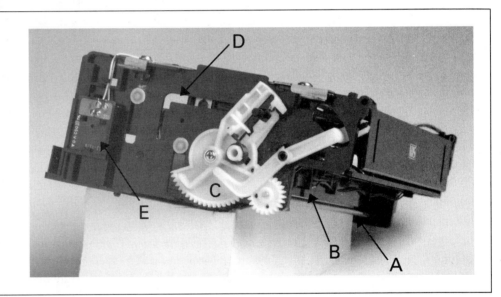

**FIGURE 5.4** Typical front loader assembly, bottom and left side view. A: Transverse shaft  B: Record-safety switch  C: Lifter arm/cam gear  D: L-shaped slot  E: End-of-tape phototransistor circuit board (supply side)

Of course there are several variations among VCR models in the way front loaders are made and operate, but they are all quite similar to the one just described. You can easily determine how a particular model works by hand cycling and watching.

To hand cycle, unplug the VCR and manually turn either the worm gear or pulley *on the motor*, whichever the particular VCR has. **Apply finger power only at the motor shaft!** Directly turning a loader gear or other component could easily result in part breakage. *Never force the mechanism when hand cycling;* the motor shaft should turn easily in either direction as long as the cassette tray is not at either end of its travel.

It is often easier to hand cycle a motor that has a worm gear on its shaft by turning the gear with the tip of a scribe or small screwdriver, especially in some models where the motor shaft is hard to reach with finger tips.

Some loaders have latches that lock the movable tray in its Eject position when no cassette is inserted. Look for these if you feel resistance when hand cycling after the cassette tray has moved only a short way inward. You can either push in on the latches, or insert a cassette, which unlocks the tray from the loader frame. Don't use force!

On some loaders there is an additional switch, usually a leaf type, in the top center of the frame, just behind the compartment door. This cassette compartment switch tells system control when a cassette is in the loader in its Eject position. You still need to push gently forward on the cassette a short way to just begin to move the cassette tray, which then transfers another switch to energize the load motor. That is, *both* switches need to be transferred before system control starts the load motor.

Rather than a leaf-type switch that is operated by the cam gear, other style loaders have two small microswitches. One signals system control when a cassette is pushed gently into the tray, moving it to the rear a small amount. Depending on the model, the other switch tells syscon when the cassette tray is all the way out in the full eject position, or all the way down in the cassette loaded position, so syscon can stop the cassette load motor. It's also possible for a front loader to signal its positions with an optical sensor, typically consisting of two infrared (IR) LEDs, an encoder disk, and two IR phototransistors. An optical encoder is described in the Chapter 3 Appendix.

At this point you should have a fairly good understanding of how front loaders basically work. If not, take the time to review before going on. The following are some problems associated with front loaders, along with probable causes:

- Cassette won't load, load motor won't turn at all when cassette is pushed gently forward.
  — Defective motor or bind in mechanism that motor torque can't overcome.
    Determine if the motor is being energized by placing a voltmeter across the two terminals on the rear or side of the motor, and then pushing gently in on the cassette until cam Switch #1, the Cassette-insert switch, transfers. Most cassette load motors are 12 VDC.

    On some VCRs, syscon shuts Off power if cam Switch #2 doesn't transfer within a few seconds of cam Switch #1. It senses something is wrong, and shuts Off the VCR. That is, Switch #1, cassette-insert, transfers when a cassette is pushed gently into the tray, energizing the motor. Switch #2 should transfer shortly thereafter, indicating loader has moved cassette in and down to reel table.

    Similarly, other VCRs will power down if the cassette-in switch on the reel table doesn't transfer within about five seconds of the cassette motor energizing at the start of a cassette load cycle.

    Some VCRs reverse the cassette load motor to eject a cassette if system control doesn't get a signal from either the cassette-in switch or a loader switch indicating that the loader has completed a cassette load cycle. The VCR may or may not power down, depending on how the system control microcomputer chip was programmed at time of manufacture.

  — Cassette tray not completely in its Eject position for some reason. Cam Switch #1 tells system control that the tray is *not* at Eject/home position, and system control won't energize the load motor.
  — Cam Switch #1 defective or not transferring when tray moves slightly in as cassette is inserted.
  — Cam Switch #2 defective, or transferred when it shouldn't be, indicating that the cassette tray is *already* down on the reel table. System control won't energize motor in load direction if it "thinks" the cassette tray is already in the loaded position.

— Cassette compartment switch found on *some* loaders (located top center behind compartment door) is defective or out of adjustment. Should transfer when cassette is placed in loader compartment. System control won't energize the cassette load motor if this switch isn't transferred.

— Cassette-in switch transferred with no cassette on reel table. Check indicator/display panel for presence of cassette-loaded icon. If syscon senses a cassette is already loaded, it won't energize the motor to load yet another cassette. Of course this situation—cassette-in switch transferred *and* cassette tray in Eject position—should never occur. The cassette-in switch could be jammed or broken. System control may power Off the VCR under this abnormal situation.

- Cassette tray moves partway to loaded position, but stops before cassette is fully seated on reel table. VCR will most likely go into auto shutdown.
  — This failure is often caused by one or more of the small vertical tabs on the floor of the cassette tray becoming bent inward. This can happen if a cassette is inserted with too much force, which can happen when the VCR is powered Off or unplugged. The VCR won't load the cassette, so someone pushes harder, bending the tabs on the tray floor.

    Once the tabs are bent, it's easy for a cassette to ride up *over the top of them*, instead of just pressing against them. This happens when the cassette is first inserted into the loader compartment. Usually the tray starts to move under motor power a short way, and then jams.

  — A slipping belt on the cassette load motor pulley can cause the tray to stop moving somewhere in its travel between full eject and load positions. A squealing sound may be heard from the slipping belt.

- Cassette tray seems to load cassette, but then motor reverses and ejects cassette. VCR may power down, depending on model.
  — Loader cam Switch #2 defective or not transferring when cassette tray is in fully loaded position.
  — Cassette-in switch not transferring at end of load cycle. Notice whether the front panel cassette-loaded icon displays at the end of the load cycle; it should!

    In these two cases, system control senses that something is wrong and reverses the load motor to eject the cassette. Both cam Switch #2 and the cassette-in switch should transfer at the completion of the cassette load operation.

  — Cassette not fully seating on reel table.

    This can be caused by a foreign object preventing the cassette tray and cassette from descending all the way, and thus there is no transfer of the cassette-in switch.

    Another possibility is that some transport component has become loose and mispositioned, preventing the cassette from seating properly. For example, the end-of-tape sensor light post that protrudes through the hole in the middle of the cassette could have shifted position due to a loose mounting screw. Now the cassette bottom hits the light post.

- Cassette fails to eject.
  — Tape unload operation not complete.

    This is probably the most frequent cause for a VCR not ejecting a cassette. In fact, the cassette load motor never even energizes.

    System control monitors the mode, or cam, switch/sensor in the undercarriage, which signals positions of the tape load mechanism and other transport components, as described in Chapter 3. If syscon detects that the shuttle assemblies and P-guides are *not* in the fully retracted position—back inside the cassette shell—then it will not eject the cassette.

    Ejecting a cassette with tape pulled out its front door is not too good. Therefore the system control microprocessor prevents this from occurring

by not energizing the cassette load motor. *Tape* load and unload problems are covered in the next section of this chapter.
— Many conditions described can also cause failure to *completely* eject a cassette. Something might be jamming the loader, or a drive belt may be slipping. Check operation of loader switches.

If a cassette is jammed in the loader, the system control microprocessor in most VCRs will shut down the power supply after about five seconds, since it doesn't get any signal from a switch on the loader that the mechanism has reached its eject position.

As you can see, all sorts of different problems can cause the same symptom. *Careful, thorough* examination of the loader assembly while hand cycling usually will disclose any mechanical problems. Look for:

- A loose or slipping belt around motor or drive pulley
- Foreign object in loader or jamming gears. It doesn't take but a sliver of something to jam gears if it gets between teeth.
- Broken parts: gears, switches, lever arms
- Bent parts: cassette tray flat springs and sheetmetal projections or tabs. Flat springs hold cassette, which rests against vertical tabs on tray.
- Loose parts: motor mount, switch bracket
- Missing or weak spring; spring unhooked at one end
- Missing C-clips or E-clips: these hold gears, pivot arms, and the like.
- Left and right side gears out of time with each other. It is possible for gears to slip one or more teeth, so that the lifter arm/cam gear on one side of the loader is ahead of or behind the cam gear on the other side. Gear teeth slippage itself is most likely caused by some other problem, like someone forcing a cassette into the machine or by another fault in the loader assembly
- Broken or pinched wire
- Poor solder connection.

Almost all cassette load and eject problems, as with the majority of VCR malfunctions, are mechanical in nature. Most cassette jamming problems are due to bent tension springs or projections in the cassette tray area. These are usually damaged by someone shoving a cassette in crooked or with too much force.

Many cassette trays have been damaged by someone forcing a cassette in when VCR power was turned Off. Since the cassette load motor won't energize and pull the cassette in, they push harder, often bending the small projections on the tray against which the cassette rests. Several newer VCRs take care of this by powering themselves up when a cassette is inserted and pushed forward gently against the cassette tray. If power is Off, system control turns power On when the cassette-insert switch, loader cam Switch #1, transfers.

The bottom line is, "Suspect a mechanical problem before going after an electronic malfunction."

One final word about cassette loaders. Sometimes it looks like a loader has correctly loaded a cassette: the cassette-loaded icon displays, and tape even loads and plays. But cassette load may still be flawed, which can cause tape edge damage, picture and sound quality problems, or both. This can happen if the cassette isn't firmly held down on the four locating studs in the transport by the loader assembly.

You can easily check for this. After a cassette loads, push down near all four corners of the cassette. It should *not* move downward at all, but be rock solid. If you detect some give or springiness to the cassette, check the leaf springs on the roof of the cassette tray. They should hold the cassette firmly down against the four transport locating studs. It could also be that the cassette tray itself is stopping its downward travel a little too soon, keeping the cassette raised slightly off the transport. Check for obstructions, binds, or adjustment of loader cam Switch #2.

## About Switches

As you've just read, there are several switches that must transfer correctly, *and at the right time,* for successful cassette load and eject operations. You probably noticed that we used the word *transfers* instead of stating that a switch is "On" or "Off," or that it is "closed" or "open," or that a switch "makes" or "breaks" contact. This is because one VCR model may have a normally open (N.O.) cassette-in switch whereas another model has a normally closed (N.C.) switch. The same is true for other switches. The end result is the same: system control "knows" when a switch transfers or changes state.

Actual switches may be encapsulated microswitches or leaf-type switches. The latter may be totally exposed or enclosed. You can't do anything about a microswitch that isn't working reliably, other than replace it. But often you can fix a malfunctioning leaf switch. Figure 5.5 illustrates a leaf switch.

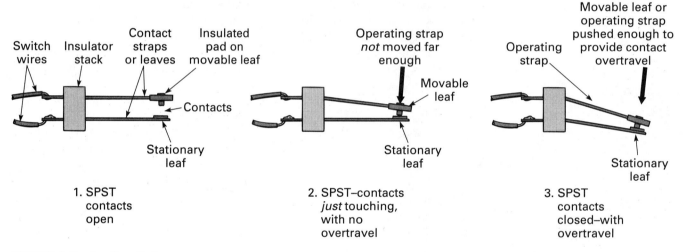

**FIGURE 5.5** **Leaf switches**

Poor leaf switch operation can usually be cured by cleaning the switch's contact surfaces. Sometimes a leaf switch gets bent or damaged in some way. Careful bending, or forming, of the thin metal strips can often restore operation. Contacts should have sufficient overtravel, or contact tension, when they come together. That is, the movable leaf should not just barely touch the stationary leaf, but should exert enough pressure to move it slightly. This ensures good conductivity with low contact resistance.

## About VCR Microprocessors

By now you are aware that the system control microprocessor is the boss when it comes to VCR operations. It senses machine conditions from the various switches and sensors throughout the loader and tape transport. It also works with another microprocessor, the *timer/operations microcomputer,* which handles front panel switches, buttons, and display, and unattended recording. These two microcomputers, *system control* and *timer/operations,* communicate back and forth with each other to handle all VCR operations. System control operates the several transport motors, presents information back to the timer/operations microprocessor to display on the indicator panel, and even powers down the VCR when it detects that an operation has not gone correctly. For example, if the loader somehow jams partway through a cassette load cycle, system control reverses the load motor to eject the cassette. If the jam prevents the cassette from fully ejecting, system control powers down the VCR.

Every once in a great while, though, the system control microprocessor or the timer/operations microprocessor can become "confused." Perhaps syscon is not sure about machine status; it may have a conflict that its factory programming cannot handle, and so it refuses to do anything or behaves in a bizarre fashion. Maybe the VCR goes into rewind right after a cassette loads, or the VCR locks up, refusing to respond to any front panel buttons. Perhaps the timer/operations microcomputer fails to power On the VCR for unattended recording when it should.

A glitch on the AC power line or a flaky switch contact might be the cause of the VCR locking up or misbehaving. What to do?

These *rare* occurrences are usually cured by totally resetting the microprocessors. "How do you do that?" you ask. Simple! Just unplug the VCR's line cord *and leave it unplugged for about 10 minutes.* Sometimes you'll find that everything works fine when you plug it back in. Unplugging the VCR *is required!* Simply turning it Off with the Power button won't do, because the system control and timer/operations microprocessors remain powered as long as the VCR is receiving line voltage, regardless of whether it's On or Off.

By the way, other microprocessor-controlled home appliances can misbehave occasionally, especially after a lightning storm, or when control panel buttons have been accidentally hit very rapidly in succession. Removing AC input power for several minutes from the entire appliance often corrects the problem by resetting the microprocessors.

## TAPE LOAD AND UNLOAD PROBLEMS

First, a quick review of what happens when tape loads and unloads. At the completion of cassette load, roller guide P-2 is positioned inside the cassette shell, behind videotape, near the supply reel. Roller guide P-3 is similarly inside the cassette shell about halfway between the supply and take-up reels. Figure 5.6 shows how things look at the completion of *cassette load,* before tape load begins. For better visibility, the loader assembly, shield plate, and cassette door have been removed.

To load tape, the tape load, or cam, motor is energized. This operates undercarriage mechanics that move shuttle assemblies with guides P-2 and P-3 out along tracks on each side of the video drum. These guides pull about 13 inches of tape from the cassette, thread it through the transport, and wrap it halfway around the video drum. Guides P-2 and P-3 come to a precise stopping position at the ends of their tracks against V-stops or V-mounts. Figure 5.7 illustrates.

Most older VCRs load tape only after Play or Record mode is selected. Some newer models load tape whenever a cassette is loaded and VCR power is On. Other models load tape only halfway after cassette load completes, and complete the load cycle when Play or Record mode is selected.

We'll broadly define tape load and unload problems as occurring whenever roller guides P-2 and P-3 fail to go to either extreme on their tracks *when they should.* As with cassette load and unload problems, almost all tape load/unload problems are caused by some mechanical malfunction, usually in the undercarriage, including:

- Slipping belts
- Broken, bent, loose parts
- Binding parts

**FIGURE 5.6** Completion of cassette load, before tape load begins.

- Meshed gears or gear segments that have slipped one or more teeth (gears are out of time with each other)
- Weak or unhooked springs
- Defective or maladjusted mode or cam switch/sensor.

As for other VCR mechanical areas, careful and thorough visual observation of gears, cam wheel, pivot arms, and linkages that work together to move the two roller guides back and forth along their tracks will often locate the problem. Depending on the failure, you might look at load and unload cycles under power *or* while hand cycling with the VCR unplugged.

When hand cycling, *apply finger power only at the tape load motor shaft!* Load/cam motors have either a worm gear or pulley and belt to drive the mechanism. Avoid forcing the mechanism while hand cycling, especially at the limits of travel; part breakage could result. (These same cautions apply when hand cycling the cassette load motor.)

Chapter 5 Troubleshooting Loader and Transport Malfunctions ■ 119

**FIGURE 5.7** Tape load complete; roller guides firmly seated in V-stops.

Meshed gears that get out of time with each other by slipping one or more teeth are usually the result of another problem, such as a foreign object jamming the mechanism. For example, suppose a tiny tot "feeds" a small toy car to the VCR. The toy might jam one of the shuttles as it moves along its track to the tape load position. This might cause a gear in the undercarriage to slip timing.

Some gears have timing marks to line them up with each other at the right spot. A dimple, small hole, or other mark next to a tooth on one gear is aligned with a similar mark on another gear or gear segment with which the gear meshes, as illustrated in Figure 5.8.

If you find that a gear has slipped timing, look for damage to gear teeth or a bent bracket on which the gear is mounted. Gears do not just slip timing for no apparent reason. It will be necessary to remove a gear or gear segment to re-time them without forcing. Gears mount with C- or E-clips, O-rings, or one or two plastic teeth at the center of the gear that latch into a groove around the shaft. Chapter 6 describes gear mounting in more detail.

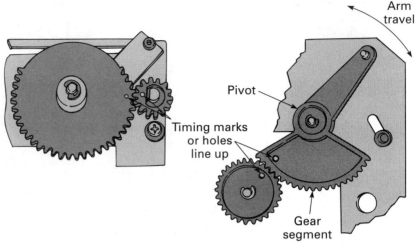

**FIGURE 5.8  Gear timing marks**

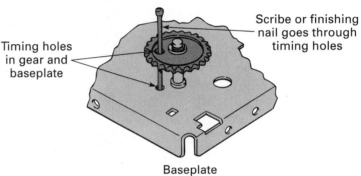

There are so many different undercarriage mechanical systems for loading tape that it would be very difficult to describe even a typical arrangement. However, you should have little, if any, trouble figuring out how things work while looking closely and operating the mechanism under power or while hand cycling. Most designs use several gears or gears along with a cam wheel and cam follower to move the shuttle assemblies and P-guides along their tracks.

Following are three sample VCR undercarriages. Figure 5.9 shows one of the simpler designs. Although this one has a cam wheel, it is not used for tape load/unload. A few gears turn two large ring gears in opposite directions to move the two shuttle assemblies. The upper ring gear is directly above the visible lower ring gear. The load motor on top of the transport drives the black cam wheel (A), which turns the white intermediate gear (B), also called an idler gear, meshed with the lower ring gear. Two small reversing gears (D) transfer motion from the lower to the upper ring gear; the two large ring gears (C) always rotate in opposite directions.

One end of a pivoting linkage arm is attached to the circumference of each ring gear, the other end connects to a shuttle assembly. When the ring gears turn, the two shuttles are pulled back and forth in their tracks between tape load and unload (home) positions.

Figure 5.10 is another undercarriage, somewhat more complicated than that in Figure 5.9. The load motor itself is on top of the transport, and drives the load mechanism through two belts. Most of the gears and linkages to move the shuttles are out of view, beneath the large operations drive gear pulley. Basically, when the load motor turns the cam wheel, the cam follower arm pivots as it rides along a track, or groove, in the cam wheel. A gear segment at the other end of the cam follower pivot arm turns two gears, which then move two shuttle linkages.

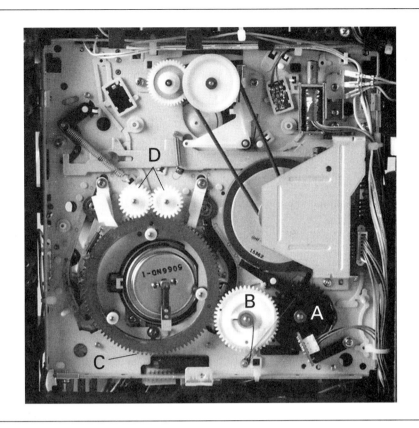

**FIGURE 5.9** Undercarriage example #1. A: Cam wheel B: Intermediate or idler gear C: Ring gears D: Reversing gears

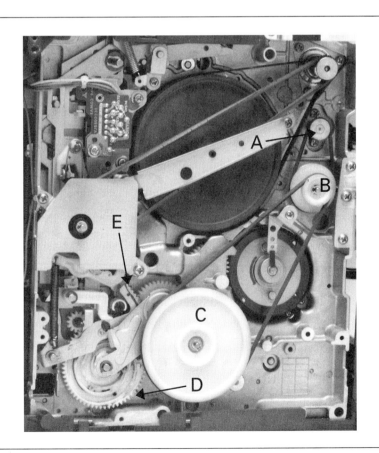

**FIGURE 5.10** Undercarriage example #2. A: Load motor pulley B: Intermediate pulley C: Operations drive gear pulley D: Cam wheel E: Gear segment part of cam follower arm

The undercarriage in Figure 5.11 works similarly to that in Figure 5.10 when it comes to tape loading and unloading. However, this undercarriage is much more open; you can really see what's going on and how parts interact with each other. Notice that the shuttle assemblies (H) have only about 1/2-inch or so of travel to go before seating in the V-mounts at each side of the video drum. When the load motor operates, rotary motion is transferred through intermediate gears to the cam wheel gear and mode sensor. A cam follower on one side of a pivot arm rides in the cam wheel groove, or track. As the cam turns, the arm pivots, moving the gear segment on the opposite end of the arm. This motion rotates the load arm gears, moving the shuttle linkage arms and shuttles.

**FIGURE 5.11** Undercarriage example #3. A: Load motor and worm gear B: Intermediate gears C: Cam wheel gear D: Mode switch/sensor E: Gear segment F: Load arm gears G: Shuttle linkage arms H: Shuttles I: Drum motor

Don't be concerned if you don't fully understand how a particular VCR mechanism works. As long as you have a general idea, that's great! Several different manufacturing designs accomplish the same end result, whether it's loading tape, engaging the pinch roller with the capstan, or braking spindles in Stop mode. Mechanical operations are not difficult to figure out when you actually have a particular machine in front of you, and can experiment with it and watch it operate. Hand cycling is a great way to see how things work.

## Mode Switch or Cam Position Sensor

Chapter 3 describes different types of mode switches and how they work. Essentially, a mode switch lets the system control microprocessor know the position of tape transport components that are operated by the tape load/cam motor. Knowing this, system control determines when, and in which direction, to energize transport motors in response to front panel pushbuttons.

For example, suppose a cassette has just been loaded and the user hits Play. System control energizes the tape load motor in one direction to move roller guides P-2 and P-3 out from inside the front of the cassette shell toward their V-mounts. The mode switch is linked to the same gear train and cam as the shuttle assemblies. When the P-guides have reached their V-mounts, the mode switch, or cam sensor, sends a signal to system control. Syscon then de-energizes the cam motor.

In a similar manner, when Stop is pressed, system control energizes the tape load motor in the opposite direction. When the P-guides are back inside the cassette shell, the mode switch indicates this and system control stops the mode motor.

You see, the mode switch tells system control where transport parts are positioned so it can determine in which direction to run the mode motor, and when to stop it. When P-guides fail to go all the way forward into their V-mounts, or don't retract all the way back inside the cassette shell, there is usually either some mechanical problem in the transport, *or* system control is not getting correct information from the mode switch.

A common tape load problem occurs when system control doesn't get a signal from the mode switch indicating that the P-guides have extended or retracted far enough. For example, a cassette is loaded and you hit Play. Syscon energizes the load motor and the P-guides move out, but system control never gets a signal from the mode switch that they have reached their V-mounts. Depending on the program put into the microprocessor chip when it is manufactured, syscon may retry the operation one or more times and then shut down if incorrect status is received from the mode switch, or syscon may shut down the VCR after the motor has been energized for a certain number of seconds, in the absence of a mode-switch signal to turn the motor Off.

Once it is determined that there is no mechanical problem, like a slipping belt or binding gear, then the mode switch is highly suspect for tape load/unload problems. Mechanical contact mode switches can cause intermittent failures due to flaky contacts. High contact resistance can be caused by cigarette smoke or other airborne contaminants. Cleaning with contact cleaner normally corrects this. If problems persist, the best thing to do is replace the switch.

Cam position sensors with IR LEDs and phototransistors are much more reliable than mechanical switches. Nevertheless, dirt on the surface of an LED or phototransistor *can* cause intermittent problems. There could also be a small piece of fuzz or glob of grease partially blocking a hole in the encoder disk.

Normally you won't have to adjust the position of the mode switch unless its mounting screws have come loose. If P-guides consistently stop short, you might try repositioning the switch slightly. Photo type sensors usually cannot be adjusted.

Frequently, bizarre transport behavior, unexpected VCR shutdown, or shutdown right after power On are caused by mode switch or cam sensor problems. Again, mechanical switches are more prone to failure than optical sensors.

Don't forget that quick play VCR models go through the tape loading cycle—threading tape through the transport and around the spinning video drum—when *cassette loading* completes, without having to press Play. With tape already loaded, playback begins as soon as the Play button is pressed, whereas other VCRs have to first go through the tape load cycle, which takes a few seconds. Many quick play VCRs keep tape loaded against the spinning drum as long as power is On.

The downside is that rotary heads wear prematurely in models that keep tape loaded, even in Stop mode, as long as power is On. Some recent quick play VCRs have required rotary head replacement much earlier than would normally be necessary for the number of hours the units were actually playing and recording. It is advisable to Eject the cassette or power down a quick play VCR when not playing or recording to prolong rotary head life. (You might also want to discuss this with your customers.)

Some other models half-load tape until Play or Record mode is entered. A small loop of tape is drawn out of the cassette by guides P-2 and P-3, but tape is not wrapped a full 180 degrees about the video drum. This reduces the time between when Play is pressed and a picture appears, while placing less wear on VCR parts and on the videotape than full tape loading.

Just be aware of these variations. It is easy to think that a VCR is malfunctioning when the shuttles move out on their tracks as soon as a cassette is loaded, when this *might* be perfectly normal for a particular VCR model. If in doubt as to what a particular VCR *should* do, consult the manufacturer's service literature. The user's manual may also give you a clue as to what is normal operation for a given VCR. If it talks about the VCR going into Play immediately or without delay, then you can be sure it's normal for this VCR to keep tape loaded whenever it is powered up.

## TAPE MOTION PROBLEMS

When we say *tape motion,* we mean physical movement of videotape between supply and take-up reels through the tape path during play, record, forward scan, reverse scan, and slow-motion video. We also include videotape movement when tape is unloaded during fast forward and rewind.

Tape movement is most critical during playback and record. Many more transport components are involved with moving tape than during fast forward or rewind. Because the transport handles tape in exactly the same way during record as during playback, any tape motion problems affecting playback will also affect record. Taking care of tape motion abnormalities in play will therefore ensure that tape movement in record is correct.

Before concerning ourselves with tape motion during play, though, it is far better to first find out if a VCR is functioning as it should be in fast forward and rewind. Always take care of the simple problems first; quite often this corrects what appear to be more complex problems. In other words, before tracking down a fault that shows up while playing a tape, first check how the unit fast forwards and rewinds an entire T-120 cassette. (Besides, failure of a VCR to rewind, even though it does everything else all right, can pile up rewind fees at the local video rental store!)

### Fast Forward/Rewind Trouble

When checking a VCR's fast forward and rewind operation, use a T-120 cassette rather than a T-60 or T-30. A standard T-120 is more of a challenge because of the greater amount of tape that must be handled, which requires more spindle torque during fast forward and rewind. A healthy machine should fast-forward a full supply reel onto the take-up reel *and* rewind a full take-up reel back onto the supply reel.

Tape motion should be smooth between reels, with no hesitation, jerkiness, or unusual sounds, like squealing, coming from the machine. You'll normally be able to hear any hesitation or abnormal slowing down during these operations. Check that the index counter increments or decrements smoothly, with no hesitation.

Demand on the transport is greatest when the reel that is pulling tape in is nearly full. If there's insufficient supply or take-up reel torque, this is when it is most likely to show up. The supply reel slows way down, hesitates, and it seems like it's just not going to make it. Sometimes it doesn't, and it simply stops. On many VCRs, system control powers down the unit when this happens. Here's why:

1. Syscon doesn't receive pulses from the take-up reel sensor; this means the reel has stopped turning.
2. But end-of-tape is *not* detected at either the supply or the take-up reel, *and* the unit is still in Fast Forward or Rewind mode.
3. There is a conflict between conditions A and B. If the machine is in Fast Forward or Rewind mode, and end-of-tape *isn't* detected, the cassette reels *better be turning,* or there's a problem. Syscon senses a problem and goes into shutdown.

Proper operation is for system control to detect take-up reel sensor pulses throughout fast forward and rewind. Syscon then detects end-of-tape at one reel or the other when light shines through the transparent leader, and puts the machine in Stop mode.

## Rewind and Fast Forward Problem Areas

The most common rewind or fast forward problem is not enough spindle torque to pull in a full reel of tape. This is almost always caused by:

- Slipping belt(s)
- Slipping idler wheel
- Slipping clutch.

In some cases, cleaning the rubber idler wheel tire and the plastic lower flanges on supply and take-up spindles takes care of rewind and fast forward problems. A slipping belt or clutch in the undercarriage can also cause low spindle drive during fast forward and rewind. Taking a close look at both sides of the transport while in either mode should readily disclose the source of drive problems. If a rubber belt or idler wheel tire is hard and dried out, glazed, or cracked, the part should be replaced. Rubber parts need to be "lively" to grip pulley and drive surfaces without slipping. A good rubber belt should snap back when stretched, not just gradually return to its pre-stretched length.

Some idler wheels have separate rubber tires that can be replaced. In other cases, the entire idler wheel is replaced. On many VCRs, the idler and clutch are replaced as a single assembly; individual component parts are unavailable. Often this is the best way to go anyway. It is frequently easier to replace the entire clutch/idler wheel assembly than several separate parts.

You can use your fingers to determine where transport slippage is occurring. For example, operate the unit with a videocassette test jig and grab the supply spindle during rewind to prevent it from turning. You can then see where something is slipping. You will probably need to rotate the take-up spindle manually at the same time to prevent system control from powering down in the absence of take-up reel sensor pulses. Be careful not to touch rubber drive tires, belts, and spindle drive surfaces with your fingers; body oil that you deposit will only make matters worse.

Figure 5.12 shows a VCR that has a separate motor, rather than using the capstan motor, to drive supply and take-up spindles. An idler wheel contacts a knurled collar, or hub, on the motor shaft, at about the 5 o'clock position of the idler wheel (A) in the photo. Depending on the direction of motor rotation, the idler wheel swings left or right so its tire contacts the lower flange of the supply or take-up spindle.

After tape has been *fully* rewound on the supply reel or wound on the take-up reel, a glance through the transparent plastic cassette windows can reveal how well the VCR is performing fast forward and rewind. Tape edges should be uniformly flat across the tops of either reel, with few circular ridges; two or three is generally no cause for concern. *Many* ridges—rings of tape wrapped at different heights—indicate nonuniform take-up tension on either supply or take-up reel. This can occur if there is insufficient or nonuniform back tension during rewind or fast forward.

**FIGURE 5.12 Idler wheel (A) drives supply or take-up spindle**

For example, if during rewind, take-up reel back tension, or soft braking, is weak as the supply reel pulls in tape, tape can sometimes reel out *too* rapidly, causing a loose tape wrap on the supply reel. Similarly, if tape holdback tension is too great, tape will be wound too tightly. Uneven back tension causes tape to wind unevenly on the supply reel. Multiple tape edge circular ridges or rings are a good indication that this has occurred.

Cleaning supply and take-up spindle flanges usually takes care of incorrect and irregular back tension. Worn or glazed soft brake felt pads or some defect in the mechanical linkage that activates them can also cause problems, but the majority of fast forward and rewind difficulties are due to low drive torque, not incorrect back tension. Figure 5.13 shows the soft braking operating arms and felt pads on a typical VCR. Note that the supply spindle soft brake pad is actually a section of the back tension band. Plastic arms operated by a cam

**FIGURE 5.13 Soft braking of supply and take-up spindles**

linkage in the undercarriage move the soft brakes against the appropriate spindle flange during fast forward and rewind. That is, soft braking is applied to the supply spindle during fast forward, and soft braking is applied to the take-up spindle during rewind.

To summarize, fast forward and rewind problems are most frequently caused by a slipping rubber idler tire or drive belt, or both. Cleaning these components and drive surfaces often is all that is necessary to restore operation. Worn or dry rubber components should be replaced. Cleaning and buffing old rubber will sometimes work for a short period of time, but once rubber has lost its liveliness, these parts should be replaced for long term reliability. Repairing fast forward and rewind problems will often clear up record and playback difficulties.

## Playback Tape Motion Problems

As mentioned earlier in this section, repairing playback tape motion problems also takes care of tape movement difficulties during record, because mechanically the two modes are identical. It probably comes as no surprise by now that careful, thorough examination of videotape as it winds through the tape path can identify various tape motion abnormalities. But what *are* tape motion abnormalities? In short, anything *that is not* absolutely consistent and smooth tape movement, from the supply reel through the tape path to the take-up reel, is a condition that needs correction for quality video and audio reproduction.

Tape motion during play should be observed with a known good scratch tape—one that has no physical imperfections such as edge curling or crinkles. The tape should first be *limbered* by being wound and then rewound end-to-end with fast forward and rewind. This limbering prevents adjacent layers of tape from sticking to each other on the supply reel, which can happen with a tightly rewound tape that hasn't been used in a while.

Examine all points along the tape path, shown in Figure 5.14, for tape motion oddities while the VCR is playing the scratch tape. Look for the following tape motion irregularities:

**FIGURE 5.14** Tape path through transport

### *Height variations*

Tape should stay at a constant height through each area of the transport. That is, the top edge of the tape should not wander up and down at any point along the tape path.

Good points at which to observe tape height are at the top of the full erase head and Audio/Control head.

### *Edge curling*

There should be no visible curling or wrinkling of either the top or bottom edge of the tape at any point in the transport. Observe all tape guides with shoulders for tape riding onto the shoulders; it should not.

Tape should be straight up and down across its 1/2-inch width as it passes between tape guides, rollers, and stationary heads.

### *Tension variations, jerkiness*

Tape should remain uniformly taut throughout the tape path, with no evidence of tape going slack, then taut, and then slack again.

A back tension arm or impedance roller that oscillates or quivers back and forth is an indication of uneven tape tension. Both supply and take-up reels should turn smoothly, with no hesitation or unevenness of rotation.

Be sure tape has first been limbered, as previously described.

### *Fluttering*

Overall tape motion in the tape path, from supply reel to take-up reel, should be just as smooth as silk, with no visible flutter or unevenness.

All this can be summed up by saying that as you look closely at tape moving through the tape path, *it should appear as if the tape is hardly moving at all!!* There should be no detectable motion of any kind, at any point along the tape path, other than tape moving smoothly from left to right. Carefully examine each portion of the tape path to detect any fluctuations or irregularities in tape motion.

Tape motion problems may sometimes be easier to see during Forward or Reverse Scan modes, when tape is traveling faster through the tape path. Good illumination is necessary to see small defects in tape motion. But one word of caution here: A bright incandescent light can cause problems if it shines on phototransistors, like the tape-end sensors. Ensure that photo sense areas are not exposed to high-intensity light during problem diagnosis; strange system behavior might result!

There are two critical areas to observe as tape is moving through the tape path: video drum and capstan/pinch roller.

## Observing Tape Around Video Drum

The lower, stationary portion of a video drum has a small ledge that guides the bottom edge of videotape as it wraps around the drum. Figure 5.15 shows this protruding ledge, called a *rabbet*. Notice that the rabbet is at an angle to the drum itself, to precisely position tape on its path around the drum.

When tape moves around the wide drum, its bottom edge should *just rest against* the ledge, or rabbet, on the lower, stationary cylinder. As you look straight down at tape moving around the drum, you should see the shiny edge of the rabbet all along the bottom edge of the tape, from where it enters the drum until it exits, as shown in Figure 5.16.

Tape should not curl over the edge of the rabbet at any point, nor should it be above the ledge, so you see any part of the drum above the ledge. Precise positioning of tape on the drum ledge is controlled by the height of entrance roller guide P-2 and exit guide P-3. Height adjustment of these guides is described in Chapter 7.

**FIGURE 5.15** Tape guide ledge called a rabbet on lower, stationary portion of video drum

**FIGURE 5.16** Edge of rabbet visible at bottom edge of videotape

## Observing Tape at Capstan and Pinch Roller

This is a vital area, because the pinch roller pressing tape against the rotating capstan shaft is what pulls tape through the entire tape path. Look especially for any tendency of the tape to angle, or skew, upward or downward as it passes between capstan and pinch roller; tape height should be constant as it passes the capstan. Check that tape is not touching the bearing at the bottom of the capstan shaft, which can happen if tape skews downward. There should be no tape crinkling.

Inspect tape as it exits the capstan and pinch roller, headed back into the cassette. There should be no tendency for the tape to adhere to the capstan or rubber pinch roller. Tape flutter as the tape exits the capstan can be caused by a dirty or gummy pinch roller, a dirty capstan, or both, to which tape tends to stick.

A dirty capstan or pinch roller, or both, are major causes of one form of a VCR "eating" tape; in this case, tape usually wraps around the capstan. In severe cases, tape wraps around the capstan many times until there's a tight wad. However, in most VCRs the absence of take-up reel pulses causes syscon to drop the VCR out of play before the wad gets too large. Tape that becomes damaged like this should be discarded. Attempting to run it through a VCR is only asking for further damage. Another primary cause for a machine eating tape in this manner is insufficient take-up reel torque, which lets tape spill between capstan and cassette. Tape ultimately gets entangled with the capstan, or pinch roller, or both. (These same difficulties, by the way, also affect audio cassette decks.)

Never hesitate to replace a rubber pinch roller. It causes more tape motion problems than any other single tape path component. Rubber hardens and doesn't grip the tape adequately, or it gets gummy and tape then tends to stick to it. With age, pinch rollers can wear unevenly, applying unequal pressure at the top and bottom of the capstan. This causes tape to skew (that is, to angle upward or downward instead of traveling a straight path). Pinch rollers can also develop hard spots and other surface abnormalities, which are often difficult to see or feel but which result in uneven tape motion.

Before adjusting transport items—unless they are obviously loose and out of adjustment, or have been replaced because of breakage—replace the pinch roller. Reconditioning a pinch roller, such as with light buffing with fine-grain sandpaper, can sometimes extend the life of a pinch roller, but for professional VCR repair, replacing the roller is the best alternative in the long run.

In addition to a worn pinch roller, dirty transport components are probably responsible for the majority of VCR problems, including tape motion difficulties. Cleaning and general VCR maintenance procedures are covered in Chapter 6.

## SUMMARY

We began this chapter discussing a *systematic approach* to diagnosing a problem. Although there is no *one* correct way to go about determining what's wrong with a machine, a few basic techniques are effective. These include starting on any problem with a relaxed, confident attitude.

Often, an incorrect diagnosis of a malfunction results from not taking the time to fully investigate. For VCR repair, a full analysis means testing as many machine modes of operation and features as possible, and noting what works, what doesn't, and what is observed to happen. This is far better than making an assumption about what's failing too early in the game.

In general, go after the simple things first. Many times the more complicated problems will be resolved when a seemingly minor failure is fixed. Replacing a worn belt that slips so badly the machine can't rewind an entire T-120 cassette may very well take care of intermittent loss of color or picture instability when playing a tape.

A systematic approach includes careful and thorough observation, *and* taking notes about what you see and hear. This is true not only as you *start* troubleshooting but at *every step of the way* until repairs are complete. For example, suppose a customer's complaint is: "My VCR eats tapes, also turns Off for no reason." While putting the VCR through its paces to see what works and what doesn't, you notice that the index counter hesitates while a tape is playing, just before the machine automatically shuts down. Make a note of this! Low take-up reel torque may be causing both the client's reported problems *and* the hesitant index counter.

Look before you leap. Look carefully for those tinted screws, usually pinkish-red or blue, before taking out every screw in sight to remove a PCB or front loader. *Be patient and gentle!* If that circuit board won't pivot out easily, there's probably a good reason. Look for the screw or latch you missed, or the cable that's snagged, before applying force. VCRs are delicate, with many fragile plastic parts, tiny cables and connectors, and miniature hardware. If you go at things as if you were dismantling a diesel tractor engine, you're likely to break all kinds of stuff.

Stay organized. Keep parts separated in numbered containers during disassembly, along with notes on what's what. This could very well prevent you from cracking a circuit board during reassembly. For example, if you put a long screw where a shorter one is supposed to go, it may run into a board or other component beneath the screw hole that you can't readily see. It takes a little extra effort and time up front, but it's well worth it in saving time and eliminating frustration later on.

Cassette load and unload problems were discussed. Recurring front loader malfunctions are due to parts that get bent or broken when excessive force is used to insert a cassette. A slipping load motor pulley and flaky cam gear switch contacts are also frequent problem areas.

When a cassette is inserted into a VCR, the front comes up against two, sometimes three, upright tabs or projections on the floor of the cassette tray. Gently pushing the cassette in a little further moves the tray inward a small amount. This transfers the cassette-insert switch, which signals system control to energize the cassette load motor. The loader moves the cassette in and down along an inverted L-shaped path.

The cassette tray, or basket, must hold the cassette firmly down on the four reel table locating posts at the completion of cassette load. It is not doing this properly if the cassette feels springy when you push down anywhere along its surface after it loads. Check for bent or broken hold-down or leaf springs on the roof of the cassette basket. Perhaps system control is stopping the cassette load motor too soon, or there's a bind or foreign object preventing the cassette tray from going down all the way.

After cassette loading, tape loading occurs in many VCRs when the machine goes into Play or Record mode. However, some VCRs fully load or half-load tape immediately following cassette loading. During a complete tape load cycle, roller guides P-2 and P-3 thread a loop of tape through the tape path, wrapping it halfway (180°) around the video drum. At the end of tape loading, roller guides P-2 and P-3 must be fully seated in their respective V-mounts on each side of the video drum. Frequently, tape load problems are caused by slipping belts and flaky mode switch contacts.

Failure of a VCR to eject a cassette can be caused by tape not fully *unloading*. System control doesn't receive a signal from the mode switch or cam position sensor that the movable P-guides are back inside the cassette shell, and so it won't energize the cassette load motor. Mechanical slippage or a faulty mode switch are common causes.

Understanding *how* a particular VCR loads a cassette and loads tape is usually necessary to fix problems in these areas, especially intermittent ones. Hand cycling the mechanics a few times by rotating the cassette load and tape load motor shafts with the VCR unplugged, is a good way to figure out how things are supposed to work. Be gentle when hand cycling: don't turn either motor beyond the limit of travel of the mechanism, or if you feel resistance. Both motors should turn very easily.

Smooth, uniform tape motion throughout the tape path is essential to high-caliber recording and playback. Carefully examine videotape as it moves past tape guides, the video drum, stationary magnetic heads, and capstan and pinch roller to discover any abnormality that can cause poor picture, or sound, or both. The tape should have straight edges, without curling or wrinkling, as it passes transport components. There should be no jerkiness, wavering, or flutter at any point along the tape path.

Critical observation areas are where tape wraps around the video drum and where it passes between capstan and pinch roller. A small ledge, called a rabbet, on the lower, stationary portion of the video drum keeps tape in precise alignment around the drum. The bottom edge of tape should just touch the small ledge, not curl over its edge or ride above the ledge. Proper tape positioning is determined by the height of drum entrance roller guide P-2 and exit guide P-3.

A worn, dirty, or glazed rubber pinch roller is probably the single most common cause of a VCR "eating" tapes, where tape wraps around the capstan. A defective pinch roller can cause many other tape motion problems as well. Pinch roller replacement is advised as a first step, *before* you adjust other tape path components in an attempt to correct tape movement abnormalities.

Using a high-intensity light to scrutinize tape as it moves through the tape path can show up minor imperfections that can have detrimental effects on reproduced picture and sound. However, take care that bright incandescent light, which has significant infrared content, doesn't shine on end-of-tape photo sensors. The VCR might unexpectedly stop, rewind, or even shut down when bright light hits the transport phototransistors.

## SELF-CHECK QUESTIONS

1. In a sentence or two, describe what is meant by a systematic approach to troubleshooting.
2. Describe a solenoid and give one example of where one might be used in a VCR.
3. Many VCRs have a 12 Volt DC cassette load motor. What difference, if any, is there between how this motor is energized when a cassette is being *ejected* compared with when a cassette is being *loaded*.
4. List the steps necessary to remove a *typical* front loader assembly from the tape transport.
5. What does the term *hand cycle* mean? Describe how you would hand cycle a typical front loader.
6. Describe a common front loader defect that causes the cassette tray to move a short way into the VCR and to then jam after a cassette is pushed into the compartment opening.
7. Describe what a *dimple, small hole,* or >-shaped mark on a gear or gear segment most likely means.
8. When checking a VCR's Fast Forward and Rewind operation, what should you use for the test videocassette? Why?
9. What are common causes for a VCR to have problems Rewinding or Fast Forwarding tape, or both?
10. Describe correct tape motion around the video drum with respect to the *rabbet*.

# CHAPTER 6
# HOW TO PERFORM VCR MAINTENANCE AND COMMON REPAIRS

Keeping a VCR alive and well means, in large part, keeping it well maintained—properly cleaned and lubricated. In this chapter we discuss what materials and tools are required and how to go about cleaning and lubricating transport components. Following appropriate techniques during routine VCR maintenance is important; applying the right cleaning solution in the wrong manner can severely damage some delicate VCR parts. Overlubricating, using the wrong lubricant, or lubricating something that is not supposed to be can cause more harm than good.

Many malfunctioning VCRs are easily corrected with a few uncomplicated, common repairs of the tape transport. Cleaning tape path components often restores full operation! General cleaning and VCR repair procedures are described within this chapter, with more specific mechanical *adjustments* covered in Chapter 7.

## CHAPTER OBJECTIVES

Upon completing this chapter, you should be able to:

1. List basic hand tools and supplies needed to maintain a typical VCR.
2. Describe what materials to use and how to clean and lubricate tape transport components.
3. Describe special precautions to take when demagnetizing and cleaning rotary tape heads.
4. Describe common procedures for servicing a VCR, such as how to remove and install C-clips, time gears, and replace a pinch roller.

In this chapter, the first things covered are the tools and supplies used for the bulk of all VCR maintenance and repair. The remainder of the chapter describes how to clean and lubricate a VCR, along with some of the more common repair procedures.

## TOOLS AND SUPPLIES

Fortunately, you don't *need* to spend a fortune on tools and supplies for the majority of VCR repair and maintenance procedures. A few common hand tools, which you may already have, and a small stock of cleaners and lubricants will suffice.

Even though you don't need *many* hand tools, you do need the *right* tools for the jobs you'll be performing. Using a Phillips screwdriver that is the wrong size, or one that has a ruined tip, can easily gouge out screw heads and make your life miserable. Having the appropriate, quality tool on hand when you need it cuts down on wasted time, frustration, and damaged parts.

You can get most tools at a hardware store or discount home improvement depot. Some tools, like a multimeter, special VCR alignment screwdrivers, and most supplies, you will probably have to get at an electronics supply house. Radio Shack carries nearly everything you could need in the way of *general* tools and supplies for VCR repair, but there are many other sources, including several excellent mail-order electronics outlets, some of which carry VCR-specific tools and replacement parts. Shop around for the best value.

Here are the suggested tools you probably should have. You don't necessarily need all of them! As far as hand tools, you can get by reasonably well for most VCR repairs with only a #1 Phillips screwdriver and a long-nose pliers. (See Figure 6.1.)

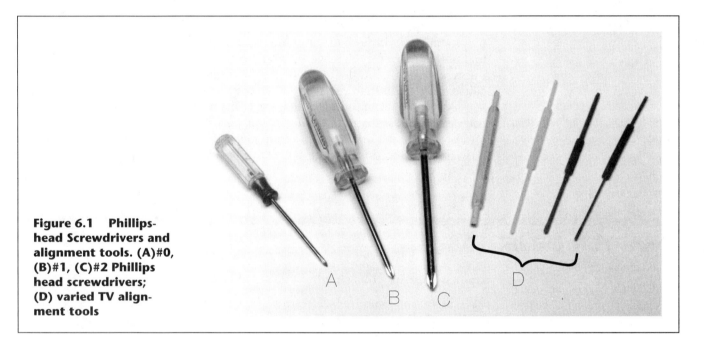

Figure 6.1 Phillips-head Screwdrivers and alignment tools. (A)#0, (B)#1, (C)#2 Phillips head screwdrivers; (D) varied TV alignment tools

## Hand Tools

- Phillips or cross-slot screwdrivers
  — #0, #1, and #2 sizes (#0 is smallest tip)
  Standard, stubby, and 10- or 12-inch stem length
  — Most Phillips screwdrivers have the size stamped somewhere on the handle or along the shaft. It usually looks something like **#0** or **#1**.
- Flat-blade or slotted screwdrivers
  — small and medium size tips (sizes 4 & 5)
  standard, stubby, and 10- or 12-inch stem length
  — set of jeweler's screwdrivers
- Torques drivers (also called Torx)
  — T-15 and T-20 sizes
  (for screws with six-pointed, star-shaped head cavities)
- Fiber or plastic alignment screwdriver(s) or set of TV alignment tools
- Duck-bill pliers (See Figure 6.2.)

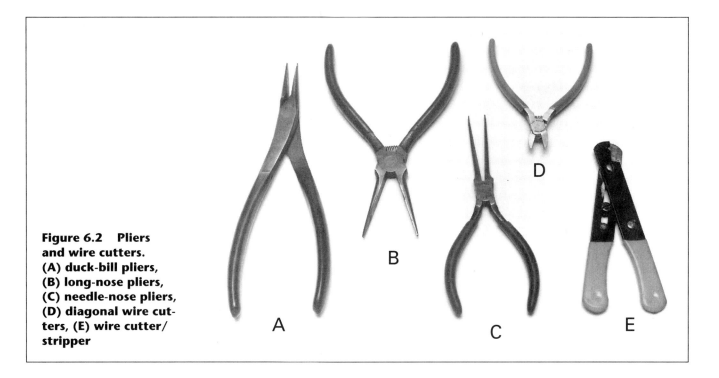

Figure 6.2 Pliers and wire cutters. (A) duck-bill pliers, (B) long-nose pliers, (C) needle-nose pliers, (D) diagonal wire cutters, (E) wire cutter/stripper

- Long-nose pliers (approximately 5")
- Needle-nose pliers
- Diagonal wire cutters, 4-inch
   (may also want 6-inch cutters)
- Wire cutter/stripper

When purchasing diagonal cutters, long-nose pliers, and needle-nose pliers, check that the pivot is solid, with no wobble. Close the jaws and hold them up to the light. You should not see any light between the two cutters of diagonal cutters or between the jaws (at least not at the tips) of long- and needle-nose pliers.

Check that long- and needle-nose pliers have some ribbing or serrations on the inside surfaces of their jaws, especially near the tips. This helps them to grip small parts.

- Nutdrivers
   — 7/32, 3/16 and 1/4-inch, or a set
- Small adjustable wrench
- Allen wrenches or hex keys
   — 0.9mm and 1.5mm, or a set
- Metal scribe
- Small claw-type parts grabber
- Dental mirror

Allen wrenches, also called hex keys, tighten or loosen small set screws, which often lock down an adjustment. Figure 6.4 shows what the head of an Allen screw looks like.

In many VCRs, an Allen screw at the base of roller guides P-2 and P-3 must be loosened before their height is adjusted. The Allen screw is tightened to hold the adjustment.

The sharply pointed metal scribe in Figure 6.3 is useful for scratching, or scribing, around an adjustable bracket or position-sensitive part before loosening its mounting screws. When reinstalling the part or its replacement, line it up with the scribed line; it will be quite close to its correct position. Scribed lines also help determine parts orientation during re-assembly.

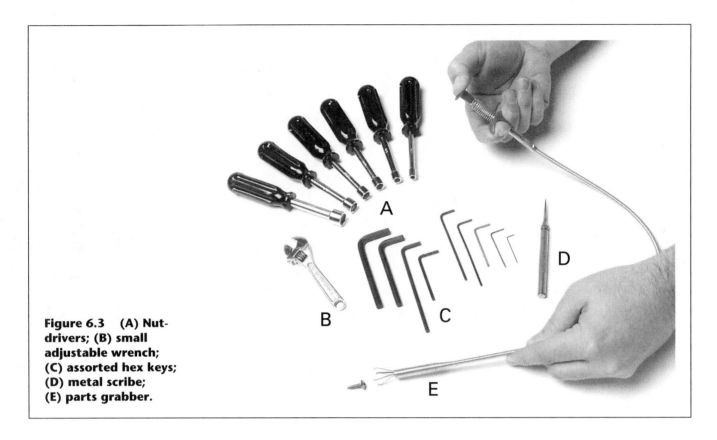

**Figure 6.3** (A) Nutdrivers; (B) small adjustable wrench; (C) assorted hex keys; (D) metal scribe; (E) parts grabber.

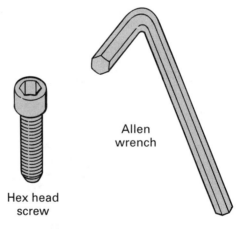

**Figure 6.4** Allen or Hex Screw head

    The parts grabber in Figure 6.3 is usually associated with auto mechanics. Some electronics and hobby supply houses carry smaller claw-type grabbers which are very similar, and better suited for VCR work.

    A small dental mirror is ideal for inspecting tape motion through the transport, especially past the upper collars or flanges on tape guides and past the audio/control (A/C) head. Most electronic supply outlets have inexpensive inspection mirrors with plastic handles that are well suited for this. Avoid a metal dental mirror, which could accidentally cause a short circuit during inspection.

- Soldering iron (20-30 watt)
  — soldering iron stand
  — small gauge 60/40 solder
  — solder-wick (desoldering braid)
- Multimeter and probes

(Chapter 11 has descriptions of soldering tools and techniques. How to select and use a multimeter—also called a *VOM*, for volt-ohm-milliammeter—is covered in Chapter 9.)

You may have noticed that the preceding tools list says "Phillips or cross-slot" screwdrivers. Although cross-slot screws *look* like Phillips-head screws, with a +-shaped head slot, there is a slight difference in the pitch and size of the slot. Most Japanese Phillips-head screws are actually cross-slot screws. A Phillips screwdriver normally works just fine, but especially for a cross-slot screw that has really been torqued down during VCR manufacture, a cross-slot driver is a better fit, with less chance the head will get gouged out. Larger electronic parts supply houses often carry cross-slot screwdrivers, sometimes known as *"Pozidriv."*

There are several "nice-to-have" tools that make service work easier. The parts grabber is an example. These are great when a screw rolls into an area you can't reach with your fingers. Phillips and flat-blade screw starters are also handy, but a magnetized screwdriver can often be used instead to place a screw or lift one from a confined space once it's loose.

You can purchase magnetized screwdrivers or make your own, which may not be as strong, but for small screws in a VCR they will work fine.

To magnetize a screwdriver, take a *strong* permanent magnet and place one pole or end against the screwdriver shaft, up near the handle. Draw the magnet down along the shaft and past the tip, keeping the magnet in contact with the driver all along the way, as illustrated in Figure 6.5. Repeat this procedure about six times, rotating the shaft a little each time. The magnet that forms one half of a magnetic cabinet latch works well for magnetizing a screwdriver.

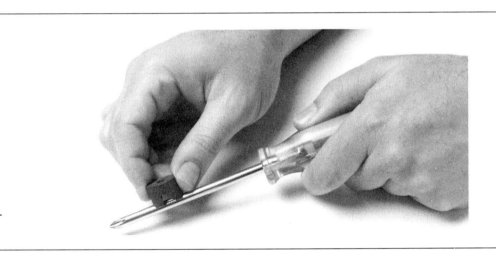

**Figure 6.5 Magnetizing a screwdriver**

A *magnetic* parts picker-upper is a convenient tool for retrieving a metal clip, spring, screw, or other small ferrous part. You can make one by affixing a small permanent magnet to the end of a stick with epoxy cement. You may also purchase this tool; home improvement, hardware, and hobby stores often carry magnetic parts retrievers, which are similar in size to a large ballpoint pen.

> CAUTION: Be careful not to place magnetized tools near videocassettes or anything else that may be adversely affected by a magnetic field.

You might add to your tool kit a 4- or 6-inch forceps, one or two hemostats, small locking pliers (generally known by the trade name "Vise Grips"), and a small vise or clamp for holding parts.

Forceps are tongs or pincers, similar to tweezers only larger. They are convenient for picking up and placing small parts. Hemostats are clamplike instruments with two jaws, similar to longnose or needlenose pliers, but with finger holes like a pair of scissors. They are convenient for clamping small mechanical parts when placing them in confined areas. Common hemostats are available with 4- or 6-inch jaws, and with either straight jaws or jaws at a 30-degree angle. The 6-inch angled ones are probably the most useful. Forceps and hemostats are carried by some electronic supply houses, but are readily obtainable from medical supply stores.

You can make a poor person's hemostat by placing a big, strong rubber band around the handles of a longnose pliers to squeeze the jaws together, but they won't lock onto a part as effectively as hemostats.

One or two small metal files are worthwhile for smoothing the occasional sharp edge, such as on sheetmetal in the loader assembly. A high-intensity light and small flashlight are helpful for critical inspection of tape movement and while servicing a VCR in general.

Another inexpensive tool worth considering is a 6-inch steel rule with small graduations, like sixty-fourths of an inch or tenths of a millimeter. Measure and jot down the height or position of a transport component, such as a reel table spindle or Audio/Control (A/C) head, as closely as possible before removing it. With this information you can install a part so it is close to the original height or position. For adjustable components, this makes overall adjustment easier; you start "in the ballpark," to use a common American expression.

It's best to buy good quality tools. Saving a few bucks by purchasing cheaply costs more in the long run, as well as frustrating you along the way. This is especially true for Phillips screwdrivers; *get the best you can afford!* Inexpensive drivers usually have poorly formed tips or tips that aren't hardened; they quickly become damaged, especially when trying to remove a screw that refuses to budge. A screwdriver with a messed-up tip will cause you nothing but grief. Throw it away!

Sooner or later you'll find one or more screws in a VCR that are difficult to loosen. First, use the right size screwdriver with a good tip. Keep the screwdriver straight in relation to the screwhead slot; even a slight angle can damage the screwhead. Then put plenty of force on the screwdriver as you try to break the screw loose.

You may have to really lean on the screwdriver to prevent the tip from riding up out of a screwhead slot. Once the screw has broken loose, ease up on the pressure. Most gouged-out screwheads result from using the wrong size or a damaged screwdriver, or insufficient force.

You can also purchase a paste-like product in a tube that increases the grip-strength between a screwdriver and screw, such as *Drive Grip*™ from ND Industries.

Other specialty tools that you might want or need for certain VCR repair and adjustment procedures are described in later sections of this book. Again, it is not necessary for you to buy all these tools at once. But when you *do* purchase, select quality tools, not the ones on the bargain table or the cheapies at the flea market.

## Supplies

As with tools, you may not need everything listed here, but the basic cleaners, swabs, and lubricants are required to properly maintain and service a

VCR. Some items shown in Figure 6.6 might be more appropriately described as tools, but are listed here because they are used in conjunction with supplies, or are specific to VCR repair and maintenance.

### *Tape head cleaner*

Here we're talking about liquid tape head cleaner that comes in a bottle or spray can, *not* the various VCR cleaning cassettes on the market. Automatic head-cleaning cassettes may be fast and easy, and they *can* do the job, but they can also cause problems. First, a little more information about these cleaner cassettes as a class, and why you should *not* rely on them to do the job for you.

There are two basic types of cleaning cassettes: dry and wet. Some dry cleaning cassettes are abrasive to rotary heads, causing wear of the delicate head tips. Wet types, in which a cleaning solution moistens a fabric tape, can damage a rotary head or change its alignment. The biggest danger with wet cleaning cassettes is that they can leave the surface of the video drum damp. If videotape is loaded before the drum dries thoroughly, the tape adheres to the drum. The term to describe this is *stiction*, a combination of the words "sticking" and "friction." In severe cases, tape stiction can destroy a rotating head.

Neither dry nor wet automatic cleaning cassettes are as effective as manually cleaning tape path components. They can also jam in the transport, possibly damaging the VCR. So, purchase tape head cleaner that comes in a bottle or spray can. Some tape head cleaners are specifically labeled for video heads or for VCRs; other tape head cleaners will generally work just as well. Although you can use other products, such as alcohol, for cleaning tape heads, it is safer to use cleaners specifically labeled for tape head cleaning.

### *Cleaner/Degreaser*

This product normally comes in a spray can, and is good for general cleaning of all areas inside a VCR. There are many different brand names of quality cleaner/degreasers available at electronic parts outlets. These products often specify that they leave no residue, which is important. Some degreasers don't completely evaporate, and thus leave a residue film. Automotive cleaners/degreasers may not be suitable for this reason; check the label.

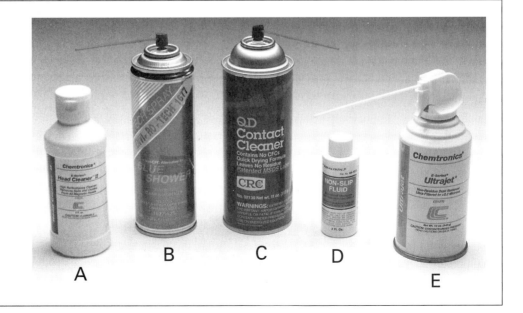

**Figure 6.6** VCR cleaners. (A) tape head cleaner; (B) spray cleaner/degreaser; (C) spray contact cleaner; (D) rubber cleaner and rejuvenator; (E) compressed air.

Alcohol *can* be used for general cleaning, and even tape head cleaning, but solutions labeled specifically for head cleaning are better for magnetic tape heads, even though they cost considerably more. Care must be taken when cleaning printed circuit boards with alcohol. Over application can soften the protective coating on some boards, giving them an overall tacky or sticky feel, which only attracts dust and other contaminants. In fact, with all cleaning products, it is wise to use no more than is necessary to get the job done. When in doubt about how a product may react with some VCR part, it is best to test it on a small area first.

Denatured alcohol available at building and medical supplies dealers is a satisfactory all-around cleaner/degreaser. *Everclear*, sold at liquor or package stores, may also be used. These products are nearly 100 percent pure alcohol. Some parts suppliers carry *isopropyl alcohol* that is nearly 100 percent pure. This is an excellent general purpose cleaner; it leaves no residue and is preferable to other alcohols, such as 91 percent rubbing alcohol. Rubbing alcohol from the drug store *may* be used as a distant second choice, but read the label to make sure it is no less than 90 percent isopropyl alcohol. Some rubbing alcohols have as much as 30 percent water! You don't want water causing oxidation inside a VCR. If you choose alcohol as a cleaning and degreasing agent, get as close to 100 percent pure as possible.

Never use silicone spray for VCR repair work.

### *Electrical contact cleaner*

This product, which usually comes in a spray can but is also sold in bottles, is designed specifically for cleaning electrical contacts, removing oxidation film and restoring contact surfaces. Contact cleaners are especially good for connections carrying very low signal voltages, such as:

- Rotary tape heads (actually, connection to rotary transformer windings in stationary part of video drum)
- A/C head
- Full erase head
- RF connectors: back panel, tuner, RF modulator, and VHF antenna switcher

Products labeled *"TV Tuner Cleaner"* or *"Tuner/Control Cleaner"* usually contain a lubricant in addition to the cleaning agent. This is fine for those old mechanical turret tuners, and for switches and potentiometers (volume/tone controls); the lubricant retards contact surface wear. But for other applications, the lubricant just attracts dust. Never use tuner cleaner with a lubricant for general VCR cleaning, especially on rubber parts.

### *Rubber cleaner/revitalizer*

Read the label and make sure this product specifically says it is suitable for pinch rollers before you use it on a pinch roller! Popular product trade names that have come into general usage are Re-Grip™ and Non-Slip™. Several other brands are also available. These are fine for all rubber parts, *except the pinch roller,* unless stated otherwise on product labeling or application sheet.

---

CAUTION: Many rubber cleaners and rejuvenators deliberately leave a tacky film to help old, hardened rubber idler wheel tires and belts grip better. This may be fine for idlers and belts, but not for pinch rollers! Tackiness here could cause tape to stick to the roller and wrap around it.

A product with a label that states it cleans and revitalizes rubber and removes glazing on rubber belts and idler wheel tires is probably OK for pinch rollers too, but keep in mind the potential problem of using some of these products on a pinch roller. You're better off replacing the roller.

### Compressed air

Great for blowing out dust and debris from inside a VCR. This product comes in a variety of spray cans, available at electronic suppliers and computer stores. Use only spray cans recommended for electronic or computer equipment. Others may develop a high static electrical charge as the spray leaves the nozzle, which can damage static-sensitive electronic components. It is advisable to wear a grounded conductive wrist strap when using spray products, and at other times as well, while servicing electronic equipment that contains integrated circuits.

### Acetone

Acetone is a *powerful* solvent, and may be successful in cleaning an extremely dirty or clogged rotary head when other methods fail. Use all other cleaning methods described in this chapter first; *acetone should only be used as a last resort!* Acetone can be purchased in some drug stores and in building supplies and paint stores. Read package labeling before using.

> DANGER: *Acetone can cause breathing problems!* DO NOT USE ACETONE in closed shops or unventilated areas. Provide plenty of fresh air circulation. Read and heed safety warnings printed on product labels.
> CAUTION: *Acetone should rarely be needed.* It is destructive to plastic!! It melts it. If you must use acetone, use extreme care not to drip it on anything plastic or nylon.

### Lightweight grease

This product, shown in Figure 6.7, comes in a small squeeze tube or jar, and is sometimes called *phono lube.* It should *not* be a petroleum-based lubricant, which can attack and deteriorate plastic or nylon gears and cams. If in doubt, read the label to see if it is all right for use on plastics.

Silicone or lithium-based products are fine. Molylube™ and Lubriplate™ are lightweight greases that are safe for use on plastics, and work well on metal surfaces too. Most electronic suppliers have nonpetroleum-based, lightweight lubricants suitable for tape decks and VCRs.

### Lightweight oil

Select sewing machine or general purpose household oil. Oil designed for musical instruments is also excellent. *DO NOT EVER use a spray lubricant inside a VCR.* A small syringe is a must for applying just the right amount of oil to just the right spot. You can buy high-grade, light machine oil in containers with a needle tip applicator at most electronic supply and hobby stores. (See Figure 6.7.)

### Cleaning swabs

There are three basic types of cleaning swabs available (See Figure 6.8.):

Foam- or sponge-tipped
Chamois cleaning sticks
Cotton-tipped.

142 ■ *Practical VCR Repair*

> CAUTION: Never use a cotton swab or cloth on rotary heads! Fibers can get caught in a head and pull it out of alignment or break the delicate head tip.

Foam-tipped swabs, also called sponge-tipped, are fine for general VCR cleaning as well as for rotary heads. *Lint free* cloths are also recommended for cleaning most parts of a VCR, but should not be used on rotary heads. Chamois swabs are the *best choice* for cleaning rotary heads. Foam- or sponge-tipped swabs may be used if you are careful not to use too much pressure. Some of these swabs have a layer of foam over a cotton swab. If the foam wears thin or tears, a cotton fiber could snag a rotary head, damaging it. Virgin chamois, purchased at an auto parts store, can be used as an alternative to chamois swabs. Simply cut a piece into small squares, about an inch or so square. Cotton-tipped swabs, including those often called by the trade name *Q-Tips*, are considerably less expensive than foam-tipped or chamois swabs and work fine for cleaning just about everything in a VCR, *but should never be used on rotary heads!*

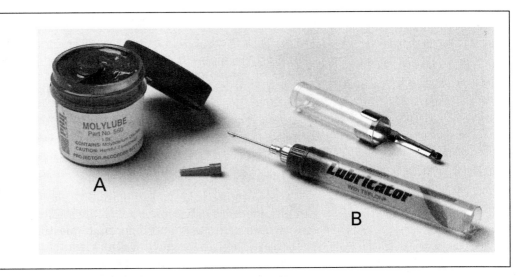

**Figure 6.7** VCR lubricants: (A) lightweight grease and (B) lightweight oil in syringe applicator.

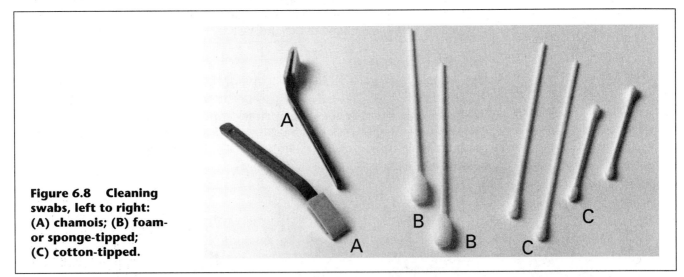

**Figure 6.8** Cleaning swabs, left to right: (A) chamois; (B) foam- or sponge-tipped; (C) cotton-tipped.

> CAUTION: Although cotton swabs are the least expensive cleaning sticks, and may be used on most VCR parts, you must be certain there are no stray cotton fibers sticking to anything when you're done. Loose fibers can cause all sorts of problems if they get caught in tape transport mechanics, especially in the tape path. For this reason, foam-tipped swabs could well be worth their higher cost.

### *Tape head demagnetizer*

It is better to purchase a demagnetizer designed for video heads instead of using one for audio tape decks. Some of the latter demagnetizers may produce an alternating magnetic field powerful enough to shatter a delicate rotary head if the tool tip gets too close. These types are fine for demagnetizing all other parts along the tape path, including stationary heads. See Figure 6.9.

Read the directions that come with the unit, and the information later in this chapter, before using a demagnetizer.

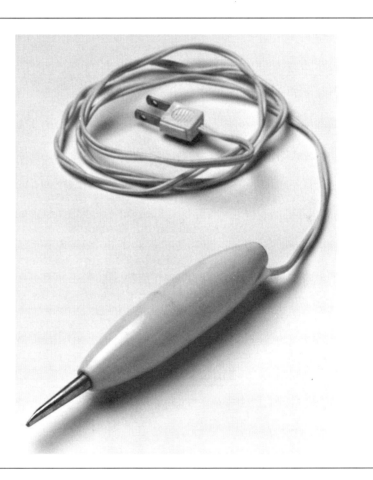

**Figure 6.9 Tape head demagnetizer**

> CAUTION: Never allow a demagnetizer tip to touch a tape head. Delicate rotary heads can shatter if a demagnetizer contacts them!

### Miscellaneous tools and supplies

- Small paint brush: helps remove dirt from corners and crowded areas
- Toothbrush: aids general cleaning. May clean a clogged rotary head gap; use *very carefully,* moving brush only horizontally with *light* pressure
- Fine sandpaper: for light buffing of glazed rubber parts, like idler wheel tires, and for cleaning stubborn corrosion on electrical contacts. Use sparingly!
- Epoxy cement: for repair of broken plastic or metal parts
- Plastic modeling putty, available at hobby stores and associated with plastic models: for repair of cracks and digs in plastic VCR cabinet
- Mild general purpose household spray cleaner: for outside of VCR cabinet. *Never use on VCR interior.* Apply on cloth or paper towel first; don't spray it directly on a VCR. It could seep inside and cause a short!
- Shop rags and disposable lint-free cloths, such as Kim-wipes, a product trade name of Kimberly-Clark.
- Notebook: keep a log of troubles you fix with VCR model numbers, part numbers, suppliers, and related data; could save you time when another VCR of the same model comes in for repair with a similar symptom.

### VHS cassettes

- Good-quality, commercially reproduced T-120s: movie or music video
- Scratch cassettes: good-quality blank or non-commercially recorded T-120s in excellent mechanical condition
- Videocassette test jig (sometimes called a dummy cassette).

A few premium-quality commercially recorded cassettes are essential to check out a VCR, and can even be used for some alignment procedures rather than expensive manufacturer's test and alignment tapes. Select major movies or popular music videos; these are generally well reproduced on top-quality commercial tape duplicators. Music videos with a lot of high-frequency content are ideal for aligning linear audio heads, in the absence of a special alignment tape.

Make sure these tapes are recorded in Hi-Fi Stereo so you can use them when working on Hi-Fi Stereo VCRs. Audio on these tapes is also recorded on the Right and Left channel linear audio tracks near the top edge of the tape, so they play back on non-Hi-Fi mono or stereo machines.

Chapter 4 contains information on scratch cassettes and how to make a videocassette test jig.

Now that you know the tools and supplies frequently used in VCR maintenance and repair, we'll move on to explain how to use them.

## CLEANING, LUBRICATION, AND INSPECTION

A large number of VCR problems are corrected by cleaning and *minor* lubrication. Doing the right things with the right tools and supplies is important. Otherwise the results could quickly become worse than if nothing were done at all. This is especially true when cleaning rotary heads, a subject covered later in this chapter.

Over time an accumulation of ferric oxide will be deposited on parts with which videotape comes in contact. Ferric oxide is the magnetic material that is bonded to the tape surface. It's really a form of rust. Parts that have more friction with the tape, such as capstan and pinch roller, generally accumulate more tape residue than other tape path components.

Dirt on a stationary tape head and in the microscopic gap of a rotary head reduces the head's ability to reproduce a good signal. This causes muddy sound, pictures with lines, loss of color, and other defects—even total loss of any recognizable image. (A typical VHS video head gap is 0.3 microns [0.0003 mm or 0.00001 inch].)

Tape deposits and dirt like cigarette smoke film on the capstan and pinch roller can cause tape to stick to these parts. This is frequently what has happened when a VCR "eats" a tape. Dirt on other transport components can cause tape mistracking, rewind failures, and erratic transport behavior.

> WARNING/CAUTION: Whenever you use chemicals, cleaners, and related products ensure there is adequate ventilation with plenty of fresh air circulation. *Do not use products in an enclosed shop area.* Avoid breathing fumes as much as possible.
>
> Immediately stop using the product, leave the shop area, and get outside fresh air at the first sign of breathing difficulty, lightheadedness, or dizziness. Avoid skin contact with the product. Wear eye protection, especially when using sprays, which can splatter. Always read thoroughly and follow directions for use and safety notices printed on the product package. Never mix two cleaning agents.

## General Cleaning

First, dirt, dust, cigarette ash, and perhaps part of Junior's peanut butter and jelly sandwich have to be cleaned out of the VCR.

Work on a clean, well-lit surface with enough room to easily turn the VCR around and upside down. Eject any cassette that may be in the VCR. Unplug the VCR and leave it unplugged for several minutes.

Take the top and bottom covers off. In most cases, you can leave the front panel on, unless it has to be removed before you can take off the bottom cover.

Put on your ESD wrist strap and connect the alligator clip on the wire to a metal bracket or the baseplate of the tape transport assembly. If you don't have a wrist strap, at least discharge any static electricity on your body by touching a ground point, such as the center cover screw of an AC outlet. Otherwise, touch the metal transport assembly.

Swing open any circuit board covering either side of the transport. Remove any shield plate over the rear portion of the transport.

If the inside is really dirty, and there's a layer of dust and crud on everything topside, a small vacuum cleaner is handy to clear it away. Use a small paint brush to dislodge debris in small recesses and tight areas. Be sure not to touch the video drum with the vacuum cleaner wand.

After vacuuming up all you can, use a can of compressed air to drive out any remaining debris. Aim the spray can nozzle so dirt blows *out* of the VCR, not further into it. Keep the spray can upright; otherwise propellant may also come out. (Don't worry if it does—it shouldn't hurt anything.)

This general cleaning should suffice for most VCRs that are kept in a fairly clean environment to begin with, and which aren't more than a few years old. Units that have been in dirty surroundings, such as where there is a lot of cigarette smoke or in a kitchen where frying is frequent, or VCRs that are many years old will likely require additional cleaning. Here's how to determine if additional cleaning is warranted, and how to go about it.

Take a clean cotton swab and wipe it for about an inch along the top surface of a circuit board or the transport baseplate. If the cotton comes away grimy and discolored, a further cleaning step is in order. Airborne contaminants can often conduct electricity to some extent, which can interfere with VCR performance. It's time to get going with that spray can of cleaner/degreaser.

- Ensure the VCR is at room temperature. Avoid spraying cleaners on warm or hot components, such as an integrated circuit; rapid cooling can crack parts, or can produce thermal stress that will shorten their lives.
- Be absolutely certain that the VCR is unplugged before using any cleaners. Don't turn the VCR On until several minutes after using any cleaner to allow for complete evaporation.
- Do not smoke or have any open flame in the area.
- Provide proper ventilation when using cleaners. A small fan at one end of the work area is a good idea! There should be fresh air circulating in the work area.

Put the VCR on its side so that cleaning solution drains out of the machine. Use paper towels or shop rags to catch the runoff and to wipe up spills. Spray cleaner/degreaser liberally on areas to be cleaned. Most spray cans come with a plastic nozzle extension tube that can be attached for directing solution into hard-to-reach areas. The extension tube also helps put cleaner exactly where you want it. Of course, you can also spray a small amount of cleaner into a paper cup, dip in a swab, and then swab the area to be cleaned. See Figure 6.10.

**FIGURE 6.10 Cleaning inside VCR with spray cleaner/degreaser**

Avoid touching components on circuit boards with your fingers. Be very careful with paper towels and swabs around circuit boards so you don't accidentally snag a component and break it off.

Spray cleaners/degreasers, alcohol, Freon TF, and trichloroethylene can be safely used on just about everything in a VCR, from plastic parts to circuit boards to tape guides. They can clean rubber parts, but also tend to dry the rubber out. If you must use one of these in place of a rubber cleaner, apply it sparingly. Don't let cleaner solution contact rubber any longer than necessary to clean the surface. Wipe the rubber part with a lint-free cloth or dry swab right after cleaning with these products. Always provide plenty of fresh air circulation when using chemicals, avoid breathing vapors, avoid skin contact, and read and follow directions and warnings on product labels.

Now that general topical cleaning has been done, we'll describe how to clean specific parts of the tape transport, namely tape guides, video drum, stationary heads, capstan, and pinch roller. Use a high-intensity work light for this.

It's difficult to say how often parts in the tape path should be cleaned. This depends to a great extent on the quality of tapes that are run through the machine, how they've been stored, and the cleanliness of the environment surrounding the VCR. Off brand videotape may be manufactured with an oxide coating that easily flakes off. Tape like this *can* clog rotary heads in a single pass.

Inexpensive cassettes may be manufactured with tape that is not uniformly 1/2-inch wide. Tape that is slightly wider can quickly leave deposits on tape guide shoulders and causes mistracking. These tapes may also have ragged edges that literally saw a groove into tape guides and other tape path components. Such tape-edge serrations may be too small to easily see with the unaided eye. Bargain tapes are generally no bargain!

Both VCRs and cassettes are affected by their environment. Cassettes that have been stored carelessly, and have picked up cigarette ash or other debris, will transfer this dirt to the tape path. A VCR that is stored or operated where there are high levels of airborne contaminants will require more frequent cleaning. Areas with smog can be unhealthy for VCRs as well as air-breathing creatures.

So, how often should you clean the tape path? Probably every 400 to 500 hours of operation in fairly ideal surroundings, using quality, name-brand tapes. Clean the tape path whenever the playback picture deteriorates and the manual tracking control is not effective in producing a good image. Any evidence of tape damage is certainly reason to inspect and clean the tape path.

Most manufacturers recommend cleaning after about 500 hours of tape path operation. Again, some VCRs keep tape loaded or partially loaded as long as power is On, for instant or very quick response when Play is pressed. These machines may need more frequent cleaning. It is best to eject the cassette or power these VCRs Off whenever a tape is not actually being played or recorded, to reduce rotary head wear.

### Demagnetizing Tape Path

Although not absolutely necessary, demagnetizing tape heads and other tape path components at the same time they are cleaned is beneficial.

Over time, residual magnetism builds up on tape heads and other ferrous parts in the tape path that come in contact with videotape, such as tilt poles, capstan shaft, and tape guides. Tape heads and transport components become very weak permanent magnets. Magnetized tape heads become less sensitive to recorded high frequencies. Residual magnetism on a tape path component also partially erases any tape that runs past it, especially the high frequencies. Routine demagnetizing of the tape path helps prevent these ills.

A demagnetizer specifically designed for video heads is preferred over one for audio tape heads, but the latter can be used with care. Some demagnetizers for audio tape recorders may produce a magnetic field strong enough to shatter delicate rotary heads. Read and follow instructions for the specific demagnetizer you are using. If you're not sure what you are doing, it's probably better not to attempt demagnetizing rotary heads.

Clear the work area of any magnetic tapes and cassettes before turning On the demagnetizer. Partial erasure could easily occur if they are in the vicinity. Plug in the demagnetizer while it is several feet away from the VCR.

***DO NOT TOUCH anything*** with the demagnetizer tip. This is easier said than done. The demagnetizer *is* an electromagnet, and therefore it is attracted to iron and steel parts. You could bring the tool close to a part, not intending to touch it, but all of a sudden the magnetic attraction slams the tool tip against the part. This can scratch the piece like an electric engraver, since the electromagnet is turning On and Off 120 times a second. A delicate rotary head could easily be destroyed!

Wave the tip of the tool *slowly* in small circles around the face of the full erase head, the face of the A/C head assembly, the capstan shaft, and the tape guides. To be safe, keep the tip one inch away from rotary heads, unless tool instructions state otherwise. After demagnetizing the entire tape path, *slowly* move the tool several feet away from the VCR before shutting it Off. If you turn the tool Off while it is close to the VCR, the collapsing magnetic field can remagnetize the parts you just demagnetized.

> Reminder: Wear a grounded conductive wrist strap to protect sensitive electronic components from electrostatic damage.

### Cleaning Tape Guides

Saturate a clean swab with cleaning solution; head cleaner is fine! Wipe all tape guide and inertia roller surfaces (see Figure 6.11). On tape guides with upper and lower shoulders or collars, pay particular attention to areas where the top and bottom edges of the videotape contact the guides, right where the shoulders begin. This is where tape residue buildup is most likely.

Do not touch any part of the tape transport that contacts videotape with your fingers. Doing so leaves an oily film that attracts debris and damages the tape.

In rare instances, you may come across a very stubborn bit of tape residue in the corner of a tape guide, where the top or bottom shoulder begins. If re-

**Figure 6.11** Cleaning tape guides. Saturate cotton or foam swab with cleaning solution. Wipe surfaces of all tape guides contacted by videotape.

peated wipes with a swab saturated in cleaner solution fail to remove the dirt, use a wooden toothpick saturated with solution to dislodge the material. ***Never touch the surface of tape guides, or anything else in the tape path, with a metal tool!*** *Any slight scratch will damage videotape.* Do not lubricate tape guides in any manner.

# ROTARY HEADS

The most critical items that require regular cleaning are the rotary heads; more so than any other VCR component, there is a right way to clean them. Doing things incorrectly here could easily damage delicate tape heads!

## Cleaning Rotary Heads

Rotary heads are cleaned with chamois or foam-tipped swabs only; *never use cotton swabs.* It is all too easy for cotton fibers to get caught on a rotary head and pull it out of alignment or break it. Foam- or sponge-tipped swabs are a second choice after chamois cleaning sticks. Just be careful; some of these swabs have a layer of cotton underneath the foam tip. Too much pressure, and the foam or sponge can wear through, exposing rotary heads to cotton fibers that can tear them out!

This is how to properly clean rotary heads (see Figure 6.12):

1. Saturate a new chamois cleaning stick with head-cleaning solution.
2. Place the chamois surface flat against the upper, rotating portion of the video drum, on a spot where there is no tape head.
3. Apply light to moderate pressure on the chamois stick.
4. Slowly rotate the upper cylinder counterclockwise several revolutions using your fingertips on the *top* of the drum. Never touch the sides of the drum! Do not touch the fine head wires at the top of the drum.
5. As you rotate the drum, *very slowly* move the swab upward on the rotating surface, maintaining pressure. Never move a swab vertically over the head tips; that can damage the heads. Move the swab only when there is no tape head beneath it.
6. Repeat these steps several times. Keep chamois flat against cylinder surface.

Saturate the chamois stick again and swab the lower, stationary section of the video drum. You should not be able to see a trace of tape oxide anywhere on the video drum after proper cleaning. Inspect for oxide trapped in the five anti-stiction grooves in the upper cylinder. Make sure the ledge, or rabbet, on the lower, stationary part of the drum is clean and bright.

There may be a very few times when the preceding rotary head cleaning procedure will still not restore a clear picture. It could be that one or more video heads is clogged with a magnetic tape particle. Typically, a dirty video head produces only a partial picture, or hardly any at all if it is extremely dirty or clogged. First, repeat the cleaning steps several more times, saturating the swab with solution each time. If it still seems that a rotary head gap may be clogged, attach the nozzle extension tube to a can of spray cleaner and aim the spray directly at each head. Wear eye protection to avoid any splatter from getting into your eyes. Do not touch the heads with the tip of the nozzle. Spray each head for about a second. Avoid getting excess cleaner inside the drum, which could get into motor bearings and cause premature failure.

If you still don't get a good picture, it could be that the rotary head tips are worn down too much, or there still could be a bit of tape oxide clogging a head gap. You've got little to lose by trying one more cleaning technique. *The fol-*

**Figure 6.12 Cleaning rotary heads. Saturate chamois stick, rotate drum CCW, move chamois upward only when no tape head is beneath it.**

*lowing procedure may itself damage a rotary head if extreme care is not taken!* If it doesn't work, you may have to replace the rotary head assembly; do all you can first!

Saturate the bristles of a clean, soft toothbrush with cleaning solution. Hold the drum so it won't turn. Wipe *gently* across each rotary head with an upward angle of about 30 to 45 degrees. Move the brush from one side of the head slot to the other, so just the tips of the bristles pass over the head tip. **Do not scrub up and down!** Repeat this *several times* for each head, keeping the bristles saturated with solution. Be careful that brush bristles don't fling cleaning solution into your eyes.

*Use acetone only as a cleaning agent of last resorts.* Test first to see if it softens or "eats" the brush bristles, plastic cleaning sticks, or foam swabs on whatever you may be using. Be very careful not to get acetone on plastic parts, which it will attack. Make sure not to breathe fumes; provide a good supply of fresh air in the work area.

Sometimes, even after all this cleaning, a clear picture is still not forthcoming. Occasionally, just playing a videotape for a while will restore a good picture. Just be sure that cleaning solution has completely evaporated before loading tape.

Only one spot on the video drum should be lubricated: the ground strap contact that touches the end of the motor shaft. Clean off any old grease and place a small dab of grease beneath the contact. Check that the contact is centered on the motor shaft, with sufficient tension to make good electrical contact.

## Replacing Rotary Heads

Individual rotary heads are not replaced: the entire rotary head assembly is replaced as a unit. Replacement head assemblies can be purchased from several electronics suppliers besides the original manufacturer.

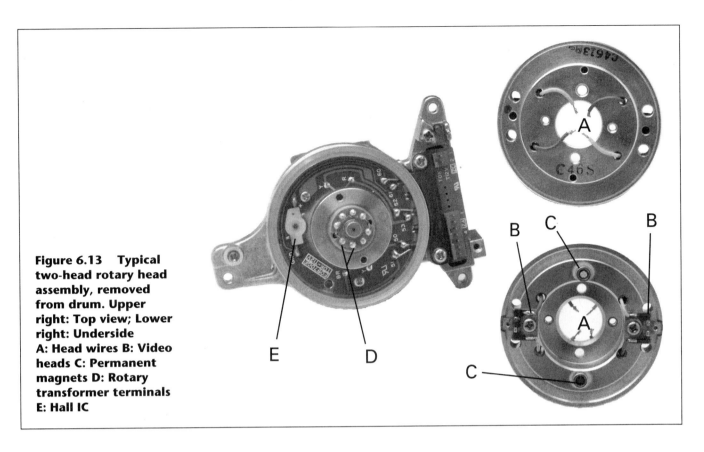

**Figure 6.13** Typical two-head rotary head assembly, removed from drum. Upper right: Top view; Lower right: Underside
A: Head wires B: Video heads C: Permanent magnets D: Rotary transformer terminals E: Hall IC

Figure 6.13 shows a rotary head assembly for a two-head VHS VCR. When the assembly is installed, two wires from each video head are soldered to rotary transformer terminals on the rotating drum assembly. This part of the transformer spins with the rotary heads, and transfers signals between the heads and the lower, stationary half of the rotary transformer.

Notice the two permanent magnets (C) on the underside of the video head assembly in Figure 6.13. As the assembly rotates, they pass over a Hall effect integrated circuit in the lower, stationary portion of the drum. A Hall IC operates somewhat like a phototransistor, but it turns On in the presence of a changing magnetic field instead of a beam of light.

Hall IC output goes to the VCR servo circuit, which determines drum *speed* and rotary head position. Speed is indicated by the drum frequency generator (FG) signal, so correct drum rpm can be maintained. Head position is indicated by the pulse generator (PG) signal. The 30Hz head switching pulse that selects which video head output is seen during playback is derived from the drum PG signal.

Although the two heads may look identical, they are not. Each is manufactured with the head gap tilted 6 degrees from vertical. (The term *azimuth* describes head gap tilt with respect to its path along the magnetic tape. Head A is tilted −6° and Head B is tilted +6°.) The important thing to know is that the two heads have different colored wires to identify which is which. If the head wires are not soldered to the proper rotary transformer terminals when the head assembly is replaced, the VCR will not play back *any* previously recorded tapes.

Normally, replacing the rotary head assembly should be about the last thing done to restore a good picture. First, make sure the problem isn't elsewhere, or just a clogged head. Mechanical and electronic adjustments should first be checked according to the manufacturer's service manual, unless the sheer number of hours of VCR service indicates that the heads are likely worn out.

Replacing the rotary head assembly is only part of the job. Sometimes, electronic and mechanical adjustments must be made after replacement.

Unless damaged, rotary heads should normally last for several thousand hours of operation. A VCR used an average of one hour every day of the year should still have remaining rotary head life after ten years of operation.

Tips of new rotary heads protrude beyond the surface of the cylinder in the neighborhood of 45 microns, or about 0.0018 inch. When tips wear down to somewhere around 5 or 10 microns protrusion (0.0002" to 0.0004"), head replacement is required because they no longer maintain sufficient contact with videotape.

The only realistic way you can accurately measure head tip protrusion is with a micrometer and a holding jig designed for use on VCRs. Tentel Corporation markets a VCR head tip protrusion gauge for around $745. (The company also has available an instructional videotape that shows how to use the gauge, as well as other VCR service gauges that they carry.) As always, take extreme care when working near rotary heads with the gauge, which has to actually contact a delicate head tip to take a reading.

Replacing a rotary head assembly is not terribly difficult, but you have to be very careful when unsoldering the old head wires from the rotary transformer terminals and when soldering on the new head wires. Too much heat, too much solder, or both, and you could destroy the rotary transformer; this usually requires replacement of the entire video drum assembly, a costly part. Take care not to damage the delicate head tips when installing the new assembly.

Following is a *generalized* procedure for replacing a VHS rotary head assembly. Refer to the service manual for the VCR model on which you are working for instructions on parts removal and replacement. Service literature also contains steps to perform mechanical and electronic adjustments. Information on manufacturer's service literature is in Chapter 18.

Figure 6.14 is a typical video drum assembly installed in a machine. This one has a spindle ground strap on the top, whereas some are on the bottom. Some rotary head assemblies have a separate shield plate, like the one pictured here. It may *not* be absolutely necessary to remove the shield plate, but it makes unsoldering the head wires easier.

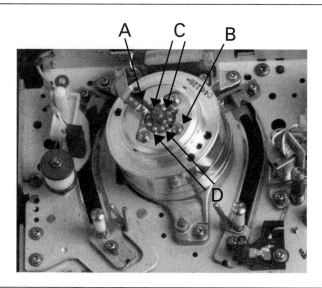

**Figure 6.14 Typical two-head video drum assembly. A: Spindle ground strap B: Shield plate C: Head wires D: Rotary transformer terminals**

## VHS Rotary Head Assembly Replacement

1. Sketch a diagram and take notes before you begin. Make additional sketches as needed during disassembly and document each step you take along the way.

2. Remove the video drum ground strap, if it's on the top. Put mounting screw in numbered container, or screw it back into the drum assembly for safekeeping.
3. Remove the two screws holding the upper drum shield plate, if present. Place screws in a numbered container and make a note of the container number and parts description.
4. Mark the colors of the head wires on your diagram, and note which wire is soldered to which rotary transformer terminal. Each head has different colored wires.

    Look for a dab of paint or some other marking near where the head wires are soldered so you can properly orient the replacement head assembly. If there is no paint spot or identifying mark of any kind on the part with the rotary transformer terminals, make a mark of your own with an indelible pen or scribe. *You must solder the replacement head wires to the exact same terminals to which the original head wires were soldered.*
5. Unsolder the two wires from each head. Do this as quickly as possible, using just enough heat, so that rotary transformer terminals and insulation around them are not damaged. Soldering and unsoldering techniques are described in Chapter 11.
6. Remove the two screws securing the rotary head assembly to the upper part of the rotary transformer.
7. Lift off the old rotary head assembly. You may need a head puller tool to remove the head assembly if it doesn't easily lift off. *Never* pry off a head assembly by twisting a screwdriver between the upper and lower cylinders. A video head puller works similarly to a wheel or pulley puller, with which you may be familiar, and is available from suppliers of VCR parts. Figure 6.15 is a photo of the same video drum with the rotary head assembly removed.
8. Clean exposed areas of the video drum with alcohol or other suitable cleaning solution.
9. **IMPORTANT:** *Do not touch any tape head on the new rotary head assembly. Be extremely careful of delicate, protruding head tips* from the moment you open the box containing the replacement assembly. *Never clean a rotary head assembly with a cloth.* Parts distributors will not refund any money for heads damaged through mistreatment or unprofessional handling.

    Orient the new rotary head assembly so the colored head wires line up with the same rotary transformer terminals as the old part. Press the assembly into place. Mount the assembly with the two screws removed in Step 6. Tighten the screws alternately—first one a little bit, then the other—until both are snug. Do not overtighten.
10. Solder the head wires on the new part to the same rotary transformer terminals as the assembly you removed. Take care not to use too much solder or too much heat. Terminals and surrounding insulation on the upper section of the rotary transformer are small and delicate.

    Perform the operation as quickly as possible. Soldered connections should look trim, bright, and shiny, not grainy or with excess solder. Check carefully that you have secure connections, with no solder bridges between terminals.

    Clean away any rosin from around the rotary transformer terminals. You can purchase rosin flux remover at electronic supply stores.
11. Replace the shield plate on top of the rotary head assembly (if there is one).
12. If you previously removed a drum spindle ground strap, clean the contact surfaces of both the ground strap and the drum shaft with alcohol. Apply a dab of grease to the end of the drum shaft. Reinstall the ground strap. The

**Figure 6.15** Video drum with rotary head assembly removed.
A: Rotary transformer, rotating half with head wire terminals
B: Rotary transformer, stationary half C: Hall device, to create drum frequency generator (FG) and pulse generator (PG) signals
D: FG/PG signal connector E: Video head signal connector

strap should be centered over the drum shaft with light downward force to make good electrical contact. Contact tension should not be excessive.

13. Clean the entire outer surface of the video drum using a chamois swab saturated with cleaning solution. Rotate the upper portion of the drum with your fingers while holding the swab against the drum surface. Do not move the cleaning stick vertically across the rotary heads.

14. After the rotary heads are replaced, other areas of the VCR *may* require adjustment. For example, the height of roller guides P-2 or P-3, or both, may need *very minor* adjustment to obtain the best playback image. It is also possible that the horizontal position of the Audio/Control head might require *slight* adjustment for correct tracking with the new assembly. Consult the manufacturer's service manual for specific information to replace the rotary head assembly and to perform mechanical and electronic adjustments after the new part is installed.

15. Make a note of the date and what parts were replaced and tape this to the bottom cover of the VCR. Make sure the note doesn't block any ventilation holes.

## STATIONARY TAPE HEADS

Clean the surfaces of the full erase head and the Audio/Control/Audio erase (A/C) head assembly with a swab saturated in cleaning solution. Do not touch head face surfaces with your fingers.

It is rare for a VCR to require replacement of the full erase head or A/C head assembly unless the machine is in very heavy service. Wear of those heads so extreme as to require their replacement would be well after rotary heads have already been replaced. Unusually high supply spindle back tension can accelerate wear of all tape path components, however; so can bargain cassettes that have rough tape surfaces.

Worn stationary tape heads will actually be cut-in along the 1/2-inch videotape width, where tape has worn away the head face. Loss of high-frequency response from the linear audio head—when height and azimuth are adjusted correctly—indicates that tape wear has widened the head gap. Sometimes a worn head will easily clog with tape oxide, requiring repeated cleanings. A/C head replacement is indicated under these conditions.

Follow instructions in the manufacturer's service manual for replacing and adjusting any tape path component. Usually there is no adjustment at all for the full erase head, but the A/C head assembly has four adjustments:

- Height: vertical position with respect to tape path
- Azimuth: side-to-side tilt or angle from vertical
- *Tilt*: top-to-bottom; forward or backward leaning
- Horizontal position: distance from video drum.

Often, adjustment screws and nuts have a dab of lock paint on them. Don't loosen these screws and nuts unless you really need to when replacing a part. Usually these factory adjustments won't change during the life of the associated part.

In general, do not remove any VCR component until you have the replacement part in hand. Sometimes you'll get a larger assembly containing the part you wish to replace as a subassembly, and it might be easier to remove and

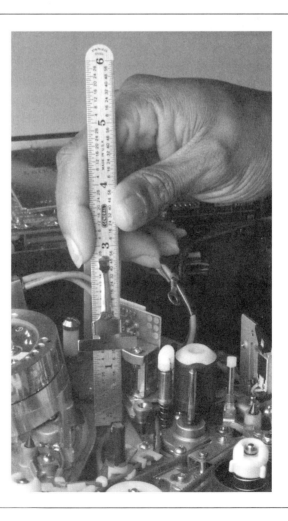

**Figure 6.16** Measuring height of A/C head assembly. A steel ruler with a sliding pocket clip and crossbar is practical for making accurate height and distance measurements.

replace the larger unit. Also, your memory of how things reassemble will be much fresher if you've just taken out a worn part. As always, take notes, scribe lines, mark parts, and make a sketch before taking things apart, which simplifies parts replacement and reassembly.

It is helpful to take accurate measurements of the A/C head height and position before removing it. Count the number of turns of any adjustment screw or nut that must be removed, and jot it down. This information will help you get the new part in *rough* alignment without having to use a manufacturer's alignment jig.

A six-inch steel rule, with graduations in tenths of a millimeter or sixty-fourths of an inch, is an inexpensive yet worthwhile tool for making small transport measurements. For example, measure from the transport baseplate to the top of the A/C head, as shown in Figure 6.16, before replacing the part.

Tape path adjustments are covered in Chapter 7, including those for the A/C head.

## CAPSTAN AND PINCH ROLLER

The capstan and pinch roller are critical tape path components; they must be kept in good condition so a VCR doesn't mangle or "eat" tapes. Cleaning the capstan shaft and lubricating its bearing are important parts of their routine maintenance.

### Cleaning the Capstan

Clean the exposed capstan shaft using a swab moistened with alcohol or cleaning solution. Do not let liquid run down the shaft and into the bearing, which could cause the bearing to seize up.

You may find a black ring of hard tape oxide near the top or bottom of the capstan shaft, or in both places. These black rings are where videotape edges are pressed against the shaft by the pinch roller. In fact, if the VCR hasn't been serviced in a while you probably *will* see this condition. Often these tape oxide rings are difficult to remove. Scrub them *many times* using swabs moistened with cleaning solution. It takes some effort. Never put excessive force against the side of the capstan shaft, which could bend it.

If the oxide still stubbornly clings, use a wooden chopstick, toothpick, or popsicle stick saturated with cleaning solution to break up the black deposit. *Never use an emery stick, sandpaper, file, or anything metal against the capstan shaft.* Repeat the cleaning with a moistened swab. Keep at it until there isn't a trace of black deposit on the capstan. The shaft should look bright and shiny, top to bottom and all the way around.

### Oiling the Capstan Bearing

This requires placing a drop or two of oil on the sintered bronze bearing at the base of the capstan shaft on top of the transport.

There will usually be a plastic or fiber washer at the base of the capstan shaft, right over the bearing. It prevents oil from migrating up the shaft and keeps dirt out of the bearing. This washer might be very thin and transparent; it is sometimes difficult to see that it's even there! Work the protective washer up the capstan shaft about half an inch and apply oil to the bearing with a needle-tip applicator. Do not overlubricate; two small drops are enough. Try not to get any oil on the capstan shaft above the bearing. Slide the protective washer back down the capstan shaft so it covers the bearing. Wipe any oil from the capstan shaft with a dry swab or lint-free disposable cleaning cloth. Again clean the exposed capstan shaft using a swab moistened with cleaning solution.

Most newer VCRs have direct drive capstans, where the motor is actually part of the capstan itself. For this type, the preceding steps are all that is usually necessary.

Many older VCRs have an *indirectly-driven* capstan: a separate motor drives the capstan flywheel with a rubber belt. The bottom of the capstan shaft usually rests on a plastic *thrust bearing* on a metal bracket beneath the flywheel. Take the bracket off—it usually mounts to the undercarriage with two screws—remove any belts around the flywheel, and then pull the entire flywheel and capstan shaft out of the machine. Don't lose the protective washer covering the bearing on top of the transport or any spacer washers that may be between the top of the flywheel and the capstan bearing in the undercarriage.

**NOTE:** If you see tape oxide deposits on the capstan shaft, be sure to clean this off *before* pulling the capstan out. This prevents contaminating the sintered bearing.

With the capstan and flywheel completely out of the VCR, lubricate the bronze bearing directly above the flywheel, as well as the one on top of the transport. Two drops on each from a needle-tipped applicator should be sufficient.

Clean the bottom end of the capstan shaft and the thrust bearing on the removed bracket. Reassemble, making sure any spacer washers are in place above the flywheel before sliding the shaft back through the bearings. Replace the protective washer over the capstan bearing on top of the transport.

## Cleaning the Pinch Roller

Clean the surface of the rubber pinch roller. Be careful if using rubber cleaner or revitalizer solution that doesn't specifically state that it may be used on pinch rollers. Some rubber rejuvenating liquids make rubber tacky so belts and idler wheel tires grip better. This tackiness is undesirable on a pinch roller, because it can cause videotape to cling to the roller.

Wet a cotton- or foam-tipped swab with rubber cleaning solution or alcohol and lightly scrub the pinch roller. Repeat with two or more swabs, wiping rather than scrubbing, until the swab stays clean, with little or no black rubber residue. If swabs continue to get quite black, this usually means the rubber has deteriorated; the roller should be replaced.

Rubber parts normally become hard and lose their resiliency and grip as they age, but sometimes, especially in a contaminating atmosphere, rubber becomes soft and tacky. In the latter case, usually no amount of cleaning helps.

There is no lubrication procedure for the pinch roller.

Inspect the pinch roller and replace it for any of the following conditions:

- Ridges or grooves at the top or bottom where videotape edges contact the roller
- Swelling or warping of the roller. Sides should be perfectly straight up and down, with no bulges or irregularities.
- Gummy or tacky surface
- Hard, shiny rubber surface
  Push in with your fingernail at the top of the roller. You should feel a slight give or yield in the rubber. It should *not* be hard as a rock.
- Cracks in rubber
- Roller wobbles on shaft, doesn't turn freely, or has gritty feeling when turned by hand

These last conditions indicate that the roller bearing is either worn or has started to seize up.

Often a pinch roller will have a shiny black glaze on its surface, even after being scrubbed with a cleaning swab. This can cause tape to skew or may allow tape to slip past the capstan and pinch roller. If none of the conditions in the preceding list are also present, the roller can be reconditioned by removing the thin shiny layer. Remove the roller and buff its surface *very lightly* with fine sandpaper. Just remove the shiny glaze. Overdoing things here can easily make the roller out of round. Clean the roller before reinstalling it in the machine.

If you have any doubt about the condition of the pinch roller, replace it. This is especially true for commercial repairs. Coaxing a little more life out of a pinch roller in one's own machine is understandable, but replacing a marginal pinch roller in a repair customer's VCR ensures it won't be back soon for pinch-roller-related problems.

### Pinch Roller Removal and Replacement

Rubber parts are the most frequently replaced VCR items. The rubber pinch roller is a prime candidate for replacement on older and high-use machines.

On most VCRs, you'll find a round, white plastic cap over the top of the pinch roller, as shown in Figure 6.17. On some models, the plastic cap itself is all that holds the roller onto its stationary shaft. Other VCRs have a Phillips head screw beneath the cap that holds the roller onto its shaft; the cap keeps dirt out of the bearing well.

**Figure 6.17 Prying off pinch roller plastic cap**

Remove the pinch roller cap by carefully prying it off with flat-blade screwdrivers. Start with two small jeweler's screwdrivers opposite each other and pry up as you work your way around the entire circumference of the plastic cap. On models where the cap itself holds the roller on its shaft, it is usually quite snug and somewhat difficult to remove. You may have to use increasingly larger flat-blade screwdrivers to work the cap off. Prying upward with two screwdrivers together helps prevent breaking the cap. Work the roller cap off a little at a time. Rushing the job and excessive force will break the plastic cap. Be careful not to place lateral pressure against the roller, or you could bend the pinch roller bracket; pry straight upward.

On VCRs where the pinch roller is attached with a screw you may be able to pull the dust cap off easily with your fingers, placing fingernails beneath the edge. If it doesn't readily lift off, do not pull harder. This could break the cap and bend the pinch roller bracket; pry the cap off instead.

Remove the screw underneath the cap, if there is one, and lift the roller off its shaft. As soon as you do, look carefully at the top and bottom ends of the roller; often one end has a different inside diameter than the other, or the bearing inside is closer to one end than the other. Note this so you can replace the roller properly. Do not loose any spacer washers that may be between the screw head and top of the roller bearing, or on the bottom of the bearing.

Before installing a pinch roller, clean the shaft with a swab moistened with cleaning solution. If you use a cotton swab, be sure there are no fibers clinging to the shaft when you put on the roller.

After pinch roller and cap have been reinstalled, check that the roller rotates freely. Clean the roller and capstan to remove any oil that may have been deposited from your hands during the procedure.

## BELTS, IDLERS, AND PULLEYS

Clean and inspect all rubber belts, idler wheels, and pulleys. Remove rubber belts and clean them with rubber cleaner and revitalizing solution. Clean the rubber tire on any idler wheel. Idler wheel rubber should give or yield slightly when you press it with a fingernail. Lightly scrub rubber parts until the swab stays moderately clean, without picking up any dirt or loose black rubber.

Clean belt pulleys, which often develop rubber deposits in their grooves. Use a wooden toothpick to break up stubborn residue if necessary.

Inspect rubber belts and idler wheel tires for signs of rubber deterioration, such as hardness or cracking. Rubber should be "lively," that is, a rubber belt should quickly snap back after you stretch it a little and let go. If a belt slowly resumes its pre-stretched length, the rubber is aged and the belt should be replaced. As with pinch rollers, a slight hard, shiny surface on an idler wheel tire may be removed with light buffing, but replacement is the better long-term fix.

On some idler wheels, just the tire itself can be replaced. Simply roll the old tire out of the groove on the edge of the idler wheel, working your way around the circumference. Put the new tire on the same way, making sure it is uniform all the way around the idler wheel. Clean the tire. Other idlers are replaced as an assembly, often with the slip clutch included in the replacement part.

There are so many varieties of idler assemblies that no single method applies to their removal and replacement. What needs to be done is usually quite obvious after examining the part and surrounding tape transport area. A slit washer, C-clip, or compression ring often mounts an idler wheel assembly to a transport shaft.

Some VCRs have one or more rubber *cog belts*. A *cog belt* is flat, with regularly spaced ridges running perpendicular to the length of the belt, as shown in Figure 6.18. A cog belt pulley has similar ridges that engage the belt cogs. Inspect these belts for dryness and cracking, damaged cogs, and frayed edges. Replace any belt having any of these conditions. Cog belts do not stretch like some other belt types, but should still be somewhat pliable, not hard, brittle, or stiff.

Cog belts do not slip the way flat, round, square, and V belts can. They are used instead of gears to keep mechanisms in time with each other, and thus are also called *gear-type* (GT) or *timing belts.*

Clean cog pulleys of any belt deposits.

Some idler wheels, pulleys, and reel spindles have sintered bronze bearings. Place a single drop of oil directly on any bronze bearing with a needle-

**Figure 6.18 Cog belt and pulley**

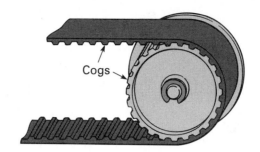

tipped applicator. Lightly oil pivot points of metal brackets and linkages, such as the pinch roller mounting arm. Avoid overlubrication!! Clean up any lubricant that gets where it shouldn't.

Many people new to small machine maintenance tend to use too much oil and grease. Just a little is sufficient. Excess lubricant on a rotating part easily gets flung to where it isn't needed, causing more harm than if it hadn't been applied in the first place. Lubricants also attract dust and dirt when over applied. Be especially careful to not get grease or oil on rubber parts, felt-lined tension band and brake pads, or on spindle surfaces that these parts contact.

Never use a spray lubricant, such as WD-40, anywhere within a VCR.

### Reel Spindles and Brake Bands

Slacken the tension band around the supply spindle and manually hold back all supply and take-up spindle brake pads, so the spindles are completely free to rotate. Then spin the spindles by hand. They should turn freely, with no

**Figure 6.19 Cleaning lower flange or rim of take-up spindle**

binds or drag. If not, remove the spindle and clean the shaft and spindle bearing. Place a drop of oil on the sintered bronze spindle bearing and apply a light film of oil to the shaft. Replace the spindle.

Clean the lower flanges or rims of both spindles with alcohol or other cleaning solution. On many models, spindles have two lower flanges. One flange is driven by the idler wheel tire; the other serves as a brake drum. This flange may have one or more of the following felt-lined parts pressing against it:

- Tape holdback tension band
- Soft brake band or brake pad
- Stop brake band or pad.

Rotate each reel spindle by hand as you swab all the way around the flanges, as shown in Figure 6.19. Don't touch flange surfaces with your fingers, which deposits body oil.

Inspect felt pieces on the tension band and brake pads. Parts with felt that is shiny, glazed, or oily, or where felt is missing or deformed, should be replaced.

## CAMS AND GEARS

Clean out any old, dried grease from cam wheels and cam gears, and from along the two movable P-guide shuttle tracks. Inspect for broken or cracked parts and any small foreign object that might be between gear teeth. Apply grease sparingly. A light film is all that is required. There should be no gobs of grease anywhere when you finish.

The general rule is to apply *oil* lightly to parts that rotate or pivot. Put oil where the parts meet, such as a movable arm around a pivot shaft. Place *grease* on parts that roll or slide against each other. This includes gear teeth and cam surfaces. As a rule, do not put grease where there is no evidence of original lubrication. For example, Teflon and Nylon gears are self-lubricating and require no grease in most applications. Excessive lubrication only attracts dirt, which can clog a mechanism.

In many VCRs, *gear timing* is critically important. *Never remove any gear, gear segment, cam wheel, cog belt, or other component that has interlocking teeth or pieces without first making a sketch. Then, take an indelible pen and mark parts where they come together.* For example, draw a line across two white plastic gears at the point where the teeth on the gears mesh. This can spare you much grief later on when you go to put things back together: refer to your sketch and line up the marks you previously made on the parts.

Many gears, gear segments, and cam wheels have a timing mark somewhere on them. This may be a dimple, ridge, small hole, or some other mark stamped or molded into the part, such as ▶ or >. Look for these marks! Make a note of their relationships, or locations, on parts that mesh with each other. Frequently two gears are *"timed"* to each other by positioning them so the mark on one gear is exactly opposite the mark on the other gear.

### C-clips, Compression rings, O-rings, and Slit washers

Several methods exist for attaching gears, pulleys, spindles, pivot arms, and other components to shafts and axles on the tape transport. They are all quite similar. Frequently a small metal clip or retaining ring keeps a gear or movable bracket in place.

Three common gear retainers are C-clips, E-clips, and compression rings. These are also called C- and E-rings and compression fasteners. They are flat pieces of sturdy sheetmetal which spring closed onto a shaft. C- and E-clips

**Figure 6.20  C-clip, E-clip, and compression ring**

C-clip    E-clip    Compression rings

normally seat into a small groove around the shaft, whereas a compression ring is used when there is no groove in the shaft. Figure 6.20 shows what these fasteners look like. C-clips and E-clips are very similar. Most people call them all C-clips, as we will in the remainder of this section.

Whenever you remove or install these type fasteners, wear eye protection. Because they spring closed, it is easy for them to fly away as you expand them to take them off a shaft. Cupping a shop rag or paper towel around a clip during removal and replacement is a good idea. It helps prevent these small parts from flying away and becoming lost.

To remove a C-clip, put the blade of a small flat-blade screwdriver in the opening *between* clip and shaft. Then turn the screwdriver to cam the clip out of the groove and off the shaft, as shown in Figure 6.21. You may find it helpful to use a second screwdriver to coax the clip off the shaft.

Use long-nose pliers to replace a C-clip. Place one plier jaw on one side of the shaft and the other on the outside of the clip, with the open end of the clip

 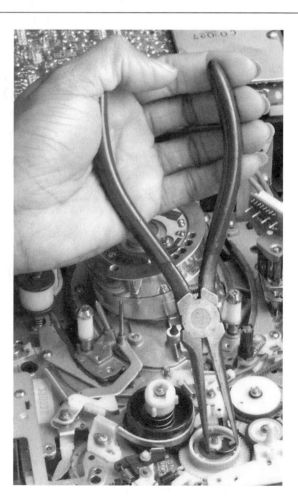

**Figure 6.21  Removing and installing C-clip**

pressed into the shaft groove. Apply even pressure on the plier jaws until the clip snaps into the shaft groove.

A compression ring is usually used on a shaft that *does not* have a groove cut into it. Nearly the entire inner circumference of the fastener grips or bites into the shaft to hold it in place. This type of fastener is harder to deal with than C-clips. Take precautions against eye injury and trap any fastener that could easily fly off in a piece of cloth cupped around the work area.

Figure 6.22 shows how to remove a compression ring fastener with a retaining ring pliers designed for the purpose. It is well worth purchasing one; however, you may not be able to find one suitable for very small compression rings. You can also use a flat-blade screwdriver to spread the ring, opening it up slightly so it doesn't bite into the shaft. Then slide the ring up and off the shaft. To reinstall, spread the clip open with a screwdriver blade, position it over the shaft, and remove the screwdriver. This sounds easier than it really is. Just be patient with compression rings.

**Figure 6.22** Removing/installing compression ring (Cup a shop rag around the fastener during either procedure)

Special C-clip removal/installation tools and a variety of retaining ring pliers are available from electronic supply houses. They make the job of taking off and putting on these tiny metal fasteners much easier and safer than using a screwdriver and regular pliers. Their use is recommended but not absolutely necessary.

Although different varieties of C-clips and compression ring fasteners are common, some VCR models use two other types of fasteners: O-rings and slit washers. An *O-ring* is nothing more than a small rubber donut that firmly grips a shaft, holding a moving part in place. Use a small screwdriver to roll an O-ring up off a shaft. Simply push an O-ring back onto the shaft to replace it.

A *slit washer* is usually made of thin nylon or plastic. It is simply a washer with a cut, or slit, and looks like that shown in Figure 6.23. The inside diameter of a slit washer is somewhat smaller than the outside diameter of the shaft on which it is used, and so it grips the shaft when installed. Remove a slit

**Figure 6.23** Plastic slit washer

washer by uncurling it from the shaft with a small screwdriver, needle-nose pliers, or both.

C-rings, E-rings, O-rings, and slit washers usually seat into a groove around the circumference of a shaft or pivot stud. A compression ring is normally used where there isn't any groove around the shaft or stud. With compression rings especially, it is important to leave a small amount of space, called *wink*, between the bottom of the compression ring and the gear, spindle, or other part it holds on a shaft. This wink, or free play, prevents the fastener from binding against the moving part. In other words, when installing a compression ring, don't place it so far onto the shaft that it rubs against a moving part beneath it.

Most electronic parts outlets have variety kits of C-rings, compression rings, O-rings, and slit washer fasteners. Because these parts are easily lost, deformed, and broken, a reserve supply of an assortment of these fasteners is a great convenience, if not a necessity.

There is still at least one other method for holding a plastic or nylon part onto a shaft. Two or more "claws" at the center of the gear or wheel may engage a circular groove in the shaft, holding the part onto the shaft, but allowing it to rotate. Figure 6.24 illustrates.

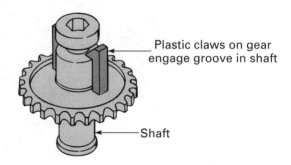

**Figure 6.24  Plastic gear with retaining claws.**

Carefully spread the claws *just enough to clear the shaft groove* while removing the gear, or the claws may break off. When replacing this type of gear, place it over the shaft and press down evenly and firmly; the claws should expand and then seat themselves in the groove. You may have to spread the claws slightly as you press the part onto the shaft.

## SWITCHES, CONNECTORS, AND SENSORS

Most of the following procedures do not need to be done each time a VCR is cleaned or has routine maintenance, but several VCR problems can be corrected by cleaning various switches, electrical connections, and sensors. Some of these procedures may clear up intermittent failures or flaky VCR operation. If the machine has been operated in a dirty environment with significant airborne contaminants, then consider performing cleaning in addition to what has already been described. Surface grime and debris on printed circuit boards (indicated by a very dirty cotton swab after wiping across a PCB) or visible corrosion indicate these additional steps may be warranted.

> CAUTION: Do not use cleaner while the VCR is plugged into AC power! Especially when using alcohol or other flammable cleaner, be certain that solution has completely evaporated and any fumes dissipated before plugging in the VCR.

## Cleaning Switches

Switch cleaning is not usually part of routine VCR maintenance unless the unit is very dirty or has problems.

Not all types of switches can be cleaned. Sealed pushbutton microswitches, normally for front panel controls, do not require any attention. Open leaf-type switches are most susceptible to dirt or corrosion; their contacts should be cleaned. Leaf or strap type switches are often found in the front loader mechanism and on the reel table. They typically indicate:

- cassette present in loader compartment
- cassette loader cam position (like the cassette-insert switch)
- cassette down on reel table (cassette-in switch)
- presence or absence of cassette record safety tab.

> CAUTION: Unplug the VCR before cleaning electrical connections! Be certain cleaning solution has completely evaporated and any fumes have dissipated before plugging the VCR back in.

To clean leaf switches, attach the extension tube to a can of spray cleaner and direct spray at the switch contacts. Manually operate the switch open and closed as you do this.

Tuner cleaner that has a lubricant is okay to use for cleaning switch contacts, but try to avoid overspray on nearby components. The lubricant will only attract dust.

Also effective for cleaning leaf switch contacts: saturate a small strip of white bond paper with cleaning solution. Pass the strip back and forth a few times between switch contacts while *gently* holding the contact straps together against the paper, as illustrated in Figure 6.25. Suitable paper for this is typewriter or personal computer printer paper; do not use newspaper, paper towels, or glossy paper. Be careful not to damage contact straps when you do this.

After you've cleaned leaf switch contacts, check that they have sufficient *overtravel* when transferred by whatever causes them to move in normal VCR operation. For example, the movable contact strap of a normally open cassette-in switch should touch the stationary contact strap and move it slightly when a cassette comes to rest on the reel table. This ensures a good electrical connection. If there is insufficient contact overtravel, carefully form one or both contact straps with a needle-nose pliers to obtain the desired overtravel when the switch transfers.

**Figure 6.25 Cleaning leaf switch contacts**

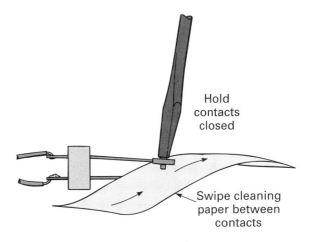

On VCRs with a mechanical mode switch or cam position sensor, spray contact cleaner or cleaning solution into any switch opening. TV tuner cleaner is fine to use here. Put a plastic extension tube on the spray can nozzle to direct the product into the switch interior. With some switches you can carefully pry up a plastic cover just a crack to get spray on the contacts inside. Operate the switch back and forth a few times by *hand cycling* the cam mechanism through *all* modes, then respray the switch with contact cleaner.

Normally you will not need to clean slide switches or potentiometers. Front panel slide switches often select recording speed, input source (Tuner/Line), Normal or Hi-Fi Audio, and Dolby On/Off. Potentiometers, called *pots* for short, are variable resistors frequently used for adjusting playback tracking, picture sharpness, and headphone volume. A pot may be a rotary or slide-type control.

If there is evidence that these controls are working poorly—when you have to fiddle with them or they are electrically noisy—spray tuner cleaner into any small opening on the switch or pot. Operate the switch or control back and forth several times and spray again.

### Cleaning Electrical Connections

Electrical connections do not normally require cleaning, except for very dirty machines or where there is evidence of corrosion or severe contamination. Hard-to-diagnose, intermittent problems also warrant cleaning VCR connectors. Spray electrical contact cleaner specially formulated for the task is the best choice, but you can also use cleaner/degreaser, or alcohol on a swab.

For external connectors, like A/V Line In and Line Out RCA jacks and F-terminals, swab or spray the connectors. Plug a cable in and out several times to clean the internal contact surfaces.

Inside the VCR, the most critical electrical connections are those with very low signal levels. These include connections at the RF tuner, Ch 3/4 modulator/antenna switcher, and at tape heads.

On many VCRs, the tuner and RF modulator are connected to each other by a short length of coax cable with an RCA plug at each end. Disconnect this cable and thoroughly clean both RCA jacks and cable plugs. Plug and unplug the cable several times after applying cleaning solution to clean internal contact surfaces.

Make certain the outer shell of any RCA plug firmly grips the outside of the jack into which it is plugged. Form the plug shell segments with long-nose pliers if there is low contact tension, as shown in Figure 6.26. The idea is to

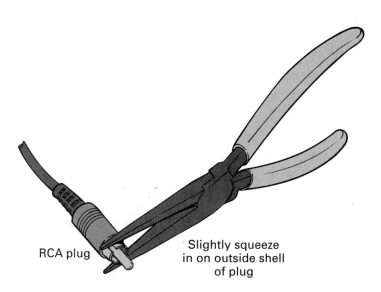

**Figure 6.26 Forming RCA plug shell for firm contact**

bend, or *"form"*, the shell contacts inward just enough so the plug grips the jack firmly.

Locate the multi-conductor connectors that plug into the video drum assembly, A/C head assembly, and full erase head assembly. There may be more than one at each location. Mark them with an indelible pen so you know their orientation when plugged in. Most connectors are designed so you cannot plug them in backward, but it doesn't hurt to take this additional precaution.

> CAUTION: Be certain the VCR is unplugged before cleaning electrical connections! Allow cleaning solution to evaporate and any fumes to dissipate before reconnecting VCR to AC power.

Carefully unplug connectors *one at a time* and spray both connector halves with contact cleaner. On many VCRs, it is possible to unplug two connectors that are near each other and accidentally swap them when replugging. Marking connectors, taking notes, making a sketch, and having only one connector apart at a time helps prevent mistakes.

Be extremely careful when pulling connectors apart that you *pull on the connector body only, never* on the wires themselves. These connectors and wires are delicate! If necessary, use a small flat-blade screwdriver to pry connector halves apart, or a long-nose pliers to pull them apart. Be gentle! Some connector shells have small plastic latches that lock the two halves together.

Rotary transformer connectors are usually on the upper side of the tape transport, at the rear of the video drum. There may be a separate metal shield covering the connector(s). Connections for the drum motor are sometimes on the undercarriage side of the drum. Figure 6.27 shows two rotary transformer

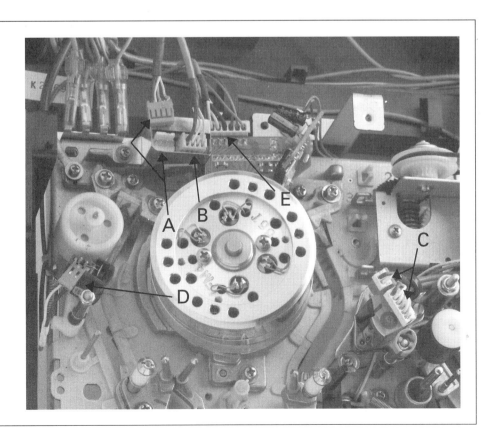

**Figure 6.27** Video drum and tape head connectors. A: Video heads (red connector body) B: Hi-Fi audio heads (yellow connector body) C: Audio/Control head D: Full erase head E: Drum motor drive and FG and PG signals

connectors on the video drum; one is for the two video heads, the other for the two Hi-Fi audio heads. These connectors are wired to the lower, stationary half of the rotary transformer, which in this machine has four windings, one for each rotary head.

A/C and full erase head connectors are normally at the top of each of these assemblies. Trace these cables, and those from the rotary transformer, to where they plug into printed circuit boards; clean connector contacts at the board ends as well.

Unplugging and plugging A/C head connectors *could* change the alignment of the assembly, so don't clean these connections unless you have indications of a problem. If you *do* work with A/C connectors, support the head assembly with one hand while *gently* removing and when re-inserting them. Use as little force as necessary so you don't twist the head mounting bracket or otherwise knock the heads out of alignment. Without safeguards, you could bend the head bracket so slightly that you wouldn't even realize it, but this would upset head alignment and degrade performance.

On many VCRs, rotary head signal connectors are underneath a metal shield plate at the drum, or at the circuit board, or both. Shield plates may be held on with one or more screws or may consist of a snap-fit metal cover. In some cases, a snap-fit cover may even be soldered in place. If so, do not unsolder the cover unless you are fairly certain there is a problem with the connector underneath.

With connector halves separated and cleaned, you should see bright, shiny contact pins. If pins look dull or corroded, further cleaning is necessary. Plug and unplug the connector several times, with applications of cleaning solution. Clean pins with solution on a swab. For stubborn cases, wipe pin surfaces with a toothpick or pencil eraser and again clean with solution. Severe contact corrosion may require light wiping with fine sandpaper, but this should be done *only as a last resort.*

You'll find many, many multi-conductor connectors on circuit boards throughout the VCR. It is not necessary to clean all these unless there is evidence of corrosion. For bizarre and intermittent problems, cleaning and reseating all connectors *may* restore stable operation.

Please be very gentle and careful when unplugging and reseating multipin connectors. Circuit boards are easily cracked and wires easily broken with carelessness and excessive force. Support the top of a circuit board with your hand near a connector when unplugging. Support the board underneath a connector when replugging. This helps prevent the board from flexing, which might create hairline cracks in the printed wiring patterns.

Again, connector contact cleaning is normally not necessary, except when corrosion or intermittent problems are present.

## Cleaning Sensors

Here we are talking about optical sensors. Cleaning switch-type sensors has already been covered, and magnetic sensors do not normally require any cleaning.

Ordinarily you will not clean optical sensors during routine VCR maintenance. However, VCR malfunctions can be caused by dirty or obstructed optical sensors. For example, dirty reflective surfaces on the underside of a take-up reel spindle could cause weak signals to be sent to system control when the take-up reel turns. This might result in system control unloading tape and shutting Off the VCR when a tape is playing, if it "thinks" the take-up reel has stopped turning when it should be revolving to pull in tape.

Lightly wipe the surfaces of the following photo sensor components with a cotton swab moistened with cleaning solution:

- End-of-tape light post
  — Sticks up into center of cassette
    IR LED on all but early VCRs, including many top loaders, where light source is small incandescent or "grain-of-wheat" lamp
- Supply tape-end phototransistor
- Take-up tape-end phototransistor
    These may be mounted in the front loader assembly or on the reel table.
- Take-up and supply reel sensors
  — IR LED
  — Phototransistor
  — Reflective surfaces on spindle bottoms
    Sensor often consists of an IR LED and phototransistor mounted on a very small circuit board on the underside of the transport, beneath the take-up spindle. Alternating reflective and non-reflective segments on the bottom of the spindle cause the phototransistor to turn On and Off as the spindle rotates.

    Depending on the model, it may be easier to remove the small circuit board *or* the spindle itself to clean sensor components. In either case, swab the LED, phototransistor, and reflective segments on the underside of the spindle, as shown in Figure 6.28. In this model, it was easier to remove the spindle rather than the small circuit board.

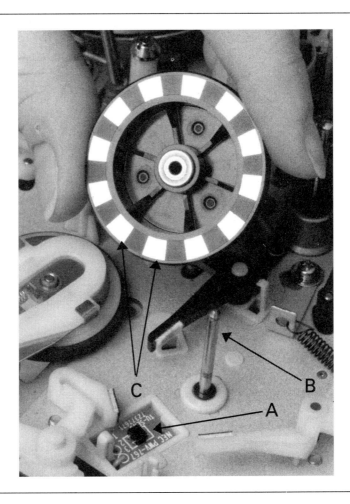

**Figure 6.28 Reel sensor components.
A: IR LED/phototransistor device B: Take-up spindle shaft
C: Reflective segments on bottom of take-up spindle**

You might get fooled! Some VCRs have a magnetic reel sensor, using a Hall effect IC. This type sensor is difficult to distinguish from an optical reel sensor before you take the spindle or circuit board off. Once these components are separated, you won't see any reflective segments on the spindle bottom; instead it will have a magnetic ring around the bottom. Instead of an IR LED and phototransistor on the small circuit board, there will be a tiny integrated circuit that looks similar to a transistor. Cleaning this type of sensor will not restore operation if reel motion is not being sensed.

Many VCRs have only a take-up reel sensor.
- Miscellaneous photo sensors

Some VCRs have IR LED/phototransistor detectors in place of a mechanical mode switch or instead of front loader cam switches. Besides cleaning the LEDs and phototransistors, clean any debris that may be blocking an aperture hole in an encoder disk or cam wheel associated with these sensors.

## SUMMARY

You may be surprised to learn that the majority of VCR malfunctions can be repaired by following the procedures in this chapter. Cleaning the entire tape path where videotape contacts tape guides and tape heads is often all that is needed to make a VCR perform like new.

All product label directions and warnings should be read and followed when using cleaners, degreasers, and other chemicals. Adequate fresh air ventilation must be provided for safe use. Avoid breathing fumes and skin contact when using these products.

A few hand tools and appropriate supplies are needed to perform most VCR maintenance and common repairs. It is much better to buy a few quality tools than many inexpensive ones, which can actually damage hardware and make the job more difficult. This is especially true for Phillips head screwdrivers; bargain drivers usually do not have hardened tips, and they may have poorly formed tips. Using such a screwdriver can gouge out screwheads in the VCR.

Rotary heads are the most critical components that require regular cleaning, using a chamois-tipped swab saturated with head-cleaning solution. Hold the swab flat against the upper portion of the video drum while slowly rotating the upper cylinder CCW by hand. Avoid any up-and-down movement of the cleaning stick across rotary heads. *Never use cotton swabs or cloths on rotary heads;* cotton fibers can snag the heads, pulling them out of alignment or breaking them.

Foam-tipped or cotton swabs moistened with head cleaner or other non-residue solution are adequate for cleaning all other tape path components. Lint free cloths may also be used. If alcohol is used as a cleaning solution, it should be *at minimum* 90 percent pure. Isopropyl alcohol at nearly 100 percent purity is available from electronic suppliers and is an adequate all-purpose VCR cleaner. Denatured alcohol near 100 percent pure is also all right to use. Overuse of alcohol may attack the protective coating on printed circuit boards. It is best to test a cleaner by applying it to a small area before using it, to make sure it won't attack the part.

Though not essential, demagnetizing the tape path can improve picture detail and high frequency audio response. A demagnetizer should be turned On and Off only when it is several feet away from the VCR. Never touch any tape head or other part with the demagnetizer tip, but wave the tip around in small circles *near* the part to be demagnetized. Be especially careful not to touch rotary head tips. Do not use a strong demagnetizer not designed for video heads too close to rotary heads.

Rubber parts are frequently the cause of tape transport problems. Belts age and stretch, idler wheel tires develop a surface glaze, and pinch rollers harden, swell or otherwise become deformed. Replacement is the best long-term solution, and should be done for commercial repairs when rubber shows signs of deterioration. Wiping rubber parts with rubber cleaner/revitalizing solution can often restore operation. Unless a solution specifically states that it can be used on a pinch roller, avoid applying it to this part. Some rubber cleaners and rejuvenators make the rubber surface sticky, which could cause videotape to adhere to the rotating roller.

One of the most frequently replaced VCR parts is the pinch roller. This critical part can cause tape damage and uneven tape motion if it has hardened with age or become deformed.

A dirty capstan shaft is a common cause for mangled tapes and tape motion problems. Difficult-to-remove tape oxide deposits on the capstan shaft can be removed with a wooden chopstick or toothpick and cleaning solution. Never use a metal implement to clean the capstan or any other part in the tape path.

Tape transport alignment and adjustments usually do *not* need to be changed unless parts that videotape actually touches are replaced. An exception is the pinch roller, which can be replaced without any adjustments being made. Screws that have a dab of lock paint on them should not be disturbed without good reason!

Tape transport lubrication is applied *very sparingly!* One drop of lightweight oil placed on bronze bearings and pivot points is generally sufficient. A needle-tipped applicator is ideal for getting just the right amount to just the right place.

Dirty, hardened grease should be removed with swabs or lint-free cloths. *Nonpetroleum-based*, lightweight grease is then applied in a *thin film* on parts that slide or rub against each other, such as the two tape load guide tracks and shuttle assemblies. Cams, cam followers, and gears should be cleaned of old grease before applying a small amount of new grease. There should be no globs of excess grease. Nylon gears do not require lubrication of any sort. If there is no evidence of original grease, do not grease Teflon or Nylon parts.

Before removing gears, gear segments, cam wheels, and interlocking parts, make a sketch. Mark a line with indelible pen where gears mesh with each other and where parts interconnect. Look for timing marks on moving parts, and make note of where they are in relationship to each other. This is all helpful in reassembling parts correctly.

Many tape transport components, such as gears, spindles, and pivot plates, are secured with several different types of fasteners. These include C-clips, E-clips, compression rings, O-rings, and slit washers. Eye protection should be worn when removing or replacing most of these retainers, as they can easily "fly." Cupping a cloth around them during removal or installation helps prevent loss. Electronic supply houses carry special tools designed to remove and install different types of retainers, but these are not absolutely necessary.

Cleaning electrical connections can sometimes resolve VCR problems, especially if there is evidence of corrosion or heavy contamination. Oxidation and corrosion of connectors carrying *very low level signals* will normally cause problems before contamination of other connectors. Low signal levels are at magnetic tape head, tuner, and RF modulator/antenna switcher connections. Clean these first. If they appear to be contaminated, then other connectors in the VCR should be cleaned also. Intermittent problems can be caused by dirty or corroded electrical connectors, by poor switch contacts, and by dirt film (such as from tobacco smoke) on optical sensor components.

## SELF-CHECK QUESTIONS

1. What is the smallest size Phillips-head screwdriver that you should have for VCR repair activities? How can you tell the size of a screwdriver when you select one off the rack to purchase?
2. When selecting diagonal wire cutters at a hardware store, what check should you perform to ensure that it is a quality tool?
3. Describe how you might use a metal scribe when removing an adjustable bracket on a VCR transport baseplate.
4. Explain why it might be smart *NOT* to apply rubber cleaner or rejuvenating liquid to a VCR's pinch roller unless product labeling specifically recommends its use on pinch rollers.
5. State two precautions that should be taken when using liquid or spray cleaner, degreaser, and alcohol on a VCR.
6. Describe an acceptable procedure to dislodge stubborn caked on tape oxide on a tape guide.
7. Explain why you should not unplug a 120V tape head demagnetizer while it is in the tape transport area.
8. Describe the correct procedure for cleaning rotary heads.
9. Why is it important to correctly orient the head assembly with the rotary transformer terminals when replacing a two-head rotary head assembly on a video drum?
10. State the particular precautions you should take unsoldering old head wires and soldering new head wires to rotary transformer terminals when replacing a rotary head assembly.
11. What is one cause for *premature* wear of the A/C head? What is an indication of A/C head wear?
12. What adjustments are required when replacing the A/C head assembly?
13. Describe one suggested procedure to perform when removing the A/C head.
14. Describe the complete bearing lubrication procedure for an *indirectly driven* capstan.
15. List four conditions that would cause you to replace a pinch roller if any one condition was present.
16. Describe the procedure for removing a pinch roller in a typical VHS VCR.
17. Describe how you would determine if a rubber belt in a VCR, such as between a separate capstan motor and capstan shaft flywheel pulley, should be replaced.
18. What would you do to the supply and take-up spindles on the tape transport during routine maintenance?
19. What two precautions should you take when removing C-clips, E-clips, and compression rings?
20. Describe the routine maintenance procedure for a mechanical type mode switch.

# CHAPTER 7
# HOW TO ALIGN TAPE PATH AND MAKE ADJUSTMENTS

For a VCR to reproduce quality audio and video, videotape must move through the transport positioned correctly in relationship to the tape heads. This ensures that recorded tracks on tape line up with the tape head gaps. Also, tape tension must be sufficient to keep videotape in intimate contact with magnetic tape heads.

The position of videotape with respect to tape heads is called *alignment*. When tape passes over tape heads during playback so they read the best signal possible, we say the heads and tape are in alignment. If tape is not in correct alignment, the picture can jitter, lose color, have horizontal noise lines running through it, or be totally absent, and audio can be weak or sound muddy.

Several tape guides, which you learned about in Chapter 3, are responsible for positioning tape precisely around the drum and at the right height throughout the tape path. Four A/C head adjustments align the audio, control, and audio erase head gaps to the videotape.

Tape holdback tension during play and record is determined by the amount of drag from the supply spindle. If tape tension is too loose, tape won't have adequate contact with rotary and stationary tape heads. This can cause the picture to be alternately stable and unstable, perhaps at regular intervals from a few seconds to many seconds. A picture with lines and dashes running through it, or one that bends or is indistinct, can also be caused by low tape tension. Audio may not be clear; it might sound like it is wavering in and out, as if you were moving a volume control up and down. Tape tension that is too high can cause unstable pictures and audio that warbles, and will accelerate wear of all tape heads.

In an ideal world, all tape transport alignment and adjustments would be made to the manufacturer's specifications as described in the service manual for a particular model VCR. These procedures usually require using several mechanical jigs and gauges, plus advanced electronic test equipment, along with special alignment tapes. Although this may be the perfect way of repairing a VCR, the high cost of all the tools required will make these ideal adjustments prohibitively expensive in many cases, for both repair shop and consumer.

For example, a tape tension gauge costs as much as $400. An oscilloscope for observing electronic waveforms could cost up to $1,000. The good news is that with some knowledge *and experience,* you can effectively accomplish most VCR tape path alignments and adjustments without high-priced equipment.

The intent of this chapter is to provide you with the knowledge and a few techniques to perform the majority of VCR mechanical transport adjustments without needing specialized tools and equipment. After studying this chapter and getting some hands-on practice, you can easily move up to using VCR test instruments at a later time, or when they are affordable. VCR manufacturers' service literature and instructions that come with various gauges and testers will guide you in their usage.

*174* ■ *Practical VCR Repair*

## CHAPTER OBJECTIVES

**Upon completing this chapter, you should be able to:**
1. Explain how to determine if supply reel back tension requires adjustment, and describe how to make the adjustment.
2. Describe how to adjust the height of supply-side guide pole P-1 and take-up side guide pole P-4.
3. Describe how to align videotape with the video drum by adjusting the height of drum entrance roller guide P-2 and exit guide P-3.
4. Describe how to align the A/C head assembly.
5. Describe how to check for and perform pinch roller alignment with the capstan shaft.
6. Describe how to set supply and take-up spindle height.
7. Describe how to check for and correct take-up spindle torque that is too high or too low.
8. Describe how to align a mode switch.

Before beginning this chapter, we suggest that you take just a few minutes to review Chapter 3. Skimming through the material and looking at the figures once more should make the following descriptions easier to understand.

### Before You Adjust

It should be stressed at the outset that the tape path cleaning procedures described in Chapter 6 should be thoroughly performed before making *any* tape path adjustment. Also, damaged or worn parts along the tape path should be replaced prior to adjusting the transport. There is no sense in attempting to correct tape mistracking caused by a deformed or hardened pinch roller by adjusting the height of one or more tape guides. Any success will most likely not last. The pinch roller should be replaced first.

Carefully examine tape motion through the entire tape path, as described in Chapter 5, before adjusting anything. Part replacement and adjustments

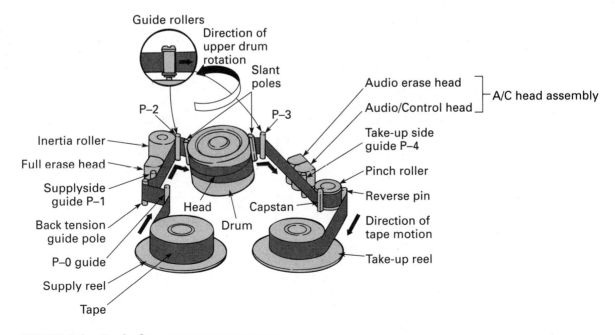

**FIGURE 7.1   Typical tape transport system**

should first be made so that videotape moves smoothly through the tape path. Once tape motion is correct, then minor additional adjustments may be needed to produce quality audio and video on the TV or monitor during playback.

Next we look at common tape transport adjustments, working from the left side of the reel table to the right side. Figure 7.1 illustrates the major tape path components addressed in this chapter.

## TAPE TENSION

Recall from Chapter 3 that supply reel back tension is provided by a felt-lined band around a lower flange, or rim, of the supply spindle. Tape travels over a movable back tension guide pole after leaving the cassette. When tape is too slack, the guide pole moves to the left, tightening the spindle band and thus taking up the slack. Conversely, the guide pole moves to the right, loosening the band, if tape tension is too great. Figure 7.2 shows the related parts in a representative VCR. For this photo, some of the tape and the entire cassette have been removed, but guides P-2 and P-3 are in the tape-loaded position at each side of the drum. The front loader has also been removed.

The particular VCR model in Figure 7.2 has a stationary tape guide, or pin, in the tape path just before the tension guide pole; not all models have this additional guide. This pin, post, or guide pole may be referred to as the *P-0 guide*. In some models, a portion of the tape tension regulator (for example, the pivot plate spring) is on the underside of the transport.

First, check the operation of the tension regulator, without loading a cassette. Block light to the two end-of-tape sensors and hold down the cassette-in switch; you could use a videocassette test jig. Then press Play. The drum should spin and the two movable P-guides should move to the tape-loaded position on each side of the video drum. On VCRs with a supply reel motion sensor you will probably have to manually rotate the supply spindle to prevent syscon from unloading tape.

> NOTE: Some VCR manufacturers' service manuals have information on operating in Play mode without a cassette loaded. This "dummies up" signals to syscon. For example, one or more jumper wires between circuit board test points and ground may deactivate tape-end sensors, and allow Play without reel motion sensor pulses and without the cassette-in switch being transferred.
>
> You will understand and be able to use jumper wires and perform some electronic tests after completing this book, especially Chapters 8 and 10.

Check how freely the back tension guide pole moves to the right and left. It should be spring-loaded nearly to the left edge of the tape transport baseplate, well left of center of supply-side tape guide P-1. You should be able to move the back tension guide pole to the right an inch or more. The guide pole pivot bracket should not bind on anything. Also check that the back tension pole is straight up and down, perpendicular to the baseplate, not bent at an angle. A bent pivot plate bracket could cause the plate to bind, the pole to be crooked, or both.

Next, play a T-120 scratch cassette to check for proper tape tension. Recall that a *scratch cassette* has good-quality, undamaged tape, but is not a premium-quality commercially recorded movie or music video. You will carefully observe two items:

1. The position of the movable back tension guide pole in relationship to the stationary *supply-side tape guide,* P1.
2. Tape contact with the full erase head on each side of the vertical center of the head face.

In this particular model (Figure 7.2), the supply-side tape guide P-1 is a nonrotating post just before the full erase head. On some models, the supply-side tape guide is considerably larger and rotates, looking more like a small thread spool, and is still referred to as the *P-1 guide* on most VCR models.

**FIGURE 7.2 Representative tape tension parts and nearby components**

This is what to look for while playing a T-120 cassette:

- Back tension guide pole should be just barely to the left of center of the supply-side tape guide, as you look in from the front of the VCR.
- Tape should firmly contact the supply-side tape guide, wrapping partway around it, as illustrated in Figure 7.2.
- Tape should wrap symmetrically around the face of the full erase head. That is, tape should contact as much of the head face to the left of vertical center as on the right.

If the back tension guide pole is far to the *left* of center of supply-side guide P-1, tape tension is most likely too loose. The back tension guide pole is probably at, or nearly at, the position it assumes when Play is simulated, or "dummied up," without tape loaded. Tape will also be wrapped further around the left side of the erase head face than on the right side.

In contrast, if the back tension guide pole is to the *right* of center of the supply-side tape guide, this means that tape tension is probably too tight. Tape will likely not cover the left side of the erase head face as much as on the right side.

Assuming that the supply spindle reel flange has been cleaned and that the tension band felt is in good shape, as described in Chapter 6, incorrect tape tension may be caused by a weak spring on the pivot bracket. In some designs you can reposition the spring to compensate. Tape tension on many models is adjusted by repositioning one end of the spring. On other models, the tension band itself is adjusted where it mounts to the baseplate. As shown in Figure 7.2, the single mounting screw is loosened, the band made tighter or looser around the spindle flange, and then the screw is retightened.

Before loosening the tension band mounting screw, use a scribe or indelible pen around the tension band bracket to mark its present location on the baseplate. Most of the time you will need to move the band bracket only *a very small amount* to achieve correct tape tension. After making any adjustment, check tape tension again while playing a T-120.

On most VCRs, the back tension guide pole should be just left of center of supply-side guide P-1, by around 1/64 to 3/16 of an inch, as you look straight back into the transport. Tape should contact the full erase head an equal amount on each side of vertical center.

On many VCRs, the above procedures are adequate for checking and adjusting tape tension. You can also make minor tape tension adjustments by observing playback on a TV or monitor. Play a high-quality, commercially recorded cassette. Manually move the back tension guide pole a *small* distance left *and* right of its normal rest position, for example 1/32-inch, with your fingertip while playing a tape. Hold it in each position for several seconds. The picture should stabilize, with no jitter or lines.

If very minor movement of the back tension guide pole either left or right causes picture problems to remain after several seconds, this may mean that tape tension is marginal, and should be adjusted. Depending on the particular model, you may have to reposition one end of a spring attached to the back tension bracket, or you might have to adjust the mounting position of the tension band. Most VCRs have some way of changing spring force on the back tension guide pole pivot plate, or bracket. An example is shown in Figure 7.3. Tape is removed, but movable guides P-2 and P-3 are in the tape-loaded position.

If *both* spring tension and tension band mounting are adjustable, make adjustments to the spring. This is usually less critical than changing the tension band mounting position. As a VCR ages, you will usually need to slightly increase spring tension to compensate for worn tension band felt and a weakened spring.

Be careful! A very small change in spring tension or tension band mounting position can cause a large change in tape tension. Make adjustments in small steps. Observe the location of the back tension guide pole in relationship to the supply-side tape guide, and watch the playback picture after each incremental adjustment. Do not adjust tape tension too high, or it will cause accelerated tape head wear.

Tape tension that is too high may appear to clear up a problem that is really the result of misalignment somewhere else in the tape path. Listen for any sound of straining from the video drum motor, which can normally be heard if tape tension is excessively high. It usually sounds something like a small buzz saw coming from the drum, not from the loudspeaker. Watch especially for tape edges climbing over the shoulders of tape guides with tape tension that is too high. It is better to have tape tension a little on the low side than too great, as long as a good quality picture is reproduced.

**FIGURE 7.3** Back tension bracket spring adjustment

After you've made any tape tension adjustment, play a quality, commercially recorded T-120 cassette for several minutes, both at the beginning and near the end of the tape. Tape motion should be smooth, with no visible tape undulation at any point along the tape path. Observe that the TV picture is stable, with no noise lines, spots, or dashes on the screen, and that there is no wow or flutter in the linear audio.

Precise tape tension adjustments can be achieved using a tape tension gauge called a Tentelometer, available from Tentel Corporation. A back tension cassette gauge may also be used for this purpose. Adjust tension to the manufacturer's specifications, usually somewhere between 25 and 30 gram centimeters (gcm).

For VCRs having a direct drive (DD) supply spindle, without a mechanical tension regulator band, consult the manufacturer's service manual for checking and adjusting tape tension. It will normally need an electronic adjustment, such as on a potentiometer that controls the amount of current to the DD motor.

## SUPPLY-SIDE GUIDE POLE

Refer again to Figure 7.1. Supply-side *guide pole P-1* is between the back tension guide pole and full erase head. Don't confuse this with the P-2 roller guide on the supply-side of the video drum. The supply-side tape guide may be a small-diameter, nonrotating pole, or a roller guide having a larger diameter.

VCRs with a large-diameter supply-side roller guide just before the full erase head normally do not have a separate inertia, or impedance, roller after the head. Refer back to Figure 7.3. One roller guide serves both functions, positioning tape at the right height and smoothing out minor fluctuations in tape tension.

Height of the supply-side tape guide is adjustable, but should not need adjustment unless it is replaced. Its height is set at the factory. Precise height adjustment requires using a manufacturer's guide pole adjustment plate. The most common arrangement is for the guide to be mounted over a vertical shaft with a spring beneath it and a *nylon lock nut* on top, as illustrated in Figure 7.4.

Rotating the lock nut clockwise lowers the guide, whereas counterclockwise rotation raises it. A nylon lock nut is similar to a conventional metal nut, but has a captive nylon insert above the metal threads on the nut. The nylon grips the screw threads tightly, preventing the nut from working loose or turning after it has been adjusted.

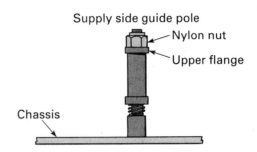

**FIGURE 7.4 Supply-side guide pole P-1**

To check supply-side guide pole height, play a good-quality T-120 scratch cassette and carefully observe the tape as it passes the tape guide. Look at the top and bottom edges of the tape where it contacts the guide. A bright light and dental inspection mirror are helpful. Tape should not crinkle, bunch up, or ride over the edges of the guide shoulders. If necessary, *very slowly* turn the adjustment nut on top of the guide CW to lower the guide, or CCW to raise the guide, until you observe correct tape motion past the guide, as illustrated in Figure 7.5.

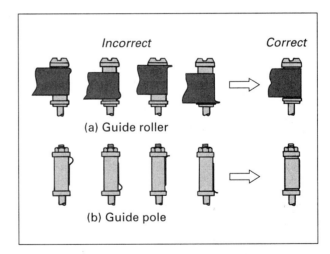

**FIGURE 7.5 Guide roller and guide pole adjustment** *(Courtesy of JC Penney)*

> NOTE: If operating with the cassette loader removed, ensure that the cassette is held down on the four transport positioning studs. It is necessary to place a weight on top of the cassette. Check that there is no give or springiness when you press down at each corner of the cassette.

Be sure to turn the adjustment nut just a little at a time! Then watch the tape for about 20 seconds or so to ensure that it is riding correctly over the guide pole. Just 1/8 of a full turn can be enough to make a significant change in how videotape passes the guide.

On some VCRs, supply side guide P-1 must be taken apart to replace the full erase head. This is because the guide pole is also used as a pivot for the erase head bracket. Take careful note of the position of the upper and lower flanges or collars, center sleeve, spring, and the nylon lock nut that make up the guide pole, so you can reassemble it correctly. Count the number of full and partial turns CCW required to remove the nut. When re-assembling, turn the nut CW onto the threaded shaft the same amount. This should get you in the ballpark for correct guide pole height. Then fine-tune the adjustment while playing a T-120 cassette, as described earlier.

There is no adjustment for the full erase head and inertia roller. Check that the erase head bracket pivots freely and that the inertia roller turns freely.

## ROLLER GUIDE POSTS

Refer again to Figure 7.1. Height of both drum entry roller guide P-2 and drum exit roller guide P-3 must be precise to properly position tape around the video drum. These guides will not normally require adjustment unless the lower drum assembly or shuttle assembly components have been replaced.

Check that the sleeves of both roller guides turn freely. If one or the other does not, replace both assemblies. The rationale for replacing both is that if one is worn sufficiently to require replacement, then the other one is probably equally worn. Replace only one at a time; adjust it as described in this section, and then replace the other one. Adjustment can be much more difficult if both roller guides are severely out of adjustment, which would be the case if you replaced both at the same time. If a roller doesn't turn freely, you might also disassemble, clean, and reassemble it. *Do not lubricate!*

Before adjusting roller guides P-2 and P-3, load tape or dummy up the tape load operation with a videocassette test jig and observe the two guides. They should both be *fully seated* into their respective V-stops. You should not detect *any* movement of either guide as you gently press it downward from the top or gently push it forward into its V-stop. If the guide moves, there is most likely a bind somewhere in the loading mechanism, or the shuttle assembly itself is hitting some part of the V-stop block. Correct whatever it is that is preventing the guide from fully seating in its V-mount before making any guide height adjustment.

Guide P-2 and P-3 height adjustment is made in two steps: first by critically observing videotape as it passes around the drum and over the guides, and then by making fine adjustments while looking at a picture reproduced on a TV or video monitor. On most VCRs, you will need a special notched adjustment driver to change the height of the video drum roller guides. Figure 7.6 shows what one of these adjustment drivers looks like, along with the top of a typical roller guide. Many units also require a similar tool to adjust the tapered nut for lateral positioning of the A/C head assembly.

Adjustment drivers are available from VCR manufacturers and electronic supply houses, especially those carrying VCR replacement parts. At nearly $10. apiece, these drivers cost more than flat-blade screwdrivers, but shop around—

> NOTE: Other items in the tape path, especially supply side guide P-1 and take-up side guide P-4, should be checked for proper adjustment before changing the height of video drum entry and exit roller guides P-2 and P-3.

**FIGURE 7.6** Split-slot adjustment nut and driver. Drivers for split-slot adjustment screws and nuts are also called RCA-type positioning tools or RCA adjustment tools.

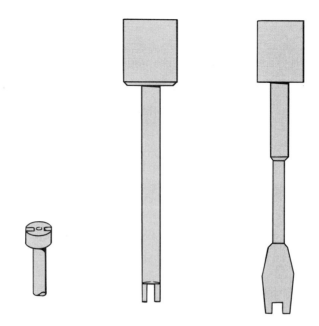

you may do better. Some suppliers have kits of special VCR tools, including several sizes of notched RCA-type drivers, that are attractively priced in comparison with the cost of individual tools.

You could take a conventional flat-blade screwdriver, clamp it blade up in a vise, and use a small metal file to make your own RCA-type adjustment tool. For *this* purpose it is better to buy a cheap screwdriver *without* a hardened tip. It will be easier to file, and adjustment screws or nuts on which this tool is used don't require much force.

On some VCRs, roller guide height is adjusted with a hex head or Allen screw at the top of the roller, rather than a split-slot nut, as illustrated in Figure 7.6. Common hex head sizes are 0.89mm, 1.27mm, and 1.5mm.

> CAUTION: Before adjusting either roller guide, a small setscrew at the base of the roller must first be loosened, as illustrated in Figure 7.7.

To check and adjust drum entrance and exit guides, load and play a good-quality scratch cassette. Use a bright light and inspection mirror to observe the bottom edge of the tape as it rides around the ledge, or rabbet, on the lower, stationary portion of the video drum. As described in Chapter 5, the bottom tape edge should just touch the ledge on the drum, without crinkling, curling over the edge, or riding above the ledge. At the same time, tape should run smoothly over both roller guides without edge crinkle, bunching up, or riding over either the top or bottom shoulder of the guides.

If adjustment is needed, loosen the small setscrew at the lower rear of the roller guide with the correct size Allen wrench. Many VCRs require a 0.89mm wrench. This is quite small! Be certain the wrench is in good shape, and place it fully into the screw head. Then turn the tool gently. An Allen wrench of this size is easy to break if you are not careful. Throw away any that have rounded edges, as they will gouge out small setscrews and make them nearly impossible to remove.

Rotate the setscrews counterclockwise approximately one turn, or less. On some VCRs you can get at the setscrews through access holes in the V-mounts while tape is loaded. In other designs, tape must be unloaded before you can loosen the setscrews in the base of the roller guides. On some models,

**FIGURE 7.7 Roller guide assembly. Most roller guides have a small hex head setscrew on the back side of the base that must be loosened before guide height is adjusted.**

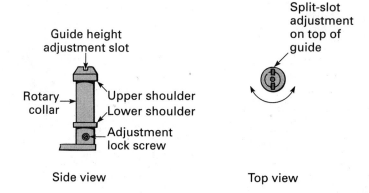

the roller guide and slant pole are on a common plate that is mounted to the shuttle assembly with a single screw. *Do not loosen this screw!* It sets the position of the slant pole, and will usually have lock paint around the screwhead.

> Do not loosen any screw associated with either V-mount or V-stop!

After you loosen the setscrew on guide P-2 or P-3, you may notice that tape turns the entire guide during play, rather than just the central roller sleeve. If so, snug up *ever so slightly* on the setscrew, just enough so you can still adjust guide height, but moving tape won't turn the whole guide.

Make roller guide height adjustments *in very small steps*, turning the adjustment screw on top of each guide no more than 1/8 of a turn at a time. Look at how the lower tape edge rides along the drum ledge and across the drum entrance and exit guides.

Unless you've replaced a broken roller guide or the video drum assembly, you will normally not need to turn either adjustment screw very far at all. If you do, there's probably something else wrong in the tape path or elsewhere in the transport. It is a good idea to make note of how far you turn each guide adjustment screw, down to the smallest part of a full turn as possible. Or you might note how many degrees of clockwise or counterclockwise rotation you move an adjustment screw. That way you can always get it back to where it was when you started.

You will sometimes have to go back and forth between the two roller guides several times, adjusting each *a very slight amount*. Adjustment of one roller guide can affect adjustment of the other. Observe tape motion for at least half a minute after making any adjustment, to make sure tape is positioned properly. Use a high-intensity light and mirror to inspect the bottom edge of the tape. It should be riding correctly all along the drum ledge or rabbet. In addition, tape must be properly positioned between the upper and lower shoulders of each roller guide.

Be patient when making these two adjustments; they are very critical to quality video reproduction. It normally takes a while to get everything right.

Once you are satisfied that tape is aligned correctly with the drum and roller guides, play a high-quality, commercially recorded cassette, observing the reproduced picture on a video monitor or TV. Be sure the manual tracking control is at its center (detent or notched) position.

> NOTE: On VCRs that have two tracking buttons with an arrow on each rather than a rotary tracking control, push both buttons simultaneously for the center of the tracking control range.

If there are horizontal lines on the screen, try turning one or the other, or both, roller guide adjustments *an extremely small amount* to rid the picture of lines or other video noise. This should not require more than 1/4 turn of either. In most cases, minor adjustment of entrance guide P-2 height will eliminate lines at the top portion of the picture; adjusting exit guide P-3 takes care of lines or dashes toward the bottom of the picture. Again, make sure that tape is positioned properly along the drum ledge and past the two guides.

It may be that the control head is slightly out of adjustment and is causing tracking problems, with horizontal lines at the top or bottom of the screen. If you cannot adjust these out with roller guide adjustments, try moving the tracking control to clear up the picture. Move the tracking control back and forth throughout its range; note where the picture is clearest. There will most likely be a range of control movement over which the picture is clear, or is the clearest it can get at this point. For now, set the control at the center of the range that produces the best picture. Then go back and adjust the height of the roller guides. Control head alignment is covered later in this chapter.

When all looks well, gently lock down the roller guide setscrews. *Do not overtighten!* This could strip out a setscrew or even change the adjustment you spent so much time achieving. Check again that tape motion is correct and that the picture is stable and noise-free. If locking down a setscrew upsets the alignment, slightly loosen the setscrew *just enough so you can turn the roller guide height adjustment.* Tweak the height adjustment as needed and reset the setscrew. Only very small height adjustment should be required at this stage.

At some time you may come across a roller guide where the bronze, lower portion of the post containing the setscrew has broken loose from the metal shuttle mounting plate. The two parts are normally held together by press fit (see Figure 7.8). You can tell if the press fit is loose by pulling up or pushing down slightly on the roller post, or trying to turn it. It should not move up and down nor turn in its mounting plate. If the guide post *does* move up and down or turn in its base plate, the assembly should be replaced.

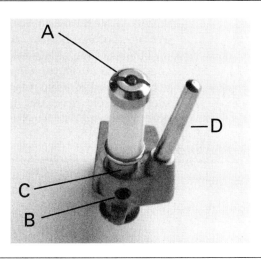

**FIGURE 7.8 Roller guide and slant pole assembly. A: Roller guide height adjustment B: Setscrew C: Bronze portion D: Slant pole**

You *may* be able to take the assembly out and repair it by running a bead of epoxy cement around the base of the bronze post where it fits into the base plate. Usually a C-clip or a small screw, or both, in the undercarriage mounts the shuttle assembly containing the roller guide and slant pole. On some models, you may have to remove a plastic rail that runs along both sides of the track in which the shuttle moves back and forth.

Be sure to clean the area where the two metal pieces come together with alcohol before applying cement. Make certain you do not get cement on the threads of the small setscrew.

If you don't get the post cemented in just the right position, you may never be able to obtain a correct height adjustment for the roller guide. Here is what you can do:

1. Loosen the setscrew and turn the guide height adjustment to its midpoint, halfway between all the way down and all the way up.
2. Raise the post in its mounting plate so that the guide is clearly higher than it should be.
3. Apply a *small* bead of epoxy cement around the bronze base of the guide pole, where it fits into the base plate.
4. Reinstall the roller guide assembly in the VCR before epoxy cement sets up. Be careful not to get epoxy cement anywhere else.
5. Play a commercially recorded tape, and lightly tap the post down just a little at a time until a *reasonable* picture appears. It may not be perfect, but don't go any further.
6. Unplug the VCR without unloading tape, so you don't disturb the position of the post. Wait a suitable time for the epoxy to completely cure (24 hours recommended).
7. Proceed to adjust the guide height as previously described.

## V-MOUNTS

Do not even *think about* touching the V-mount screws! There should be no reason to change V-stoppers for the life of any VCR. V-blocks are positioned with a special manufacturer's adjustment fixture. However, on some models the V-mounts are part of the lower video drum assembly, and will be replaced if the larger assembly is replaced. V-stop mounting screws will have a dab of lock paint on them.

In general, be sure you know what you are doing before loosening any screw in a VCR that has lock paint on it. Some screws mount components to the transport; others perform an adjustment.

Before loosening a paint-locked mounting screw, scribe around the part or take precise measurements to locate the exact position of the part, or do both. For paint-locked adjustment screws, jot down the exact amount you turn them, and in which direction. These steps will help you get a part closely repositioned or an adjustment near its original setting.

## A/C HEAD

After passing around the drum and exit guide P-3, videotape next passes over the Audio/Control head assembly, which also contains the audio erase head. (Some manufacturer's literature refers to the A/C head as the ACE head.) The A/C head has three, four, or five separate magnetic tape head gaps, depending on whether the VCR has stereo or mono linear audio, and, on stereo units, whether one linear audio track can be recorded while retaining the other track.

The same audio and control heads are used for record and playback. Figure 7.9 shows the face of a stereo A/C head and the videotape tracks. This head assembly has separate linear audio erase heads, which enables one audio channel to be recorded while saving the other audio channel and video. Some VCRs with stereo linear audio have only a single erase head gap that covers

Chapter 7 How to Align Tape Path and Make Adjustments ■ 185

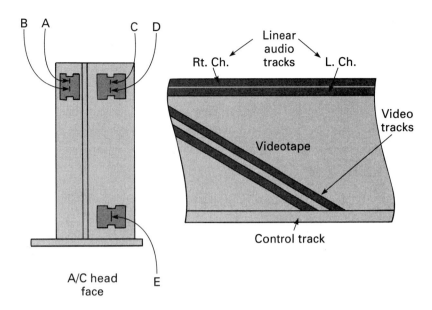

**FIGURE 7.9** Stereo A/C head assembly and tape tracks. A/C magnetic tape head gaps: A: Ch 1/R Ch linear audio erase B: Ch 2/L Ch linear audio erase C: Ch 1/R Ch linear audio Rec/PB D: Ch 2/L Ch linear audio Rec/PB E: Control track logic (CTL) Rec/PB

both audio tracks. On those VCRs, rerecording one linear audio channel while retaining the other is not possible.

As described in Chapter 3, there are four A/C head assembly adjustments:

- Height
- Tilt (top to bottom; forward or backward leaning)
- Azimuth (side-to-side angle)
- Tracking (lateral distance from video drum).

You might want to review the material in the Chapter 3 Appendix for descriptions of these adjustments.

As illustrated in Figure 7.10, on a typical VCR the A/C head is on a small metal plate. This head plate then mounts onto a lower bracket with screws that have springs beneath them. The head plate can pivot back and forth and from side to side on a raised metal dimple at the front of the lower bracket. Loosening or tightening adjustment screws in the head plate pivots the head back and forth and from side to side, changing A/C head tilt and azimuth, respectively.

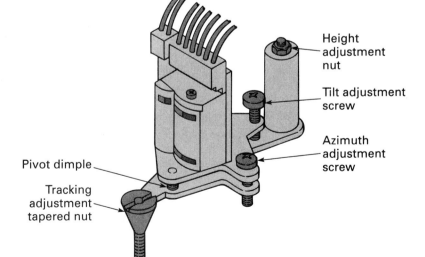

**FIGURE 7.10** Typical A/C head assembly mounting and adjustments

The lower bracket pivots on a vertical post. Head height is adjusted with a nylon lock nut on top of the post, in the same manner as the supply-side guide pole, previously discussed. The entire A/C assembly is spring-loaded against the side of a tapered nut. Turning the nut up or down changes the distance of the assembly from the video drum slightly, which affects mechanical tracking.

For other A/C head mounting schemes, take a minute or two to study the mounting hardware. This should reveal which screw or nut makes which adjustment. You can also refer to the manufacturer's service literature for A/C alignment. Some A/C head adjustments are made with hex head rather than Phillips head screws.

As with other transport components, there is usually no reason to make any A/C head adjustment unless the assembly has been replaced or has been subjected to force, such as unplugging or plugging in the head connector. Adjustment screws normally are paint-locked. Don't be too quick to adjust the A/C head without good reason. Adjustment of take-up side guide pole P-4 should be checked before adjusting the A/C head.

Low volume and muddy-sounding linear audio indicate that the A/C head may be out of alignment, assuming the head surface is clean. Horizontal lines that float vertically up the TV screen and the inability to adjust tracking could mean that the control track is not being played back correctly, because A/C alignment is off.

Ideally, A/C head alignment is done with a special alignment test tape and an oscilloscope according to the manufacturer's service literature; however, the method described here usually gives satisfactory results if done carefully.

Select a high-quality, commercially recorded movie or music video. This will be used throughout the alignment. It should be recorded in stereo if the VCR on which you are working is stereo.

Remember to make notes of precisely how much you turn each adjustment screw or nut so you can return it to its original position if you need to.

### A/C Tilt Adjustment

*Tilt* describes the top-to-bottom angle of the face of the head assembly. The face of the head should be vertical, perpendicular to the baseplate. That is, the head should not tilt forward or backward. If the head is tilted forward, it causes tape to skew downward. This can be seen as the bottom tape edge wrinkling or bunching up at the bottom shoulder of the drum exit roller guide or at the take-up side guide P-4 bottom shoulder, or both. Conversely, if the head is tilted backward, tape skews upward, causing the top edge to wrinkle or bunch at the top shoulder of one or both tape guides.

Press Play and inspect a scratch videotape as it runs from drum exit guide P-3, past the A/C head, to take-up side guide pole P-4. As you look straight down at the top edge of the tape, you should see a single black line; the tape should not be twisted. The top edge of the tape should be directly over the bottom edge. Also check that neither tape edge is wrinkling, bunching up, or riding over the edge of either shoulder on the take-up side guide pole.

If a tilt adjustment is necessary, turn the adjustment screw very slowly as you look at the tape. Adjustment is correct when there is no twist in the tape between drum exit guide P-3 and the A/C head *and* tape is centered on take-up side guide P-4.

### A/C Height Adjustment

If you look straight into the face of an A/C head, you should be able to see the linear audio head at the top and the control head on the bottom—something like the ones illustrated in Figure 7.9. These head areas on the A/C head surface usually look like small silver rectangles. You will not be able to see the actual head gaps themselves; they are too small. The aim of adjusting A/C height is for the top and bottom of the videotape to just cover the audio *and* the control track heads.

Perform coarse head height adjustment by inspecting videotape as it passes over the head. A bright light and handheld inspection mirror are musts. Adjust head height slowly until you see just a sliver of the silver audio head *above* the top edge of the tape, and an equal size sliver of the control head just *below* the bottom tape edge. Tape should definitely *not* touch the head baseplate nor the top of the assembly; either of these indicates that head height is way out of tolerance.

## A/C Azimuth Adjustment

Fine head height and azimuth adjustments are done while monitoring audio from the linear audio heads when playing a commercially recorded tape. If the VCR is a Hi-Fi model, be sure to operate the front panel audio selection switch to the Normal or Hi-Fi Off position. Otherwise you will monitor audio from the spinning Hi-Fi audio heads instead of from the linear audio heads. If there is a switch labeled Stereo/Main or Stereo/Mono, place the switch in the stereo position. If there is a Dolby NR switch on the front panel of the VCR, place this in its Off position.

> NOTE: Some Hi-Fi Stereo VCRs have only monophonic linear audio. A single head gap records a monophonic track, which is both left and right stereo channels combined. This head plays back both stereo channels on a prerecorded stereo tape together, in mono.

You can choose one of several methods to monitor the linear audio signal from tape. The objective is to maximize audio signal strength during playback by adjusting head height, tilt, and azimuth. Here are some suggestions for monitoring.

Monitor with a set of headphones. Many VCRs have a headphone jack on the front panel, usually a 1/8-inch phone jack that will accept a personal-stereo-type headset. You can also use headsets with a 1/4-inch phone plug with a 1/8-inch male to 1/4-inch female adapter.

If the VCR has no headphone jack, you can connect the Audio Out line jacks to a stereo receiver or tape deck with a headphone jack. Connect one or two RCA audio cables to the Aux In or Tape In jacks on the receiver or tape deck. If using an audio tape deck, you'll probably have to put the machine in Record mode to get headphone output from the input signal from the VCR. Just plop in an audio cassette that has the record safety tab in place, go into Record, and hit the Pause button on the audio deck.

Although you can usually do a more effective job wearing headphones, you can also listen to playback from a TV speaker or audio system loudspeaker. Whether wearing headphones or listening to a speaker, pay particular attention to the high frequencies in the audio. These high frequencies occur in the "S" and "Z" sounds in human speech; brushes on drumskins, cymbals, triangles; and the various synthesizer sounds in many dance, hip-hop, and other popular music videos. Playing a music video that has a lot of zing and sparkle in the high end is a good choice.

In some cases, you might be able to visually monitor the audio signal strength, as well as aurally. Some VCRs, usually Hi-Fi Stereo models, have two rows of LEDs that are peak audio level indicators. Most professional VHS decks have VU meters (VU stands for volume unit). Either will indicate maximum audio signal while you adjust the A/C head.

By routing line level audio signals to an audio tape deck, you can observe the peak level LEDs or VU meters on *that* machine as you adjust A/C head height, tilt, and azimuth. Again, you want to *maximize* the readings, keeping left and right channels fairly equal, as you make fine adjustments to head height, tilt, and azimuth. You should still listen, but the peak reading LEDs or VU meters will make it easier to see when maximum volume is reached. Analog VU meters are better for this than peak reading LEDs, because the latter do not show small differences between signal levels as well. LED meters light up in signal level steps, rather than indicating continual changes as a meter needle does.

You may be able to measure audio signal levels from the Line Out or headphone jack with *some* multimeters, especially digital volt meters and amplified, high input impedance analog meters. The latter are sometimes called FET multimeters, where FET stands for field effect transistor. They generally have an input impedance of 10 megohms or more. This type of meter is a solid state version of the older vacuum-tube voltmeter (VTVM). Put the meter on (probably) its lowest AC voltage scale, usually 1.5 or 2.0 volts full scale.

Yet another alternative is to connect an AC voltmeter across the speaker terminals of the TV or audio system to which the VCR is connected. Don't be concerned if you don't understand anything about multimeters or voltmeters at this point in your study. Chapter 9 covers this topic.

It is rather difficult to use a meter or LED-type level indicator to adjust for highest signal strength while playing a music video or commercial movie, because the program itself is constantly changing in volume. What is needed is a *steady* audio signal. Fortunately, many commercially recorded movies have just this very thing at the end of the movie. It is usually several minutes of a steady 1,000-Hz sine wave audio test tone. The tone, if present, normally comes on a short time after the last of the credits have rolled off the screen and the screen has gone blank. If you can locate a videocassette with this tone, it is adequate for adjusting A/C head height, tilt, and azimuth.

Listen for the tone to become clearer and louder, and have more presence and "air" as you make A/C head adjustments. Then observe the meter or peak reading LEDs to fine-tune for maximum signal level. You'll find a meter more sensitive for determining the best adjustment position than your ears alone.

One at a time, very slowly turn the A/C head azimuth and height adjustment screws as you listen. You should find a narrow range of azimuth adjustment over which the sound gets louder; it seems to open up, becoming more airy, with greater clarity in high frequencies. Now fine-tune head height to maximize high-frequency volume. Go back and forth between the two adjustments to obtain the best high-frequency response and the highest overall signal level.

On VCRs with linear stereo, you may not be able to maximize signal levels on both channels at the exact same position with the height screw, and perhaps not even with the azimuth adjustment. If this is so, adjust for the best compromise between the highest level on each channel.

You might find a range of height adjustment over which there doesn't seem to be much change in what you hear, but the sound gets softer and less clear on either side of this range. In this case, position the height adjustment at the center of the range that gives the best clarity and volume. Go back and fine-tune the azimuth adjustment. Make adjustments slowly! Listen carefully for the best high-frequency response and the loudest signal.

When you are satisfied with both head height and azimuth adjustments, go back and tweak the tilt adjustment to see if any improvement can be made in high-frequency response and signal level. Make sure the tape is still straight between drum exit roller guide and A/C head, and that tape edges don't curl around shoulders on the take-up side guide pole P-4.

Go back to the height and azimuth adjustments again and see if any improvement can be made. At this point, any refinement may be hard to detect. Only very small amounts of tilt, height, and azimuth adjustment should be required at this stage. Adjust all three for best high-frequency response and highest overall volume level.

With the linear audio head(s) properly adjusted as just described, the control track record/playback head and audio erase head(s) should also be in correct alignment.

Once adjustments are optimized, lock-paint the adjustment screws with fingernail polish or enamel hobby paint. Do not get paint into screwhead slots, which could make things difficult at some later time. Don't overdo it with the lock paint!

### A/C Tracking Adjustment

The purpose of the A/C tracking adjustment is to center the electronic tracking control so it can compensate for timing differences in tapes recorded on a variety of other VCRs. A/C head tracking alignment is done with a quality, commercially recorded cassette. On most VCRs, an RCA-type adjustment driver, as previously described for the roller guide posts, is needed to turn the split-slot A/C tracking adjustment tapered nut.

On VCRs with automatic tracking, place the front panel Tracking selector switch to the Manual or Auto Off position when checking or adjusting A/C tracking.

Set the manual tracking control to its center, detent position. Press both tracking buttons together on units with two tracking pushbuttons. Play the tape while observing the TV or video monitor.

Assume for now that you have a good picture, with no lines running across the screen. Operate the front panel tracking control slowly in one direction until noise lines just start to appear. Note the amount you turn the control, or the length of time you push a tracking button, until this happens. Now return the control to its center position. Next, operate the control in the opposite direction, again noting the amount of rotation, or length of time, for lines to appear.

If A/C head tracking position is correct, the amount you operate the manual tracking control to either side of its center position, to where the picture deteriorates, will be about the same. However, if you operate the control just a small amount in one direction before the picture goes bad, but quite a ways in the other direction before lines appear, this means A/C tracking needs adjustment.

However, you may not get a good picture even with the front panel tracking control centered. Operate the control over its entire range. If you don't get a good picture somewhere, either there is something else out of whack with the VCR, or A/C head tracking is way out of alignment. Suspect the latter *only if the A/C assembly has been replaced or the cone-shaped adjustment nut has been inadvertently turned.* In either case, center the manual tracking control and move the A/C tapered adjustment nut up or down until a clear picture appears. Then proceed as follows.

Should you decide that A/C tracking adjustment is needed, so that commercially recorded tapes produce good pictures in the middle of the manual tracking control range, here's what to do:

1. Operate the manual tracking control to wherever the best picture displays.
2. Move the control one way or the other until you see several lines running through the screen; *note in which direction you move the control.*
3. Turn the tapered (cone-shaped) A/C adjustment nut up or down until the picture noticeably improves.

> NOTE: Some older VCRs do not have a tapered nut to adjust A/C tracking. Instead, two screws mount the entire assembly to the transport baseplate. Loosen these screws and shift the entire assembly closer to, or further away from, the video drum. It is much more difficult to make this adjustment, so don't be too eager to tackle it unless the assembly is being replaced.
>
> There's something else that may work. Inspect the component side of the circuit boards. Look for a small, round adjustable resistor with the word **TRACKING** printed near it. Most often it will be about 1/4-inch in diameter, have three leads soldered into the board, and a small screwdriver slot on the top. It will usually have a component identification also printed on the board, something like **VR305**. The VR stands for *variable resistor*. There will probably be several similar variable resistors, called *potentiometers* or *pots*, nearby.
>
> Note the position of the tiny slot on the tracking trimmer pot. Use a plastic alignment screwdriver and turn this pot one way or the other while playing a tape to center the tracking.

4. Operate the manual tracking control again to get a good picture. Note the direction in which you move the control.
5. Decide if a good picture appears closer to or further from the center position of the manual control, as compared with Step 1.

Depending upon that answer, take path A or B:

**A:** If the preceding steps move tracking *closer* to, but not quite on center, repeat them. Move tracking control *in the same direction* as noted earlier until lines appear. Then turn the adjustment nut *in the same direction* until lines disappear. Repeat if necessary until a good picture displays with the front panel tracking control centered.

**B:** If Steps 1–5 moved tracking *further* from center, move the tracking control to where a good or improved picture displays. Then move it *in the opposite direction* from the way you moved it in Step 2. Turn the adjustment nut *the other way* until the picture clears up. Repeat as necessary until a good picture displays with the front panel tracking control centered.

This sounds much more complicated than it really is. Once you are actually working on a VCR, you just go back and forth between the manual tracking control and the A/C adjustment until a commercially recorded tape plays with the tracking control centered. When you have centered A/C head tracking, dab locking paint on the top threads of the tapered nut to hold the adjustment.

## TAKE-UP SIDE GUIDE POLE

This tape guide, called *guide P-4*, is adjusted identically to the *supply-side* guide pole P-1, previously covered. (Refer back to that section.) As with the supply-side guide, height is factory set and should require adjustment only if the guide pole is replaced. Ideally, this is done with a manufacturer's jig or height adjustment plate. If needed, height adjustment of this guide should be done before adjusting video drum guides P-2 and P-3 and before adjusting the A/C head assembly.

Many VCRs have a plastic cap over the adjustment nut at the top of this guide pole, which supports the cassette door in its open position. Simply pull the guide pole cap straight off to get at the nut.

## PINCH ROLLER

There is *no* adjustment screw or nut for the pinch roller, but you may have to perform an adjustment. The pinch roller mounting bracket, or pivot plate, may become bent so that the pinch roller surface is no longer parallel with the capstan shaft. It is vitally important that these two components be parallel. If the top and bottom of the pinch roller apply unequal pressure against the capstan, tape skew usually results.

First, check the pinch roller itself for any surface deformities, bulging, hardness, or stickiness. Replace the roller if it exhibits any of these symptoms. Next, manually cycle the cam mechanism with no cassette in the machine. After roller guides P-2 and P-3 have seated into their V-blocks, continue to slowly hand cycle in the same direction while watching the pinch roller. The pinch roller should start to move in toward the capstan.

Stop hand cycling when the pinch roller is about 1/4-inch away from the capstan.

Hold a small piece of white paper or a business card vertically between the capstan shaft and guide pole P-4, as shown in Figure 7.11. With bright illumination, look between the pinch roller and capstan at the white paper from the right side of the roller.

**FIGURE 7.11** Checking parallelism of pinch roller and capstan shaft. With cam mechanism hand cycled so pinch roller almost touches capstan, check that roller and capstan shaft are parallel.

Slowly hand cycle to bring the roller closer to the capstan. When the roller *almost* touches the capstan, you will see the paper as a thin white vertical line between pinch roller and capstan shaft. Stop hand cycling. This line should be equal in thickness from the top of the roller to the bottom. If the white line is thinner at the top or bottom, this means that the roller bracket is slightly bent or twisted.

This is also a good way to detect any swelling or irregularities in the roundness of a pinch roller. With the roller nearly touching the capstan, rotate it through 360 degrees; the vertical white line should stay at a constant width from top to bottom. If the line width varies, replace the pinch roller.

If this check indicates that the roller bracket is bent, carefully take a pair of pliers and form the bracket so the pinch roller is parallel to the capstan. Do not use too much force and overform the bracket. Normally only a very small twist is needed to set things straight.

## SUPPLY AND TAKE-UP SPINDLES AND BRAKES

Normally, not much goes wrong with supply and take-up spindles themselves. However, the back tension and forward torque are important. Supply spindle back tension and its adjustment have already been discussed at the beginning of this chapter. Ideally, spindle back tensions and torques should be checked and adjusted with a special torque gauge designed for VHS VCRs. The cost of a VHS dial torque gauge can be as high as $300.

Correct take-up torque on most VCRs is approximately 150 gram-centimeters (gcm) during play. Fast forward and rewind torque is around 400 gcm. Braking torque of around 450 gcm is usually acceptable for the supply spindle in the clockwise direction and take-up spindle going counterclockwise. In the opposite direction, these spindles should have braking torques near 100 gcm. You should consult directions accompanying the dial torque gauge and VCR manufacturers' service literature to determine how to check spindle torques with a gauge.

Without a torque gauge, you can, over time, get a *rough feel* for what are acceptable and unacceptable spindle torques. Simply grab a spindle hub at the top while operating a well-functioning machine in Play, Fast Forward, and Rewind modes with a videocassette test jig loaded. Turn the spindles in both directions during Stop to get a feel for braking torques. After a little experience doing this on several VCRs, you'll develop what might be called "calibrated" fingers; you will be able to tell when spindle torque or tension is too high or too low.

Forward torque of the take-up reel is controlled by the slip clutch in many VCR models. With *too much* torque, the take-up reel pulls on the tape excessively. This can cause either the top or the bottom edge of the tape to curl over P-4's shoulders, damaging or even folding the tape edges. Too much tape tension can also stretch videotape or put creases in it.

Replacement, adjustment, or rebuild of the take-up clutch is called for when forward torque on the take-up reel is too high. On a few models, the clutch is adjustable and you may be able to obtain correct take-up torque this way. Some clutches can be rebuilt by replacing worn felt washers between the clutch disks.

On many models, the clutch and idler wheel are not available separately; instead, both are replaced as a single unit. Even on units where the clutch and idler are separately replaceable, it is a good idea to replace both at the same time. If either part has deteriorated enough to require replacement, the other is most likely close to the end of its useful life as well.

Check take-up reel forward torque with a T-120 scratch cassette. The take-up reel should pull-in tape smoothly, with no hesitation or jerkiness. Check at the beginning of the cassette and also at the end, when the take-up reel is nearly full. Insufficient take-up torque can cause tape to stall, and then wrap around the capstan. Auto shutdown during Play or Record may also indicate insufficient take-up reel torque. If the reel stops turning, reel sensor pulses are no longer sent to system control, which unloads tape and powers down the unit.

Low take-up torque can be caused by slipping belts, excessively slipping take-up clutch, slipping idler tire, or a combination of all three. The most common cause is a slipping idler wheel. Cleaning and rejuvenating belts and idler tires and cleaning the spindle flange, or rim, surfaces can often cure the problem. (How to perform these procedures is covered in Chapter 6.) Hardened, gummy, or cracked rubber parts should be replaced.

Here is one way to check for *excessive* forward take-up reel torque. Play a T-120 scratch cassette. Take a screwdriver and *gently* push back on the pinch roller bracket *near its base*, as shown in Figure 7.12. Force the bracket back far enough so the pinch roller *just* stops turning—no further. If take-up spindle torque is excessively high, the take-up reel will still tend to pull-in tape with the pinch roller disengaged from the capstan. Perform the test at the beginning of the cassette, when there is little tape on the take-up reel.

**FIGURE 7.12** Disengaging pinch roller from capstan during play. Excessive take-up reel torque will tend to still pull in tape with the pinch roller disengaged.

A squeaking or squealing sound as videotape moves between capstan and pinch roller also indicates excessive take-up torque; tape is literally being pulled from between capstan and pinch roller. In some cases, you may even hear a rather raucous squeal from the loudspeaker, along with linear audio, caused by very rapid fluctuations in tape speed. Sometimes the audio sounds jerky, with high pitched "chirps" as tape lurches past the A/C head. You may see that videotape looks stretched or twisted just before it enters the cassette shell—this is another sign of high take-up torque. Again, a faulty take-up slip clutch is usually the cause for excessive torque on the take-up reel.

You can easily detect low braking torque. Use a T-120 cassette and go back and forth among all operating modes—Play, FF, REW, and Stop—at both the beginning and end of a cassette. For example, switch between FF and REW without going to Stop first. Look for any tendency of the tape to spill into the transport or inside the cassette shell. This indicates poor spindle braking. Poor braking is usually caused by weakened brake lever springs, deteriorated felt brake pads or bands, oil on spindle rim brake surfaces, or a combination of these.

VCRs with direct drive reel spindles usually control the various spindle torques electronically, by varying the amount of voltage to the spindle motors.

Some VCRs have an electromagnetic clutch as part of each spindle. Essentially an electromagnet pulls the spindle down against a felt washer. With no voltage applied to the clutch coil, the spindle is free-wheeling. With a small amount of DC applied, the spindle clamps the felt washer a little bit, adding some drag, or back tension, to the spindle. Higher DC voltages produce more drag, and even higher voltages brake the spindle. For both types of VCRs, consult the manufacturer's service manual to make electronic spindle torque adjustments.

### Spindle Height

As discussed in Chapter 3, correct spindle height is vital. If it is too high or too low, the tape reel inside the cassette will not float between the floor and ceiling of the cassette shell, but will rub.

On most VCRs without direct drive, spindle height is determined by several small shim washers underneath the spindle. If there is a C-clip, compression ring, O-ring, or slit washer near the top of the spindle shaft, this is most likely how spindle height is set. Shim washers normally come in two or more precise thicknesses, such as 0.5, 0.13, and 0.25mm. Stacking a combination of these shim washers on the stationary shaft elevates the spindle to the proper height. *Do not lose these shim washers or swap them with other washers!*

If you remove a spindle of this type, be careful not to lose any shim washers; they often stick to the spindle bottom when you take it off. Immediately after taking a spindle off its shaft, turn it over and check for any shim washers on its underside. Place these wayward washers back on the spindle shaft.

On some direct drive models, the spindle is fastened to the rotating motor shaft with two small setscrews. Before ever removing a spindle of this type, measure precisely the height of the spindle above the reel table with a small ruler. Measure from the baseplate to the top of the lowest flange, or rim, on the spindle. Jot this measurement down so you can reinstall the spindle at the exact same height.

Other direct drive VCRs have an intermediate gear between the spindle motor and spindle. Spindle height is usually set with shim washers as previously described.

Manufacturers' service manuals specify spindle height for each VCR model. Many manufacturers have a jig available for precisely setting spindle height. Spindle height dial gauges are also available from VCR test instrument companies, but these can cost several hundred dollars.

## MODE SWITCH

Recall that the mode switch tells system control the position of the cam mechanism: roller guides extended or retracted, pinch roller engaged, Pause, and so on. Many VCRs have an optical mode switch. This consists of an opaque plastic wheel with small holes and slits in it, called an *encoder disk*, that is geared to the cam mechanism. Two or more IR LED and phototransistor pairs then detect the position of the mechanism by which phototransistors turn On when IR light shines through the encoder disk.

For this type of mode switch, you usually cannot adjust the small circuit board holding the LED and phototransistor pairs. Instead, the encoder disk itself must be timed to the mechanism. As illustrated in Figure 7.13, encoder disk timing is often accomplished by aligning a small hole near the center hub

Chapter 7 How to Align Tape Path and Make Adjustments ■ 195

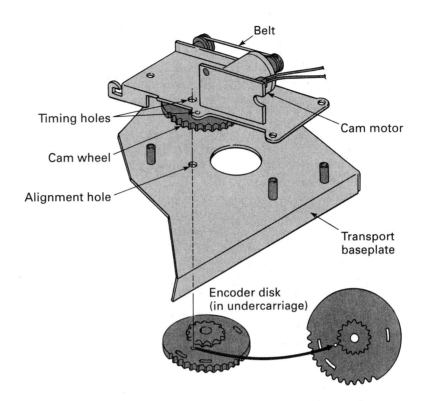

**FIGURE 7.13** Timing a mode switch encoder disk. Holes in encoder disk, baseplate, and upper cam wheel all line up. Mechanism in Stop or home position.

of the encoder disk with a hole in a cam wheel and a hole in the transport baseplate. A timing pin, such as a medium-size finishing nail, should pass through all three holes when the mechanism is in the Stop or home position, with tape fully unloaded.

Similar timing arrangements are possible. An encoder disk may have a dimple, or a small line, or an arrow-shaped mark on its surface that must line up with a hole or corresponding mark on a gear or cam. Segment gears and large-diameter ring gears in loading mechanisms customarily have timing holes or marks to line them up with the transport baseplate and the gears that drive them.

Usually you should hand cycle the tape load mechanism to the fully unloaded, or Stop, position before removing any of its parts. It is imperative to thoroughly document how parts mesh or hook up to each other prior to disassembly. Sketch a diagram detailing parts' positions. Mark where parts come together with an indelible felt-tipped pen if you don't see obvious timing marks or holes. Doublecheck your work before taking things apart. This is absolutely essential if you don't have a service manual; it can be *extremely difficult* to get gear trains, cam wheels, and associated parts back together correctly without marks showing where they mesh or connect and without a faithful drawing.

> CAUTION: Failure to correctly time, or align, tape load and cam mechanism gears and related components can result in part breakage, electrical component damage, or both. Check your work carefully when reassembling and check by hand cycling prior to operating the mechanism under power.

Many VCRs have a mechanical mode switch, mounted to the undercarriage with two screws. These screws will probably be paint-locked. A cam-driven linkage arm typically moves the mode switch lever to different positions as the tape load mechanism operates. There will usually be a small arrow head or

**FIGURE 7.14** Mechanical mode switch timing. Line up > mark on switch body with corresponding mark on sliding switch lever. Engage cam linkage arm with switch lever when mechanism is in Stop (tape unloaded or home) position.

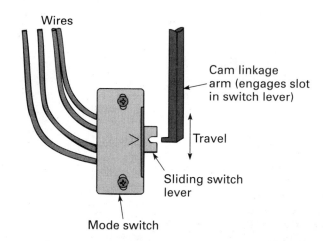

>-shaped timing mark on the switch body, as illustrated in Figure 7.14. This timing mark indicates the position for the sliding switch lever when the mechanism is in its Stop, or tape unloaded or home, position, when the cam linkage arm is engaged with the switch lever.

To *time* (another term used for *aligning*) the mode switch, operate the cam mechanism to the tape-unloaded, Stop position, with roller guides P-2 and P-3 inside the cassette shell. (You don't actually need a cassette or jig loaded.) Engage the cam linkage arm with the switch lever, slide the switch body back or forth until the mark on the switch body points to the corresponding mark on the movable switch lever, and then tighten the switch mounting screws.

Intermittent tape load problems, such as failure to completely load or unload tape, can be caused by poor electrical contact within the mode switch. If thoroughly cleaning switch contacts with spray cleaner doesn't correct the problem, the mode switch should be replaced. In rare cases, slightly moving the mode switch one way or another from its correctly timed position *may* resolve cam mechanism problems.

After replacing or aligning a mode switch, operate the mechanism *several times* through its different modes. Check that video drum entrance and exit guides P-2 and P-3 fully seat into their V-blocks at the completion of tape loading.

Load a scratch cassette that has had its door removed. Load and then unload tape. Check that guides P-2 and P-3 are back on their tracks, inside the cassette shell, when tape is unloaded. They should be in back of or behind the videotape (toward front of VCR) about 3/16 of an inch, so tape doesn't catch on them when the cassette ejects. Also check that the pinch roller comes into full contact with the capstan. Check all other cam-operated modes for the particular model VCR.

## SUMMARY

This chapter described how to make several mechanical adjustments along the tape path. Proper positioning, or alignment, of videotape with tape heads is important for quality audio and video reproduction. Working from the supply reel through the tape path toward the take-up reel, the first important adjustment affects tape back tension furnished by the supply spindle.

Holdback tension keeps tape in proper contact with tape heads, and is usually provided by a felt-lined band wrapped partway around a flange on the supply spindle. A back tension guide pole monitors tape tension and regulates how tightly the band grips the spindle flange as tape runs through the tape path.

Tape tension on most VCRs is correctly adjusted when the back tension guide pole is just left of center of the supply-side tape guide. Video and audio should be stable, with no streaks or noise spots in the video and no wavering of the sound. Tape holdback tension is adjusted on most models by changing the position of a spring attached to the back tension guide pole pivot plate.

Height adjustment for both supply-side (P-1) and take-up side (P-4) tape guides is factory set, and should normally not need adjustment unless a guide is replaced. Adjustment is accomplished by turning a nylon lock nut at the top of the guide post, in most VCRs. While playing a T-120 cassette, adjust each tape guide so tape is centered between the upper and lower guide shoulders. Guide height is correct when the top and bottom edges of videotape do not crinkle, bunch up, or ride over the edge of either guide pole shoulder. If necessary, these guides should be adjusted prior to adjusting drum entry guide P-2, exit guide P-3, and the A/C head assembly.

Each video drum guide (P-2 and P-3) has a small setscrew at the base of the guide post that locks the height adjustment. Setscrews most often have a hex head. A special RCA split-slot adjustment driver is required to make the height adjustment on most roller guides, although some are adjusted with an Allen wrench. An RCA-type adjustment driver resembles a flat-blade screwdriver with the center of the blade cut out.

Video drum roller guides P-2 and P-3 are adjusted so that the bottom edge of videotape just grazes the ledge, or rabbet, on the lower, stationary portion of the video drum. The bottom edge of the tape should not crinkle, ride over the edge of the rabbet, or ride above the drum ledge. Tape should also be centered between the upper and lower shoulders of each guide and produce a stable, noise-free TV picture. Slight adjustment of the entrance guide normally clears up lines or noise at the top of the picture; exit guide adjustment often takes care of lines at the bottom of the screen.

There is no V-stop adjustment. Under normal circumstances, there is no reason to loosen V-stop mounting screws, which are paint-locked. With some VCR models, V-mounts are part of the video drum assembly, and as such are replaced whenever this larger assembly is replaced.

There are four adjustments of the Audio/Control/Audio erase (A/C) head assembly:

**Tilt: forward or backward leaning:** Adjust so head surface is perpendicular to transport base. Tape between drum exit guide and A/C head should be straight, not twisted. Tape should not wrinkle, bunch, or ride over top or bottom collars on take-up side guide pole P-4. Fine-tune along with azimuth adjustment.

**Height:** Adjust so audio and control tape heads are equally covered by top and bottom of tape, respectively. Use bright light and inspection mirror. Fine-tune along with azimuth adjustment.

**Azimuth: side-to-side head tilt:** A/C head gaps must be perpendicular to tape. Adjust while playing commercially recorded tape to obtain best high-frequency linear audio response and strongest signal. Listen on headphones or loudspeaker. May also observe strength of 1,000-Hz tone at end of movie on LED peak level readouts, VU meters, or on low voltage range of AC voltmeter. Fine-tune tilt and height and repeat azimuth adjustment for clearest sound.

**Tracking: centers good picture range of manual tracking control by changing A/C head distance from video drum:** Observe picture while playing quality, commercially reproduced video cassette. Adjust tapered nut up or down until good picture deteriorates when tracking control is moved roughly the same amount either side of its center position. RCA split-slot adjustment driver required on most VCRs.

Although there is no adjustment for the pinch roller, its mounting bracket can become slightly bent. This causes the roller to make uneven contact with the capstan, which in turn causes tape to skew up or down. Tape damage can easily occur. To check for a bent roller bracket, hand cycle the cam mechanism until there is a gap of about 1/16-inch between capstan shaft and pinch roller. Look at the vertical space between capstan and roller. It should be uniformly wide from top to bottom.

If the bottom or top of the pinch roller is closer to the capstan shaft, gently bend or form the roller bracket with a pair of pliers until the roller and capstan are parallel with each other. This inspection technique is also great for revealing swelling and eccentricity of the pinch roller as you rotate it.

Ideally, supply and take-up spindle forward and braking torques are checked and adjusted with a special VHS dial torque gauge. A practical substitute approach used by experienced technicians is to determine whether torques are approximately correct by grabbing the spindle hubs while the VCR is in Play, Fast Forward, and Rewind modes. Similarly, braking torque can be roughly judged by manually turning spindle hubs CW and CCW while in Stop mode.

Take-up tension that is too low is routinely caused by a dirty spindle flange, dirty or deteriorated rubber idler tires and belts, or by a take-up clutch that is slipping too much. Tape spills, eaten tape, and auto shutdown can be caused by low forward torque on the take-up reel spindle.

*Excessive* take-up tension can cause tape to climb over one of the take-up side guide pole (P-4) shoulders. High spindle tension may pull tape past the capstan and pinch roller, creating an erratic chirp sound from the speaker along with program audio. One check for high take-up tension is to push back on the pinch roller bracket until the roller just disengages from the capstan while in play. Tape should not continue to be pulled in by the take-up reel. Excessive take-up spindle torque is usually caused by a faulty take-up slip clutch.

Precise spindle height is necessary. On most VCRs where spindles rotate on a stationary shaft, spindle height is determined by two or more shim washers beneath the spindle. Shim washers usually come in two or more thicknesses. Spindles of this type are held on the shaft with a C-clip, O-ring, compression ring, or slit washer. It is easy for a shim washer to adhere to the bottom of a spindle when the spindle is removed. It can then fall off and get lost.

On VCRs where the spindle shaft rotates, such as in some direct drive models, the spindle fastens to the shaft with two setscrews. For both types of spindles, carefully measure and jot down the height between transport baseplate and top of the first spindle rim, or flange, before removing the spindle. A special VHS spindle height dial gauge or manufacturer's spindle height jig may be used to set spindle height to specification.

There are generally two types of mode switches, also called cam position sensors. In one type, an opaque encoder disk with holes rotates between pairs of IR LEDs and phototransistors. Light shining through holes in the disk indicates the cam position. Encoder disk timing is obtained by aligning a small hole or timing mark on the disk with a similar hole or mark in another gear or cam. An encoder disk timing hole may also have to line up with a timing hole in the transport baseplate.

The other type of mode switch is mechanical. A sliding switch lever is moved back and forth to different positions by a linkage arm connected to the cam mechanism. Correct mode switch alignment results when a timing mark on the switch body points to a corresponding mark on the switch slider.

Timing both mode switch types is done with the VCR in the Stop or home position; roller guides P-2 and P-3 should be inside the cassette shell, about 3/16-inch away from the back side of the videotape.

It is essential to carefully sketch and mark the location of gears, cams, gear segments, and linkages operated by the cam motor before disassembly. Note in particular timing marks and timing holes. Timing marks usually line up at the Stop position. Component damage could occur if cam mechanism components are not correctly timed to each other.

Do not be too quick to loosen adjusting screws that have a dab of lock paint around them. These factory adjustments normally don't need to be changed unless an associated part is replaced because it is broken or worn. It may be difficult to reestablish factory adjustments without special alignment jigs or gauges.

Whenever any adjustment screw or nut is turned, be sure to make a note of the exact amount you turn the screw or nut, and in what direction. In this way, you can refer to your notes and return the adjustment to where it was before you began, if necessary. Clockwise or counterclockwise rotations should be noted in eighths of a turn.

## SELF-CHECK QUESTIONS

1. What is the correct position for the back tension guide pole with Play dummied up in a typical VCR? That is, the tape load cycle is complete with guides P-2 and P-3 seated in their V-stops, but with no tape actually loaded?
2. What is the correct position for the back tension guide pole when playing a tape on a typical VCR?
3. What are some typical causes for low tape holdback tension in an old VCR that you take in for repair?
4. Describe how you would check the height of supply-side guide pole P-1. What would you do to change the height of P-1 slightly?
5. You are going to check the tape path by playing a scratch cassette while the front loader is removed from the VCR. What must you do after manually placing a cassette on the reel table and before pressing Play?
6. What would you do to eliminate a few horizontal lines or dashes near the *bottom* of the picture when playing a commercially recorded movie? Assume that the manual tracking control has been adjusted for the best picture possible.
7. Suppose you work on a VCR that has horizontal lines across the picture. You find everything in good order, except for the height of drum entrance roller P-2. You change its height by two turns and now get proper tape motion around the drum and guides and a beautiful picture. Two weeks later the same VCR is back. You find that P-2 once again needs adjustment. What would you suspect the problem might be?
8. What is the largest number of head gaps that an A/C head assembly could have? What are they? Where are these gaps located on the face of the assembly?
9. What happens when you make the mechanical tracking adjustment of the A/C head?
10. On a typical VCR, what A/C adjustments are located on the head mounting plate itself?
11. What must you do on a Hi-Fi Stereo VCR when aligning the A/C head?
12. Describe how you would use a commercially recorded music video to align the A/C head.
13. Describe how you would use recording level meters or an LED bargraph display on an audio cassette deck to align the A/C head assembly.
14. What should you do on a VCR with automatic tracking when adjusting mechanical tracking of the A/C head?
15. Describe how to check alignment of the capstan shaft and pinch roller.
16. Explain how to develop what the textbook calls calibrated fingers for estimating spindle torques.
17. What would it be wise for you to do when taking apart any portion of the mode or tape load mechanism?

# CHAPTER 8
# UNDERSTANDING BASIC ELECTRONICS

*Practical* knowledge of electronics is required to troubleshoot and repair the many circuits in a VCR. In this chapter you will learn how electricity flows through a circuit, starting with a simple flashlight.

"But," you're thinking to yourself, "I'm studying how to repair VCRs, not flashlights!" You're right. However, there's a lot you can learn about fundamental VCR circuits by first seeing how a flashlight works. From there, you can more easily understand electrical properties: voltage, current, resistance, and power, along with their relationships in a circuit. Next you will see how these electrical units of measure are calculated for simple series and parallel circuits.

This chapter is not going to teach you the world of electronics. It *will*, however, give you an understanding of the basic electronics used in VCRs. Then you can diagnose many VCR electronic failures. You will also need to understand the material in this chapter before you can effectively use a meter to test VCR circuits and components.

Two basic types of electric current, direct and alternating, are present in a VCR; both are described, with examples of where each is used. The chapter concludes with a brief overview of analog and digital circuits, both of which are found in a VCR.

If you already know something about electricity (even if it's only a vague memory from a high school class), that's fine; it will help. But whether you don't know a volt from a dune buggy, or you are highly knowledgeable about circuits, we hope you'll find this material fairly painless and highly relevant to your VCR repair career.

## CHAPTER OBJECTIVES

**Upon completing this chapter, you should be able to:**
1. Describe electron flow in a simple DC circuit.
2. Describe the components and operation of a single-cell flashlight, including electron flow.
3. Give a one-sentence definition for each of the following electrical terms:
   - Insulator
   - Conductor
   - Semiconductor
   - Voltage
   - Current
   - Resistance
   - Power
   - Impedance.
4. Describe the relationship between voltage, current, resistance, and power in an electrical circuit. Given values for any two in a circuit, determine values of the other two.

5. Discuss what is meant by alternating current (AC), direct current (DC), analog, and digital. Give an example of where each is found in a VCR.

The emphasis in this chapter is on practical, day-to-day electronics. Electronic theory is only mentioned. You don't need to be an electronics engineer to be successful at VCR repair. For students who want to learn electronic theory, there are many excellent books available from libraries and bookstores with more in-depth electronic discussions.

A handheld calculator is helpful during portions of this lesson.

In Chapters 9 and 10 you'll see how to perform electrical measurements and component tests with a multimeter, and you'll learn how various electronic components work in VCR circuits.

Some of the following explanations are broadly stated to help newcomers to electronics get up to speed quickly and understand overall concepts. Those of you already knowledgeable about electronics will recognize this material as foundational rather than comprehensive.

## INTRODUCTION TO ELECTRONICS

Everything is made up of atoms. Atoms themselves are made up of three parts: protons, neutrons, and electrons. Electrons have a negative electrical charge, protons have a positive charge, and neutrons are neutral, having neither a positive nor a negative charge. Electrons orbit around the protons and neutrons that make up the nucleus of an atom, just like the Earth orbits about the sun. The combined negative charge of all electrons in an atom equals the positive charge of the nucleus; thus, the atom itself has no charge.

Electrons are what electricity and electronics are all about. When an energy source excites certain types of materials, one or more electrons literally fly out of the orbit of one atom and move to an adjacent atom. A fundamental law of physics is that opposite charges attract. When a negatively charged electron flies away from an atom, this leaves the atom with a positive charge. This positive charge then attracts negatively charged electrons from adjacent atoms. And that's really what electricity is: a flow of electrons from atom to atom to atom through some material, like a copper wire.

But what excites the material in the first place to make an electron fly out of an atom's orbit? The two main types of exciting forces we use most often are chemical and magnetic. In batteries, a chemical reaction causes an atom to give up one or more electrons, called *free electrons*. Electricity in our homes is produced by generators. An electric generator produces a flow of free electrons when a wire physically moves within a magnetic field.

That's about as far as we're going to go with atomic, electron, and magnetic theory. All you really need to know right now is:

1. In electronics, we're talking about *controlling* the flow of electrons, and
2. Electrons, having a negative charge, move toward something with a positive charge.

## ELECTRON FLOW

To further understand electricity and electron flow, look at a simple single-cell flashlight, as illustrated in Figure 8.1. Here the battery is the chemical energy source for moving electrons. Because of a chemical reaction inside the battery, many electrons move from atom to atom to atom within the battery until they figuratively pile up at one end of the battery. This gives that battery

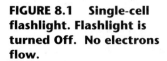

**FIGURE 8.1 Single-cell flashlight. Flashlight is turned Off. No electrons flow.**

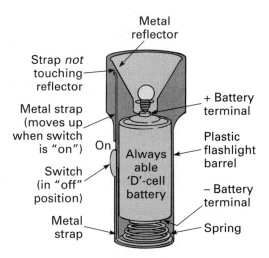

end a negative charge, so that end is called the *negative terminal*. The atoms from which these electrons came now have a positive charge, which is at the other end of the battery, its *positive terminal*.

When the flashlight is turned Off, electrons piled up at the negative end of the battery have nowhere to go, so there is no flow of electrons. When the switch is moved to its On position (Figure 8.2), the movable metal strap touches the metal reflector into which the flashlight bulb is secured. Electrons race out of the negative terminal of the battery and travel through the metal spring, up the metal strap to the switch. They continue through the switch on the movable strap to the metallic reflector and into the screwthreads of the bulb. Electrons then go through the bulb's thin filament and out the base of the bulb to the positive battery terminal. As you will soon learn, electrons have a somewhat difficult time going through the bulb's filament, causing it to get so hot it produces visible light.

Remember that opposite charges attract. It is the positive charge at one end of the battery that attracts the electrons from the negative battery terminal. Along the way, electrons move from atom to atom through the metal parts of the flashlight and the bulb. Considered together, the battery, copper straps, and flashlight bulb form a *circuit*. A circuit is simply a pathway for electron flow. Electrons always move from negative to positive in a circuit.

Materials through which electrons can easily move are called *conductors*. Most metals conduct electricity very well. Silver is the best conductor, but cop-

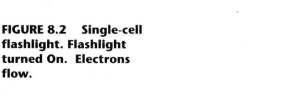

**FIGURE 8.2 Single-cell flashlight. Flashlight turned On. Electrons flow.**

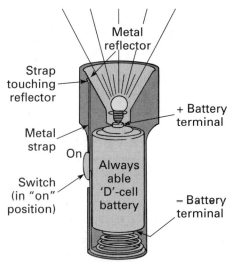

per, because of its much lower cost, is the most frequently used electrical conductor. The vast majority of house wiring, for example, is copper wire. Gold and aluminum are also good conductors. The better a material conducts electricity, the less electrical loss there is. For most wires, we want *minimum* electrical loss.

Some low-level signal connectors in VCRs and other electronic equipment are gold-plated; gold is a very good conductor *and* it doesn't oxidize or corrode. Oxidation and corrosion don't carry, or conduct, electricity very well, especially weak signals.

In the flashlight, the spring contacting the negative end of the battery, the copper straps, and the metal reflector are all good conductors of electricity.

Some materials are not nearly as good electrical conductors as copper. Electrons can travel through those materials, but not easily. However, sometimes this is very desirable! When electrons have a difficult time traveling through a poor conductor, the material heats up. Electrons have a very hard time getting through the filament of the flashlight bulb, so the filament gets very hot. The heating wires in an electric toaster are made of metallic alloys, such as nickel and chromium, which don't conduct electricity very well, so they get red hot when electrons move through them.

A very common electronic component, the *resistor*, is frequently made from carbon graphite, which is a mediocre conductor of electrons. Resistors are found in virtually every electronic device, and are the most numerous component in electronic circuits; a VCR may have well over one hundred resistors. They are used to *control* the amount of electrons flowing in different parts of a circuit. Figure 8.3 shows several carbon composition resistors. You will read much more about resistors later in this chapter and in Chapter 10.

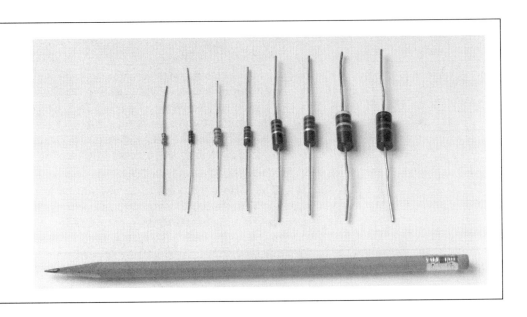

**FIGURE 8.3 Carbon composition resistors. The larger a resistor is physically, the more electrical power it can handle.**

*Insulators* are materials through which electrons *don't* travel. Plastics, rubber, glass, porcelain, ceramics, wood, and fabrics are good insulators. A plastic flashlight barrel is an insulator. Air is also a good insulator, for the most part. When the switch on the flashlight is Off, or open, there is air between the end of the movable switch contact and the metal reflector; no electrons flow. Sometimes, though, as with lightning, air can become a conductor.

Semiconductors fall somewhere between being good conductors and good insulators. Semiconducting materials are germanium and silicon. In day-to-day

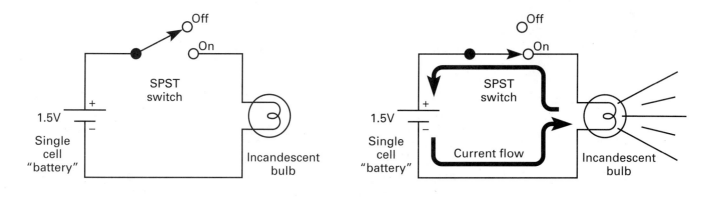

**FIGURE 8.4** Single-cell flashlight schematic diagrams.

language, *semiconductor* ordinarily refers to a device such as a transistor, diode, or integrated circuit chip that has been manufactured from silicon or germanium. These devices have unique characteristics for controlling the flow of electrons through them. These electronic components are described in Chapter 10.

Figure 8.4 has schematic representations of our single-cell flashlight. A schematic diagram is much easier to draw and work with than a drawing of the actual flashlight components. The diagram on the left shows the switch in the *open* or Off position; on the right it is *closed* or On.

When there is no complete path through which electrons can flow, as in the schematic diagram with the flashlight switch Off, there is an *open circuit*. A complete path for electron flow, like the flashlight with switch On, is called a *closed circuit*.

An electrical circuit has three basic parts: *source, path,* and *load.* These parts of a circuit, along with the components in a single-cell flashlight, are:

- **Source**: provides electrical energy to the circuit
  — single-cell battery
- **Path**: conductors to carry electrons
  — copper straps, closed switch
- **Load**: uses energy supplied by the source
  — flashlight bulb.

**The following point is important to keep in mind.** No matter how complicated an electronic circuit may be, or how bewildering the schematic diagram seems, it is easier to understand when you trace the flow of electrons from the source, along a circuit path, through a load, and then back to the source, just like our flashlight. An electronic circuit in a VCR may actually be composed of many smaller, separate circuits that together make up the larger circuit.

Often there is only one source of energy, but many paths and many loads. Consider the AC wiring in your house or apartment. The common energy source is the electricity coming into the electric company meter. From there, different paths carry, or conduct, electricity to the kitchen, living room, and bedroom. These paths consist of fuses or circuit breakers, copper wires in the walls, wall switches, and electric cords on appliances. And there are all sorts of different loads: light bulbs, TVs, toasters, hair dryers.

A similar situation exists in a car. A 12-volt battery is the common source. Circuit paths include fuses in the fuse block, dashboard and steering column switches, and many individual conductors, or wires, in the wiring harness. Electric loads include the engine ignition system, headlights, radio, and windshield wiper motor.

Just as in the flashlight circuit, electron flow can be traced from the source, along a circuit path, through the load, and back to the source, as in Figure 8.5. This is a simplified schematic of a small part of an automobile's wiring, showing circuits for the interior lights and windshield wiper motor. Note the schematic symbol for electrical ground; it looks like a downward pointing arrow, connected to the negative battery terminal. This is the metal frame of the car. Anywhere on a schematic where you see two such symbols, visualize a wire conductor between the two. The metal car frame itself is an electrical conductor.

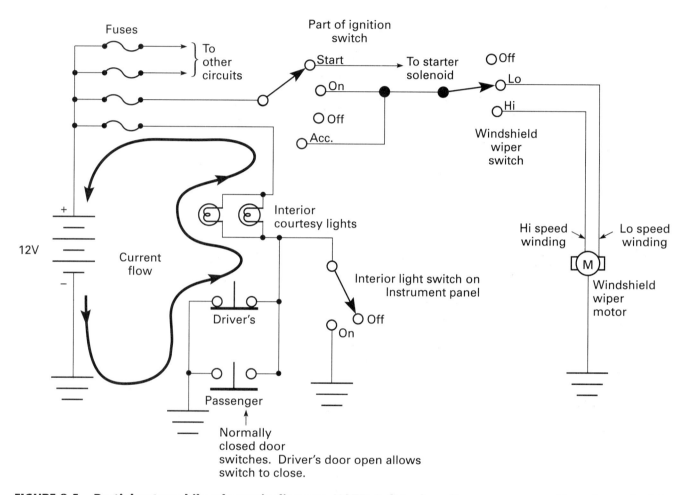

**FIGURE 8.5** Partial automobile schematic diagram. NOTE: A dot where lines cross means there is an electrical connection between wires. There is *no electrical connection* where lines cross without a dot.

Here is how we trace the flow of electrons in the automobile interior light circuit. The driver's door is open.

1. From the negative battery terminal (source) to ground (car frame)
2. From ground, through the door switch that is closed (driver's door open)
3. From the other side of the driver's door switch, along wiring harness conductors to one side of each interior light bulb
4. Through the filaments of both interior lights (the load)
5. From other side of light bulbs, along wiring harness conductors to the fuse
6. Through fuse to positive battery terminal.

Notice that it doesn't matter whether the ignition switch is On or Off for the interior lights to work. With both driver and passenger doors closed, the interior lights will be out; they will illuminate if the panel switch is moved to its On position.

A portion of an automobile electrical system was used here to trace current flow, because many readers already have some familiarity with car wiring. The same principles apply to DC circuits in a VCR or other electronic device.

Throughout this text, circuit operation is discussed using the *electron theory for current flow,* which states that electrons go from the negative terminal of a battery or power supply, through circuit components, and back to the positive supply terminal. Conversely, *conventional current flow theory* is electricity flowing from the positive supply terminal, through circuit components, back to the negative side of the supply. Some people prefer to use conventional current flow, especially in electronic schematics where several electronic components contain an arrow. Electron flow is against the arrow, whereas conventional current flows with the arrow. You will see more about such schematic symbols in Chapter 10. My suggestion is that you stick with electron flow ( ⊖ to ⊕ ), and just get used to electricity flowing against the arrows in electronic wiring diagrams. Electron current flow seems to be more widely accepted, and is considered technically correct.

## ELECTRICAL UNITS OF MEASURE

Now that you know how to trace electron flow from source, along a circuit path, through a load, and back to the source, we can go on to describe electrical units of measure. To understand how electronic circuits work and to be able to test circuits and components with a meter, you will need to know what electrical units of measure are.

> Remember that this overview of electronics is intended to help you build VCR troubleshooting skills. If these concepts are all new to you, don't panic. The next few chapters will offer examples of how to use what you are learning. The big picture may seem fuzzy now, but as you continue, you'll begin to understand how to use these concepts to troubleshoot electronic circuits in VCRs.

There are three basic characteristics, or conditions, in any electrical or electronic circuit: *voltage, resistance,* and *current.* We can also describe the amount of "work" being done in a circuit. This we call electrical *power.*

To help understand electricity, and these electrical units of measure, we'll use water flowing from a garden hose as an analogy to electricity flowing through a circuit.

When you connect a garden hose to a faucet, you are concerned with how much water you are going to get out of the hose. The amount of water that comes out is controlled by two conditions:

1. How far you turn the faucet On. At a trickle, the water has little pressure. When the faucet is turned full On, there is much more water pressure.
2. Hose diameter. A small-diameter hose will restrict the flow of water more than a large-diameter hose.

In the water-and-hose situation, increase either of the first two conditions, and the amount of water flow also increases. Decrease either water pressure or hose diameter, and you decrease the amount of water flowing from the hose.

With electricity, *voltage* is similar to water pressure. It is the electromotive force (EMF) behind the stream of electrons that constitutes a flow of electricity. With higher water pressure, more water flows. With higher voltage, more electrons flow.

### *Voltage is electrical pressure or force.*

The size of a hose is likewise similar to the amount of resistance in an electric circuit. A large-diameter hose offers little resistance; lots of water can flow. Similarly, a large-diameter wire offers less resistance to electron flow than a small-diameter conductor.

With more resistance, a lesser amount of water or electricity flows. A small-diameter hose delivers less water. A small-diameter wire can handle less electricity than a larger one. Recall that plastic and rubber are good insulators; they strongly resist electron flow. We can therefore say that insulators have an extremely high resistance to electron flow. A large-diameter copper wire is a good conductor of electricity; it has very little resistance to the flow of electrons.

### *Resistance is opposition to electron flow.*

The amount of water flowing is dependent upon water pressure *and* the degree of restriction offered by the hose. In an electric circuit, the voltage and amount of resistance determine the electrical current, that is, the quantity of electricity flowing in the circuit. Number of drops of water equals amount of water. Number of electrons equals electrical current.

### *Current is the amount of electron flow in a circuit.*

The higher the current, the more electrons that flow through a circuit.

Perhaps a further explanation will help. Let's put the end of our garden hose into a bucket. For a particular faucet setting and a particular hose size, we get six ounces of water every second.

### *How could we increase the amount of water flow?*

We could turn the faucet On further; this would increase water pressure. Or we could use a hose with a larger diameter; water flow would be less restricted. Or we could do both: increase water pressure *and* reduce flow resistance.

### *In an electrical circuit, how do we increase current flow?*

Electrical pressure, or voltage, could be increased, or electrical resistance could be reduced, or both. (See Figure 8.6.)

With electricity, the counterpart of the amount of water flowing is *current;* we talk about current flow in a circuit. The electrical parallel to water pressure is *voltage,* while the equivalent of restriction to water flow is electrical *resistance.* Understanding the concept of voltage, resistance, and current, we next describe the units of measure for each and how they relate mathematically in an electrical circuit.

Voltage = Pressure
Resistance = Restriction
Current = Amount

**208** ■ *Practical VCR Repair*

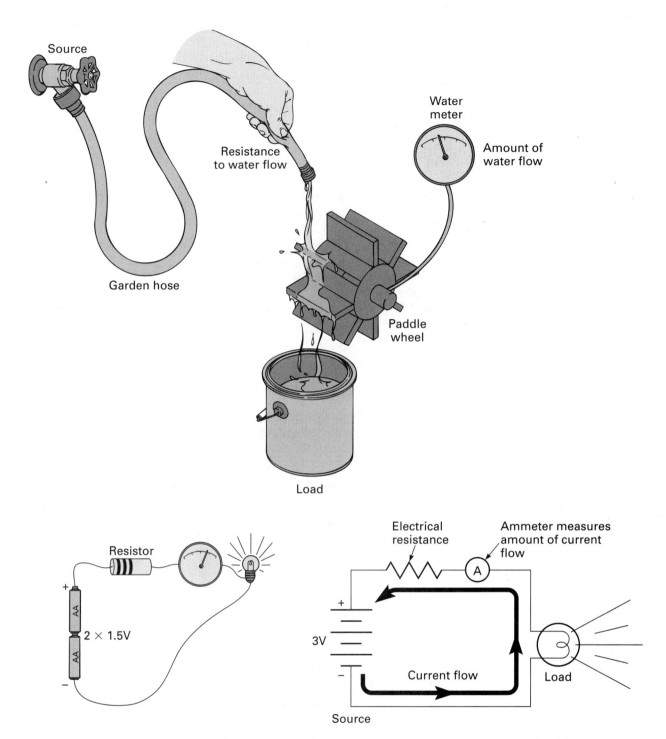

**FIGURE 8.6** Water and electrical flow. Increasing water pressure, reducing hose restriction, or both increases water flow. Increasing voltage, reducing resistance, or both increases electrical current flow.

### Voltage

- Unit of measure is the *volt*, abbreviated V
- Symbol is either E (for electromotive force) or V (this textbook uses E)

### Resistance
- Unit of measure is the *ohm*
- Symbol is R or Greek letter Ω (omega)

### Current
- Unit of measure is the *ampere,* abbreviated Amp or A
- Symbol is I

A German scientist named George Ohm determined how to relate the values of voltage, resistance, and current in an electrical circuit. You've probably heard of it; it's Ohm's Law.

## OHM'S LAW

Ohm's Law states that the voltage across any part of an electrical circuit is equal to the current flowing through that part of the circuit, in amperes, multiplied by the resistance of that part of the circuit, in ohms:

$$E = IR \quad \text{Which is the same as E = I times R.}$$

For example, if we knew that 3 amperes of current were flowing through an electrical resistance of 5 ohms, we could determine the voltage drop across the resistance, like this:

$$E = IR = 3 \times 5 = \textbf{15 volts}$$

So the voltage across the resistance would be 15 volts. We also call this the *voltage drop* across the resistance.

Resistors are electronic components that restrict the flow of electrons. You might want to look back at Figure 8.3 to see what common carbon composition resistors look like. There are probably more resistors of this type in most electronic devices than any other type of component. A VCR could easily have more than 100 resistors.

Resistors perform two basic functions in electronic circuits:

- Provide a voltage drop
- Limit the amount of current flow.

Actually, any resistor performs *both functions at the same time,* but often it is helpful to think of a particular resistor as *predominately* performing one or the other function in a given circuit. Until you do VCR troubleshooting for a while, you might not recognize whether a particular resistor serves primarily as a voltage-dropping resistor or a current-limiting resistor. Don't be concerned about this! The difference is not always distinct, something like the difference between a pot and a pan in the kitchen, but now you can recognize the terms voltage-dropping and current-limiting. The main point is to get the big picture of how a resistor behaves and its effect in an electronic circuit.

Resistors come in many ohmic values, from fractions of an ohm all the way to several million ohms, or megohms. Later on in this chapter you'll learn about electrical power. For a given type of resistor, like carbon composition, the larger the resistor is physically, the more power it can handle.

For now, just realize that resistors have a specific value in ohms. For example, you can readily purchase resistors with resistance values of:

10, 33, 47, 68, 100, 220, 470, and 680 ohms
1000, 1.2K, 2.2K, 3.9K, 4.7K and 5.6K ohms
10K, 22K, 47K, 100K, 470K, 1meg, and 2.2meg ohms

The K stands for 1,000, so a 3.9K resistor is 3,900 ohms. Meg is similarly used to mean one million. A 2.2meg resistor is 2,200,000 ohms. This could also be stated as 2.2 megohms. Note that there are many other standard value composition resistors manufactured; the above list is just to give you an idea of typical resistor values.

The schematic diagram symbol for a resistor is like a squiggle, something like this: —⋀⋁⋀—. You'll see resistors represented by this symbol during this course and throughout your electronics career. On a schematic diagram, the value of the resistor in ohms or a number beginning with an R, like R3, R27, or R305, is printed adjacent to each resistor symbol. The latter is a reference number; the R stands for resistor. You look elsewhere on the schematic or in a parts list to find the actual value for R3, R27, or R305.

See if you can determine the voltage drop across resistor R1 in the following circuit.

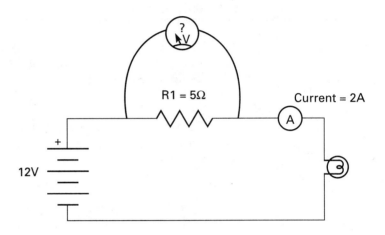

**FIGURE 8.7 Calculating voltage drop. The ammeter measures current flow in the circuit. What is the voltage across resistance R1?**

Plugging the values for current and resistance into Ohm's Law formula $E = IR$, we have:

$$E = 2 \times 5 = \mathbf{10\ volts}$$

The voltage drop across resistor R1 is 10 volts.

All components in Figure 8.7 are connected in what is called a *series circuit*. Each component is connected end-to-end with its neighbor components, like individual links in a chain. Series circuits are explained further, later in this chapter.

There are two electrical laws for series circuits:

- The amount of electrical current flow is the same through each and every component in a series circuit.
- The sum of all voltage drops in a series circuit is equal to the source voltage.

With this information, can you determine how much voltage is across the light bulb in the Figure 8.7?

Sure! It's 2 volts. Since the source voltage is 12, and 10 volts are dropped across resistor R1, the difference of 2 volts must appear across the lamp.

Using Ohm's Law we can also solve for the value of resistance if we know the value of voltage and current in a circuit:

$$E = IR$$

Swapping sides of the equation gives us:

$$IR = E$$

Dividing both sides by I gives:

$$IR/I = E/I$$

E/I is the same as E ÷ I. The I over I portion on the left cancels out, so, when simplified, this equation becomes:

$$\mathbf{R = E/I}$$

Resistance equals voltage divided by current.

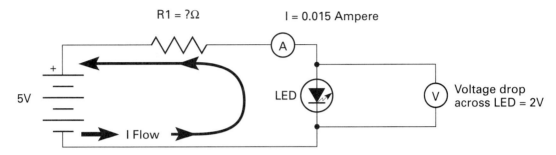

**FIGURE 8.8    Calculating resistance.  What is the resistance of R1?**

Refer to Figure 8.8. What is the value of the resistor required so that a two-volt LED drawing 0.015 ampere can be operated from a supply voltage of 5 volts?

This *could* be a real VCR repair situation! Suppose a small carbon resistor, like R1, has broken off a printed circuit board and is missing. Most common LEDs operate at around 2 volts and draw about 0.015 ampere. Knowing this information, and measuring the supply voltage at 5 volts, you could determine the approximate value of a replacement resistor.

Use Ohm's Law formula: R = E/I

But before you can use Ohm's Law, stop for a moment. Did you notice that you were given the voltage across the load, at the LED, of 2 volts? Since the supply voltage is 5 volts, the resistor has to reduce this to the 2 volts required by the LED. So the resistor has to have a 3-volt drop across it. Therefore, 3 volts is the voltage you plug into Ohm's Law formula. Also plug in the value of current, which is 0.015 ampere. Then solve for R.

$$R = E/I$$
$$R = 3.0 \div 0.015 = \mathbf{200\ ohms}$$

Soldering a 200-ohm resistor into the circuit will get the LED working again.

Actually, the nearest standard value resistor available is either 180 or 220 ohms. We'd probably select a 220-ohm resistor, since this would cause a little less than 0.015 ampere to flow through the LED, whereas a 180-ohm resistor would allow more than 0.015 ampere to flow. It's safer for the LED to have less current flowing through it.

If you had trouble following the last two sample problems with Ohm's Law, please do not go any further in this chapter at this time. Take the time to go back and reread the material. Use a calculator or pencil and paper to determine the unknown values of voltage and resistance.

You can also use Ohm's Law to find out how much current is flowing through part of a circuit if you know the voltage and resistance of that part of the circuit. For example, what is the maximum current drawn by a 12-volt DC motor in a VCR front loader assembly? The motor has a measured resistance of 25 ohms when disconnected from the circuit.

First, we need to solve for current, I. Starting with Ohm's Law:

E = IR     which can be re-written as:

**I = E/R**     substituting the known values for voltage (E) and resistance (R) in this formula gives:

I = 12 ÷ 25 = **0.480 A**, or about 1/2 ampere.

This is fairly typical for the maximum amount of current drawn by small DC motors used in VCR front loaders and tape load mechanisms. (If you are very knowledgeable about electricity and motors, then you might know that in actuality a DC motor of this type normally draws less than its maximum current when it is running. Maximum current flow occurs only when the motor is stalled.)

Where else might you use Ohm's Law? Suppose that a particular VCR frequently blows its power supply fuse. The fuse always seems to blow when the cam motor is moving to load or unload tape, but not every time. Is it a binding mechanism that is causing the problem, an intermittent short to ground in a cable from the mode switch, a partially shorted cam motor winding, or some other problem?

You decide to check the motor first. You're not sure what voltage the cam motor is supposed to be; most are around 12 volts, but without the schematic it's best to check. Measure the voltage going to the cam motor when you press Play to load tape. You read 11.2 volts on your voltmeter. Luckily, the fuse doesn't blow before you get a voltage reading. (A voltmeter is the function of a multimeter which measures volts. You'll learn how to take measurements for all multimeter functions in Chapter 9.)

Next, unplug the VCR and measure the resistance of the motor winding. You read about 4 ohms.

Plug the two values into the Ohm's Law formula to solve for current:

$$I = E/R$$

Substituting measured values gives:

$$I = 11.2 \div 4.0 = 2.8 \text{ A}$$

This is a rather high current for this type motor. As we saw in the previous example, most motors of this type draw somewhere around 1/2 ampere, or 0.500 A. At 2.8 A, this motor is drawing about five times what it typically should. Most likely there is a partial short in the motor winding.

So what's a short? *A short is an electrical shortcut:* an electric circuit is shortcut around the intended load, causing excessive current flow. A short has very low or zero resistance to current flow. The formula I = E/R says that if resistance goes down, which would happen if there were a partial short in the motor winding, then current goes up. We say that current (I) and resistance (R) have an *inverse relationship*. If one goes down, the other goes up.

The excessive current drawn by the motor sometimes causes the power supply fuse to blow. You can be fairly sure this is a 12-volt motor from the 11.2 volts you read across the motor. But because the motor draws more current than it should, the power supply voltage goes down from 12 volts to nearly 11 volts. The supply just can't provide the current demanded by the partially shorted motor.

In this case the motor must be replaced. You get a new motor and measure its resistance before installing it. The ohmmeter reads 25 ohms; certainly, this will draw less current than the partially shorted motor, which measured 4 ohms. With the new motor installed, you measure 12.3 volts across it when it is running. That's more like it! The power supply fuse no longer blows.

Knowing how to calculate circuit values with Ohm's Law is a necessity in nearly all electrical and electronics troubleshooting and repair work at some time or another. Here again are the three Ohm's Law formulas:

> Ohm's Law Formulas
>
> E = IR   Voltage equals current multiplied by resistance
> I = E/R  Current equals voltage divided by resistance
> R = E/I  Resistance equals voltage divided by current

Also remember that in a series circuit, where components connect end-to-end like links in a chain:
- The sum of all voltage drops equals the source voltage.
- Current is the same value at any point in the circuit.

Earlier, a short was described. A defective component with very low resistance, like the DC motor that had much lower resistance than it should, is said to be shorted or partially shorted. The opposite of a short is an *open*. Whereas a short is a very low resistance where too much current can flow, an open is an extremely high resistance, or a resistance having an infinite number of ohms. The $\infty$ symbol means infinite or infinity. No current can flow in an open circuit.

> SHORT: 0 $\Omega$, maximum current flow
> OPEN: $\infty$ $\Omega$, no current flow

Suppose a blob of solder accidentally drops across two adjacent wiring patterns on a printed circuit board. That would cause a short circuit. A cracked circuit board, where a printed wiring trace is broken, would cause an open circuit.

As you will learn in Chapter 10, several basic component tests you can perform with an ohmmeter indicate a defective component if a short (0 ohms) or an open ($\infty$ ohms) exists. (An ohmmeter is the function of a multimeter which measures ohms.)

## Ohm's Law Exercise

If this is your first exposure to electronics, then you might not feel entirely comfortable with these concepts yet. That's okay! You will pick them up in time. To help you along, here are exercises to work through on paper.

With the information you've learned so far, you can determine the unknown values (?) of voltage, current, and resistance in Table 8.1 for the circuit in Figure 8.9. Important additional information is included in the solutions that follow these problems.

First, try to determine the unknown value in each problem on your own. Then be sure to read the solutions carefully. (Here's a hint: you don't need to use the Ohm's law formula in problems A and B to solve for the unknown.)

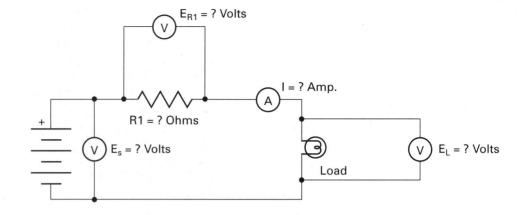

**FIGURE 8.9** Series circuit for Ohm's Law exercise in Table 8.1.

These symbols are used in Table 8.1:

$E_S$ = source voltage, same as battery voltage in Figure 8.9.
$E_{R1}$ = voltage across resistor R1
$R_{R1}$ = resistance of resistor R1
$E_L$ = voltage across the load, a light bulb
$I$ = electrical current in the circuit

Table 8.1 Ohm's Law exercise

| PROBLEM | $E_S$ | $E_{R1}$ | $R_{R1}$ | $E_L$ | $I$ |
|---|---|---|---|---|---|
| 1. | ? | 18V | — | 20V | — |
| 2. | 120V | 70V | — | ? | — |
| 3. | — | ? | 270Ω | — | 0.16A |
| 4. | — | 87V | ? | — | 3.00A |
| 5. | — | 5V | 150Ω | — | ? |
| 6. | 24V | — | ? | 9V | 0.125A |

## Ohm's Law Exercise Solutions

### Solution for 1

Here we are given the voltage drop across resistor R1 and the voltage drop across the light bulb load; we are asked to determine the battery, or source, voltage.

In a series circuit, the sum of all individual voltage drops equals the source voltage. So, the source voltage must equal the voltage drop across R1 (18V) plus the voltage drop across the light bulb load (20V):

$$E_S = E_{R1} + E_L \text{ equals } 18 + 20 = \textbf{38 volts}$$

### Solution for 2

Here we are given the source voltage (120V) and the voltage drop across resistor R1 (70V); we are asked to determine what the voltage drop is across the light bulb load.

As in Problem 1, the total of all voltage drops in a series circuit is the same as the source voltage.

Therefore, the voltage across the light bulb is the 120V source voltage minus the 70V drop across resistor R1, which is:

$$120 - 70 = \textbf{50 volts}$$

### Solution for 3

Here we are told that 0.16 ampere of current is flowing through the circuit. Because it is a simple series circuit, we know that 0.16A is flowing through resistor R1 *and* the light bulb load. Current flow from the battery is also 0.16A.

Resistor R1 has a value of 270 ohms; we are asked to find the voltage drop across the resistor.

Using Ohm's Law formula E = IR, we simply substitute the values of current (I) and resistance (R) to determine the voltage across R1:

$$E_{R1} = 0.16 \times 270 = \textbf{43.2 volts}$$

Note that in this case we did not even know the battery voltage, nor did we need to. What could we assume about the battery voltage? Well, it would have to supply at least 43.2 volts to provide the 0.16 amp of current through the 270 Ω resistance.

For this problem we really don't know much about the light bulb, other than 0.16 amp of current is flowing through it.

One other thing while we are working this problem: One-one thousandth of an ampere is a *milliamp* (mA). That is, 0.001A equals one milliamp or 1mA. Most electronic circuits have electrical currents that are less than an ampere, so milliamp is frequently encountered as the unit of current in electronic circuits. Some examples:

- 0.250A = 250 milliamp = 1/4 ampere
- 1/2A = 500mA = 0.5A
- 0.16A = 160mA
- 0.020A = 20mA
- 1/3A = 333mA = 0.333A

### Solution for 4

Given the voltage drop across resistance R1 of 87 volts, which has 3 amps of current flowing through it, we are asked to calculate the resistance of R1.

Using Ohm's Law formula R = E/I we substitute for the values of voltage (E) and current (I) to solve for $R_{R1}$:

$$R = 87 \div 3 = \textbf{29 ohms}$$

Note again that we don't know the source voltage, but it must be at least 87 volts. We also do not know anything about the light bulb, other than it has 3 amps of current flowing through it. Depending upon the particular light bulb and its filament resistance, the bulb might not even be glowing *or* it might be brilliantly lit.

The important thing here is to consider just one component's resistance and the voltage drop *across the same component* when working with series circuits. Many electronics novices get confused on this issue. For example, they might use the source voltage to determine the resistance value of R1 in exercise 4. You were not given the source voltage, so you probably avoided this pitfall.

### Solution for 5

Here we are given the voltage drop (5V) across resistance R1, which has a value of 150 ohms. We are asked to find the amount of current flowing through R1. Because this is a series circuit, the same amount of current flows through each and every circuit component.

With Ohm's Law formula I = E/R, we substitute values for the voltage (E) across R1 and the resistance (R) of R1 to determine the current (I) through R1.

$$I = 5/150 = 5 \div 150 = \textbf{0.033A} \text{ or } 33\text{mA}$$

### Solution for 6

For this last exercise we are given the source voltage (24V), the voltage drop across the bulb (9V), and the circuit current (125mA). We are asked to determine the value of resistance R1.

Well, by now we know to use Ohm's Law formula R = E/I.

But wait! If we find the value in ohms of R1 using the voltage across some *other* component in the circuit (the bulb), then we'd get the wrong answer. We must first determine the voltage drop across R1, and use *that* value for E in the formula R = E/I.

This is easy. Since our source voltage is 24 and 9 volts is measured across the light bulb, this must place 24 minus 9 or 15 volts across R1. We can now solve for the value of R1:

$$R = E/I$$

Substituting 15 for E and 0.125 for I gives:

$$R = 15 \div 0.125 = \mathbf{120\ ohms}$$

One other pitfall to be aware of when using Ohm's Law formulas: always be sure that the units you are working with are volts, ohms, and amperes. Quite often electronic circuits have current readings in milliamps. These values have to be converted to ampere before using with volts and ohms in a formula.

For example, in previous Exercise 6, if we used 125mA in the formula R = E/I instead of 0.125 amp, we would have:

$$R = 15/125 = 0.12\ \mathbf{ohms} \ldots \text{which is 1,000 times smaller}$$
$$\text{than the correct value of } 120\Omega.$$

If you had any difficulty understanding the solutions to Exercises 1 through 6, now would be a good time to go back and review the material presented so far in this chapter before going on. Work the exercises on paper using a calculator.

## Electrical Power

So far we've discussed three units of measurement in an electrical circuit: voltage, resistance, and current. Another circuit concept with which you should be comfortable before troubleshooting and repairing electrical or electronic gear is *power*.

*Electrical power is the amount of work performed over time.* Often, this work is in the form of heat, as with the resistance wires inside an electric toaster. Going back to our garden hose and water analogy, electrical power is comparable to the *volume* of water in the bucket.

It is fairly easy to understand that the volume of water flowing into a bucket from a hose in a given period of time is related to the water pressure and the number of drops, or amount, of water flowing at any instant. Increase the water pressure or the number of drops, or both, and the total volume of water delivered to the bucket will increase.

The same basic principle holds true in an electric circuit. If the voltage or current, or both, increase, then electrical power increases. Whenever there is electrical current and voltage in a circuit, electrical power is produced.

| WATER | ELECTRICITY |
|---|---|
| Pressure | Voltage (E) |
| Restriction to flow | Resistance (R) |
| Amount of flow | Current (I) |
| Volume of water | Power (P) |

The unit for electrical power is the *watt,* symbolized by the letter W. You are most likely familiar with a 60-watt electric light bulb.

Just as voltage, resistance, and current in a circuit have relationships that can be expressed with formulas, so too can electrical power. Power is symbolized in formulas by the letter P. Electrical power in watts is determined by multiplying the current flowing through a circuit component in amperes by the voltage across the component in volts. That is:

$$P = IE$$

In a common flashlight with two D-cells, the flashlight bulb draws about 500mA. What is the wattage rating of the bulb?

We'll figure that each fresh D-cell battery supplies 1.5 volts. The two batteries in series in the flashlight gives us a voltage source of 3.0 volts. So:

E = 3.0 volts
I = 0.5 amp    (remember, 500mA equals 1/2 amp)
         therefore:
P = 3.0 × 0.5 = **1.5 watts**

Okay, how much current does a 60-watt table lamp draw from a 120-volt household outlet? Answer:

P = IE         Swap items on each side of equal sign to get:
IE = P         Then, dividing both sides by E yields:
IE/E = P/E     Simplifying gives:
I = P/E        Substituting known values for P and E:
I = 60 ÷ 120 = **0.5A**

A standard 60W bulb draws 1/2 ampere of current from the wall outlet.

Besides P = IE, there are two other formulas you can use to determine electrical power consumed, or dissipated, by a component. When you don't know *both* the current *and* voltage, but you do know the resistance *and* either I *or* E, then use one of these formulas:

$$P = I^2R$$
*or*
$$P = E^2/R$$

The little $^2$ after the I and E means that the value for current or voltage is *squared,* which means multiplied by itself. For example, how much power is dissipated by a 470Ω resistor that has 25mA of current flowing through it? Answer:

P = I²R        Substituting values:
P = 0.025 × 0.025 × 470 = **0.29 watt**

Remember that a resistor reduces current flow through a circuit, but it also must dissipate electrical power. Resistors of a single ohmic value are constructed in several different physical sizes and power ratings specified in watts to dissipate the power required by a circuit.

Resistors dissipate power as heat. The ability to dissipate heat keeps a resistor from blowing up. For example, a 1/2-watt resistor can dissipate more heat than a 1/4-watt resistor, even if they have the same ohmic value. Typically, for safety reasons, you install a resistor with double the wattage rating needed by the circuit to ensure that the resistor can stand the heat without destroying itself.

To repeat, the wattage or power rating of a resistor is an independent feature of its construction. Therefore, any resistor you install must have (1) the correct ohmic value and (2), as a separate matter, the correct power rating. A resistor's ohmic value reduces current flowing through the circuit to the required amount and the power rating handles the heat the resistor dissipates.

> **POWER FORMULAS**
> $P = IE$
> $P = I^2R$
> $P = E^2/R$

Most of the electric power we use generates heat. In an incandescent electric light bulb, much more power goes to produce heat than light, about 80 percent. Florescent lamps are more efficient; a larger percentage of the power they consume goes into producing light than for an incandescent bulb, but they also generate heat. An electric motor produces quite a bit of heat as well as shaft horsepower.

We know that any time there is voltage and current flow, there is power. For electronic components, most of the power consumed produces heat. A TV set that consumes 420 watts of power converts very little of this power into light energy from the picture tube or acoustic energy from the loudspeaker. Most of the power produces heat.

Electronic engineers must determine how much power a resistor in a circuit will dissipate so a resistor that is large enough to handle the heat produced can be specified. Most resistors in electronic circuits do not consume much power; 1/4-watt and 1/2-watt resistors are adequate for the majority of circuits. Often times, 1/6W, 1/8W, and even 1/10W resistors are used in VCRs and other electronic circuits.

How much power is consumed by a 680Ω resistor that has 4.7 volts across it? Because you know two circuit values, you can calculate either of the other two. Answer:

$P = E^2/R$     Substituting with known values:
$P = 4.7 \times 4.7 \div 680 = 22.09 \div 680 =$ **0.032W**

A 1/4-watt resistor (0.25W) would be plenty adequate for the circuit, nearly eight times the power rating that is actually needed.

Suppose you've bought a pack of two red light-emitting diodes (LEDs) at an electronics store. Component data on the back of the package says that the voltage across the diode should be 2.0V and that the absolute maximum current that it can handle is 20mA.

You want to hook the diode up to a 9-volt rectangular transistor-radio battery. You also realize you'll have to connect a resistor in series with the LED to drop the voltage down to 2 volts across the diode. Since 20mA, or 0.020A, is a *maximum* current rating, you decide to select a resistor so that about 12mA of current actually flows through the LED, so the LED isn't operating at its maximum rating.

What are the calculated values in ohms and watts for the resistor you need?

First, we'll find the ohmic value required of this resistor. If 2 volts is needed across the LED and the power source is 9V, then there will be 9 − 2 volts or 7 volts across the resistor:

$$E = 7$$

Our design current is 12mA, so I = 0.012. Solving for R:

$$R = E/I = 7 \div 0.012 = \mathbf{583\Omega}$$

Now we'll determine how much power the resistor will dissipate:

$$P = IE = 0.012 \times 7.0 = \mathbf{0.084W}$$

The nearest standard resistor value readily available is 560Ω. A 1/4W resistor (0.25W) would easily handle the power requirements. Of course a larger resistor, such as a 1/2W resistor, would also be fine to use.

---

After you finish reading the next chapter, you might want to purchase a few items and experiment with electronics on your own. Start with a 9V battery, an LED, and a 560Ω 1/2W resistor. Short test leads with an alligator clip at each end are convenient for connecting the three components. Since Radio Shack stores are in just about every city, here are their stock numbers:

| | |
|---|---|
| Red LEDs (2) | 276-041 |
| 560Ω 1/2W Resistors (2) | 271-020 |
| Alligator clip test wires (10) | 278-1156 |

All parts, including a 9V battery, should cost about $6.00. One of the LEDs and one of the resistors you can even use to build an electronic infrared (IR) detector, which is described in Chapter 11.

Just don't connect the LED directly to the battery. Without the current limiting resistor, it will be destroyed instantly!

---

How else might you benefit from knowing power formulas?

Suppose mom calls on the phone and asks for some help. She says a fuse for the living room frequently blows when she's ironing at night, watching TV. She's worried that there might be something wrong with the wires in the wall or with the fuse box in the basement. Knowing you were studying electrical stuff, she thought she'd ask your advice before calling in an expensive electrician. You decide to take a crack at it. Besides, Mom's baking your favorite apple pie.

At Mom's you unscrew the 15-ampere fuse she says has been blowing. You then go about the house checking what appliances, lights, and outlets aren't working now because of the unscrewed fuse. Next you make a list of items on the 15A circuit along with the power rating for each, found on the bottom or rear panel of most appliances. Here is your list:

- L.R. floor lamp        100W
- Two L.R. table lamps   150W (75W each)
- D.R. chandelier        240W (six 40W bulbs)
- TV                     385W
- VCR                    42W
- Electric clock         2W
- Hallway ceiling light  75W
- Front porch light      100W
- Driveway floodlight    150W

Adding all the wattages you get 1244 watts, ...and wonder whether this is enough to blow a 15A fuse.

With the formula P = IE you solve for I, the current drawn if all lights and appliances were On at the same time:

P = IE          Changing equation sides:
IE = P          Dividing both sides by E:
I = P/E         Substituting known values:
I = 1244 ÷ 120 = **10.4 amps**

You use 120 for the voltage, because 120V is the nominal household voltage. You could have used 117 or 110, but even with 110 in your calculation, the current is only 11.3A, still under the 15A rating of the fuse.

You think, "No, this shouldn't cause a 15-amp fuse to blow." But then you remember Mom saying something about ironing. You ask her where she plugs it in and, sure enough, it's the same outlet that the living room floor lamp is plugged into. Checking the rating plate on the iron you discover it consumes 850 watts.

Adding 850 to the 1244 watts of everything else on the circuit gives a total of 2094 watts.

Using 2094 you again solve for current:

$$I = 2094 \div 120 = \mathbf{17.45 \text{ amps}}$$

"Yes," you reason, "this amount of current *would* blow a 15-amp fuse."

You tell Mom to either not iron at night with all the lights and TV On, or move the ironing board to the other side of the living room where there's another outlet not on the same circuit. Also tell her the apple pie was just great!

So where might you need to use Ohm's Law and power formulas when working on a VCR?

Imagine you are repairing an older VCR. You've determined that the tape load DC motor is defective. It appears to have an open motor coil; you can't measure any resistance with your ohmmeter across the two motor terminals and the motor just won't run. You know the motor drive electronic circuit is working, because you measure about 12-volts DC at the motor when you press Play. It's time to replace the motor.

Try as you may, you can't seem to locate an exact replacement motor. But then you remember an old, broken, battery-operated toy dune buggy your little brother gave you to fix. It was smashed so badly it was hopelessly beyond repair. You kept it, though, thinking maybe you could use its DC motor.

You dig out the dune buggy. Its motor looks almost identical to the one you need for the VCR. It even looks like it will mount okay. The drive pulley from the VCR motor even fits the shaft of the dune buggy motor!

But wait! The dune buggy was powered by three AA batteries in series. "Huummm, let's see. That would be three times roughly one-point-five volts per battery for a total of four-point-five volts," you say to yourself. You're right! This is a 4.5-volt motor and the VCR needs a 12-volt motor. Wiring the 4.5-volt motor directly into the VCR would most likely burn out the motor, destroy the motor drive electronic components, or both.

As you start to throw the old toy back into the junk box, you remember something: "If a resistor is wired in series with a load, some of the supply voltage will appear across the resistor, and some of the supply voltage will appear across the load." "That's it!", you think to yourself, "all I've got to do is find the right value resistor and I can use the four-point-five-volt motor in the twelve-volt VCR circuit." You decide that with what you know about Ohm's Law and power formulas you can figure out what value resistor to use.

You jot down a little sketch of what you hope to accomplish. It looks like Figure 8.10. You're really right on track; the correct value resistor will decrease the voltage going to the motor to 4.5 volts.

Looking at your sketch, it appears you've already figured out the first step: you know that with the 12-volt supply, the resistor must have 12V minus 4.5V or 7.5 volts across it. That is, with 7.5 volts across the series resistor, the remaining 4.5 volts will be applied across the motor.

To go any further, however, you realize that somehow you must determine how much current will be flowing in the circuit. You decide to determine how much current the 4.5V toy motor draws, so you connect it to 3 AA-cell batteries, with a milliammeter connected in series between the batteries and the motor. (A milliammeter is the function of a multimeter that measures current in milliamps. You'll learn how to take measurements for all multimeter functions in Chapter 9.) Your hookup looks like Figure 8.11.

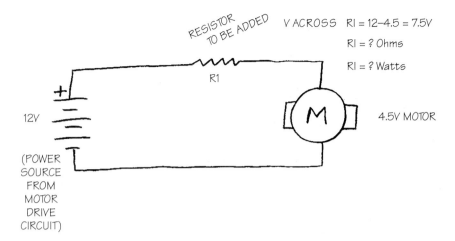

**FIGURE 8.10** Adding a resistor to 4.5V motor in 12V circuit.

You put the milliammeter in series with the motor and battery source. Current is the same at all points in a series circuit, so it doesn't matter on which side of the batteries or motor that you connect the meter, as long as it is in series. You place a little drag on the rotating motor shaft with your finger tips to simulate the motor moving the tape load mechanism in the VCR. The milliammeter indicates about 600 milliampere or 0.6A.

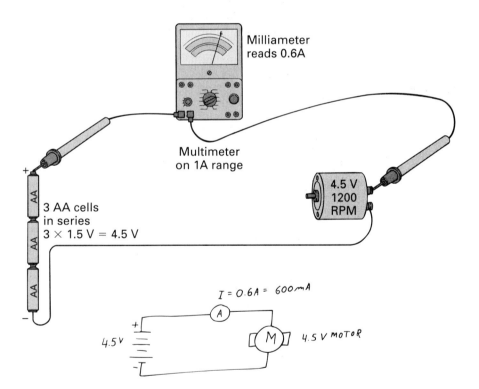

**FIGURE 8.11** Measuring current drawn by 4.5V motor.

At this point you know that the added resistor should have about 7.5 volts across it when the 4.5-volt motor draws about 0.6 amps of current. With this information, you determine the value in ohms for the resistor. A resistor that provides a specific amount of voltage drop, as in this circumstance so a 4.5-volt motor can operate from a 12-volt supply, is called a *voltage dropping resistor*. To determine its value:

$E = IR$      So:
$R = E/I$      Substituting 7.5 for E and 0.6 for I:
$R = 7.5/0.6 =$ **12.5 ohms**

That's it! A 12.5-ohm resistor should do the trick. You then remember electrical power. If there is 0.6 amp of current flowing through a 12.5-ohm resistor, the resistor is going to consume power, which means it will heat up. How much power does the dropping resistor need to dissipate? Answer:

$P = I^2R$     Substituting 0.6 for I and 12.5 for R:
$P = 0.6 \times 0.6 \times 12.5 =$ **4.5 watts**

So, you would purchase a resistor that was close to 12.5 ohms and which had a power rating of at least 5 watts.

As mentioned earlier, normally for circuits that are powered for any length of time, a resistor with at least twice the power handling capability of that actually required should be used. This conservative approach allows the resistor to run cooler, which means it won't burn out. For this particular application, however, the cam motor is operated for only short periods of time, so a 5-watt resistor would suffice. But under other circumstances it would be prudent to wire in a 10-watt resistor or one with even higher power handling capability instead of the 5-watt resistor.

Don't be overly concerned if you don't understand everything about the previous VCR repair. How to use a multimeter to take voltage, resistance, and current readings is covered in the next chapter of this book. You will learn more about resistors and LEDs during Chapter 10 on electronic components.

You may be concerned about remembering the three Ohm's Law formulas and the three power formulas. Actually, you only need to remember *just one* Ohm's Law formula and *just one* power formula. If you know these two, you can always figure out all the rest by algebraic manipulation and substitution with the two formulas.

For example, the only two formulas I bother remembering are:

1. **E = IR**
2. **P = IE**

I use two different ways for remembering them. For the Ohm's Law formula, the letters E, I, and R are in the same order as in the alphabet. You can probably guess how I recall the power formula: Mom makes a great one with apples!

If you need to find R when you know E and I, just manipulate the one Ohm's Law formula like this:

| | |
|---|---|
| E = IR | So: |
| IR = E | Divide both sides by I: |
| IR/I = E/I | Simplify; I/I cancels out, leaving: |
| **R = E/I** | |

Likewise if R and E are known and you need to calculate I:

E = IR
IR = E
IR/R = E/R
**I = E/R**

Knowing one power formula, you can determine the other two by substituting the appropriate Ohm's Law formula and then simplifying.

For example, suppose you need to determine the power dissipated by a certain value resistor that has a known voltage across it, but you don't know I. P = IE won't do, but it will get you there with a little work. Here's how:

| | |
|---|---|
| P = IE | Since I is unknown we need to get rid of it and substitute something that I is equal to. As we just determined above: |
| I = E/R | So we substitute E/R for I in the P = IE formula: |
| P = E/R × E | . . . which is the same as: |
| P = E × E/R | . . . which is: |

$P = E^2/R$

Similarly, you can get the formula to calculate power if you know the current and resistance, but not the voltage:

| | |
|---|---|
| $P = IE$ | E is unknown, so substitute what E is equal to: |
| $E = IR$ | Substitute IR for E in the $P = IE$ formula: |
| $P = I \times IR$ | ... which is the same as: |
| $P = I^2R$ | |

Knowing *how* to use Ohm's Law and power formulas is really more important than actually memorizing them. You can always look up a formula. The formula chart in Figure 8.12 is ideal for determining which formula to use for finding Power (P) in watts, Current (I) in amperes, Voltage (E), and Resistance (R) in ohms.

To use the formula chart, start near the center for the quantity to be calculated:

- P  for power in watts
- I  for current in amperes
- E  for voltage in volts
- R  for resistance in ohms

Based on whatever two electrical values are already known, select the formula to use from the three chart sectors adjacent to P, I, E, or R.

If you need to determine resistance when you know the power and voltage, start at the large R near the center circle of the chart. From the three formulas in adjacent sectors, locate the one that contains both knowns, P (power) and E (voltage). In this case the formula in the sector between 6 o'clock and 7 o'clock is the one containing the known quantities. For example, what is the resistance of a 120-volt electric iron that draws 850 watts? From the formula chart in Figure 8.12:

| | |
|---|---|
| $R = E^2/P$ | Substituting known values: |
| $R = 120^2/850$ | ... which is: |
| $R = 120 \times 120 \div 850 = $ **17 ohms** | |

Notice in the chart that two formulas require finding a square root, indicated by the symbol $\sqrt{\phantom{x}}$. Finding the square root of a number means determining the number that when multiplied by itself equals the number under the $\sqrt{\phantom{x}}$ sign. For example, the square root of 16 is 4, expressed as: $\sqrt{16} = 4$. Many

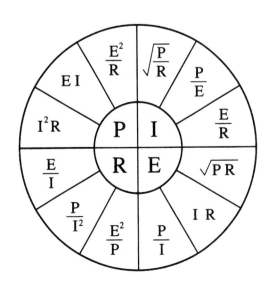

**FIGURE 8.12** Formula chart for calculating values of power, current, voltage, and resistance.

calculators have a square root function; just press the $\sqrt{\ }$ button to find the square root of whatever number is displayed.

For instance, what is the voltage drop across a 5-ohm resistor that consumes 12.8 watts of power? From the chart in Figure 8.12, select the formula that solves for E when R and P are known:

$E = \sqrt{PR}$    Substituting known values:
$E = \sqrt{12.8 \times 5}$ ... which is the same as:
$E = \sqrt{64} = $ **8 volts**

With a calculator, multiply 12.8 times 5 to get 64, then press the $\sqrt{\ }$ button to find the square root of 64, which is 8. In actual repair situations, you will very rarely need one of the formulas where the square root of a number needs to be determined.

## SERIES AND PARALLEL CIRCUITS

Whether you realize it or not, you have already studied a great deal about how electricity operates in both series and parallel circuits. Previous exercises where a battery, light bulb, and resistor are connected end-to-end, like links in a chain, are examples of series circuits. The various 120V appliances and lights on the same 15A household circuit is an example of a parallel circuit.

Here are some fundamental laws for series circuits:

---

**SERIES CIRCUIT LAWS**

- The *sum of all voltage drops* across individual circuit components equals the source, or supply voltage.
- *Current* flowing through each device is the same.
- *Total power consumed* by the entire circuit is equal to the sum of the amounts of power consumed by each component in the circuit.
- *Total circuit resistance* (external to the power source) is equal to the sum of the individual resistances.

---

Complete this exercise to fully understand what's going on with the series circuit in Figure 8.13. Be sure to refer to the above laws for series circuits. You may be able to get the correct answers without looking at the solutions, which follow. If not, be sure to carefully follow the reasoning behind the answers.

Determine the following values for the series circuit:

Total resistance:        $R_T$ = _____ $\Omega$
Circuit current:         $I_T$ = _____ A
Voltage drop across R1:  $E_{R1}$ = _____ V
Voltage drop across R2:  $E_{R2}$ = _____ V
Voltage drop across R3:  $E_{R3}$ = _____ V

### Series Circuit Solutions

Refer to the laws for series circuits, above.

First, we'll solve for the total resistance in the circuit and then for the total current flow in the circuit. Once we know the total current flow in the circuit,

Chapter 8 Understanding Basic Electronics ■ 225

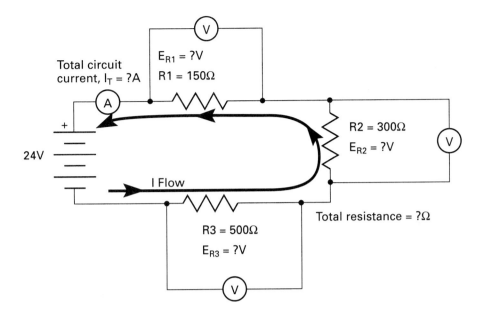

**FIGURE 8.13** Simple series circuit.

we automatically know the current through each resistance; in a series circuit they are the same. Knowing the current flow through each resistor, we can then determine the voltage drop across each resistor:

Total resistance: $R_T = R1 + R2 + R3 = \underline{\phantom{000}} \Omega$
$R_T = 150 + 300 + 500 = \mathbf{950\Omega}$

To solve for total circuit current, use the known voltage (24) and the known total resistance in ohms (950) in Ohm's Law formula:

$E = IR$     Solving for current (I):
$I = E/R$     Substituting known values:
$I = 24 \div 950 = 0.0252631 = \mathbf{0.025A}$, or 25mA

From here we multiply current flow (I) in amperes through each resistor times the individual resistance values (R) in ohms to find the voltage drop across each resistor:

$E = IR$
$E_{R1} = I_{R1} \times R_{R1}$     Substituting known values:
$E_{R1} = 0.025 \times 150 = \mathbf{3.75V}$

In like manner:

$E_{R2} = 0.025 \times 300 = \mathbf{7.5V}$
$E_{R3} = 0.025 \times 500 = \mathbf{12.5V}$

Notice that if we add the individual voltage drops across each of the three resistors we get 23.75 volts. This is essentially equal to the supply voltage of 24V. The 0.25V difference is because we rounded down current from 0.0252631 amp to 0.025A.

We can also determine the power dissipated by each resistor using Ohm's Law formula $P = IE$, substituting the value for current *through* each resistor and the voltage drop *across* each resistor.

Total power supplied by the 24V battery is equal to the sum of the amounts of power consumed by each resistor. We can also determine total power with the same formula, using the applied supply voltage in the formula $P = IE$:

$P = 0.025 \times 24 = \mathbf{0.6 \text{ watts}}$

226 ■ Practical VCR Repair

> ## PARALLEL CIRCUIT LAWS
>
> - The voltage drop across each individual circuit component is equal to the source, or supply voltage.
> - *Total circuit current* is equal to the sum of the currents flowing through each device.
> - *Total power* consumed by the entire circuit is equal to the sum of the amount of power consumed by each component in the circuit.
> - *Total circuit resistance* (external to power source) is always less than the device with the smallest resistance in the circuit, and is determined by the formula:
>
> $$R_T = \frac{1}{\frac{1}{R_1} + \frac{1}{R_2} + \frac{1}{R_n}}$$
>
> Where:
>
> $R_T$ = total resistance
> $R_1$ = resistance #1
> $R_2$ = resistance #2
> $R_n$ = other resistances

Next, take a close look at the fundamental laws for *parallel* circuits:

Don't let that complicated formula for total resistance in a parallel circuit get you down. You will hardly ever need it, but knowing what it is and how to use the formula could come in handy some day.

We'll see how drastically things change when the same value resistors from the previous, series circuit exercise are placed *in parallel* across the same voltage source, as depicted in Figure 8.14. Try solving for the unknowns.

Review the laws for parallel circuits and determine the following values for the parallel circuit:

Current through R1:  $I_{R1}$ = _____ A
Current through R2:  $I_{R2}$ = _____ A
Current through R3:  $I_{R3}$ = _____ A
Total circuit current:  $I_T$ = _____ A
Total resistance:  $R_T$ = _____ Ω

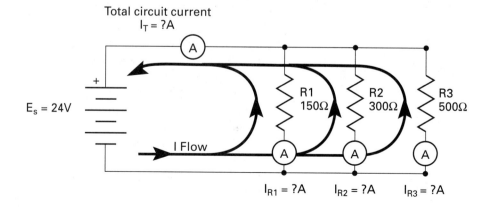

**FIGURE 8.14** Simple parallel circuit.

## Parallel Circuit Solutions

First off, notice that the voltage drop across each of the three resistors is equal to the supply voltage, 24V. Then, it's easy to solve for the current through each resistor:

$E = IR$          Solving for I:
$I = E/R$        Substituting with known values:
$I_{R1} = 24 \div 150 =$ **0.16A** or 160mA
$I_{R2} = 24 \div 300 =$ **0.08A** or 80mA
$I_{R3} = 24 \div 500 =$ **0.048A** or 48mA

Calculating the total amount of circuit current is simply a matter of adding the current through each resistor:

$I_T = I_{R1} + I_{R2} + I_{R3}$
$I_T = 0.16 + 0.08 + 0.48 =$ **0.288A** or 288mA

To solve for the value of the total resistance across the supply voltage, use a calculator and the formula:

$$R_T = \frac{1}{\frac{1}{R_1} + \frac{1}{R_2} + \frac{1}{R_n}}$$

Where:

$R_T$ = total resistance
$R_1$ = resistance #1
$R_2$ = resistance #2
$R_n$ = other resistances

In this case $R_n$ becomes $R_3$. Substituting with known values:

$$R_T = \frac{1}{\frac{1}{150} + \frac{1}{300} + \frac{1}{50}}$$

Simplifying:

$$R_T = \frac{1}{0.006666 + 0.003333 + 0.002}$$

$R_T = 1 \div 0.0119999 =$ **83.33Ω**

We could have determined the total resistance in a somewhat easier manner. Can you figure out how?

Since we know the total voltage (24V) and the total current, which we previously calculated (0.288A), we plug these values into Ohm's Law and solve for R:

$E = IR$          Solving for R:
$R = E/I$        Substituting with known values:
$R = 24 \div 0.288 =$ **83.33Ω**

With what we know, the power dissipated by each of the three resistors can be determined easily using the formula $P = IE$, where I is the current through each individual resistor and E is the supply voltage of 24V, which appears across each resistor.

Total power consumed by the circuit can be determined by adding the amounts of power dissipated by each of the three resistors, or by substituting the total amount of current drawn by all three resistors in the P = IE formula. We calculated $I_T$ to be 0.288A, so:

$$P_T = 0.288 \times 24 = \textbf{6.9 watts}$$

... quite some difference from the total power of **0.6** watts with the same value resistors *in series* across 24 volts, which we calculated a little earlier. The 150, 300, and 500Ω resistors *in parallel* have a much lower resistance to current flow than when they are in series.

Since E = IR, for a given applied voltage, if the resistance goes down, the current goes up. This equates to higher power consumption.

How did you do with the series and parallel circuit exercises? If you had trouble, and couldn't make sense out of the solutions given, you should definitely go back and review the material in this chapter before continuing. Grab pencil, paper, and calculator. Actually work through the exercises; hands-on experience makes it easier to grasp.

If you're having trouble understanding the effect of parallel resistance in a circuit, review the scenario at Mom's house and what happens when she plugs in the clothes iron. That's adding parallel resistance to the 120V, 15A household circuit.

Knowing the formulas for total resistance in series and parallel circuits can often be useful. For example, suppose you are designing or repairing a circuit that calls for a 1,000Ω, 1/2W resistor. You look all through your collection of resistors, but can't find one having these values.

Aha! You recall that resistance values in series add. You correctly reason that any two resistors whose values together equal 1K-ohms could be used in place of a single 1K resistor. Looking through your resistor collection again you find a 560Ω resistor and a 470Ω resistor. Putting these *in series* gives 1030 ohms, close enough in most cases for the 1,000Ω resistor called for.

You could also put two 2.2K-ohm resistors *in parallel,* which would give a combined resistance of 1,100 ohms, still close enough for most applications. It is easy to figure out the combined resistance when resistors in parallel are of the same value. For two or more parallel resistors, where each has the same value, the rather complicated formula for total resistance in a parallel circuit can be simplified to:

$$\textbf{R}_T = \textbf{R/n}$$

Where:

$R_T$ = total resistance
R = value of each resistor in ohms
n = the number of like-value resistors

Example: What is the value of three 5.6K resistors in parallel?
Solution:

$$R_T = 5600 \div 3 = \textbf{1867}\Omega$$

As you become more familiar with electronics, you might very well decide to design some of your own simple circuits, build a project from a magazine article, or modify an existing piece of equipment for some special purpose. Knowing how to calculate voltage, current, resistance, and power in circuits is a tool you will use repeatedly in these endeavors.

## ALTERNATING AND DIRECT CURRENT

Two fundamental types of electricity can flow in a circuit: direct current (DC) and alternating current (AC).

Batteries supply *direct current;* the electricity in your house or apartment wiring is *alternating current.* The ignition system, lights, windshield wiper motor, and radio in your car work from a 12-volt direct current system provided by the car's battery. Household appliances operate on alternating current from the electric power company.

Virtually all consumer and industrial electronic equipment works on DC. Despite this fact, power companies transport electrical energy as AC, because it can be transported further and less expensively than DC. A power supply inside electronic equipment converts the AC to DC for use by electronic circuitry. If the device is battery operated, in many cases the batteries are the entire DC power supply.

Utility companies transport power over long distances at very high voltages, in hundreds of thousands of volts. Why? Well, the formula $P = IE$ tells the story. For a given amount of power, if the voltage is high, the current is low: One ampere at 10 volts is 10 watts, just as 10A at 1V is 10W. Low current at very high voltage means smaller size wires can be used to deliver the needed amount of power. If the voltage was lower, larger, more expensive wires would be needed.

Transformers at electric company substations convert the very high voltages on long-haul power lines to lower voltages, in the thousands of volts, for distribution on regular wooden utility poles throughout the community. Transformers on these utility poles then convert the high voltage down to 120 volts for homes and small businesses. Notice that the 120-volt wires between the transformer and a building are larger than the high voltage wires running along the top of the utility poles. At the lower voltage of 120, larger wires are needed to handle the current.

Alternating current is used to distribute electric power because transformers simply do not work at all on direct current. There is no easy way to efficiently convert one DC voltage level to another, as can be done with alternating current using transformers.

Direct current always flows in the same direction. Electrons come out the negative terminal on a battery, go through a load, and return to the positive battery terminal. Each battery terminal remains negative or positive. Current flows only one way.

Alternating current reverses the direction of current flow, usually at some regular rate. In the United States, electric power alternates or changes direction 60-times each second. This is a frequency of 60 Hertz (60Hz). Fifty Hertz power is common in many other parts of the world.

Consider the two wires carrying electricity from a 120-volt, 60Hz wall outlet to your desk lamp. Sometimes one wire is negative with respect to the other; at other times the same wire will be positive when the other wire is negative. This changes back and forth 60 times a second.

If we were to look at the voltage on just one of the wires for 1/60 second, this is what we would see:

- Voltage starts at zero volts and gradually builds up to a maximum positive voltage
- From the maximum positive voltage, voltage then gradually decreases back down to zero volts
- From zero volts, voltage gradually builds up to a maximum negative voltage
- From the maximum negative voltage, voltage gradually decreases back to zero volts.

. . . and all this happens in just one-sixtieth of a second. It is one *AC cycle.* The cycle then repeats itself 59 more times during the same second.

Figure 8.15 illustrates how an AC voltage changes during two complete AC cycles.

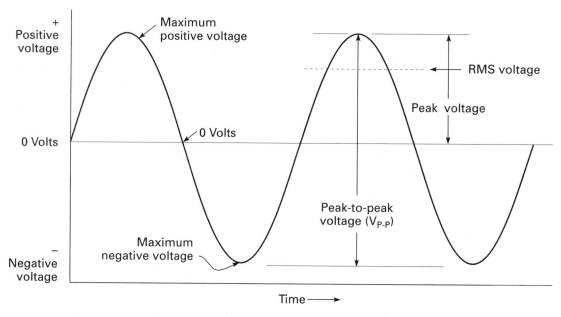

**FIGURE 8.15** Two complete alternating current cycles

Without getting too technical, the manner in which AC voltage changes during a cycle follows the trigonometric sine function related to the AC generator producing the power. And so we describe the voltage wave as a *sinewave*.

You may have *heard* 60Hz sine waves! If you touch a finger to the center pin of an RCA plug on a cable plugged into an audio amplifier input and the volume is turned up, you will most likely hear a low frequency hum from the loudspeaker. What is happening is that your body is acting like an antenna. It picks up radiated 60Hz energy from the wiring in the walls and nearby AC line cords. The amplifier increases the weak signal enough so you can hear it. Should you do the same thing with an automobile audio amplifier, you will not hear any 60Hz hum, unless there are power lines nearby.

The sinewave in Figure 8.15 depicts the AC voltage for two cycles on one of the two wires going to the desk lamp. The voltage on the other wire follows an identical pattern, but in the opposite direction, like a mirror image. That is, when one wire is at its maximum positive voltage, the other wire is at its maximum negative voltage.

As you can see from the sinewave, the voltage going to your lamp is actually zero two times during every cycle: once at the very beginning of the cycle and again half way through the cycle. Your desk lamp is really being turned On and Off 120 times each second! Also notice that the highest, or *peak*, voltage only occurs for a very small amount of each cycle.

Because AC power is actually turning On and Off, it takes a higher maximum AC voltage to produce a specific amount of power than a DC voltage, which is On all the time. This AC voltage is referred to as the *peak* voltage. Referring back to Figure 8.15, notice that the voltage between the negative and positive peaks of the sinewave is labeled $V_{p-p}$, meaning the *peak-to-peak voltage*.

An AC voltage is not normally described or measured according to either its peak or peak-to-peak voltage values. Instead, it is rated and measured at the voltage that would produce the same amount of power as a DC voltage. This AC voltage is called the root mean square (RMS) voltage. We're not going to get into why it's called that; just know that with AC, we're normally talking about *RMS voltage*.

Most AC voltmeters read RMS voltages. Usually the "RMS" is left off when describing an AC voltage. Unless specified as a peak or a peak-to-peak value, you can assume an AC voltage description or meter reading is in volts *RMS*.

Household power in the United States is 120VAC. This is the RMS, or *effective,* value of the AC voltage. The word effective means 120 volts AC placed across a resistance will produce the same amount of power as 120VDC across the same resistor. You *may* hear someone talk about RMS voltage as an average voltage, which is technically incorrect.

You will rarely need to know anything about peak or peak-to-peak (*P-P*) voltages in VCR repair, although we will talk about this some more during our discussion of power supplies in Chapter 12. Manufacturers' service literature will often have electronic adjustment procedures with pictures of signal waveforms with P-P voltage readings. This requires using an oscilloscope, which displays electronic waveforms on a cathode ray tube (CRT), similar to a small TV screen. Oscilloscope use is not covered within this textbook.

The relationships between RMS, peak, and peak-to-peak voltages are as follows:

- $V_{Peak} = V_{RMS} \times 1.414$
- $V_{RMS} = V_{Peak} \times 0.707$
- $V_{P-P} = 2 \times V_{Peak}$

What are the peak and peak-to-peak voltages of common 120VAC household power? Answers:

$V_{Peak} = 120 \times 1.414 = \mathbf{170V}$

$V_{P-P} = 2 \times 170 = \mathbf{340V}$

In Chapter 12 on VCR power supplies you'll learn where peak and peak-to-peak voltages come into play.

All basic automobile electric power is DC, supplied by the battery. In a house or apartment, all basic electric power is AC. In many electronic circuits both AC and DC voltages may be present on the same wire or component lead at the same time. For example, an output transistor in an audio amplifier may have 35 volts of DC on one of the transistor leads and 5 volts of AC, which is the music signal delivered to a loudspeaker.

## Impedance

*Impedance* is to AC circuits what resistance is to DC circuits. The unit of measure for impedance is the ohm. For a device like a resistor, heating wire, or incandescent light bulb, the DC resistance and AC impedance are identical. However, for many other components and combinations of components in a circuit, AC impedance and DC resistance can be drastically different.

This is true for electromagnets, coils, transformers, inductors, capacitors, and combinations thereof. You will read more about these components in Chapter 10. A term similar to impedance is *reactance,* which is the opposition to AC current flow caused by an inductor or capacitor. Although slightly different from impedance, you may hear the word reactance used in its place.

The main thing you should know right now is that DC resistance and AC impedance are *not the same*. But you treat impedance and resistance the same in Ohm's Law and power formulas. The symbol for impedance is Z, which is the same as R in the formulas.

See the Chapter 8 Appendix for additional information about impedance, along with some examples using Z in formulas.

## Polarity

You'll often come across the term *polarity* in association with electrical and electronic circuits. Polarity describes the direction of current flow through a device. It also describes whether a wire conductor, battery terminal, or component lead is positive, identified with a plus sign, or negative, indicated by a minus sign. Many electronic devices, such as diodes, transistors, LEDs, integrated circuits, and electrolytic capacitors, must be connected in a circuit in the right direction, that is, with correct polarity. For example, as shown in Figure 8.16, if an LED is connected backward to a DC voltage source, it will not be damaged, but it won't light.

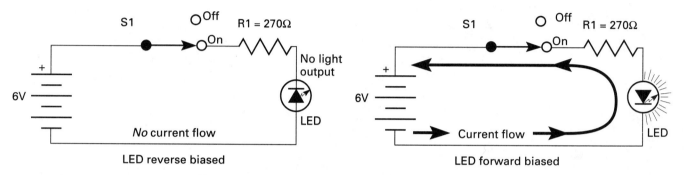

**FIGURE 8.16  Polarity of an LED in a circuit matters. The LED in the circuit on the left is installed backward, or with reverse polarity, and no current flows.**

When a power supply DC output, battery, or polarity-sensitive device is accidentally connected backward, the condition is referred to as *reversed polarity*. In many cases, reverse polarity can destroy an electronic component. For example, soldering an electrolytic capacitor into a circuit backward will cause the capacitor to heat up and perhaps even explode when the circuit is powered up. You'll learn about electrolytic capacitors and other polarity-sensitive components in Chapter 10.

AC current comes out of your wall outlet when you plug in your TV or other home appliance. To repeat, this current is 110 to 120 volts and alternates, or changes polarity, 60 times each second. These are standards for voltage and frequency in the United States. When one side of the AC line is electrically positive, the other side is negative. This changes back and forth at a 60-Hertz rate.

However, electronic circuits work with direct current. A power supply converts AC from the wall outlet to DC for use in electronic equipment, like a VCR. A power supply usually changes the voltage also. For solid state electronics, low DC voltages are required. For example, a typical VCR power supply might have the following output voltages:

- +5VDC
- +12VDC
- −30VDC
- 5.2VAC
- 5.2VAC
- Gnd/Com

The power supply output labeled Gnd/Com is the ground or common conductor for all DC voltages. In some VCRs, it may also be one side of a low AC voltage. As a common power supply output, the Gnd/Com lead is the negative terminal for positive DC voltages and is the positive terminal for negative DC voltages. For the power supply outputs just listed, Gnd/Com is the negative polarity terminal for the +5VDC output *and* at the same time the positive polarity terminal for the −30VDC output. Gnd/Com is really just a shared conductor for the various DC outputs of the power supply.

Notice that there are *two* outputs labeled 5.2VAC. "Why?" you ask.

That's not a mistake. In this case there are actually two output terminals on the power supply for this AC voltage. The polarity of one terminal alternates back and forth *with respect to the other;* there is no ground or common reference for this AC output back to the power supply. Instead, current goes out one of the wires and back on the other. This alternates 60 times a second. The 5.2 volts of alternating current is available *between the two terminals,* not between ground and either of them.

Direct current power supply outputs go to various VCR circuits. A twelve volt output frequently powers the front loader and cam motors. Higher DC voltages often supply other transport motors, a timer integrated circuit (IC) chip, and electronic tuner.

A low voltage AC output usually powers a heater in the florescent display. An AC output may also be used as a reference signal for the clock/timer. Essentially, this is how this is accomplished: Alternating current supplied by power companies in the United States has an extremely steady, accurate, and reliable frequency of 60 Hertz. First, 120VAC from a wall outlet is transformed down to 5 or 6 volts AC by the VCR power transformer. This low voltage 60Hz AC signal is next shaped into 60Hz pulses called square waves, which are then electronically divided by 60 to get 1Hz pulses, or one pulse each second. These very precise one-second timing pulses are then used to step the clock and timer circuits. (Some VCRs have a quartz crystal oscillator that provides timing pulses independently of the AC line frequency.)

Getting back to power supply outputs, the majority of VCR circuits are powered by +5 volts DC. For a particular model VCR, it is necessary to refer to the schematic diagram in the manufacturer's service literature to determine power supply output voltages, what connector pins carry the various output voltages, and where each is used in the VCR.

Just for comparison, tube-type equipment usually requires DC voltages *higher* than the 120-volt AC line. Typical voltages may be in the range of 350 to 450 volts DC (VDC). The cathode ray tube (CRT) in a TV or computer monitor usually requires at least 1,000 volts DC for each inch of diagonal screen measurement. For example, a 19-inch TV picture tube might have anywhere between 19,000 and 23,000 volts applied to it.

Once again, polarity refers to the positive ⊕ and negative ⊖ ends or leads of a power supply, battery, or electronic component. Polarity also means the electrical orientation of a device in a circuit. Installing a battery backward or wiring in a component with its ⊕ and ⊖ terminals swapped is called *reversed polarity,* which can damage electronic components and circuitry. Components that are sensitive to the direction in which they are connected into a circuit are sometimes called *polarized*.

Resistors, switches, and incandescent light bulbs are *not* polarity-sensitive; it doesn't make any difference which terminal or lead of these devices is connected to ⊕ or ⊖. Simple DC motors with only two terminals, like those typically used in VCR front loaders and cam mechanisms, are polarity sensitive. Connected to DC one way, they turn in one direction. Reversing the polarity of the applied DC voltage at the motor terminals causes the motor to turn in the opposite direction.

With AC, the term polarity can also refer to more than just the fact that the two conductors in an AC circuit change from positive to negative at some rate, like 60Hz. Most household appliances today have a polarized plug on the two-conductor line cord; notice that one blade on the plug is wider than the other. This forces you to insert the plug in a wall receptacle only one way.

Three-wire grounding plugs are also considered polarized, since you can only insert them one way. This plug polarization has nothing at all to do with which wire is positive and which is negative. The electricity is still changing polarity 60 times a second.

Plug polarization on AC appliances determines which line cord conductor is at *neutral* or ground potential and which is *hot* or 120 volts above neutral or ground. The wider 120VAC outlet contact is at earth ground potential and is called *neutral*. The narrow slot in a wall receptacle is the *hot* conductor; there are 120 volts available between hot and neutral/ground. Polarized plugs are on appliance cords primarily for safety reasons. Consider a standard plug-in incandescent lamp. The shell or threaded portion of the lamp socket is wired to the wide, neutral plug prong. Then if you are in bare feet on the cellar floor while changing a bulb and accidentally touch the threads of the bulb as you screw it into the socket, you won't get a shock. The polarized plug ensures that the threaded portion of the socket is connected to neutral, which is at the same electrical potential or voltage as the ground on which you are standing. Because the socket shell and cellar floor are at the same ground/neutral potential, you won't get shocked by touching both at the same time.

If the lamp had a non-polarized plug, you'd have a 50/50 chance that the socket shell would be connected to the hot side of the AC line, depending on which way the plug was inserted into the wall outlet. Accidentally touching the lightbulb threads in a socket shell that is connected to the hot side of the AC line could deliver an extremely dangerous, perhaps fatal, shock if there were an electrical path to ground through your body.

Any electrical connector that can be plugged in when oriented only one way is called a *polarized connector*. In general, all connectors in a VCR are polarized. It's still a good idea, however, to mark the two connector halves with an indelible pen so there is less chance of accidentally connecting them together backward. Some connectors have a small plastic lockout plug or key in the female half that prevents a pin in the male half from seating if an attempt is made to plug it in backward. These plastic plugs *have* been known to fall out when the connector halves are apart, thus allowing the connector to be plugged backward.

The relevance of this discussion about polarity will become clear in Chapters 9 and 10. If electronics is new to you, consider re-reading this discussion before moving on to those chapters.

## ANALOG AND DIGITAL CIRCUITS

It seems that we see or hear the term digital almost constantly these days. Starting with the Compact Disc (CD) in the '80s, which employs digital recording and playback, digital technology has become increasingly utilized. There's the new Digital Compact Cassette (DCC), digital telephone answering machines, digital special effects incorporated into TVs and VCRs, and the digital superhighway that will link our digital PCs and digital TVs to a world of information and services. More and more, these products are using microprocessors to control their various operations and to process analog audio and video signals.

VCRs have both digital and analog circuitry; therefore, you need a basic foundation of what analog and digital signals are. But exactly what is digital? For that matter, what is analog?

### Analog

In an *analog* circuit, a particular signal is represented or composed of many different voltage levels at numerous frequencies. An analog audio signal, for example, has extremely complex waveforms. At any instant in time, the

**FIGURE 8.17** Analog audio signal displayed on an oscilloscope

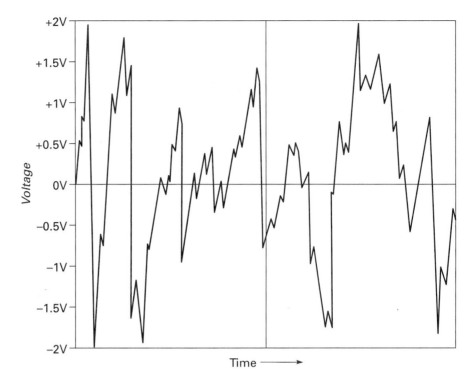

voltage of an analog music signal could be at any value over the operating range of the circuit.

For example, suppose a videocassette is playing. The signal from an RCA audio Line Output jack on the VCR is fed to an oscilloscope. We might see something similar to Figure 8.17 on the 'scope face. The horizontal X-axis is time, in this case 1/100 of a second, and the vertical Y-axis is the audio output voltage from the VCR.

Notice that the analog audio waveform in Figure 8.17 can have any value of voltage between −2V and +2V during the 1/100 of a second that was sampled. Another way to describe an analog signal is to say that it is *constantly changing in amplitude and frequency*. Amplitude or signal voltage can be at nearly an infinite number of values over time.

The volume control on your TV is an example of an analog device; the signal it controls can vary over a wide range from full Off to soft to medium to all-the-way-up loud.

Another area where the term analog is used is in reference to meters. A meter that has a moving pointer that passes over a printed scale with numbers is an *analog meter*. A speedometer with a pointer that moves in an arc past printed numbers like:

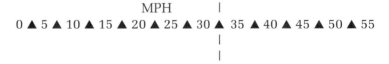

... is an analog readout of the vehicle's speed. Analog electronic test meters are similar, having a moving pointer or needle.

## Digital

Whereas an analog signal has varying voltage levels, a *digital* signal can have *only two states* or conditions, On or Off. In a circuit, digital On and Off states exist as only two voltage levels. One voltage means On, another voltage means Off.

There are several ways to describe the two possible digital states or conditions:

- On or Off
- Logical 1 or Logical 0
- High or Low
- True or False
- Yes or No

The most frequently used descriptors for the state of a digital circuit are: On and Off; 1 and 0; High and Low.

For example, in a digital circuit, an On condition may be represented by plus 5 volts and Off represented by zero volts. These are the only two voltages levels that will ever be present in the circuit. Plus 5 volts represents 1 or High; zero volts represents 0 or Low.

Think of a flashlight as a digital device: it has two possible states, either On or Off. We could say the flashlight is in a logical 1 state when it is On and in a logical 0 condition when it is turned Off.

A specified time interval during which a signal is either On or Off is called a *bit*, which stands for *binary digit*. A binary digit can have only one of two values or conditions at any time: 0 or 1. That is, a bit can be either On or Off, High or Low, True or False, 1 or 0.

Figure 8.18 illustrates 8 bits during a period of time. Each bit is identified by a number from 0 through 7. Notice which of the bits are 1 (On) and which are 0 (Off). As you can see, bits numbered 0, 2, 4, 5, and 7 are On, with a voltage level of +5V. Bits 1, 3, and 6 are Off, with a voltage level of 0V.

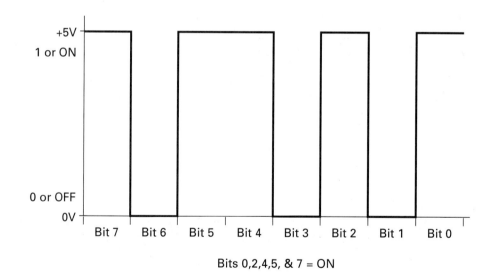

**FIGURE 8.18** Digital signal having 8 bits numbered 0 through 7

You could think of the 8 bits as eight individual flashlights labeled 0, 1, 2, 3, 4, 5, 6, 7. At a particular moment, flashlights 0, 2, 4, 5, and 7 are turned On, and flashlights 1, 3, and 6 are Off. The 8 bits taken together are called a *byte*.

A combination of bits On and Off in one or more bytes can represent an analog voltage level at a particular instant in time, for example at some point in the analog music waveform shown in Figure 8.17. This is essentially how an analog audio or video signal can be stored and handled digitally: various combinations of bits On and Off represent different voltage levels along the analog waveform. With sufficient numbers of bit combinations, all analog voltage levels can be represented digitally.

Analog signals are changed to a stream of digital bits that represent the voltage at various points along the analog waveform by an *analog-to-digital*

*(A/D) converter*. This digital data can then be stored on a CD or computer disk, or it can be modified by a digital signal processor (DSP). The digital bit stream can later be changed back to an analog signal with a *digital-to-analog (D/A) converter*. You may also see *ADC* for analog-to-digital converter and *DAC* for digital to analog converter.

An important part of a VCR is the system control microprocessor. VCR status can be expressed using 8 bits. For example, bits that are On might indicate the following to the system control microprocessor:

| On Bits | Meaning |
|---|---|
| 0 | VCR Power On |
| 2 | Front loader cam at cassette load position |
| 4 | Cassette-in switch transferred |
| 5 | Pinch roller loaded against capstan |
| 7 | Take-up reel sensor pulses being received |

This VCR microprocessor example is intended to give you a general idea of how digital signals can indicate various VCR states or operations. Chapter 14 looks at VCR microprocessors in a little more detail.

The term digital also applies to meters and readouts. Just as digital speedometers with direct-readout displays have replaced speedometers with a needle on most cars, so too have digital volt-ohm-milliammeters (VOMs) become more popular than analog meters. You'll learn about both types in Chapter 9.

This is just an introduction to analog and digital. As long as you have a broad idea of how the two types of signals differ and how they are used, that is all right at this point in your learning. How and where digital and analog signals are used in a VCR can vary greatly by model. You will need to consult the manufacturer's service manual and schematic diagrams for specifics on a particular model.

In general with VCRs, all audio and video for straightforward reception, recording, playback, and RF modulation to Ch 3/4 is handled by analog circuitry. Special effects, like picture-in-picture (PIP) and noise-free slow motion are handled by digital circuitry. Timer programming, system control, front panel display, IR remote control, and tuner channel memory are also digital in most cases.

## SUMMARY

A lot of ground was covered during this chapter, which started with a brief explanation of what electricity is: the flow of electrons through a conductor. Electrons flow from negative to positive.

For some people it is helpful to relate electricity to water flowing through a garden hose or pipe. Water pressure is similar to voltage, the amount of water flowing at any instant is analogous to electric current, and any restriction to the flow of water, like pinching a garden hose, is like electrical resistance.

Within a circuit, the relationship of voltage, current, and resistance is described by Ohm's Law, **E = IR**, where E is voltage in volts, I is current in amperes, and R is resistance in ohms. That is, the voltage across a circuit component is equal to the current flowing through the device multiplied by the resistance of the device. A closed switch should have 0-ohms resistance, so we would not expect to see any voltage drop across a switch in its On position. A defective switch in an operating circuit, where there *is* some switch contact resistance, will have a voltage drop across the switch.

Knowing that **E = IR**, we can derive the formulas for determining current (I) and resistance (R):

$$I = E/R \quad \text{and} \quad R = E/I$$

In most electronic circuits, current is expressed in thousandths of an amp, or a milliamp (mA).

Examples: 0.25A = 1/4A = 250mA; 0.018A = 18mA; 0.006A = 6mA

Power is the quantity of electricity consumed, or dissipated, by a device. Power is similar to the volume of water in a bucket.

The power in watts consumed by a device is determined by multiplying the current flow through the device by the voltage drop across it:

$$\mathbf{P = IE}$$

If any two of the units voltage, current, resistance, and power are known, the other two can be calculated with the appropriate formula. A circular chart presented in this chapter may be used to determine which formula to use, depending on what two electrical quantities are known or given.

There are two basic types of electrical circuits: series and parallel. In a series circuit:

- The sum of all voltage drops equals the applied voltage
- Amount of current is the same through each component
- The power consumed by the entire circuit equals the sum of the amounts of power consumed by each circuit component
- The total circuit resistance (external to power source) equals the sum of the individual resistances

In a parallel circuit:
- The voltage drop across each component equals the applied, or source, voltage
- Total circuit current equals the sum of the current flows through each parallel device in the circuit
- Total circuit power equals the sum of the amounts of power consumed by each parallel circuit component
- Total circuit resistance (external to power source) is smaller than the smallest parallel resistance, and is determined by the formula:

$$R_T = \frac{1}{\frac{1}{R_1} + \frac{1}{R_2} + \frac{1}{R_n}}$$

Where:

$R_T$ = total resistance
$R_1$ = resistance #1
$R_2$ = resistance #2
$R_n$ = other resistances

For resistors of the same value, this can be simplified to:

$R_T = R/n$

Where R is the value of one resistor in ohms and n is the number of like-value resistors in parallel.

There are two primary types of electric current: direct (DC) and alternating (AC). Direct current always flows in the same direction because the voltage source terminals stay constantly negative and positive. A battery is an example of a DC voltage source. Alternating current, such as household power, changes polarity of the two terminals back and forth between negative and positive many times a second. AC power in the United States changes polarity at a 60Hz rate. The waveform that describes voltage changes in an AC circuit is the *sinewave*.

The chapter concluded with a brief description of analog and digital signals. An *analog signal* is constantly changing in amplitude and frequency. The signal from a VCR audio Line Out jack is an example of an analog signal. A *digital signal* has only two discrete amplitudes or voltage levels. This is called binary, meaning two. One voltage level indicates a logical 1 or On condition; the other voltage level indicates logical 0 or Off. For example, in a digital circuit, +5VDC could represent 1 or On and zero volts could mean 0 or Off. The word *bit*, which is short for binary digit, describes the two possible states of a digital signal or circuit.

An analog signal can be represented by several bits considered together. Eight separate bits considered together are called a *byte*. Within one or more eight-bit bytes, different combinations of bits being On and Off represent the amplitude at various sampling points along an analog waveform. An A/D converter changes analog to digital, and a D/A converter changes digital signals back to analog.

Within most VCRs, analog circuits are used to receive and amplify RF signals, record and play back video and audio on videotape, and output to line level and RF jacks. Digital circuitry is often used for picture enhancement, picture-in-picture, and slow motion effects as well as the usual system control and timer/operation microcomputer functions. Digital circuits create on-screen menus, decode remote control pulses, and provide electronic tuner preset memory. Manufacturer's service literature should be consulted for specifics on a particular VCR model.

Note that there is a Chapter 8 Appendix with additional information about subjects covered in this chapter.

## SELF-CHECK QUESTIONS

1. What is meant by an electrical conductor? Give at least three examples. What is meant by an electrical insulator? Give at least three examples.
2. What are the three basic parts of an electrical circuit? Give an example of each part in an automobile or household electrical circuit.
3. Explain what is meant by electron current flow and conventional current flow. Which of these two is considered technically correct and is used throughout this VCR repair course?
4. What basic word describes each of the following characteristics of an electrical circuit? Give the unit of measure and symbol for each characteristic.
   a. Electric pressure or force
      1. Unit of measure
      2. Symbol
   b. Restriction to flow of electricity
      1. Unit of measure
      2. Symbol
   c. Amount of electric flow
      1. Unit of measure
      2. Symbol
5. State Ohm's law for electrical circuits in a sentence and as a mathematical equation.
6. What are two principal uses for a resistor in an electrical or electronic circuit?
7. What is meant by a series circuit?

8. State the two electrical laws for a series circuit.
9. What happens to the amount of electricity flowing in an electrical circuit if the electrical pressure remains the same, but the restriction to the flow of electricity goes down? State the Ohm's law equation that shows this relationship.
10. What does the symbol $\infty$ mean? How is this symbol frequently used in electronics?
11. What is the unit of measure for electrical power? If you know the voltage drop across a resistor and the amount of current flowing through the resistor, how can you determine the amount of power being dissipated by the resistor?
12. How much power is handled by a 500Ω resistor that has 12V across it?
13. You want to add a jumbo LED indicator to the dashboard of your car to tell when the brake lights are actually on. What ohmic value series resistor do you need so that the voltage across the diode will be 1.85V and the current through the LED is limited to 20mA? Assume the nominal voltage from the car's electrical system is 13.7V. How much power must this series resistor dissipate?
14. State the series circuit law for the total value of two or more resistors in the circuit. State the series circuit law for the total amount of power consumption.
15. State the parallel circuit law for the voltage drop across individual circuit components. State the parallel circuit law for the total amount of circuit current.
16. In a couple of sentences, describe what happens to the voltage of an alternating current signal during one complete sine wave cycle.
17. Describe what the term electrical polarity means. When might you need to be concerned about polarity while servicing a VCR?
18. Give a very **general** description of an analog electronic signal. Give one example of where you would expect this type signal in a typical VCR.
19. Describe a digital electronic signal. Give one example of where you would expect to find digital circuitry in a typical VCR.
20. In a VCR circuit, a particular digital signal line is described as being "high" when the Cassette-in switch is transferred. Describe what this means. What are two other ways to describe the logic high state of this circuit line?

# CHAPTER 9
# HOW TO USE A MULTIMETER

Meters are one way in which you can "see" what's going on in an electrical circuit. Voltmeters, ohmmeters, and ammeters are essential in all electronic work. They are among the simplest and least expensive pieces of test equipment available. A multimeter is as basic to an electronics technician as a hammer is to a carpenter.

By studying this chapter, you will learn how to use a multimeter to measure electrical voltage, resistance, and current. Knowing how to connect a meter and read these electrical circuit values is a fundamental skill that an electronics technician must have to perform basic tests and troubleshooting.

## CHAPTER OBJECTIVES

Upon completing this chapter, you should be able to:
1. Describe how to connect a basic volt-ohm-milliammeter (VOM) in a circuit to measure voltage, resistance, and current.
2. Describe basic differences between analog and digital multimeters.
3. Correctly read values of voltage, resistance, and current from illustrations of an analog meter face printed within the chapter.
4. Describe some of the features to look for when selecting a VOM suitable for VCR troubleshooting and repair, including an appropriate meter sensitivity specification.

## INTRODUCTION TO MULTIMETERS

From Chapter 8, you know that voltage, resistance, and current are fundamental characteristics of electrical and electronic circuits. Many times it is necessary to know what one or more of these values are in a circuit to perform problem diagnosis. For example, if the DC motor in a VCR front loader mechanism doesn't turn at all when it should, you would want to know if the correct voltage was present at the motor terminals.

You *could* use one meter to measure AC voltage, a different meter for DC voltages, another meter for taking resistance readings in ohms, and yet another meter for determining how much current is flowing in a circuit. Rather than have all these separate meters, most electronic technicians use a combination meter called a *multimeter*. This single meter can measure AC and DC voltage, resistance, and DC current in milliamps. There is rarely a need to measure AC current in consumer electronic equipment, and most electronic circuitry has less than 1 ampere of current. Many multimeters *are* capable, however, of measuring up to 5 or perhaps even 10 amps of DC current.

One or more switches on the multimeter allow the test instrument to function as an AC voltmeter, DC voltmeter, ohmmeter, and DC milliammeter. Some multimeters can also measure AC current, in milliamps and amps. Because a multimeter can measure volts, ohms, and milliamps, it is often called a *VOM*, for volt-ohm-milliammeter. (Pronounce VOM as "Vee Oh eM.")

A multimeter is connected to a circuit or component with two test leads. Usually one meter lead is black and is referred to as the *common* or *negative lead*. The other test lead is red, called the *hot* or *positive lead*. Both black and red test leads have a banana plug at one end; they plug into banana jacks on the multimeter. Various types of probe tips at the other end of the test leads make electrical contact with the circuit or component under test.

As you learned toward the end of Chapter 8, there are two basic types of multimeters, analog and digital. An *analog VOM* has a needle or pointer that moves over printed scales on the meter face for reading voltage, resistance, and current. A *digital meter* displays actual numbers, usually on a liquid crystal display (LCD) panel, and may also be called a *DMM,* for digital multimeter. A DMM is sometimes called a *DVM,* for digital voltmeter, even though it *may* also read resistance and DC current. Figure 9.1 is a photo of typical analog and digital handheld multimeters.

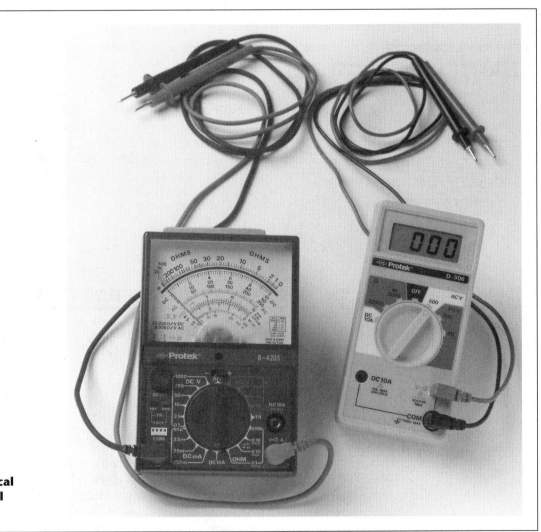

**FIGURE 9.1 Typical analog and digital multimeters**

When using a multimeter, your first concerns will be the meter's *function* and *range*. Its *function* refers to whether the multimeter is measuring AC voltage, DC voltage, resistance, or DC current. Its *range* refers to the scale at which the multimeter is measuring those electrical quantities. For example, to measure volts, you would set the multimeter to act as a voltmeter; that is *setting its function*. Then you would set the meter to one of its available voltage ranges, such as one that will measure a maximum of 50 volts. The same fundamental procedure is used to measure resistance or current. First choose the correct function, then choose the appropriate range setting, in ohms or milliamps, respectively.

Analog and digital multimeters have either a rotary selector switch or pushbuttons to select the appropriate function and range. Some DMMs are auto-ranging; they automatically select the correct range of voltage, resistance, or current. You still need to select the function.

Multimeters do not directly measure impedance.

For either type of multimeter, whether analog or digital, the *basic* operation to measure an unknown voltage, resistance, or current is:

1. Select the *function* with a rotary switch or pushbuttons on the meter:
   - AC volts
   - DC volts
   - DC resistance
   - DC current in mA (milliamps) or amps (A)
   - AC current (available on *some* meters) in mA or A
2. Select the appropriate *range* of voltage, resistance, or current to be measured, for example:
   - 2.5V, 10V, 50V, 250V, or 500V
   - R × 1, R × 100, or R × 10,000 ohms
   - 1mA, 10mA, 100mA, or 500mA
3. Connect two test leads from the meter to the circuit or component to be measured. Usually:
   - Black test lead plugs into meter terminal marked Common, Com, or ⊖
   - Red test lead plugs into meter terminal marked ⊕ or V-Ω-mA
4. Read the measured value on the meter.

With some meters, the red test lead may have to be moved to a different banana jack on the meter for some ranges. For example, a separate jack may be labeled 5A, for measuring up to 5 amperes of DC current, rather than a meter range switch making this selection.

### Safety Precautions:

1. Be conscious of any shock hazard that may exist in equipment on which you are working. American National Standard C39.5 considers any AC voltage over 30V RMS or DC voltage over 42.4V to be a possible shock hazard.
2. Do not work alone when taking measurements in circuits where a shock hazard may exist. Let someone else know that you are going to be working on equipment with a potential shock hazard.
3. Before making connections to a circuit, be aware of voltage sources and current paths. Fuses should be of the proper rating. AC equipment designed to be grounded should be properly grounded.
4. Be aware that even with equipment powered Off, dangerous voltages might *still* be present. A power supply capacitor, for

example, can store a charge after the supply is turned Off. This could be caused by a bleeder resistor across the capacitor that has opened. A bleeder resistor should normally discharge a capacitor within a minute or so.

5. Inspect test leads for damaged insulation, cracked probes, or weak connections at either end. Do not use defective test leads; repair or replace them.
6. Do not work on a damp floor or in damp shoes. The workbench surface should be dry insulating material, such as wood or Formica.
7. Do not touch exposed parts of test leads when circuit being measured is powered up.
8. Do not touch any object, such as a water pipe or metal bench, that might provide a path for current flow to power line ground or to the common side of the circuit on which you are working. Get in the habit of placing one hand behind your back when working near hazardous voltages. This can prevent current from flowing through your body from one hand to the other.

Now that you know the very basics of a multimeter, we'll describe how to use one to take voltage, resistance, and current readings. Most of the following discussions and illustrations are for analog meters with a moving pointer. These take a little more practice to read correctly than direct-reading digital meters.

There are many differences among various makes of digital and analog multimeters. Always study the instruction booklet that comes with a particular meter to ensure you are using it properly and safely.

## HOW TO TAKE VOLTAGE MEASUREMENTS

Recall from Chapter 8 that voltage is present *across* a component in a circuit. To find the voltage of a battery, the black meter lead is placed on the negative battery terminal and the red lead on the positive terminal. To determine the voltage drop across a resistor in a circuit, the black test lead is connected to one resistor lead and the red test lead is connected to the other resistor lead.

In general, an analog VOM uses the same printed scales on the meter face for both DC voltages and DC current, and has somewhat different scales for AC voltages (and AC current, on some meters). Check the face carefully when reading an analog meter to make sure you are using the correct scale. AC scales are often printed in red, with all other scales in black.

For both voltage and current, 0V and 0mA is indicated when the meter pointer is at the left end of the scale. The numbers at the right end of the printed scales are the highest voltage or current that can be read with the meter range switch in a particular position. This is referred to as the *full scale voltage* or current.

For example, if the meter selector is set to the 250V position, 250 volts is the full scale or maximum voltage that can be read. This corresponds with the 250 at the far right end of one of the meter scales, as in Figure 9.2. Likewise, if the selector switch is set to 2.5V, the 250 at the right end of the meter scale indicates the full scale value of 2-1/2 volts. In this case, the reading on the meter scale is divided by 100 to determine the actual voltage being measured.

**FIGURE 9.2 Analog VOM scales.** *Courtesy of Simpson Electric Co.*

What is the full scale voltage when the meter function/range switch is set to 500V? The full scale voltage is 500 volts, the highest voltage that can be measured with the range switch in this position. You would read the meter scale that has 50 at the right end, and multiply by 10.

For example, suppose the meter's selector switch is set to 500V. What is the voltage being measured when the meter pointer comes to rest over the scale line halfway between 30 and 40 on the 50V scale, as in Figure 9.2?

Because the meter pointer falls over 35, and the range switch is set to 500V, multiply the meter reading by 10. The answer is 350 volts.

To take a voltage measurement, plug the black meter lead into the meter jack labeled Com., Common, or ⊖. Plug the red test lead into the jack marked ⊕ or V-Ω-mA.

Before taking any measurement with an analog meter, check that the meter is *mechanically zeroed.* With the meter leads not touching each other, nor connected to any circuit or component, the meter pointer should fall directly over the left end of the scales. If it does not, use a small screwdriver and very slowly turn the adjustment screw at the bottom, center of the meter face. This screw is generally plastic. Turn it a small amount one way or the other until the meter pointer is over the 0 volt and infinite ohms (∞) markings at the left end of the meter scale.

Set the function switch on the meter to either AC or DC for the voltage you intend to measure. Some VOMs have a separate switch to select AC or DC, whereas on others the range switch has several separate positions for AC and DC voltages.

In addition, some meters also have a polarity switch with two positions, usually labeled −DC and +DC, or just marked ⊖ and ⊕. What this switch does is to electrically swap the black and red test leads inside the meter. Normally you leave this switch in the +DC position, where the black lead connects to the negative side of a DC voltage and the red lead connects to the positive side. This switch is handy in many cases, such as when measuring a minus power supply voltage in a VCR. (More on this in just a while.)

Once you've selected AC or DC, select the full scale voltage. Unless you absolutely know approximately what voltage you expect to read, it is *always* prudent to put the meter on its highest voltage range. Once you have made the connection with the two test leads, you can switch the meter to a lower voltage range to obtain a more accurate reading. If you don't start at a high voltage range, you could damage or destroy your meter, the circuit under test, or both.

Suppose you want to measure the voltage at the +5V output of a VCR power supply (PS). You consult the schematic diagram and find that PS ground is on connector P101, Pin #1 and the +5V output is on P101, Pin #7. You confidently set the meter switch or switches to measure DC at 10V full scale. You connect your black test lead to P101-1 and the red lead to P101-7.

You look at your meter and see that the meter pointer has slammed all the way over to the right end of the meter, well past the full scale marking of 10. To your dismay, you notice that the delicate meter pointer is actually bent a little! You hurriedly switch the meter to the next higher, 50V, position.

Now you see that the meter reads about 48 volts on the 50V scale. What happened? You check where you placed your meter leads and discover you accidentally probed P101 pin #8 with the red lead instead of Pin #7. Looking at the schematic, you see that +48V is present on Pin #8.

You should have started with the meter set to 250V (or even 500V). With the meter set to 250V, the meter pointer would be over on the left side of the scale, *below* the 50-volt point on the 250V scale. Since 50V is the next lower full scale range, you know it is safe to move the meter to the 50V range. You do so and observe that the meter reads 48 volts. Since you expected to read about 5 volts, you check your leads and find your mistake in probing the wrong pin on P101. *By starting with the meter set at a higher voltage than you expect to read and then going to a lower range, you prevent damage to the meter.*

To repeat: It is always a good idea, when taking voltage or current readings, to place the meter in its highest range before making connections with test leads. Then you can switch to a lower range to obtain a satisfactory reading.

Your objective is eventually to find a meter range that brings the needle closest to the right end of the meter scale, *without going past the right end of the scale*. An analog meter is most accurate at the right side of the scales.

Suppose you take the same voltage measurement at power supply connector P101, but this time the meter pointer moves *to the left* of the *left end* of the meter scale. This indicates reverse polarity: the black meter lead is connected to a positive DC voltage with respect to the red lead. If the meter has a DC polarity switch, simply move the switch to its −DC position. This electrically reverses the black and red leads inside the meter. Now the meter reads up scale. If the meter doesn't have a polarity reversing switch, swap the positions where the black and red probes connect to the circuit.

*Never* swap meter leads at the meter test jacks while the other ends of the leads are connected to a circuit under power. You could get a shock if you touch a banana plug while swapping. Also, the banana plug might accidentally slip from your fingers, come in contact with a ground point or some other circuit component, and "fry" the circuit you are probing!

Look at Figure 9.3. How would you connect a multimeter to measure:

- Battery voltage?
- Voltage drop across the LED?
- Voltage drop across resistor R1?

First, determine the battery voltage. Because the power supply is a battery, set the meter to read DC voltage. Because you don't know what the battery voltage is, set the voltmeter to a higher voltage than you would expect to read, like the 50V range.

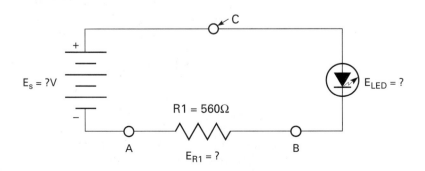

**FIGURE 9.3 Reading circuit voltages. Where to connect a voltmeter?**

Connect the black meter lead to point A, the red lead to C.

Point A could be the actual negative terminal of the battery or the lead on resistor R1 that connects to the negative battery terminal, or at any noninsulated point along the way. For example, the actual circuit may be on a printed circuit board. In this case, the black meter lead *could* probe any noninsulated point along the wiring pattern on the PCB between the negative battery terminal and the resistor lead. Likewise, point C could be the actual positive battery terminal or the lead of the LED connected to the positive battery terminal.

You see that on the 50V scale the meter pointer is somewhat lower than, or to the left of, 10. You can therefore switch the meter to the 10V range. You do, and read the battery voltage as 8.8 volts. If the meter reads backward in the first place, either change the polarity switch on the meter (+DC/−DC) or reverse the black and red probes where they connect to the circuit.

To measure the voltage drop across the LED, connect the black lead to point B and the red lead to point C. You observe that the needle moves only a very small amount to the right of the 0 volt line. This must be a low voltage, so change the meter from its 10V range to its 2.5V range. Here the needle points halfway between 150 and 200. Since the voltage would be 2-1/2 volts with the pointer over the 250 at the right end of the meter scale, the actual voltage drop across the LED must be 1.75 volts.

To measure the voltage drop across resistor R1, first switch the meter back to the 10V or higher range. Connect the black lead to point A and the red lead to point B. You read 7 volts.

Notice that with all three readings:

- The black meter lead is placed closer to the negative terminal of the battery than the red lead

  *and*

- the red lead is connected closer to the positive battery terminal than the black lead.

Unless a meter polarity switch is in the −DC position, the black lead probes the more negative side of the circuit being measured.

Most, *but not all,* DC voltage readings that you take when working on a piece of electronic equipment are *with respect to ground,* or a common point. That is, one meter lead is connected to ground when taking a voltage measurement. Unless otherwise noted, all voltages printed on a schematic wiring diagram are with respect to ground. This means circuit ground, chassis ground, or circuit common.

Schematic wiring diagrams use three different symbols for circuit common. These are shown in Figure 9.4, along with a partial automobile circuit using a symbol for electrical "common." A schematic diagram will normally use only one of the common or ground symbols, but some electronic equipment has a chassis ground that is electrically isolated from common; this would be indicated with separate symbols for each. A common that is separate from chassis ground is often called a *floating ground* or *isolated ground.*

Ground is normally the common point for all voltages. In almost all cases, metal chassis components are at ground potential. In a VCR, the metal transport baseplate is at chassis ground potential. *Voltages in different VCR circuits may be either positive or negative with respect to ground.*

When measuring AC voltage, first set the meter function switch to AC Volts. As when measuring DC voltages, start with the highest voltage range, then switch to a lower range if the measured voltage is within a lower range on the particular meter. With AC, you don't need to be concerned with polarity. Black and red probes can be connected either way across the voltage to be measured, and the meter will always read up scale.

Be extremely careful should you measure AC line voltage, such as at a wall receptacle. Use proper probe tips with metal ends that are long enough to

**248** ■ *Practical VCR Repair*

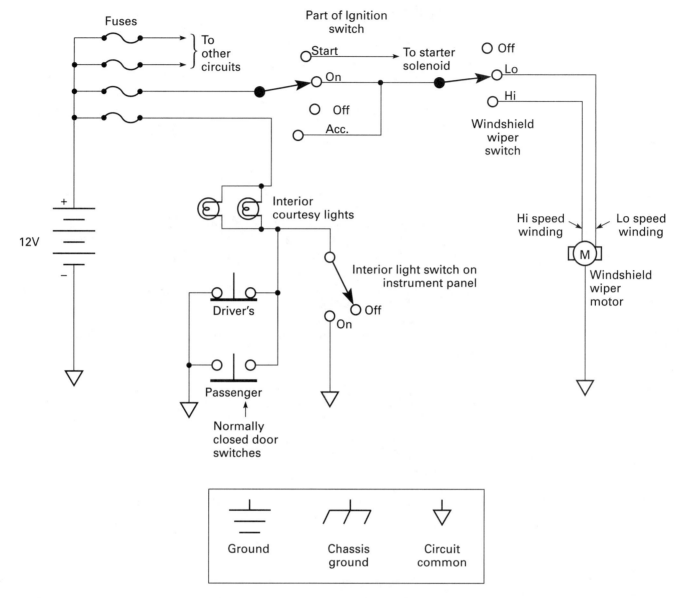

**FIGURE 9.4** Schematic symbols. NOTE: A dot where lines cross means there is an electrical connection between wires. There is *no electrical connection* where lines cross without a dot.

reach the outlet contacts. Do not use paper clips or other makeshift methods to make an electrical contact. Never leave probes attached to high voltages after a reading has been taken. Remember to keep one hand behind your back when working near hazardous voltages.

When making voltage measurements, check that the meter is *NOT* set to a resistance or current range, which could damage the meter or the circuit under test.

## HOW TO TAKE CURRENT MEASUREMENTS

In most electronic troubleshooting and repair, you will rarely take current measurements. But if you design your own electronic circuit of some type, you might very well measure current to determine the required value of components, like resistors. For example, suppose you are building an auto burglar alarm from plans in an electronics magazine. The parts list calls for a particular relay. Rather than buy the relay specified, you want to use a similar one you already have in your electronics junk box. You may have to measure the coil current for your relay to see if it may be substituted in the circuit. (Relays are covered in Chapter 10.) In any event, an electronics technician should know how to measure electrical current in a circuit.

To measure circuit current, a milliammeter is connected so that all current in the circuit flows *through the meter*. Recall from Chapter 8 that the amount of current is the same through each component in a series circuit. Therefore, the VOM itself must become a series component of the circuit in order to measure current flowing through it. This requires that the circuit be open at some point; the meter leads are then connected across the open circuit. With the meter connected across the open circuit, the meter effectively closes the circuit once again. Current flows through the circuit *and* through the meter, which measures it.

For example, if you want to see how much current is being drawn by a VCR front loader motor, disconnect one of the two wires soldered to the two motor terminals. Then connect one meter lead to the wire just disconnected and the other test lead to the motor terminal where the wire was originally connected. The milliammeter is now in series with the DC motor. All current to operate the motor must also flow through the meter.

Just as when measuring voltage, it is a good idea to start measuring current with the meter set to its highest milliamp range. This protects the meter from damage should the current be larger than you expected. For example, start with the meter set to 500mA. Then, if the meter pointer comes to less than 100mA on the 500mA scale, you can switch the meter to its 100mA range. If the pointer is now less than 10mA on the 100mA range, you can safely set the meter to its 10mA range.

What value of current is indicated by the meter in Figure 9.5?

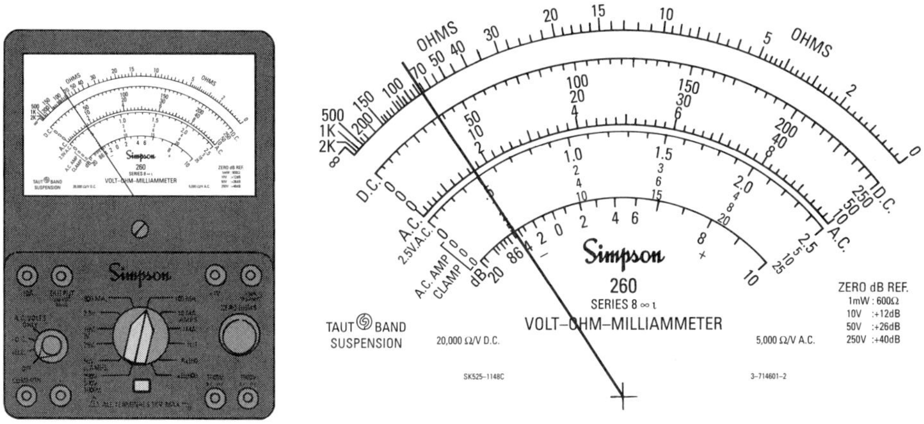

**FIGURE 9.5 VOM scales.**
*Courtesy of Simpson Electric Co.*

Because the range switch is set to 100mA, read the scale with "10" on the right end, and multiply by 10 for the full scale current of 100mA. The meter pointer is three-quarters of the way between 0 and 2 on the 10 scale, or at 1.5. Multiplying this by 10 gives 15mA. There is 15mA of current flowing through the LED in the circuit.

In this case, selecting the next lower current range, 10mA, would cause the meter pointer to slam against a small peg inside the meter, to the right, past the full scale markings. This is called *pegging the meter*. Competent electronics technicians take precautions, like beginning a measurement with their VOM set to the higher voltage and current ranges, so they rarely peg a meter.

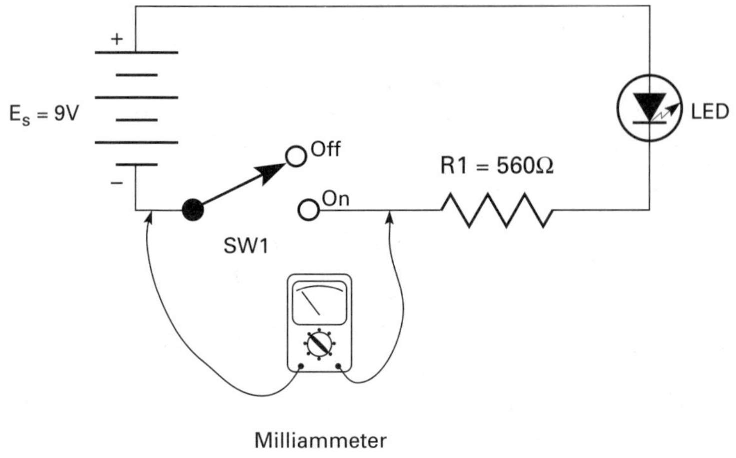

**FIGURE 9.6  Measuring circuit current. Where to connect a milliammeter?**

As when measuring a DC voltage, if the meter reads backward, operate the meter's polarity reversal switch or swap the probe ends of the two test leads.

Refer to Figure 9.6. How would you connect a VOM to measure the current flowing in the circuit?

Recall that an ammeter must be connected *in series* with the circuit, so that circuit current flows *through* the meter. This means the circuit must be discontinuous, or open, somewhere. When there is an On/Off switch in the circuit, simply placing the switch in its Off position opens the circuit, and no current flows. Then connect one meter lead to one side of the switch and the other test lead to the other side of the switch. The ammeter itself now completes the circuit: current flows through the meter and the rest of the circuit.

For the circuit in Figure 9.6, connect the black meter lead to the negative battery terminal or to the switch terminal connected to the battery; electrically, these are the same. Connect the red test lead to the other switch terminal, or to the LED lead that connects to the switch. Now, with the switch open or Off, circuit current flows *through the meter,* and the LED lights. With the switch closed or On, the meter is short-circuited; all current now flows through the switch contacts. The LED lights, but with no current flowing through the meter, it measures 0mA.

Be extremely careful that the VOM is not set to one of the resistance ranges when taking a voltage or current reading. This could easily destroy the meter. Some meters have small, fast-acting fuses that *may* protect them from damage under such an abusive condition. Always replace a meter fuse with one of the exact same rating.

## HOW TO TAKE RESISTANCE MEASUREMENTS

Taking voltage and resistance measurements are the most common uses to which you will put your VOM. Quite often, you will want to know if there is electrical *continuity* between one point and another in a circuit. In other words, is there a complete or continuous path for current flow? This is when you reach for your multimeter and set it up as an ohmmeter.

For example, you've taken a fuse out of a circuit, but you can't tell by looking at it whether it's good or not. If it's a glass fuse, you can *usually* tell if it's blown. The inside surface of the glass will have a blackened smudge, caused by the vaporized fuse wire condensing on the glass. Sometimes you can see a break in the fuse wire inside. At other times, you *can't* readily see whether a fuse is open or whether it has continuity.

Some fuse bodies are ceramic or other material where the fuse wire is not visible. Even with glass fuses, sometimes the fuse wire opens where you can't see the break. This is true for the cylindrical glass fuses most often used in VCR power supplies. What happens is that, over time, metal fatigue caused by the fuse wire warming up and cooling down when the VCR is On and Off, respectively, causes the wire to break away from one of the metal end caps. You can't see the break.

A good fuse has electrical continuity between its two contacts. It should have 0 ohms resistance. An open fuse, on the other hand, has a very high resistance, infinity ($\infty$). This can be checked with an ohmmeter.

At times you will want to know if a cable is good or not. Is there continuity on all conductors from end to end? There should be. Are two conductors shorted together? They should not be.

For example, suppose there is no VCR playback picture on a video monitor. The Video Out jack on the VCR is connected to the Video In jack on the monitor with a six-foot coax cable having an RCA plug at each end. For the cable to be good, the following conditions must be present:

- Continuity between the center pin of the RCA plug on one end of the cable to the center pin of the plug on the other end (0 ohms)
- Continuity between the outer shell of the RCA plug on one end of the cable to the outer shell of the plug on the other end (0 ohms)
- *No continuity* between the center pin and outer shell of one of the RCA plugs ($\infty$ ohms)

In a similar manner, you might want to check a multi-conductor cable inside a VCR, like one from the front control panel printed circuit board to another PCB containing the system control microprocessor chip. Unplug the cable from both boards and check for end-to-end continuity of each individual conductor at the two connector ends. A poor wire crimp inside a connector can cause a conductor to become open. This problem is difficult to see unless you use your ohmmeter.

When taking a *voltage* or *current* reading, the circuit is powered up; otherwise there would be no voltage or current to measure. But when measuring *resistance,* the device being measured *must have no voltage applied to it from the circuit.*

However, there needs to be *some* power source to move the meter pointer. The power source is one or more batteries inside the VOM. This must be the *only* source of current when taking a resistance reading or making a continuity check.

> CAUTION: Never take a resistance measurement or make a continuity check in a circuit that is powered up! You could easily damage the VOM.

Most VOMs have three or more resistance ranges. Always use the R $\times$ 1 range when checking continuity. To check for a short, such as between the shield and center conductor of the RCA cable in the preceding example, the R $\times$ 1 range is usually sufficient. However, there *could* be a high resistance, or partial short, between the two conductors. This resistance causes *electrical leakage.* You might not detect a small amount of leakage on the R $\times$ 1 range. By selecting a higher range, such as R $\times$ 10,000, you will be able to see if there is slight leakage between the conductors. There should not be.

The first step to making a resistance check is to *zero the ohmmeter.* This is different from mechanically zeroing the meter pointer (previously described). Zeroing an ohmmeter compensates for slightly different voltages as

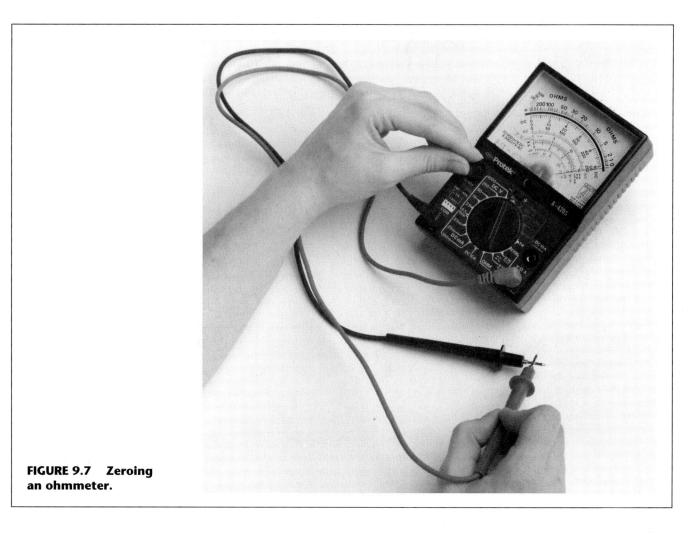

**FIGURE 9.7** Zeroing an ohmmeter.

batteries inside the meter age, and for separate batteries used on different ohmmeter ranges. This is how to zero an ohmmeter (see Figure 9.7):

1. Select the R × 1 ohmmeter range on the VOM selector switch.
2. Press the test lead probe tips *firmly together* so they are making good electrical contact with each other.
   - At this time, the meter pointer should be somewhere near the right side of the OHMS scale. It might be a little to the left or to the right of the 0 ohms mark.
3. Slowly rotate the *Ohms adjustment control* on the meter one way or the other until the pointer lies directly over the zero mark on the right end of the OHMS scale. This control is usually labeled "ZERO OHMS" or "Ω ADJ."

The meter is now ready to take continuity checks or measure low values of resistance, generally 0Ω through 1,000Ω. To measure higher values of resistance, move the meter range switch to its R × 100 or R × 10,000 position, and rezero the meter as you did on the R × 1 range. It is best to re-zero a meter each time you change resistance ranges.

There may be a time when you are *unable* to zero an ohmmeter. No matter how far you turn the zero ohms adjustment, the pointer won't move far enough to the right so it lies over the 0 ohms mark. Most likely one or more batteries in the meter are weak, missing, or dead. It is also possible your meter has a poor battery connection. Inspect battery connections for dirt or corrosion, *and* for firm mechanical contact between meter and battery terminals.

Batteries in most analog multimeters have nearly the shelf life of the particular battery. Current is drawn from the battery only while you are making the zero ohms adjustment and while making a continuity check or measuring a resistance. Even then, the current drain is very small. (For a 100-microamp analog meter movement, it is 100 microamps with the pointer over the zero ohms mark on the meter scale. That's 0.0001 ampere!) With such a long battery life, along with very low current drain, battery contacts can build up a thin oxidation layer. Cleaning battery terminals and contact surfaces will eliminate contact resistance. Clean battery terminals and connections with contact cleaner or pencil eraser, or both, until they are bright and shiny. On meters using an AA-, AAA-, or C-cell battery, gently form the meter spring terminals so they make solid contact with the battery terminals. For meters that use a rectangular 9V battery, gently close up the snap connector on the battery and on the battery clip in the meter so there is a tight mechanical fit between the two.

Also inspect meter leads if you are unable to zero an ohmmeter. Leads should fit very snugly into the meter jacks. Test lead wires should be firmly attached at probe and meter ends. A gentle tug should *not* cause the wire to separate from a probe or banana plug. If it does, the connection was poor. Resolder or get new test leads.

*It is important to zero an ohmmeter for each resistance range.* Some VOMs have a separate, higher voltage battery for high resistance ranges, such as R × 1K and R × 10K (K = 1,000). The meter may zero on R × 1, R × 10, and R × 100 ohm ranges, but a separate weak battery may prevent zeroing on the higher resistance ranges.

Make sure to read the OHMS scale on the meter face when measuring resistance, not the DC or AC scales. The OHMS scale is uppermost on the face of most analog VOMs.

In most VCR troubleshooting and repair work, you will be checking continuity, rather than taking an actual resistance measurement. For example, you want to check whether a fuse or a switch contact has continuity. With your ohmmeter set to R × 1, place one meter probe at each end of the fuse. If the meter reads 0 Ohms, there is continuity. Or connect a test lead to each of the two switch terminals to see if a good, low resistance contact is being made.

It is usually a good idea to disconnect at least one end of the component that you are checking for continuity. This electrically isolates it from other circuit components. For example, if you place your ohmmeter across a fuse while it is still installed in a VCR, you might read several ohms of resistance but not 0Ω. The fuse itself might be totally open ($\infty\Omega$), but what you are seeing is resistance in other circuit components on each side of the fuse. This alternate path for current flow is called a *back circuit*. If you aren't very careful to observe that the meter isn't 0Ω, but some other value, you might incorrectly conclude that the fuse is good, when in fact it is open! By taking the fuse out of the circuit and checking it, you'll know for sure whether the fuse is open ($\infty\Omega$) or if it has continuity (0Ω).

Following is an example of when you might use the R × 1 range of your ohmmeter to check continuity.

Suppose a customer brings in a VCR. The problem: nothing happens when the Power button is pressed. With the VCR plugged in, the fluorescent display blinks **SU 12:00 am**, so you know at least a portion of the power supply is working. You unplug the VCR, take the covers off, and check three fuses in the power supply. They all test good; that is, each has continuity (0Ω). You don't see any obvious sign of trouble. Next you decide to see if the Power-On pushbutton itself is working. Figure 9.8 is a small portion of the schematic diagram for the VCR.

Keep the VCR unplugged. Zero the ohmmeter on the R × 1 range. Connect the meter leads to each side of the normally open (N.O.) pushbutton microswitch on the front panel printed circuit board. It doesn't matter for this continuity check which meter lead goes to which side of the pushbutton.

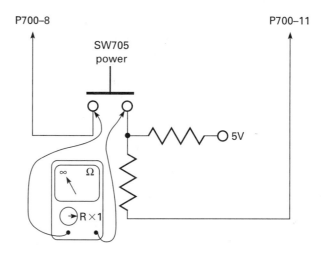

**FIGURE 9.8** Partial VCR front panel schematic. P700 is a multi-pin connector on the front panel printed circuit board. A 15-conductor cable runs between P700 and a PCB on which the system control microprocessor chip resides. SW705 is a normally open (N.O.) pushbutton; the contacts close when the button is pushed.

Without pressing the pushbutton, the measured resistance should be $\infty\Omega$, since this is a normally open pushbutton; an open has infinite resistance. You *may*, however, measure some value of resistance between $0\Omega$ and $\infty\Omega$ on the R × 1 resistance scale. Why? This is most likely resistance in a back circuit caused by the system control microprocessor. But what you are really concerned about is whether the resistance across the pushbutton goes to $0\Omega$ when the button is pushed.

You push the Power On button. The meter pointer moves up scale to about 1K ohms. You release and press the button several more times. Sometimes the meter pointer hardly moves towards 0 ohms at all, and sometimes it goes to somewhere between 100 and 1K ohms, but it *never* goes to 0 ohms. It *should* go to $0\Omega$, indicating that the pushbutton switch is closed, with no contact resistance! Apparently this pushbutton switch is defective; *it has too much contact resistance*. You carefully unsolder the switch from the circuit board, solder in a replacement, and now the VCR powers up just fine.

Now suppose you have the same model VCR, the same schematic, and a similar problem. In this case the VCR frequently will not turn On, or will not turn Off once it *is* On, without numerous depressions of the Power pushbutton. You set up your meter to measure resistance across the power switch exactly like you did for the previous problem. Without pressing the pushbutton, the meter needle is somewhere around the $1K\Omega$ mark on the R × 1 range, way down toward the left end of the scale.

Because this is to the far left of the ohms scale, you switch your meter to R × 100. You now see that the reading is 1,500 ohms. That is, the meter pointer is over "15" on the ohms scale, times the range switch setting of R × 100; 15 × 100 = 1,500.

But you're not sure whether this is some resistance in the switch itself or in a back circuit, on the other side of Pins 8 and 11 of connector P700.

To eliminate any possible back circuit from the resistance reading you can simply unplug connector P700. Refer to the schematic in Figure 9.8. With P700 disconnected, there is no alternate path for ohmmeter current to flow; the only possible path is through switch SW705. You disconnect P700. The ohmmeter now reads about $3,000\Omega$: meter pointer over "30" times 100, with VOM selector still on R × 100. You press the Power pushbutton and the meter reads 0 ohms, as it should. Releasing the button, it registers about $3,500\Omega$. You press and release the button several times. The ohmmeter goes to 0 ohms each time the button is pressed, but when it is released it never goes to $\infty\Omega$ as it should. Instead, the meter consistently reads between 2,000 and 5,000 ohms.

This particular pushbutton has quite a bit of *leakage resistance* when it is open, falling somewhere between $2K\Omega$ and $5K\Omega$. Instead of being completely open (∞Ω) when released, or in its normal condition, it has a lower value resistance. Without the switch completely opening, the system control microprocessor doesn't receive the right voltage level to indicate that the switch has been released after it is pressed.

You replace the pushbutton, and now the VCR powers up and down as it should.

Although it is rare for a switch of this type to develop leakage resistance between its contacts, it *is* possible, especially if a beverage was spilled on the VCR and it dribbled inside the switch.

A very important use of a multimeter's ohmmeter function is to check for continuity, shorts, and opens. If you had difficulty understanding the previous two VCR troubleshooting sequences, review them now, before going on in this textbook.

Sometimes you only want to confirm that there is *some* amount of resistance, less than infinity, in a particular component or circuit, without being too concerned with the actual resistance value. In many cases you won't *know* what value DC resistance a component is supposed to have, especially if it's a tape head, transformer winding, coil or inductor, or motor winding. All you really want to know is whether the component is open or not. If it *is* open, it's definitely defective. If it measures *some* value of DC resistance, it still could be defective, with a partial short in a winding. You would need to substitute a known good part to be sure.

Imagine that you have a non-Hi-Fi, monophonic VCR to repair. The customer says that when she dubs new audio onto a previously recorded tape, she hears both the original audio *and* the new audio when the tape is played back.

You check the machine, and everything seems to work OK. It records and plays back fine. It is only when new audio is dubbed over existing audio on the linear track that the original audio is not erased first. You suspect there might be a problem with the linear audio erase head, which is part of the audio/control (A/C) head assembly. The reasoning behind this: In regular recording, the full erase head erases any previously recorded audio, as well as any previously recorded video, but when dubbing new audio while retaining original video, the full erase head isn't energized. If it were, it would erase the original video, too. The original audio, therefore, must be erased by the separate linear audio erase head. But with this VCR, original audio is still heard, so the audio erase head must not be doing its job.

You disconnect the multi-pin connector at the A/C head and connect your ohmmeter leads across the two pins that go to the linear audio erase head. The measurement is ∞Ω! Even with the meter set for R × 10K, the head coil still reads open.

Upon carefully inspecting the A/C head assembly you discover a very poor solder joint where one of the connector pins is soldered to a small PC board on the back of the head assembly. Touching this up with a soldering iron and solder restores correct operation. The audio erase head now erases previously recorded linear audio when new audio is being dubbed, while retaining the original video.

Now assume the same situation, except this time when you measure the resistance of the audio erase head it reads 3 ohms. You have no idea whether this means the head is good or not, *but at least it's not open!* Even manufacturer's service literature will not specify DC resistances for motors, heads, coils, and transformers.

From this point, you decide to check continuity of the two conductors going to the audio erase head *from the other end* of the multi-conductor A/C cable. Aha! One of the small metal contacts at the other end of the cable is almost completely pushed out of the plastic connector housing. Evidently the contact didn't seat correctly in the connector body when the unit was manufactured. It finally worked itself loose to the point where electrical contact with the pin on the audio PC board was lost. In this case you discovered the problem without really needing your meter; you could readily see the pushed-back connector contact.

You can use the same general techniques to check for an open in many VCR components, such as a motor, head, or transformer winding. However, observe this precaution. If you check a tape head with an ohmmeter, demagnetize the head afterwards. The meter passes a small amount of DC through the head winding and this can magnetize the head. A magnetized head can partially erase a pre-recorded tape, and will not record or play back high-frequency detail as well as a nonmagnetized head.

## Notes on Digital Multimeters (DMMs)

Some multimeters, generally digitals, have a convenient feature: audible continuity checking. With this feature, the meter beeps when a low value resistance is across the test leads.

For example, on one manufacturer's popular DMM, any value below 20 ohms on the 200-ohm range, the lowest resistance range, sounds the beep; above 20 ohms it's quiet. The LCD panel still reads out the actual resistance value. This audible checker is convenient when testing continuity of a multiple conductor. You don't have to take your eyes off the cable ends you're probing to read the meter; a beep confirms that a wire has end-to-end continuity.

However, an audible beep continuity checker *may* be somewhat deceiving. For example, with this digital meter, you *could* have a connection with up to 20 ohms of resistance, and the meter would still beep its "Okay." There are other digital multimeters where the audible beep sounds even though the resistance is as high as 100 ohms! Be aware of this when using an audible continuity check. A switch or other connection that has 18 ohms resistance would cause many circuits to malfunction. A beep indicates that a wire or connection *is not open,* but doesn't necessarily mean it's zero ohms.

With digital multimeters, you normally don't have to worry about connecting test leads to a voltage that is higher than the range to which the meter is set. For example, if you connect the probes to 75 volts DC when the meter is set to its 2V range, the meter will not be damaged. Most DMMs display something, such as **OL** for overload or even the word **ERROR**, to indicate that the applied voltage exceeds the selected range. Check the instruction manual for your meter.

Many digital meters can display only the digit "1" in the leftmost position. DMM specifications refer to this "1" as a "half digit." DMMs like this usually display either 3-1/2 or 4-1/2 digits. Thus, if you select the 20V range, the maximum voltage a 3-1/2 digit meter can actually display is 19.99, whereas for a 4-1/2 digit meter it is 19.999. Twenty volts would indeed be an *over range* condition; you would need to select the 200V meter range to measure it.

Another nice feature of digitals is that you don't have to be concerned with DC polarity. A minus sign (−) appears on the display before the numeric digits if the voltage or current being measured is negative.

Most digital multimeters do not require zeroing on resistance ranges, and therefore have no "Ohms Adj." control on the panel. Their electronic circuitry does this automatically.

In Chapter 10 you will learn how to use the ohmmeter function of a multimeter to test several electronic components, such as diodes and transistors.

## PRACTICE EXERCISES

The purpose of this section is for you to become comfortable reading the various scales on an analog multimeter. You will be presented with several analog meter faces and asked to determine the correct reading. Following the exercises are the correct answers, with an explanation of how they are achieved.

Compared with analog meters, digitals are much easier to use; the actual value is displayed in numbers on the meter face for the function and range selected. Therefore, this practice exercise is for analog rather than digital multimeters. Follow the user's guide for a particular DMM.

Refer to Figure 9.9 for the next two questions. Be sure to note the position of function and range switches, and read the right scale on the meter face. Include in your answer the correct unit of measure: amps, milliamps, ohms, volts AC, or volts DC. What electrical value is indicated:

by pointer A? _____    by pointer B? _____

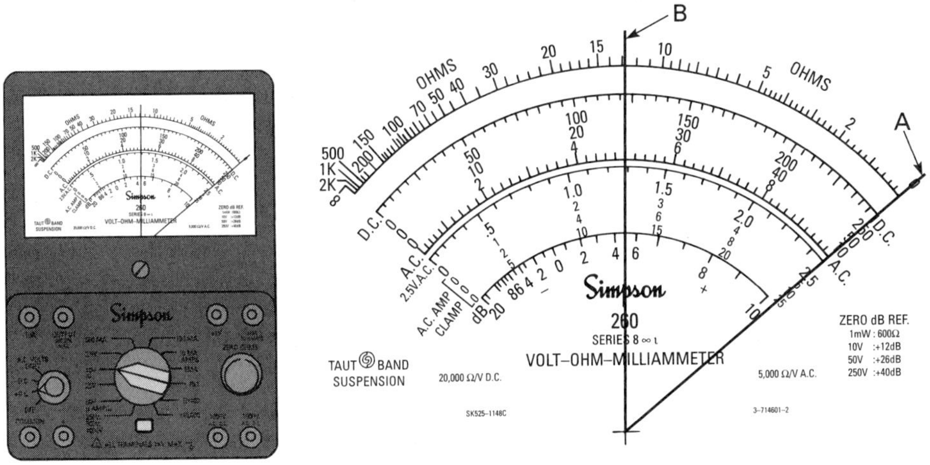

**FIGURE 9.9  Multimeter exercises A and B.** *Courtesy of Simpson Electric Co.*

Refer to Figure 9.10 for the next two questions. What electrical value is indicated:

by pointer C? _____    by pointer D? _____

**FIGURE 9.10  Multimeter exercises C and D.** *Courtesy of Simpson Electric Co.*

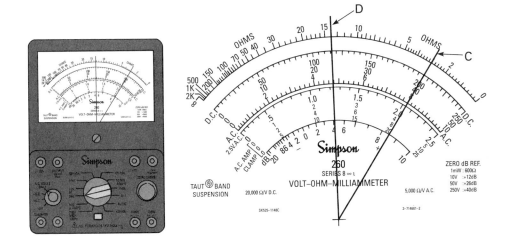

Refer to Figure 9.11 for the next two questions. What electrical value is indicated:

by pointer E? _____    by pointer F? _____

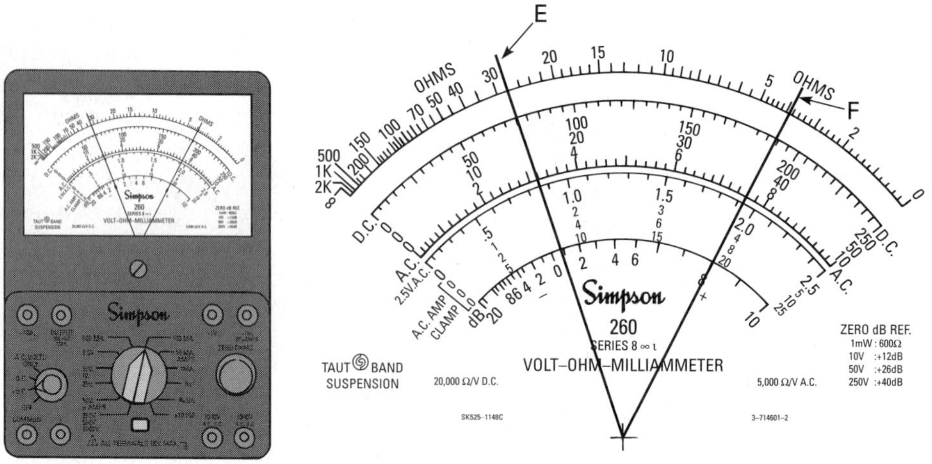

**FIGURE 9.11  Multimeter exercises E and F.** *Courtesy of Simpson Electric Co.*

Refer to Figure 9.12 for the next two questions. What electrical value is indicated:

by pointer G? _____    by pointer H? _____

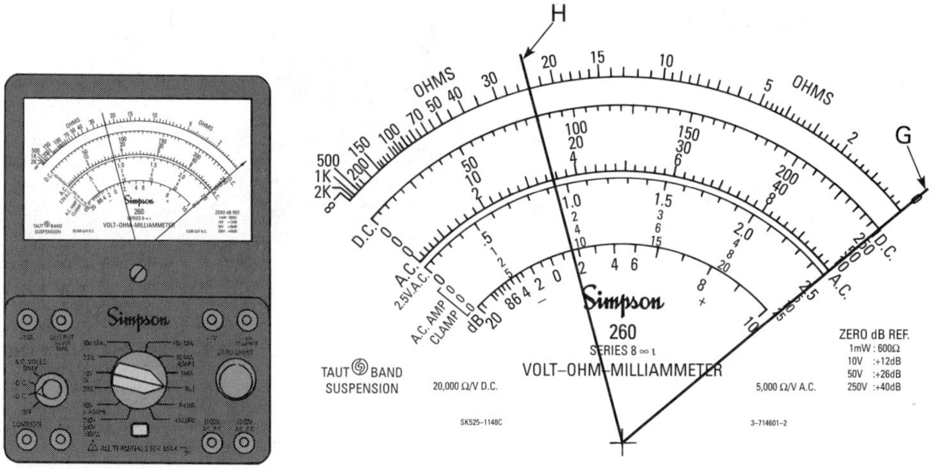

**FIGURE 9.12  Multimeter exercises G and H.** *Courtesy of Simpson Electric Co.*

Refer to Figure 9.13 for the next two questions. What electrical value is indicated:

by pointer J? _____    by pointer K? _____

**FIGURE 9.13  Multimeter exercises J and K.** *Courtesy of Simpson Electric Co.*

## Practice Exercise Answers

A: **10 VDC**    The small function switch is set to +DC and the range switch is set to 10V, which is the full scale value for this voltage range. Pointer A is at the far right, or full scale end, of the meter scales. The voltage being measured is therefore 10VDC.

B: **5 VDC**    The small function switch is set to +DC and the range switch is set to 10V, which is the full scale value for this voltage range. Pointer B is over the scale marking that is halfway between 4 and 6 on the 10-volt scale. The measured voltage is therefore 5VDC.

C: **20 VAC**    The small function switch is set to AC VOLTS and the range switch is set to 25V, which is the full scale value for this voltage range. There is no full scale value of 25V on the meter face, but there is a full scale value of 250; use this scale and divide by 10 for a full scale value of 25V.

Because this is an AC rather than a DC voltage, the AC scale *below* the numbers on the meter must be used. Notice that it is slightly different from the DC voltage scale. Pointer C is over the 200 mark on the lower AC scale; dividing by 10 gives the correct reading of 20VAC.

D: **12 VAC**    The small function switch is set to AC VOLTS and the range switch is set to 25V, which is the full scale value for this voltage range. There is no full scale value of 25V on the meter face, but there is a full scale value of 250; use this scale and divide by 10 for a full scale reading of 25V.

Because this is an AC rather than a DC voltage, the AC scale *below* the numbers on the meter must be used. Notice that it is slightly different from the DC voltage scale.

The scale line halfway between 100 and 150 must be 125. Then, each of the five scale divisions between 100 and 125 must be 5. That is, the first small line to the right of 100 is 105, the second small line is 110, the third 115, and so on. Pointer D is over the fourth small line between 100 and 150 on the 250 scale, and so is pointing to 120. The range switch is set to 25V instead of 250, so divide the scale reading of 120 by 10 to get the correct answer of 12VAC.

E: **30 mA DC**    The function switch is set to +DC and the range switch is set to 100mA, which is the full scale value for this current range. There is no full scale value of 100 on the meter face, but there is a full scale value of 10; use this scale and multiply by 10 for a full scale reading of 100mA. This is a DC reading, so use the upper DC scale on the meter, not the lower AC scale.

Meter pointer E is over the scale marking that is halfway between 2 and 4 on the 10 scale. This then is 3. Multiplying by 10 to convert to a full scale value of 100mA gives the correct answer of 30mA DC.

F: **76 mA DC**    The function switch is set to +DC and the range switch is set to 100mA, which is the full scale value for this current range. There is no full scale value of 100 on the meter face, but there is a full scale value of 10; use this scale and multiply by 10 for a full scale reading of 100mA.

Pointer F is over the mark that is two small divisions before the 8 on the meter scale. The longer line halfway between 6 and 8 on the scale must be 7. Because there are five small divisions between 7 and 8 on the meter scale, each division must be 0.2. That is, the first small line to the right of 7 is 7.2, the second division to the right of 7 is 7.4, the third is 7.6. The pointer is therefore on 7.6. Multiplying by 10 for the full scale value of 100 gives the correct answer of 76mA DC.

G:  0 ohms      Pointer G is over the 0 at the right end of the Ohms scale, and the range switch is set to R × 1. Zero multiplied by 1 is 0. The meter is reading 0 ohms. This could be a good continuity check from one end of a wire to the other end, or between contacts of a good switch in its On position.

H:  24 ohms     Pointer H is over the second division on the ohms scale between 20 and 30. As there are five divisions between 20 and 30, each division must have a value of 2. That is, the first division *to the left* of 20 is 22, the second is 24, the third is 26. With the pointer over 24, times the range of R × 1, the correct resistance reading is 24 ohms.

J:  500,000 Ω   Pointer J is over the 50 on the ohms scale. The range switch is set to R × 10,000. Fifty multiplied by 10,000 ohms equals 500,000 ohms. This answer can also be expressed as 500KΩ or 1/2megohm.

K:  22,000 Ω    Pointer K is over the first division to the left of 2 on the ohms scale. As there are two major divisions between 2 and 5, they must be 3 and 4, respectively. With five small divisions between 2 and 3, each must have a value of 0.2. That is, the first division *to the left* of 2 is 2.2, the second division is 2.4, the third 2.6, and so on. Pointer K is therefore over 2.2 on the ohms scale. Multiplying 2.2 times 10,000 ohms, since the range switch is set to R × 10,000, equals 22,000 ohms. This answer can also be expressed as 22KΩ.

## SELECTING A MULTIMETER

Several types of multimeters are available in the $20. to $35. price range. Virtually all electronics supply houses are good sources for meters that will meet your VCR troubleshooting needs. A home improvement center or hardware store *may* also carry a suitable meter, but when considering a purchase from one of these latter sources, check the voltage and resistance ranges their meters have, as well as the ohms-per-volt rating of specific multimeters.

Some multimeters are designed primarily for work with electrical wiring in a house, building, or factory. They may not have a low voltage range that will allow you to accurately measure a low voltage, such as 5 VDC. A resistance range of at least R × 1,000 is also required for electronics work, but is seldom necessary for an electrician who is checking building wiring. Be sure the meter you purchase has at minimum a 10 VDC range and an R × 1,000 range. If it has a 2.5V or 5V range and an R × 10,000 range, so much the better.

Another meter characteristic is the ohms-per-volt (Ω/V) or *input sensitivity* specification. This designates how much resistance the meter itself places into a circuit, depending on the full scale range selected, when a voltage is being measured. For example, a meter with 1,000 ohms-per-volt input sensitivity set to its 5V full scale range puts 5 × 1,000 or 5,000 ohms of resistance across the circuit being measured. In many electronic circuits, this relatively low value resistance is sufficient to cause the circuit to completely malfunction! This is obviously undesirable. By contrast, a meter with a 5V range and rated at 20,000 Ω/V puts only 100,000 ohms of resistance across the measurement points in the circuit.

So why is it better to have 100,000 ohms rather than 5,000 ohms connected across a circuit when measuring voltage? Recall parallel circuits. The lower the resistance, the higher the current flow. The lower the meter sensitivity, the lower the resistance it puts in parallel with the circuit it is measuring. And the more current the meter itself draws from the circuit, the more it can upset circuit operation. The circuit was not designed to operate with this additional, low resistance in parallel with it. Some circuits can be so loaded down in the presence of a meter's parallel resistance that they stop working properly. In contrast, a high-sensitivity voltmeter doesn't load down the circuit it is measuring, which allows the circuit to function as it should.

A meter's input sensitivity in ohms-per-volt is often termed its *input impedance*. A VOM having a 10KΩ/V input sensitivity may also be described as having an input impedance of 10KΩ. (In common practice, the term *input impedance* is generally used with digital multimeters or analog meters that have transistorized front end amplification. Such analog meters are frequently described as FET VOMs or "VOMs that work like a VTVM." FET = field effect transistor; VTVM = vacuum tube volt meter.)

For purely electrical systems, as house wiring, electric distribution systems, or nonelectronic automobile circuitry, the ohms-per-volt input sensitivity of a meter is relatively *un*important. Circuit loading by a meter having a low Ω/V input sensitivity, such as only 3,000 or 5,000Ω/V, will not affect voltage measurements of primary AC wiring in a building or on an automobile's DC electrical system. For electronics work, the opposite is true!

Once more, when you measure a voltage, the meter itself puts an additional load resistance in parallel with the circuit being measured. *Ideally,* a meter should introduce absolutely no additional resistance into the circuit. The truth of the matter, however, is that a multimeter *does* introduce resistance into the circuit under test.

The higher a meter's input sensitivity, the less current it draws from the circuit being probed. For electronics work, it is essential to use a meter with a relatively high input sensitivity so the meter won't load down circuits. A multimeter with a *minimum* of 20,000Ω/V input sensitivity on DC ranges is adequate for most routine electronic measurements you will be taking. Avoid meters with a 3K or 5KΩ/V input sensitivity. These are typically sold at home improvement, hardware, and auto parts stores. They are fine for AC power and vehicle electrical systems, but can load down electronic circuits, giving false readings.

Digital multimeters usually have a much higher input sensitivity than analog VOMs. Digital multimeters typically have an input impedance of ten megohms (10MΩ) or higher. This is not to say that the digital meter is necessarily the better meter. Again, an analog meter with a 20,000 ohms-per-volt DC input sensitivity is adequate for most VCR and general electronics work.

The *AC input sensitivity* of most analog VOMs is one half, or less, that of the DC rating. For example, a multimeter with a 20,000 ohms-per-volt rating on DC ranges may have only a 5K or 10KΩ/V sensitivity on AC ranges. This is generally alright for routine electronics troubleshooting, where you will be measuring AC voltages in a power supply rather than AC signal voltages in electronic circuitry.

Once you're satisfied that prospective multimeters have sufficient voltage and resistance ranges, and an input sensitivity of at least 20,000 ohms per volt, your next decision is analog or digital?

The cost of digital multimeters has come down to the point where they are very competitive with analog meters. As stated earlier, DMMs will almost always have a higher input sensitivity than an analog VOM.

Digital meters are more accurate than comparably priced analog meters. A typical DMM has a DC accuracy of under 1 percent, whereas a good analog meter will have an accuracy of around 2 percent to 3 percent. A meter with 2 or 3 percent accuracy is adequate for your needs.

Digital multimeters are easier to read than analogs. For a handheld multimeter, a liquid crystal display produces a direct readout in numbers, so you don't have to interpret a meter scale. Some digital meters even have *auto-ranging;* they automatically select the correct voltage range depending on the voltage being read. Digitals have either a round selector switch or a row of pushbuttons for selecting meter function and range. Pushed once, a button stays depressed and selects a function or range. Push another button and the first pops back out to its Off position. On some meters two buttons must be depressed together to select a particular function and range.

Digital meters are less prone to being overloaded and burned out by accidental misuse than analogs (for instance, if test leads are connected to 100 volts with the meter set to its 10-volt range). Digital meters are in general more rugged and dependable from the physical standpoint also. Drop an analog meter, and the delicate meter movement may be damaged. Digitals usually stand up pretty well to this type abuse. Frequently, DMMs incorporate a buzzer or beeper continuity checker. This handy feature was described earlier in this chapter.

So which type should you buy? It's really your personal choice. Whatever decision you make, be sure to thoroughly read the user's manual that comes with your meter. Working with a digital meter may take a little time to get used to, depending on the particular model and the method for switching among meter functions and ranges. For example, on one popular DMM, two buttons are used together to select among the meter's voltmeter, ammeter, and ohmmeter functions. With both buttons depressed, it's an ammeter; with button #1 in and #2 out, it's a voltmeter; with #1 out and #2 in, it's an ohmmeter. That's why it's important to read the booklet that comes with your meter.

Many technicians prefer a digital meter over an analog. Although digital multimeters are great, if I had to select *just one* meter to live with, I'd probably go with a good quality-analog. (Although you realize, I really *love* my digital and use it often!) You're probably wondering, "If digitals are easier to read, more rugged, more accurate, have higher input sensitivity, and can be purchased for comparable cost, why would anyone prefer an analog?" Frankly, there are times when an analog meter can tell you more about what's going on in a circuit. For example, you can connect analog meter leads to two ends of a printed circuit with the meter in the R × 1 position, and then tap around the circuit board to locate a crack. You can easily detect any flicker of the meter needle if the circuit pattern isn't thoroughly sound.

This could go unnoticed on a digital, or the LCD digits may change so rapidly you're not too sure what you saw. Perhaps on an analog meter, the pointer dropped all the way to infinity for just a fraction of a second, and then returned to zero ohms. This would certainly indicate a fracture in printed circuit wiring.

You might measure a voltage, and then tap around a circuit board while keeping an eye on the meter. If the meter pointer suddenly changes, up *or* down, and then returns to its original reading, this most likely says that you are tapping close to a bad solder joint, cracked PCB circuit, or a poor connection. Events that can cause a brief meter flicker on an analog meter are not always easy to interpret, or even to see at all, on a digital.

Similarly, if you measure some steady voltage with an analog, and the meter momentarily kicks up a few divisions, and then returns to the reading, you know there must be something causing impulse noise in the circuit. Intermittent and fluctuating circuit changes are usually easier to detect with the moving pointer of an analog meter compared with a digital.

Some digital multimeters have an analog bar graph, and at least one company markets a true combination analog-and-digital multimeter. It has a mechanical meter movement with pointer as well as an LCD digital readout. With these meters you get the best of both worlds, but they cost more than an equivalent analog or digital alone.

So the choice is yours. You won't go wrong with either type of meter. Most electronics novices prefer a digital, if for no other reason than that they are easier to read than an analog. Some of you will choose to use both types. Many suitable units are available for less than $50. Some meters have enhanced features, like audible continuity or a built-in transistor tester, that you may feel are worthwhile for the additional cost.

Get a good set of test leads, and possibly a spare set. Some test leads that come with lower-priced meters are fairly fragile (read "cheap"). They don't hold up well to frequent use. A high-quality set of test leads on which you can rely is a good investment.

You'll probably want more than one set of test leads for another reason: the standard leads with 1/2-inch probe tips that come with most meters just aren't adequate for all testing situations. Sometimes an alligator clip on one or both test leads is extremely convenient. For example, you can clip the black lead to a convenient chassis ground point while probing power supply outputs with the red lead.

Miniature clip-on test leads, which have tiny spring-loaded metal hooks that clamp onto a wire end or connector pin, are ideal for connecting to closely spaced component leads like those on an integrated circuit. You'll likely find this style of test lead the most useful for VCR troubleshooting, a "must have." You could easily short adjacent IC pins with a standard probe tip, possibly destroying a component or smoking a printed circuit board wiring trace. Mini clip-ons prevent this.

There are even wire-piercing probes that puncture the insulation on a wire to contact the conductor. Although you probably won't use this type of probe for VCR repairs, they are very handy when tracing automobile circuits. In short, you'll want test leads with different probe arrangements to supplement the leads that come with the meter.

You can either buy separate test leads, each set having different probe ends, or probe tip adapters. I prefer separate test leads; a slip-on adapter is just one more connection that can go bad over time. You must have excellent test lead connections for reliable multimeter measurements.

An alternative is to construct your own test leads. Electronic suppliers carry red and black test lead or probe wire, which is very flexible and holds up well to repeated flexing. They also carry a wide variety of probe tips and banana plugs for custom-made test leads.

Read the entire instruction booklet that comes with the particular meter you purchase. A well-written instruction booklet will tell you how to properly use and care for the multimeter. Depending on the meter, the manual may even describe how to make measurements other than the usual AC and DC voltage, resistance, and current. For example, the meter might also measure decibels (dB) or capacitance.

## Keep Your Meter Healthy

Get in the habit of rotating a multimeter's selector switch back and forth through all range settings or operating all pushbuttons a few times before you use it. If the meter hasn't been used in a while, it's possible for very light oxidation to form on function and range switch contacts, resulting in undesirable contact resistance. Operating the switch a few times effectively cleans off any of this oxide.

Here's a review of some multimeter do's and don'ts. These principles apply to most situations where you could have difficulties, such as burning out an analog meter movement or pegging the meter so hard the pointer gets bent. Incorrectly using a meter can also damage the circuit you are testing. For example, probing a DC output voltage of a VCR power supply while the meter is set to measure current effectively shorts out the supply. This could cause a transistor or diode in the supply to short or open.

### *Multimeter Do's*

- Check the function/range selector switch(es) before probing

    For example, don't read a voltage with the selector switch set to an ohmmeter range, like R × 1. Could be "Bye-bye meter!"

    Always start with a high voltmeter or ammeter range, and then select lower ranges after the meter leads have been connected, if appropriate, to obtain an accurate reading on the meter scale.
- Keep test leads and probes, tips, and other accessories in excellent condition
- Use the right probe for the right job. Don't go probing around a tiny connector having closely spaced contacts with a big alligator clip. You could easily short out adjacent pins, causing component damage
- Protect your meter from harsh physical abuse, especially analog meters, which have a delicate meter movement
- Read and understand the book that comes with your multimeter
- Know your meter

### *Multimeter Don'ts*

- Don't take any resistance measurement in a circuit that has any power applied
- Don't replace any meter fuse with one having a higher current rating or slower response time
- Don't leave meter leads connected to the circuit under test once you've taken your reading
- Don't connect your ammeter directly across a power source, such as a battery or power supply output. An ammeter must always be connected *in series* with the load
- Don't swap black and red test leads at the meter to reverse polarity when the other ends of the leads are connected to a live circuit
- Don't expose the liquid crystal display on a DMM to prolonged direct sunlight, which could happen if it was left on the seat of a car.

    Always treat all electrical circuits with respect! Knowing what you're doing is important. If you don't know what you're doing in areas that could prove dangerous, such as in 120-volt household circuits, then don't do it! Find out more first, ask someone who knows how to help you, or leave it to a professional in those areas.

## Know Your Meter

The more experimenting you do with your meter, the more comfortable you'll be with its operation. You can measure the resistance of a toaster or electric iron. Then use the formulas you learned in the last chapter to determine how many watts of power it consumes at 120 volts. You should come out close to the wattage rating stamped on the appliance.

Don't expect this to work out with motor-driven appliances or even incandescent light bulbs. The DC resistance of a motor is valid only for computing current drawn when the motor is first turned On. Although the actual DC resistance of a motor doesn't change when it's running, the current drawn is much less than the applied voltage divided by the resistance (I = E/R). This is just the nature of motors; they require more current to start than after they're up to speed. (Technically, a running motor produces a *counterelectromotive force (CEMF)*, which is a voltage opposite in direction of the applied voltage. CEMF reduces the current drawn by the motor.)

The resistance of a light bulb is *not* constant. When cold, the resistance value of a filament is much lower than when it's at full brilliance. Because of this, incandescent bulbs draw more current at the very instant they're turned On, until the filament comes up to temperature.

For example, take a 25W, 120V incandescent light bulb. What should the resistance of the filament be when it is powered from 120 volts? You know the values for voltage (E) and power (P). To find resistance (R), use the power formula that shows the relationship of E, P, and R:

$P = E^2/R$      Turn formula around so R is on left side of = sign:
$E^2/R = P$      Next, divide both sides by $E^2$:
$E^2/E^2R = P/E^2$      Simplifying, since $E^2/E^2 = 1$ gives:
$1/R = P/E^2$      Flip both sides of equation over to get:
$R = E^2/P$      Finally, substitute known values for E and P:
$R = 120 \times 120 \div 25$
$R = 14,400 \div 25 =$ **576 ohms**

Thus, a 120V, 25W incandescent light bulb should have a *hot* filament resistance of approximately 576 ohms.

You can also determine how much current the bulb draws when lit, using the formula:

$$I = P/E$$

Plugging in known values for P and E:

$$I = 25 \div 120 = \textbf{0.21A, or 210mA}$$

When you actually measure the cold filament of this particular bulb with an ohmmeter, though, you find that it is only 50 ohms! That means at the very instant this bulb is turned On, it draws *a lot more* than just 210 milliamps. How much more?

$$I = E/R = 120 \div 50 = \textbf{2.4 amps}$$

Of course, this higher current flows for only a fraction of a second after power is applied to the bulb, until the filament gets hot.

How much power does this 25-watt bulb actually take at the very instant of turn-On? Use the following power formula, plugging in the measured cold resistance value of 50Ω:

$$P = E^2/R = 120 \times 120 \div 50 = 14,400 \div 50 = \textbf{288 watts}$$

Again, this higher power is consumed only at the instant of turn-On. Now you know why a light bulb generally blows when it is turned On. The sudden inrush of current through the cold, low-resistance filament causes a weak spot in the filament to pop open.

If you have a multimeter, read through the booklet accompanying it. Then do some experimenting with flashlight batteries, low voltage indicator lamps, and various resistors to get the hang of how it works.

Radio Shack sells low current indicator lamps that operate on either 1.5, 6, or 12 volts. Do not try to use 12V automobile lamps, such as brake, directional, and backup lamps or headlamps; they draw too much current for use with small dry cell batteries. The indicator bulbs you purchase should draw between 25mA and 250mA.

Use some 1.5V and 9V batteries as your power source. You can get battery holders for one or more AA, AAA, C, or D cells. These let you easily make connections to battery terminals, and you can put multiple cells in series for higher voltages. For example, a battery holder with four AA cells makes a six-volt battery. Snap connectors with short lengths of wire for 9V "transistor-radio" batteries are also available at electronic parts stores. AAA cells tend to be rather expensive, considering their smaller capacity. AA or C cells would be a good all-around choice for experimenting.

Purchase several different values of resistors. For experiments with low-voltage indicator lamps, you will most likely want resistance values between just a few ohms and perhaps 1,000 ohms. Radio Shack sells two 1/2-watt carbon resistors per pegboard card for around 50 cents. The value of the resistor is printed on the top of the package, or you can use the resistor color code in Chapter 10 to determine a resistor's value. You may want to include 10Ω, 50Ω, and 100Ω wirewound resistors, either 2W, 5W, or 10W, in addition to a variety of 1/2W carbon resistors. Radio Shack has these also, priced around $1.00 for two.

Measure the current drawn by an indicator lamp with the correct voltage applied. For example, connect a 6V lamp, milliammeter, and 4 C-cells in series (4 × 1.5V = 6V). Once you measure the current drawn by the bulb at 6V, determine what value resistor you need to wire in series with the 6V bulb so you can light it with a 9V battery. Be sure to also figure how much power the resistor needs to dissipate. If it's more than roughly 0.3W, a 1/2W resistor will get pretty warm in constant use. If the value you determine is more than 1/2W, you'll need a larger wattage resistor if the circuit is to be under power for more than just a few seconds. Otherwise, the resistor will burn up!

Here are just a few things you might try. Measure values of circuit voltage, current, and resistance with a VOM:

- With a 6V battery and 6V bulb, see what happens when you connect a resistor in series with the bulb and battery. Try this with several resistors, each having a different value.
- Experiment with *two* resistors *in series*. Try this with two resistors of the same value, and then with two resistors of different values. Measure the voltage drop across each resistor.
- Try two resistors of the same value *in parallel* with each other, both in series with the battery, milliammeter, and bulb. Measure the voltage drop across the resistors.
- With your meter, verify the laws of series and parallel circuits, which were discussed in Chapter 8.
- What happens to circuit current with different values of resistors in series with the bulb? What is the voltage drop across each resistor? Across the bulb?
- How hot do resistors of different values get in various configurations?
- Do your meter readings work in Ohm's Law formulas? Remember, the *cold* filament resistance of an incandescent bulb won't work in Ohm's Law formulas when the bulb is actually lit.

Refer to Figure 9.14. What voltage do you think you would measure with meter leads connected to each side of the flashlight switch in the diagram on the left? What about the meter across the switch in the diagram on the right?

- Voltage across the open switch (L. diagram): _____ V
    Check this with an experimental hookup.
    *Why* does a voltmeter across a switch in its Off position read this value?
- Voltage across the closed switch (Rt. diagram): _____ V
    Check this with an experimental hookup.
    *Why* does a voltmeter across a switch in its On position read this value?

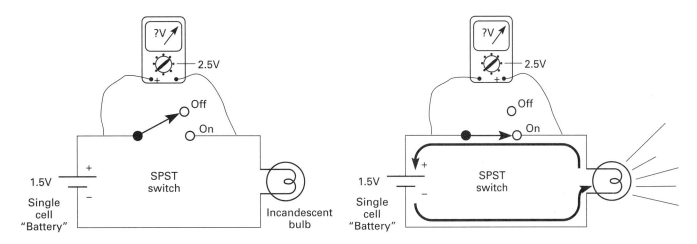

**FIGURE 9.14** Single-cell flashlight schematic diagrams

Reading all the material in this and the previous chapter—and even fully understanding it all the first time through—is fine, but until you actually experiment with some basic electronic components and a multimeter, it'll just be theoretical. Take the time and effort to get some hands-on experience to make the theory come alive and have some real meaning. And it's fun, too!

Use yourself as a resistor. Put your meter on its *highest* ohmmeter range. This should be at least R × 1,000, but if you have an R × 10,000 or higher range it is better. Just touch your fingers to the two probe tips. That is, grasp the black probe tip between thumb and forefinger of your left hand, and the red probe tip between thumb and forefinger of your right hand. Don't worry, you won't get a shock.

Remember, the only source of voltage is the 1.5-volt battery in the meter, although some meters may also have a higher voltage battery for use on high resistance ranges, usually between 9 and 22-1/2 volts. This voltage won't harm you. The current is also very low, typically 100 microamperes or less. You absolutely won't feel a thing electrically when you touch the probe tips.

While grasping the probe tips, you should see *some* reading on your ohmmeter. It's simply measuring your skin resistance at the point of contact with the probe tips, plus body resistance between your two hands. As you press harder on the probe tips with your fingers, you *lower* the contact resistance between the meter leads and your fingers. You should observe the meter needle move higher, that is, to a *lower* resistance reading. Moistening your finger tips further lowers contact resistance; the meter should indicate an even lower resistance value. If your skin is very dry, you might need to moisten your finger tips slightly to get much of a reading in the first place.

The key word is *Experiment!* Take time to work with your multimeter. Avoid any voltage checks at a 120VAC level until you're first comfortable, and confident!, with your meter skills using some flashlight batteries, a few low-voltage indicator lamps, and a selection of resistors.

Once you've learned and practiced basic multimeter skills, you may at some time want to measure the AC voltage at a wall receptacle. At *all times* be safety conscious when working around 120-volt circuits. Watch what you're doing, and don't do anything like sticking paper clips or a screwdriver blade into an AC outlet so you can make contact with your meter leads. Use meter probe tips designed to reach the inside contacts of an AC receptacle safely.

After measuring the voltage at an AC outlet in this manner, using the proper probes, always remove the probe tips from the outlet. Don't leave them hanging there stuck into the outlet after you get your reading.

Please, do your initial experimentation with a couple of batteries, low-voltage indicator bulbs, assorted value resistors, and switches. Several alligator clip test leads are a big help connecting these components for your experiments. You can get a packet of these jumper wires at most electronic supply houses fairly inexpensively. You can also purchase or make up several jumper leads with small spring-loaded hook-like clips at each end. These mini-clips are convenient for gripping small component leads, but they cost more than alligator clips.

As you practice more with your multimeter on simple circuits with these components, where there is no danger of shock or fire, basic electricity concepts and multimeter use will become *second nature* to you. You really need to reach this stage of familiarity with meter-related electronics before working on VCR circuits. When troubleshooting a VCR, you'll have enough things on which to concentrate, other than wondering how to use your meter. You need to know how first.

So enjoy working with and learning about your meter, batteries, indicator lamps, and resistors. Your gain in hands-on experience is well worth what the time, a few dollars, and your interest will cost.

## SUMMARY

A multimeter measures AC and DC voltages, DC current, and resistance. Another name for this type of combination meter is volt-ohm-milliammeter (VOM).

When taking voltage measurements, always start with the meter range switch set at a higher voltage than you expect to measure. This can prevent meter damage if, for example, you actually connect meter leads to 24V instead of an intended 5V point, with the meter set to the 10V range. Connect meter leads to each side of the circuit or component to measure voltage *across* the circuit or component.

A milliammeter must be electrically *in series* with the circuit or component to measure current flow through the circuit or component. This requires opening up the circuit at some point to insert the two meter test leads, easily done if there is a switch or pushbutton in series with the circuit. Simply place the meter leads across the two switch contacts while the switch or pushbutton is in its open or Off position. Start with the meter at its highest current setting to prevent damaging it with current that is unexpectedly higher than a lower range.

When taking a resistance measurement or making a continuity check, *all power must be removed from the circuit.* Otherwise the meter may be damaged, besides giving an erroneous reading. Zero an analog ohmmeter on each range by firmly touching the two test lead probes together and rotating the Ω ADJ. or ZERO OHMS control on the meter until it reads 0 ohms. Most digital multimeters (DMMs) have no ohms adjustment.

Use the R × 1 range for checking continuity of a device, such as a switch, fuse, or an individual wire in a cable. If a fuse is good, it has continuity and will measure 0 ohms. Use a high resistance range, like R × 10,000, when checking for leakage. Electrical *leakage* results from some value of resistance less than infinity when there should be an open, or infinite ohms.

For example, if there is leakage between two conductors in a cable, you will measure something less than ∞Ω, such as 2 megohms (2,000,000Ω). This would be the reading with the meter pointer over 200 on the ohms scale and the meter set to its R × 10K resistance range.

It doesn't make a tremendous difference whether you purchase an analog meter, one with a mechanical pointer, or a digital multimeter (DMM) with a direct readout display. For an analog VOM, be sure it has a DC input sensitivity of at least 20,000 ohms per volt. Some inexpensive meters or testers sold at hardware, home improvement, and auto supply stores have DC input sensitivities much less than 20KΩ/V. These are unsuitable for measuring in many electronic circuits because the meter itself loads the circuit down, causing the circuit to behave differently than if the meter were not connected. This meter loading effect causes erroneous voltage readings. Digital meters and FET analog meters (having a field effect transistor amplifier) usually have an input sensitivity of 10 megohms or more. This puts minimal loading on the circuit under test.

Becoming thoroughly familiar with the operation of your own meter is important. Read the instruction manual that comes with it and practice using the various meter functions with simple battery-powered circuits to become confident in multimeter usage.

## SELF-CHECK QUESTIONS

1. What can you measure with a typical VOM?
2. Describe the readouts on an analog multimeter and on a digital multimeter.
3. On a typical multimeter, where would you plug in the black meter test lead? Where would you plug in the red lead?
4. What is the purpose of a two-position multimeter switch labeled −DC and +DC?
5. What should you do with a multimeter's range switch before measuring an unknown voltage? Why?
6. Describe in general terms how you would connect a multimeter to measure the amount of current being drawn by the 12V DC cam motor in a VCR.
7. Describe the procedure for zeroing an analog ohmmeter.
8. Referring to VOM Example 1, what electrical value is the meter reading? Include the unit of measure, such as ohms or volts, in your answer.

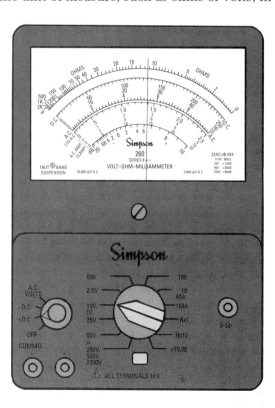

**FIGURE 9.X**  *Courtesy of Simpson Electric Co.*

9. Referring to VOM Example 2, what electrical value is the meter reading? Include the unit of measure in your answer.

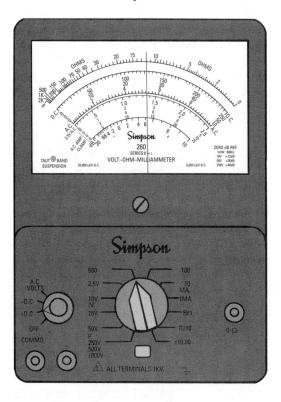

**FIGURE 9.Y** *Courtesy of Simpson Electric Co.*

10. Referring to VOM Example 3, what electrical value is the meter reading? Include the unit of measure in your answer.

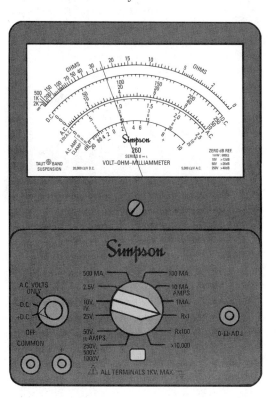

**FIGURE 9.Z** *Courtesy of Simpson Electric Co.*

# CHAPTER 10  ELECTRONIC COMPONENTS

The building blocks of all electronic circuits are the individual components. Separately, these electronic parts all control or affect the flow of electrons in some way. Connected together in all sorts of different circuits, electronic components combine their individual characteristics to make TVs, VCRs, virtual reality, stereos, computers, global communications, and many other feats possible.

To comprehend how electronic circuits work, and how they might fail, it is first necessary to know how each type of component behaves on its own. From there, the next step is to understand how various components work together in a piece of electronic equipment. One effective troubleshooting method is to test individual components in a circuit to determine the cause of a circuit malfunction.

In this chapter you will learn the fundamental operating characteristics of many individual electronic components, how they are represented in a schematic wiring diagram, and what simple tests can be performed on them with a multimeter.

## CHAPTER OBJECTIVES

**Upon completing this chapter, you should be able to:**

1. Describe the essential electrical characteristics of and uses for each electronic component in the following list.
2. Identify the schematic diagram symbol for the listed components.
3. Describe common ways in which the listed components fail and simple multimeter tests that can be performed to detect a failed component.

Electronic Components
- Wire
- Switches
- Resistors
- Transistors
- Coils
- Fuses
- Relays
- Diodes
- Capacitors
- Transformers

Before beginning this chapter, it would be very helpful to get a current catalog from an electronics supply house such as Radio Shack. Most catalogs have photos or illustrations of components described in this chapter. Then, when diodes, transistors, and other electronic parts are discussed, you can see a fairly wide variety of components that are readily available.

As you study how various electronic parts work, concentrate on the overall effect the component has in an actual working circuit. This chapter does not go into detailed VCR circuit descriptions, but provides basic electronics knowledge. With this foundation, and some experience working with simple experimental hookups (and perhaps some easy VCR electronic repairs), you can then go on to learn more about the highly complex operations of VCR electronic circuits. This you can do with technical books designed for those already comfortable with basic electronics.

In other words, if electronics is new to you, this chapter will continue to build your foundation for more advanced study; if electronics is already familiar to you, this chapter may give you a fresh perspective and a helpful review.

> NOTE: Electronic circuits in this chapter are sometimes simplified for instructional purposes. Component values and circuit diagrams are *not* intended as plans for actual working circuits.

And now to discover how electronic components work and how to test them.

## WIRE

A copper wire is probably the simplest electronic component. It carries electrons from one end to another. There is not a great deal you need to know about wires for VCR repair, other than how to check end-to-end continuity with an ohmmeter on its R × 1 range, which was described in Chapter 9. Following are a few facts about wires that you might find helpful.

The material and physical diameter of a wire determine how much current the wire can safely carry without getting hot. Wire *does* have a certain amount of resistance; the smaller the wire, the more ohms per foot it has. Most wire is made of copper. For a given wire size, it can handle more current than aluminum. Silver is the best conductor, but the cost of wire made of this metal is prohibitive for most applications.

Wire sizes are specified by an *American Wire Gauge* (AWG) number. Opposite of many things we are used to, *the smaller the AWG, the larger the wire* diameter. For example, a number ten (#10) wire going to an electric stove or clothes dryer is much larger than the #18 wire on a common table lamp. The 10-gauge wire is needed to handle the large amounts of current drawn by a stove or dryer. If 18-gauge lampcord was used, the wire would get very hot, melt its insulation, and probably start a fire.

House wiring is 12-gauge for 20A circuits and 14-gauge for 15A circuits. Larger 12-gauge wire can safely handle 5A more current than 14-gauge wire. By comparison, most wires inside a VCR are quite small, because they carry very small amounts of current compared to house wiring. Most VCR wiring, other than the 18-gauge line cord and 18- or 20-gauge wire in the power supply, is 22- to 26-gauge.

A wire can be either solid or stranded. *Stranded wire* is made up of many small, individual copper fibers twisted tightly together to form a single conductor. A given size wire can handle the same amount of current whether it is solid or stranded. The main difference is that stranded wire is much more flexible than solid.

Again, the smaller the wire gauge, the larger its diameter and the more current it can carry. This is because larger wire has lower resistance per foot. Sixteen-gauge copper wire has approximately 0.004 ohms per foot, whereas smaller 24-gauge copper wire has about 0.03 ohms per foot. Another way to compare the two: a 250-foot length of #16 wire has the same voltage drop as about 33 feet of #24, when both carry the same amount of current.

A typical printed circuit board has many short wire jumpers to connect, or bridge, between wiring patterns. These are usually designated with a J on wiring boards, like J52, but are seldom indicated on schematic diagrams. Normally these jumpers are uninsulated, and make a convenient place to attach test lead clips. Test points, short metal posts soldered to the component side of circuit boards, and designated TP203, TP204, etc., are also helpful places for attaching meter and other test equipment probes.

## Wire Failures and Checks

A *cable* or *cord* is two or more individual wires or conductors in a common jacket or grouped together. We frequently say "wire" when we really should say "cable." In rare cases individual wires or conductors in a cable can short out to each other, or have leakage resistance between them. This usually is the result of some physical damage, such as a pinched or overheated cable. You learned how to test for interconductor leakage with an ohmmeter set to a high resistance range, such as R × 10K$\Omega$, in Chapter 9.

Most problems with wires and cables are *not* caused by the wire itself, but rather by the many types of connectors, plugs, jacks, terminals, and pins to which the ends of a wire or cable are attached. One of the most frequent causes of electronic problems is not a shorted capacitor, open diode, or leaky transistor; it is a poor electrical connection. This could be a loose battery contact, bent plug, corroded pin, cold solder joint, bad switch or relay contact, or poor wire terminal crimp, to name several causes.

When troubleshooting an electronic failure, you can often save a lot of time by checking for poor connections first. Intermittent problems are frequently the result of a flaky circuit connection. Gently tapping or wiggling connectors and circuit boards will often cause the problem to either clear up or get worse, helping to locate the source of the trouble.

For example, suppose a VCR intermittently unloads tape and shuts down for no apparent reason. You suspect it may be caused by the system control (syscon) microprocessor failing to receive take-up reel pulses. You've checked that there is sufficient take-up torque and have even cleaned the reflective underside of the take-up spindle and the surfaces of the IR LED and phototransistor that make up the optical sensor. The VCR still occasionally fails.

Next you put the VCR in play and gently tap the three-pin connector at the little reel sensor PCB with the plastic handle of a small screwdriver. The machine keeps right on playing. You then tap the connector at the other end of the reel sensor cable where it plugs into a large circuit board; all is still OK. You carefully tap around the wiring side of the large circuit board itself. Sure enough, at one point as you tap the circuit board the failure recurs; tape unloads and the VCR powers down.

Following this procedure, you narrow down the cause to a small area on the large circuit board, near the system control (syscon) microprocessor chip. Close inspection reveals a hairline crack in the circuit board trace going to an integrated circuit (IC) lead. (The customer *did* say the machine had been accidentally dropped!) Soldering a short jumper wire across the tiny crack in the printed circuit trace fixes the intermittent problem.

Now if you happen to have the schematic circuit diagrams for the VCR, you could attack the same intermittent problem somewhat differently. Here's what you might do to check for reel sensor circuit continuity:

1. Unplug the VCR.
2. Set multimeter to check continuity on R × 1 scale.
3. Referring to schematic diagram,
   - connect one test lead to IR phototransistor output on small reel sensor PCB;

- connect other test lead to the sensor input pin on the syscon IC. (This is where those miniature spring-loaded hook-like probe clips work so well!)

At this point the meter registers 0 ohms; there is good continuity between sensor output and syscon input.

4. Tap around the two connectors at the ends of the cable running between the small reel sensor PCB and tap on the main board in the VCR, which contains servo, audio, video processing, and system control circuits.
5. As you tap the main circuit board, you notice that the meter needle flickers downward, to the left towards $\infty\Omega$, when you tap around one small area on the board. Visually checking this area with a strong light and magnifying glass, you discover the hairline crack in the printed wiring.

An 8× or 10× loupe, available at photography and art supply stores, is ideal for seeing small cracks in printed circuit foils, as well as for reading the fine print on boards and components.

Of course, these are just two examples of how you might track down a poor connection. Gently wiggling wires going into a multi-pin connector and cleaning any contacts that are obviously dirty or corroded can also be effective.

In the preceding scenario about the intermittent reel sensor pulses, there could be several other causes. A +5VDC line going to both the IR LED and phototransistor input, or the ground wire to the IR LED could have a poor connection somewhere along the way. The IR LED or phototransistor itself might be marginally functional. Later in this chapter, and in upcoming chapters, you'll learn some checks you can perform on these components.

## FUSES

The purpose of a fuse is to open an electric circuit when current flow exceeds a certain amount, determined by the rating of the fuse. Opening a circuit under high current conditions can save electronic components from damage and prevents overheating, which could cause a fire.

A fuse is really a very thin wire, which either melts or vaporizes when too much current flows through it. The resulting open in the circuit stops current flow. In electronic equipment, most fuses are cylindrical glass with a metal cap at each end. There are two popular physical sizes: 1-1/4 × 1/4-inch and 5 × 20 mm. The 1-1/4 × 1/4-inch size is used in many automobiles. You'll find both sizes in VCRs, but the smaller 5 × 20mm has become more common.

Both physical fuse sizes come in two basic types: fast-acting and slow-blow. The majority of electronic equipment is protected with fast-acting fuses, which open very quickly when their particular current rating is exceeded. This is important for solid state circuits and analog meter movements, which can quickly be destroyed when too much current flows through them, for even a very small amount of time. VCRs use fast-acting fuses.

Some devices, those with fairly husky motors or large capacity power transformers, often have a slow-blow fuse in the AC line. This is because inductive loads, like large motors and power transformers, require much higher current when they are first turned On than when they are operating normally. A slow-blow fuse will not open immediately if its current rating is exceeded; it allows a short-term overcurrent condition.

With glass fuses, you can generally tell whether a fuse is fast-acting or slow-blow. Fast-acting fuses have a single fuse wire running from end cap to end cap inside the glass cylinder. Slow-blows usually have a small spring soldered to one end of the fuse wire near the center of the fuse. If too much current flows for too long through a slow-blow fuse, the solder heats up, melts, and the spring literally pulls the fuse apart, opening the connection.

Never use a slow-blow fuse in place of a fast-acting fuse. It may not open up fast enough to prevent component damage under a high current condition. It's not harmful to replace a slow-blow with a fast-acting fuse, but it will probably open up unnecessarily every now and then when the equipment is first turned On.

A cylindrical fuse has its current rating stamped into one of the two metal end caps. Look for something like 1/4A, 0.75A, or 1A. A magnifying glass or loupe is a big help. Don't replace a fuse with one having a higher current rating. Doing so defeats the purpose of having a fuse in the first place.

## Fuse Failures and Checks

You learned how to check the continuity of a fuse in Chapter 9. Most frequently a fuse in an electronic circuit blows for one of two reasons: either there is a momentary power line surge, or some circuit component has shorted out or become extremely leaky.

In some cases a fuse will open up fast enough when there is a surge so that other components aren't damaged. If this is the case, replacing the fuse with one of the same type and rating restores operation. Unfortunately, a high probability exists that if a fuse blew, something in the circuit it was protecting shorted out. In this case, a replacement fuse blows right away.

For example, suppose that a VCR power supply component like a rectifier diode or electrolytic filter capacitor develops a short. This causes high current to flow in one of the secondary windings of the power transformer. This in turn produces high current in the transformer primary winding, blowing the fuse between the AC line and transformer.

The blown fuse has protected the transformer from burning out and possibly causing a fire, but the power supply component is still shorted out. Until the shorted part is replaced, a new fuse will blow.

Diodes, filter capacitors, and transformers are discussed later in this chapter and in Chapter 12.

# SWITCHES

A *switch* turns the flow of current On and Off in a circuit. There are several types of switches in electronic equipment, but they all basically do the same thing: open or close one or more circuits.

Switches are categorized according to the number of circuits they can control at one time; this relates to the number of switch *poles*. A simple On/Off switch, such as you have already seen in Chapters 8 and 9, is a *single pole switch*. Because it has only one active, or On position, it is a *single throw switch*. A common On/Off switch is thus a single pole, single throw (*SPST*) switch.

A single pole switch can also be made with *two* On positions, and is called a single pole, *double throw* (SPDT) switch. Wall switches in a house that can turn a hallway light On and Off from two different locations are SPDT switches. These are often called *three-way switches,* which is really not correct at all, although this is the accepted nomenclature used by electrical professionals. They *do* have three switch terminals, but only two positions or throws. So it depends on which hat you're wearing: for an electrician, it's a *three-way switch;* if you have your electronics technician cap on, the same switch configuration is a *SPDT*. Figure 10.1 shows how two SPDT switches can independently control a single load.

A switch can have more than one pole. A double pole, single throw (DPST) switch is electrically the same as if you took two SPST switches and

276 ■ Practical VCR Repair

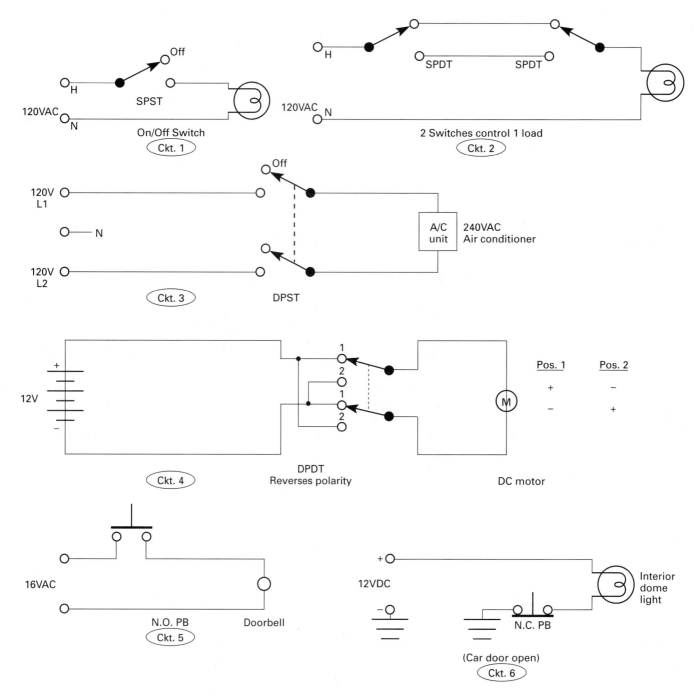

**FIGURE 10.1 Common switches.**

mechanically ganged them so they worked together. DPST switches are often used when it is desirable to turn *both sides* of a circuit Off, especially for safety reasons. A disconnect switch for a 240V air conditioner unit is an example of where a DPST switch is used, as illustrated in Figure 10.1.

Just as there is a SPDT switch, there is also a double pole, double throw (DPDT) switch. A frequent use for a DPDT switch is to reverse DC polarity. Many multimeters have a polarity reversal switch, with the two switch positions labeled +DC and −DC. This control is a DPDT switch. (In some meters, this function may be combined in a rotary switch that also sets the meter up for AC or DC measurements.)

A DPDT switch on the DC power pack for a model train changes the polarity of the DC voltage to the two track rails, thus changing direction of rotation of the DC motor in the locomotive. Figure 10.1 shows a DPDT switch wired to reverse DC polarity.

A momentary contact pushbutton is a type of switch that can be either normally open (N.O.) or normally closed (N.C.); pushing the button *closes* contacts in a N.O. pushbutton. Pressing a N.C. pushbutton opens its contacts. Releasing the button returns the contacts to their *normal* condition.

A doorbell button is N.O.; pushing it closes the two contacts, completing a circuit, and the bell rings. A pushbutton that turns On an interior light when a car door opens is normally closed (N.C.). With the door closed, the pushbutton is pushed in and the contacts are open. Opening the car door causes the switch plunger to assume its *normal* condition: contacts close, completing the circuit, and the light lights.

Don't get confused on this. You might think it is *more normal* for a car door to be closed than open, but when talking about a momentary pushbutton, the terms *N.O.* and *N.C.* mean the state of the pushbutton when its plunger or button is *not* depressed. Figure 10.1 shows both N.O. and N.C. pushbuttons.

Notice the schematic symbols for DPST and DPDT switches; the dashed line between the two operating switch arms indicates that they are mechanically joined and transfer together. Take a few moments to trace the flow of current in each of the six circuits in Figure 10.1. Mentally operate the various switches to the different positions. That is, mentally swing the switch arms with the arrow heads to their alternate positions.

Can you see how the two separate SPDT switches in Circuit #2 can each turn the light On or Off? If the light is On, either switch can turn it Off. If the light is Off, either switch can turn it On.

In Circuit #4, trace current flow through the DPDT switch in both Positions 1 and 2:

1. Start at the negative battery terminal, go through the closed switch contacts in Position 1, to one side of the motor.
2. Go through the motor, through the other set of closed switch contacts in Position 1, back to the positive battery terminal.

Now repeat this with the switch thrown to Position 2. Do you see how the voltage applied to the DC motor changes polarity when the switch is thrown?

There are many variations of switch types beyond those just described. There are switches with three, four, and more poles. Switches can be spring-loaded, so they really operate more like momentary contact pushbuttons, and there are pushbuttons that stay open or closed after you push them once, thus acting more like a switch than a pushbutton. This type of pushbutton is generally called a Push On/Push Off switch.

Switches also come in a variety of mechanical configurations. There are toggle switches, slide switches, rocker switches, and rotary switches. Toggle switches have a handle that moves the switch to its positions, like a common wall switch. Some switches have an operating lever that resembles a baseball bat, and are called bathandle switches. Slide and rocker switches are just variations in mechanical construction. On a schematic diagram, a DPDT switch will look the same, regardless of whether the physical switch in the equipment is a toggle, slide, rocker, rotary, or push/push type switch. Electrically, they all work the same way.

Nearly all VCR pushbuttons, whether on the front panel or handheld remote, are momentary N.O. Some may *appear* to function as a Push On/Push Off switch, but the On or Off *electronic latch* function is performed by system control circuitry rather than mechanically by the switch or pushbutton itself.

For example, on nearly all VCRs, Power control is a N.O. momentary contact pushbutton microswitch. Push the button once and a signal goes to the syscon microprocessor chip. Syscon sends a signal to the power supply, electronically turning On several DC output voltages. When the Power button is released, syscon maintains the Power On condition. Power is latched On, independent of the switch. With the unit On, pressing Power again sends a signal to syscon, which now removes the "Power On" or "Go" or "Start" signal to the power supply, turning Off several DC voltage outputs.

Two- and three-position slide switches are also common on VCRs, used to select recording speed (SP/LP/EP), Normal or Hi-Fi sound, Dolby NR On/Off, and input from antenna or line jacks. Refer to the schematic diagram of a particular model VCR to see what switch contacts are used.

Note that many times the actual physical switch in a piece of equipment has more terminals than are needed or used. For example, a DPDT slide switch may have only two terminals actually connected, and therefore it really functions as a SPST switch. A single part number DPDT slide switch can be wired as a SPST, SPDT, DPST, or DPDT switch. The schematic may or may not show unused switch terminals.

### Switch Failures and Checks

As you learned in Chapter 9, switch contacts can develop high resistance or fail to make any contact at all ($\infty\Omega$), when there should be 0 ohms between switch terminals with the contacts closed. In rare cases, a switch develops leakage resistance between contacts, which shows up when the contacts are open. You can check for these conditions with an ohmmeter. Be sure to account for back circuits. Most of the time it will be necessary to disconnect one or more wires going to a switch in order to electrically isolate it from other components. Then when you check for proper switch function, the ohmmeter will measure just the switch, not other components in the circuit.

## RELAYS

A *relay* is really a switch that is operated by an electromagnet rather than by someone's fingers. Instead of having a toggle handle that moves and transfers contacts, as in a switch, a relay has a magnetic coil that attracts a movable arm, called an *armature,* to which one or more electrical contacts are mechanically attached. The armature moves and contacts transfer.

Recall the operation of a solenoid from Chapter 3: a plunger, also called an *armature,* is drawn into the solenoid when its coil is energized. If a solenoid plunger were connected to the handle of a conventional toggle switch, the combination would be a relay.

Relays can have the same contact arrangements as switches, for example, SPST and DPDT. Figure 10.2 illustrates a single pole, double throw (SPDT) relay, along with its schematic symbol. This relay has a 5VDC coil. When 5 volts DC is applied to the coil, the relay armature is attracted to the coil and relay contacts transfer, turning 120V Lamp A Off and turning On Lamp B.

Relay contacts that are closed when the relay is *not* energized are called normally closed (N.C.) contacts. When the relay coil is energized, N.C. contacts open and normally open (N.O.) contacts close.

Relays are frequently used for remote control and circuit isolation. In Figure 10.2, low voltage on small 22-gauge wire remotely energizes a relay to control two 120-volt lamps. Red and green lights above a drive-through teller

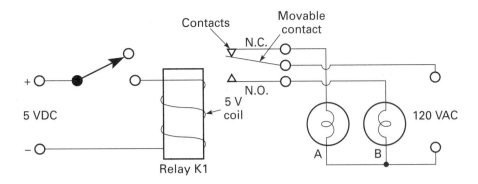

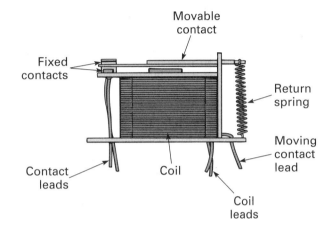

**FIGURE 10.2  Typical low-voltage relay.**

lane at a bank could be controlled in this manner, without having to actually run 120V wires to switches inside the bank. The relay would be mounted in an electrical enclosure close to the lane lights, with the SPST switch controlling the relay coil and a simple 5-volt DC power supply inside the bank. Also notice that the 120V load circuit is totally isolated from the 5V control circuit. By isolated, it means there is no path for current flow between the two circuits.

You are more likely to find relays in older VCRs than in today's VCRs. Some older VCRs have a relay to rapidly switch between the outputs of the two video heads 30 times a second as the drum spins. Operating at this rate, the relay itself sounds rather like a faint, low frequency buzzer. A few VCRs mute the audio during forward scan and reverse scan with a relay. During these modes the relay coil energizes, causing N.O. contacts to transfer to their closed condition, which shorts out the audio signal to ground so the sped-up sound is not heard.

Some four-head VCRs use a relay to select which pair of heads is connected to the video preamps in playback, depending on the tape speed (SP, LP, or EP/SLP). Usually the unused heads are shorted out by relay contacts. Newer VCRs accomplish these switching operations with solid state integrated circuits instead of relays. This is more reliable; relay contacts carrying very low signal levels, such as from magnetic tape heads, are prone to develop contact resistance, which calls for contact cleaning.

On relays where you can see and get at the contacts, contacts can be cleaned using the same procedure for cleaning leaf switches and other contacts, described in Chapter 4. Other relays are totally encapsulated; you cannot get at the contacts. They may look like a tall, *dual inline package* (*DIP*) integrated circuit. Another style of encapsulated relay consists of two contacts enclosed in a small glass cylinder, similar to a glass fuse, with an electromagnetic coil around the cylinder. When current passes through the coil, the relay contacts transfer.

This type of relay is called a *reed relay;* the contact arms inside the glass cylinder are the *reeds.* Some reed relays have multiple glass cylinders enclosed within a common magnetic coil. Each glass cylinder can be manufactured with either a N.O. SPST or a N.C. SPST contact.

## Relay Failures and Checks

The main things that can go wrong with a relay are that the coil can open up and the contacts can get dirty or corroded and develop a high resistance. Not much can be done for an open coil other than replacing the entire relay. Use your ohmmeter to check for an open coil. Some relays have the coil resistance in ohms stamped on the relay cover. You should see some value of resistance, not an open, when you check with your ohmmeter. Remember to disconnect one side of the relay coil when you do this so you are not measuring a back circuit.

A relay coil can also develop a short, especially when it is energized, or *picked,* for long periods of time. Unless you know what the coil resistance is supposed to be, a partially shorted coil will be nearly impossible to detect with a simple resistance check.

If the relay has a DC coil (most relays in electronic equipment do), you can determine what the coil resistance should be if you know the coil voltage and how much current it is supposed to draw. Some relays have this information marked on them. Just use Ohm's Law to determine the approximate resistance value of the coil. If your actual reading is more than 20 percent lower than the calculated reading, there's a good chance the coil is at least partially shorted.

A shorted coil may manifest itself as the relay failing to pick, running very hot, or burning out a driver transistor or integrated circuit. Before replacing an open or shorted relay driver, check for the possibility of a shorted relay coil.

Some relays have two coils. One, the *pick coil,* initially attracts the armature to the electromagnet pole piece. The second, the *hold coil,* then keeps the relay energized *after* it is picked. Here's why two coils are used in this manner. It takes a lot more power to initially pick a relay quickly than to keep the armature attracted after it's picked. A high current is applied to the pick coil and a much smaller current is applied to the hold coil. Once the relay is picked, usually after less than 1/4 of a second, the pick circuit is de-energized and the low-current hold coil keeps the relay energized until voltage to the hold coil is dropped.

Many DC relays have a small *diode* soldered across the two coil terminals, so it is in parallel with the coil. The diode is a small cylindrical device with a wire lead coming out each end. There will be a dot or a ring at one end of the cylinder, or the cylinder itself will be bullet shaped. The purpose of the diode is to short out, *shunt,* or *swamp* a reverse polarity voltage, called *back EMF* (electromotive force), when the coil is de-energized. A relay coil produces a back EMF that is *many times higher* than the normal forward coil voltage. This high reverse voltage can damage a driver transistor or integrated circuit. The diode across the relay coil is transparent to voltage that operates the coil, but acts like a short to back EMF, protecting other components in the circuit.

A diode connected across a DC coil is sometimes called a *kick-back diode.*

This diode can give you a bogus ohmmeter reading when you check relay coil resistance. If you happen to have the test leads connected across the coil so the voltage from the meter battery is in the same direction, or polarity, as the back EMF from the relay, the diode will short it out. It will appear as a very low resistance, and could lead you to believe the coil is partially shorted. Change the polarity switch on the meter or reverse the meter probes. Now, ohmmeter voltage is applied with the *same* polarity as the normal operating voltage of the relay, the diode appears as an open, and the resistance of just the relay coil itself is measured.

You will learn much more about diodes later in this chapter, including how to check them. If the preceding is not all that clear to you right now, don't be too anxious about it. Reread this section after you have learned about diodes.

# RESISTORS

There are two main classifications of resistors: fixed and variable. *Fixed resistors* have only one ohmic value, and unless the resistor is defective, always have the same ohmic value. The ohmic value of a *variable resistor* can change, or vary.

Most variable resistors with which you are probably familiar change their resistance when a knob is turned or a slide lever is moved back and forth or up and down. A rotary control on a lamp dimmer and a slide-type volume control on your boom box are examples of manually operated variable resistors.

### Fixed Resistors

You saw some illustrations of carbon composition resistors and a few examples of how resistors are used in Chapter 9. Carbon resistors are used extensively in electronic equipment. Most often they are discrete units, having the form of a round cylinder with an *axial lead* at each end. Electronic components that have a wire lead coming out of each end, like common fixed resistors, are described as having *axial leads*. This simply means the two leads and the component itself are in line on the same axis, like the rear wheels of an automobile on a common axle.

Axial leads on discrete resistors are bent at 90 degrees, inserted into holes in printed circuit boards, and soldered to the foil wiring pattern. When components are connected to a printed wiring board in this manner, it is called *through hole mount* (*THM*) or *pin-in-hole*. You will see this type of resistor frequently on VCR printed circuit boards. Most of the following discussion refers to discrete, axial lead resistors.

Several resistors can also be manufactured together in a single package, usually called an *R-pack*. R-packs are made with several resistor leads in a single line, called a *single inline package* (*SIP*), or in a package with two rows of inline pins called a DIP, for *dual inline package*.

Resistors are also made for surface mount technology (SMT) printed circuit boards. Here, each resistor is a small, flat, rectangular component with two flat metal tabs. The tabs are soldered to connection pads on the surface of a circuit board, instead of pin-in-hole like discrete resistors with axial leads.

### Resistor Color Code

As you know, resistors have two fundamental characteristics: ohmic value and wattage rating. Normally you will be concerned with just these two. A third specification is *tolerance,* which indicates the most a resistor's *actual* ohmic value can be expected to vary from its rated value. Most resistors have a tolerance of 10 percent, 5 percent, or 1 percent. For example, a 330-ohm, 5 percent tolerance resistor could have any value between 330 ohms minus 5 percent of this value to 330 ohms plus 5 percent, or from 313.5 ohms to 346.5 ohms.

Resistance value and tolerance of round, axial lead carbon-film and metal-film resistors are usually noted by several color bands around the insulated resistor body. Each color band represents a different digit, 0~9, or a multiplication factor, or a tolerance percentage.

Most resistors with which you will be working have four color bands. The first and second color bands represent a two-digit number, whereas the third band is a multiplier of that number. These three together determine the resistor's ohmic value. The fourth color band specifies the resistor's tolerance. For example, a resistor with a red first band, a violet second band, and orange third band has a value of 27,000 ohms:

- Red first band = 2
- Violet second band = 7
- Orange third band = × 1,000 (or × 1K)

The fourth band identifies the resistor's tolerance. Most resistors in consumer electronics equipment have a tolerance of 10 percent or 5 percent. A silver fourth band indicates 10 percent; gold means it is a 5 percent tolerance resistor.

If the resistor in our example has a 10 percent tolerance, it means that the actual resistance value could be anywhere from 10 percent less than 27KΩ to 10 percent more than 27KΩ:

$$10\% \text{ of } 27{,}000 \text{ equals } 0.1 \times 27{,}000 = \mathbf{2{,}700\Omega}$$
$$27{,}000 - 2{,}700 = \mathbf{24{,}300\Omega}$$
$$27{,}000 + 2{,}700 = \mathbf{29{,}700\Omega}$$

So, a 27K, 10% resistor could have an actual value anywhere between 24,300 and 29,700 ohms. Some circuits require a closer resistor tolerance. A 5%, 27K resistor has an actual value within ±5% of 27,000, which is between 25,650 and 28,350 ohms.

Again, most discrete resistors in VCRs and consumer electronics have either 5 or 10 percent tolerance. It is always acceptable to replace a resistor with one having an equal or tighter tolerance. Thus you may replace a 470-ohm, 10% resistor with a 5% resistor. Replacing a 5 percent resistor with a 10 percent unit may cause slight deterioration in the circuit's operation or precision, such as in a critical timing circuit or voltage reference circuit.

In some circuitry, it is necessary to have highly accurate, 1 percent tolerance resistors. These have five color bands: bands 1, 2, and 3 are the first, second, and third digits, band 4 is the multiplier, and band 5 is brown, signifying a 1% resistor. Figure 10.3 contains the color codes for 4- and 5-band resistors.

If the multiplier band is any color black through blue in the chart, simply add the number of zeroes in the multiplier to the number created by digits 1 and 2 or by digits 1, 2, and 3. For example, if the multiplier band is black, don't add any zeroes. For an orange multiplier band, add three zeroes.

Color band 1 is nearest the end of a resistor, with bands 2, 3, and 4, (and possibly 5) equally spaced to the right of band 1. Note the wide space between bands 3 and 4 on a 4-band resistor, and between bands 4 and 5 on a resistor with 5 bands. On many resistors *there is no wider spacing* between the third and fourth bands on 5 percent and 10 percent resistors, but the fourth band will be silver (10 percent) or gold (5 percent). So how will you know which end of a resistor is band 1? Well, since neither silver nor gold represents the first digit in the color code, you know to determine a resistor's value starting with the band at the other end if band 1 is silver or gold.

Not to confuse the issue, but you *may* come across a five-band resistor where the fifth band is *not* brown. In all likelihood, that resistor is military grade; the fifth band can mean different things. For example, an orange fifth band says the resistor is reliable enough for missile system electronics; yellow says reliability is sufficient for space flight systems. A white fifth band just means the component leads may be soldered. Resistors with a white fifth band are sometimes found in consumer electronic equipment and from electronic parts suppliers.

Resistors with a tolerance exceeding 1 percent, like 0.5 percent or 0.25 percent, will generally have their values printed on the resistor body. You will just about never come across these in consumer electronic equipment. Sometimes 1 percent tolerance resistors will have printed values rather than color codes.

**FIGURE 10.3** Standard color code.

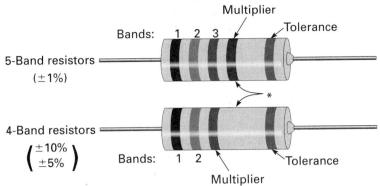

| Band 1<br>1st Digit | | Band 2<br>2nd Digit | | Band 3 (if used)<br>3rd Digit | | Multiplier | | Resistance<br>Tolerance | |
|---|---|---|---|---|---|---|---|---|---|
| Color | Digit | Color | Digit | Color | Digit | Color | Multiplier | Color | Tolerance |
| Black | 0 | Black | 0 | Black | 0 | Black | 1 | Silver | ±10% |
| Brown | 1 | Brown | 1 | Brown | 1 | Brown | 10 | Gold | ± 5% |
| Red | 2 | Red | 2 | Red | 2 | Red | 100 | Brown | ± 1% |
| Orange | 3 | Orange | 3 | Orange | 3 | Orange | 1,000 | | |
| Yellow | 4 | Yellow | 4 | Yellow | 4 | Yellow | 10,000 | | |
| Green | 5 | Green | 5 | Green | 5 | Green | 100,000 | | |
| Blue | 6 | Blue | 6 | Blue | 6 | Blue | 1,000,000 | | |
| Violet | 7 | Violet | 7 | Violet | 7 | Silver | .01 | | |
| Gray | 8 | Gray | 8 | Gray | 8 | Gold | .1 | | |
| White | 9 | White | 9 | White | 9 | | | | |

The wattage rating of a resistor is determined by its physical size. For a given type of resistor, the larger it is the more power it can handle. Figure 10.4 shows discrete, axial lead carbon composition resistors in 1/4W, 1/2W, 1W, and 2W sizes. Most VCRs use many resistors that are only 1/6W or 1/8W, which are even smaller than the 1/4W resistors in the figure.

The majority of resistors in consumer electronics are 1/10W, 1/8W, 1/6W, 1/4W, and 1/2W. It is acceptable to replace a resistor with one having a *higher* wattage rating. For example, suppose you measure a 1/6W resistor in a VCR circuit and find it to be 650 ohms, when it should be 270 ohms. Go ahead and replace this with a 270Ω, 1/4W or even 1/2W resistor. Resistors with a wattage rating of 1/6W, 1/8W, and 1/10W can be hard to find or buy in small quantities, but are available from larger electronic suppliers and mail order companies.

Recall that K after a resistance value means *kilohms* or × 1,000. A 3.3KΩ resistor is 3300 ohms. An M means megohms or × 1,000,000. A 1.2MΩ resistor is 1,200,000 ohms. (Kilo = times one thousand; Mega = times one million)

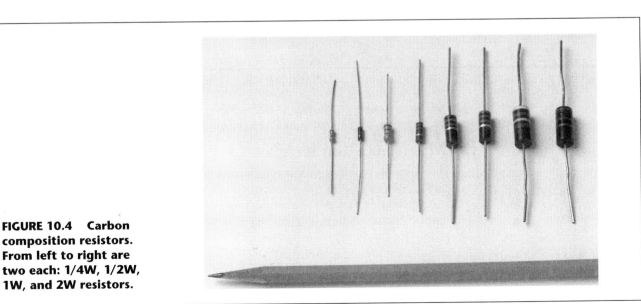

**FIGURE 10.4** Carbon composition resistors. From left to right are two each: 1/4W, 1/2W, 1W, and 2W resistors.

## Color Code Practice

Refer to the chart in Figure 10.3 to determine resistance and tolerance values for resistors having the following color code bands:

A. Red-red-black-gold
B. Orange-white-red-gold
C. Yellow-violet-yellow-silver
D. Brown-black-black-gold
E. Green-blue-gold-gold
F. Orange-orange-green-silver
G. Red-gray-violet-red-brown
H. Gold-brown-gray-blue

## Color Code Practice Answers

A.  Band 1 red = 2 for 1st digit
    Band 2 red = 2 for 2nd digit
    Band 3 black = × 1 multiplier
    Band 4 gold = 5% tolerance
    This is a 22 ohm, 5% resistor.

   Note that gold in the fourth band is *not* a multiplier for a resistor with just four bands. Had this been a 1% resistor with brown in the fifth band, then the gold fourth band would mean to multiply the digits of bands 1, 2, and 3 by 0.1.

B. 3900Ω, 5% (an answer of 3.9K ohms is also correct).
C. 470,000Ω, 10% (an answer of 470K ohms is also correct).
D. 10 ohms, 5%.
E. 5.6 ohms, 5%. (In this case, gold in band 3 says to multiply the 56 indicated in bands 1 and 2 by 0.1).
F. 3,300,000 ohms, 10% (an answer of 3.3 megohms is also correct).
G. 28,700 ohms, 0.1% (an answer of 28.7K ohms is also correct).
H. 680Ω, 5%.

   Did this one trip you up? The resistor color code is actually given *backward*. Because gold or silver can never be a first-digit color, though, you have to read the resistor from the other end: blue-gray-brown-gold. It doesn't matter in which direction a resistor is wired into a circuit; it is not polarity sensitive. Also, as you look at a resistor on a circuit board, it could be oriented in any direction. Don't expect that color band 1 is always at the left or at the top of a resistor.

Just about every electronics supply outlet sells a convenient color code guide for determining values of resistors and other color-coded electronic components. Simply turn three cardboard color wheels on the guide to match the colors of the first, second, and third bands of a component, and the guide displays the correct value. It works the other way around also, to determine code colors when you know a value. You can usually pick up one of these handy guides for less than $2.00.

## Wire Wound Resistors

Sometimes it is necessary for a fixed resistor to dissipate more power in a circuit than a carbon resistor is able to handle. In these cases, a *wire wound resistor* is called for. A wire wound resistor, which can run much hotter than a carbon resistor, is constructed of a coil of resistive wire wound on a ceramic core. Wire wound resistors, seldom used in consumer electronic circuits, are manufactured to handle from two watts up to hundreds of watts of power. If present, they are usually in a power supply or other high-current circuit, like the power output stage of an audio amplifier.

A VCR may have one, or perhaps more, wire wound resistors in its power supply, but this is likely the only place. Wire wound resistors in a power supply typically have values under 1,000 ohms, in the 2- to 5-watt power range, and are similar in overall size to a 1W or 2W carbon resistor. They are usually round, sometimes rectangular, with axial leads. Wire wound resistors do not have a color code; the value in ohms is printed on the ceramic body. It is rare for a wire wound resistor to have an ohmic value over 25KΩ.

## Variable Resistors

Potentiometers are the most common form of variable resistor in electronic circuits. A *potentiometer* (pronounced poe-TEN-chee-ahm-mitter) and called *pot* for short, is made of either carbon composition material or resistive wire with a connection at each end. (So far, not really different from a fixed-value carbon or fixed-value wire wound resistor.) But a pot also has a movable contact arm called a *wiper* that can travel along the resistive material from one end of the resistor to the other. The wiper can therefore contact, or tap, any value of resistance from zero ohms up to the maximum value of the pot.

Figure 10.5 shows several styles of potentiometers, including a dual slide type directly below the schematic symbol for a pot. The three small ones in the lower left are designed for infrequent adjustment or calibration of electronic circuits; they are called *trimmers* or *trim pots*. Use an insulated screwdriver or TV alignment tool to adjust these. Every pot has three terminals or connections, one for each end of the resistor and one for the wiper contact.

The ohmic value of a pot is the *maximum resistance* from end to end. Thus, a 10K pot measures 10,000 ohms from one end to the other. Any value between 0 and 10,000Ω can be tapped by the wiper. When a voltage or signal is placed across the ends of a pot, the wiper can therefore pick off any value of voltage or signal between 0 at one end of the pot and the maximum applied at the other end of the pot. In this way, a pot becomes a *variable voltage divider;* the wiper position determines how much voltage or signal goes on to the next part or *stage* in a circuit. For example, notice the pot with alligator clip leads at top-center in Figure 10.5. Imagine the two leads on the left connected to a

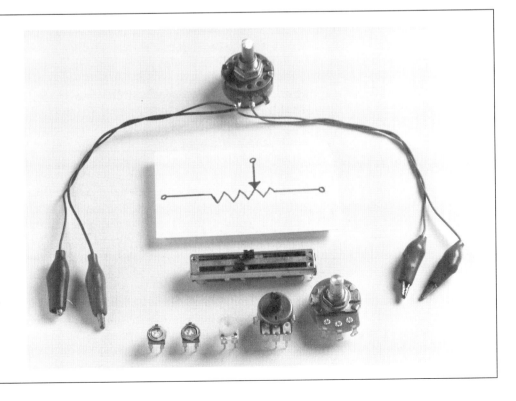

**FIGURE 10.5** Potentiometers and schematic symbol.

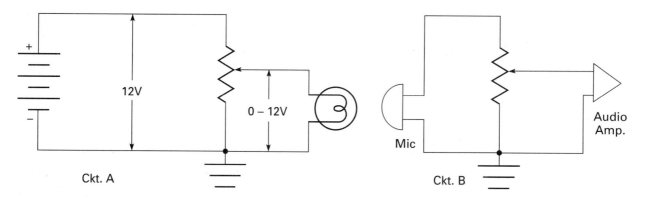

**FIGURE 10.6  Potentiometer uses.**

12-volt battery and the two on the right connected to a small 12-volt indicator bulb. As the pot's control shaft is slowly turned from one end to the other, the bulb would gradually go from all the way Off to full brilliance. The next figure shows this schematically.

Figure 10.6 shows two similar uses for a pot. In Circuit A, a DC voltage is applied across the pot. The wiper then picks off any value between 0V (bottom of pot) and the applied voltage (top of pot) to send to the lightbulb. In Circuit B, the electrical output of a vocalist's microphone is connected across the potentiometer. This pot acts as a mic volume control; the wiper position determines the amount of signal that goes on to the audio amplifier, and thus the amplitude of the singer's voice coming from a loudspeaker or being recorded.

In some circuits, instead of a signal being applied *across* a pot, a variable resistor is connected *in series* between the input source and output load. Only two of the three terminals on a pot are actually used.

A variable resistor with two terminals is called a *rheostat* (pronounced REE-oh-statt). In electronic equipment, a pot is generally used where a rheostat is called for.

Figure 10.7 shows three ways a rheostat may be depicted on a schematic wiring diagram; all three circuits are equivalent. A variable resistor is often shown as a resistor with an arrow running across it at an angle, as in Circuit C. Note that in Circuit B one end of the pot is connected to the wiper; electrically this is still connected as a rheostat rather than as a pot, because the input signal *is not* applied across the pot.

VCRs use many small trimmer potentiometers, wired either as pots or rheostats, to make fine adjustments to video, audio, and servo circuits. You'll find them soldered to one or more printed circuit boards. These adjustments are made at the time of manufacture and under normal conditions don't require routine readjustment. Consult the service manual for the particular model VCR to correctly adjust VCR trimmer pots.

Don't fool with them unless you are sure what you are doing. Carefully note the position of a trimmer pot before turning it so you can return it to its original position if necessary. One good way to do this is to unplug the VCR and carefully measure the resistance between one end of a pot and its wiper with an ohmmeter *before* making an adjustment. Mark the value down on a piece of paper. Now you can always return the trimmer to its original value.

### Resistor Failures and Checks

Most often when resistors fail they either *increase in value* or open up all together. Small resistors are physically fragile; mechanical abuse can cause a lead to pull away from the carbon material in the center of the resistor body. Overheating can also make a resistor increase in value or open.

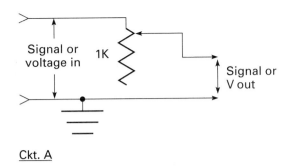
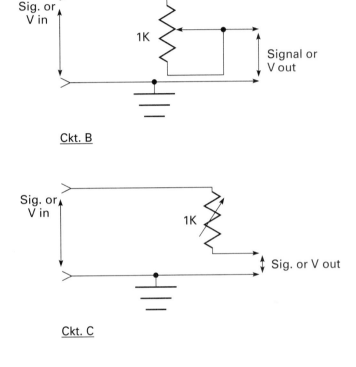

**FIGURE 10.7  Rheostats. All three circuits are electrically equivalent.**

You can normally pick out an overheated carbon or metal-film resistor visually. The resistor will have a darkened or even burnt appearance, and the color bands will have lost their vibrancy. There may even be evidence of cracking or blistering. In nearly all cases of overheating, the resistor itself is not at fault; rather, some other component, such as a capacitor, transistor, or diode, has either shorted out or has become leaky. This causes too much current to flow through the resistor, which then overheats. Generally, you will want to find and correct the cause of overheating first, then replace the resistor.

Potentiometers can increase in value or open up end-to-end, but by far the most common failure is poor contact of the wiper. You have probably experienced this with a volume control on an older radio or TV, where there is a dead spot on the control or it is "scratchy." You have to fiddle with it to get sound. This can sometimes be fixed by spraying contact cleaner into the pot and vigorously working the control back and forth several times through its entire range. TV tuner cleaner with lubricant is OK for this. If this is ineffective, the pot will have to be replaced. This is seldom a problem on VCRs, because there are no pots that get heavy usage.

You can check the resistance of a resistor with an ohmmeter. Once again, if the resistor is in a circuit, you will generally have to isolate the resistor so you are measuring only the resistor, not other components in the circuit. Always be aware of possible back circuits when performing in-circuit resistance measurements.

But if you connect your meter leads across a resistor in a circuit and it measures *higher* than it should, then you *know* the resistor is either open or has gone up in value. Other circuit components cannot possibly *increase* the value of a resistor; any back circuit could only make the resistance reading lower.

Sometimes a resistor becomes electrically *noisy* or unstable. The resistor has the right value when measured with an ohmmeter, but in an actual circuit with current flowing through it, the resistance fluctuates. If the resistance changes rapidly, the resistor is said to be *noisy*. In an audio circuit, the result is random popping, crackling, hiss, or a rushing sound. In a video circuit, a noisy resistor can cause random streaks, dots, or picture instability. Noisy resistors are rare.

An unstable resistor usually changes value slowly as it heats up. The affected circuit often works correctly for a while, but after the unit heats up, some problem appears. Unstable and noisy resistors can be very hard to locate. A can of spray component cooler is often effective. When a heat-sensitive component gets cooled off, normal operation is temporarily restored. Use component cooler carefully; spraying a hot integrated circuit could cause it to crack under thermal stress.

## DIODES

Another basic electronic component is the *diode.* There are many different types of diodes, but they all share one essential characteristic: *current flows in only one direction through a diode.* A diode acts like a one-way valve for electricity. This characteristic is important in many electronic circuits where current flow is desired when the applied voltage is one polarity, but not when it is reverse polarity.

Diodes, transistors, and integrated circuits are made of different compounds, of which silicon is the main element. Silicon, more abundant in nature than any other element except oxygen, is chiefly sand. Silicon is *doped* by combining it with other ingredients to form two different types of silicon, called N-type and P-type; each has unique electrical properties when grown into crystals. In a very simplistic sense, you can think of N-type and P-type silicon crystal as follows:

### N-type silicon

- has extra free electrons that can travel through the material
- N = negative

### P-type silicon

- is missing electrons, creating *holes;* these holes can travel through the material as the electrons move from atom to atom, like bubbles through champagne
- P = positive

Recall that electrons flow from negative to positive.

When both N- and P-type silicon are combined in a single wafer or chip, electrons will flow in only one direction, from N to P. Electrons will not flow from P to N.

The area where the two types of silicon contact each other is called a *PN junction.* A diode works as it does because it contains a single PN junction.

PN junctions can be made of germanium as well as silicon. The majority of solid state devices, however, are silicon-based.

Figure 10.8 shows PN junctions, or diodes, connected to a battery and incandescent bulb. Notice the schematic symbol for a diode above each PN junction. Notice also that the battery voltage in Circuit B is *opposite* the polarity of the applied voltage in Circuit A.

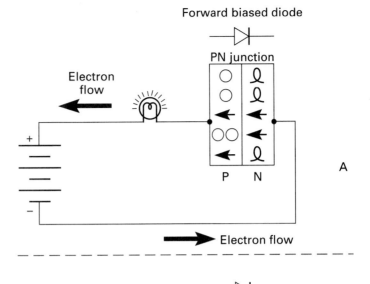
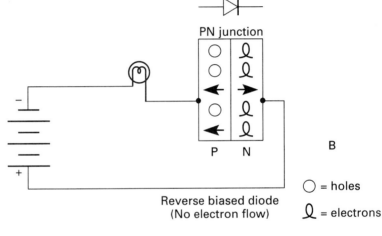

**FIGURE 10.8** Forward and reverse biased PN junction.

When N-type material is connected to the negative battery terminal, electrons flow through the diode and the lamp lights. The diode is *forward biased*. Current flows *against* the arrow in the diode symbol.

When the N terminal of the diode is connected to the positive battery terminal, the diode is *reverse biased*, and no current flows in the circuit. Electrons *do not flow* in the same direction as the arrow in the diode symbol.

A diode, unlike a resistor, therefore has directional properties in a circuit. It *does* matter in which way a diode is wired into a circuit. The P terminal of a diode is called the *anode*. The N terminal is the *cathode*.

Figure 10.9 shows two *nearly identical* circuits. Notice that the only difference between Circuits A and B is the direction in which diode D1 is wired. In Circuit A, diode D1 is *reverse biased*, which means that the anode is more negative than the cathode. No current can flow through a reverse biased diode. Because the diode is in series with the milliammeter and light bulb, no current flows through the meter or bulb. In Circuit B, diode D1 is *forward biased;* the anode is positive with respect to the cathode. Current flows through a forward biased diode, so the bulb lights and the milliammeter registers the amount of current the bulb draws at 6 volts.

In actuality, for a silicon diode, about 0.6V is dropped across the diode when it is forward biased, so the voltage across the bulb is closer to 5.4V (6V battery minus 0.6V across diode). In the majority of circuits, this forward voltage ($V_F$) drop, called the *Fermi voltage,* can be ignored, but the diode will *not* conduct until the forward bias voltage across the diode exceeds this voltage threshold.

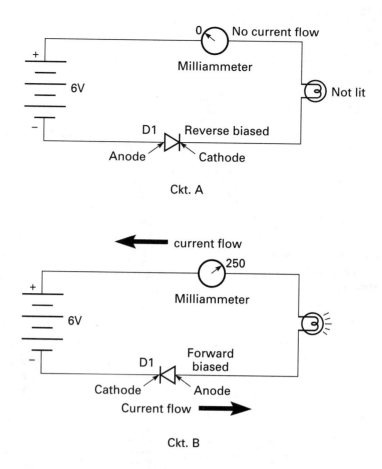

**FIGURE 10.9** Reverse and forward biased diodes. The diode in Circuit A is *reverse biased*; no current flows. In Circuit B, the diode is *forward biased*; its cathode is more negative than its anode, and current flows.

Notice in Circuit B with the forward biased diode, electron current flow is *against the arrow* of the diode symbol. Why is that? Well, in the early days of electricity, it was believed that current flowed from a positive charge to a negative charge. Schematic symbols were created with arrowheads to indicate current flowing from positive to negative. Much later, scientists discovered the electron, and that electrons travel from negative to positive. Nevertheless, the older way of thinking, called *conventional current flow,* is still used in electronic diagrams today. Another name for conventional current flow is *positive current flow,* where current flows *with the arrow* in schematic symbols. Some people prefer to follow a circuit diagram with positive current flow, from positive to negative, because it's easier to think of current flowing *with* the arrow. That's entirely OK, as long as you are consistent in using either electron *or* positive current flow when tracing flow within a circuit. This textbook sticks with electron flow, where current flow is *against the arrow.*

Figure 10.10 illustrates a half-wave rectifier. Diode D1 allows current to flow through load resistor R1 only during one-half of the AC sine wave input. This occurs because the diode is forward biased and therefore conducts or allows current to flow during only one-half of the AC cycle. The diode conducts because of the negative voltage on AC line "B," applied to D1's cathode through R1, at the same time D1's anode is positive from AC line "A." Remember that a diode conducts when its cathode is negative with respect to its anode. Notice that current flows from negative "B" through R1, against the arrow of D1, to the positive voltage on line "A."

Many newcomers to electronics have a difficult time understanding how the half-wave rectifier in Figure 10.10 works. Often the problem is understanding how the diode conducts when it *looks as if* the cathode has a *positive* voltage applied to it.

**FIGURE 10.10  Half-wave rectification.**

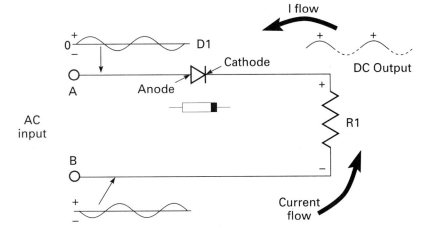

Think of diode D1 as a SPST switch that is closed only when AC line "A" is positive with respect to AC line "B." This places a positive on D1's anode and a negative voltage, through R1, on D1's cathode. When a diode's anode is positive with respect to its cathode, it is forward biased and conducts. The top of R1 is positive with respect to the bottom of R1, *but* the cathode of D1 is more *negative* than its anode, by about 0.6V, and so D1 acts like a closed switch.

You *could* think of the closed switch as letting the positive voltage get through the diode to the top of R1. (That's conventional or positive current flow.)

If you still don't see it, return to the beginning of this section on diodes and go through the material once again. You might also review alternating current and direct current in Chapter 8.

Most diodes in VCRs are axial lead components; they are for the most part cylindrical and look similar to resistors without color bands. Some diodes are barrel-shaped, and others are nearly spherical. Diode bodies are usually glass for low current signal diodes, and are plastic, ceramic, or metal for higher current diodes, such as *rectifiers* in a power supply. A *rectifier* is a device for changing AC to DC. Either the shape or some printed mark identifies the *cathode* end of a diode. Figure 10.11 illustrates several diodes. Most diodes in a VCR have their cathode ends marked with a printed band. Some manufacturers print the schematic symbol directly on the diode body.

Memory helper: The cathode end of a diode symbol looks somewhat like the letter **K**, so you might remember that this is the "Kathode."

### Some Unique Diode Types

Notice the Zener (pronounced *ZEE-ner*) diode in Figure 10.11. This type of diode has a particular voltage at which it *will* conduct when *reverse biased*. Zener diodes are available in many different voltage ratings, and are often used as voltage regulators or as voltage references. A Zener diode can provide a very steady voltage even though the supply voltage wanders all over the place. Many VCR power supplies incorporate one or more Zener diodes.

For example, a 12V Zener diode will not conduct when the reverse bias voltage across it is less than 12V. At 12V and higher, the diode conducts, just as if it was forward biased, *but the voltage drop across the diode remains at 12V*. This is called the *reverse breakdown* or *Zener voltage*.

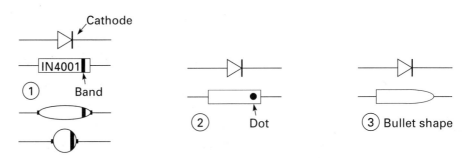

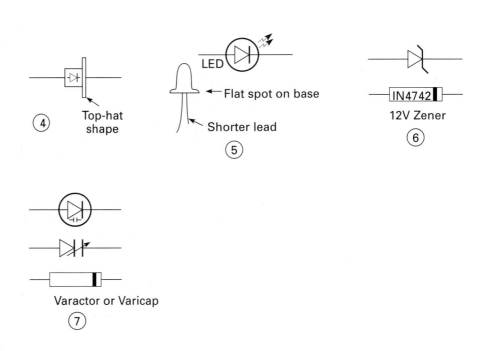

**FIGURE 10.11** Diodes.

Suppose you need stable 5VDC to power a low-current circuit, like 45mA. You want to power the circuit from either a 9V battery or a 12V automobile, which supplies about 13.8V when the car is running. A simple solution is to use a Zener diode as a voltage regulator.

Figure 10.12 is a circuit with a 5V Zener diode. Indicator lamp I1 represents the 5V, 45mA circuit you wish to power. The input voltage can vary anywhere between about 8V with a weak 9V battery to about 14V from a running automobile. Note that there *must* be a resistor in series with Zener diode ZD1 to drop the difference in voltage between the supply and Zener voltages.

**FIGURE 10.12** Zener diode voltage regulator. Diode ZD1 provides a constant voltage drop across the indicator lamp with varying input voltages over an 8–14V range.

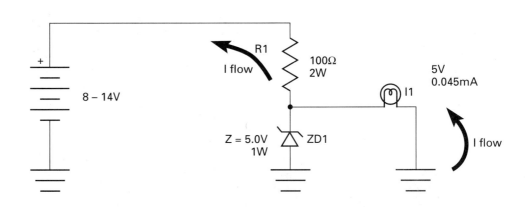

Notice that Zener diode ZD1 is *reverse biased*; its cathode is more positive than its anode. However, once the battery voltage equals or exceeds ZD1's Zener voltage, ZD1 conducts. The difference between the applied voltage and the Zener voltage of 5V is dropped across R1, leaving a constant voltage across ZD1 and the load, indicator I1, which is in parallel with it. That is, if the battery voltage is 9V, then 5V is dropped across ZD1 and 4V across R1. If the battery voltage is 11V, then 5V appears across ZD1 and 6V across R1.

This is a simple shunt-type voltage regulator, which is rather inefficient, but for low current loads is sometimes adequate. A more efficient voltage regulator can be made by teaming up a Zener diode with a power transistor. (*Shunt* is another term for a parallel component.)

Referring back to Figure 10.11, notice the varactor diode. A *varactor*, also called a *varicap*, is a diode that precisely changes the amount of electrical capacitance between its anode and cathode with a change in *reverse bias* voltage across the diode. No current flows through the diode; instead, it acts like a variable capacitor. It changes its capacitance as the DC voltage across it varies. All diodes exhibit a change in capacitance with a change in applied reverse bias voltage. (Capacitors and their uses are discussed later in this chapter.)

Varactors are used extensively in electronic tuners, replacing clunky turret-type tuners. When you rotate a small VCR tuning wheel to preset a channel, you are really turning a potentiometer that varies the DC reverse bias voltage applied to one or more varactors.

Manufacturers usually recommend that an entire electronic tuner section, or tuner PCB, be replaced if it develops a problem, rather than attempting a repair. This textbook, therefore, does not discuss varactors any further, but at least you will recognize the two symbols for a varactor when you see them on schematic diagrams.

## Other Diode Characteristics

In addition to the basic characteristic of diodes—that they conduct in only one direction—there are several other diode characteristics. These are considered the diode's electrical ratings, as follows:

### Forward voltage ($V_F$)

The voltage needed in the forward biased direction for a diode to conduct. For germanium signal diodes, this is around 0.4V. For silicon diodes, $V_F$ is normally around 0.6V. (The *forward voltage* of one or more diodes is sometimes used similarly to a Zener diode for very low voltage regulators or as a reference voltage. For example, two diodes in series, each having $V_F = 0.55V$, could provide a 1.1V reference voltage across them when forward biased.)

### Reverse voltage ($V_R$)

This is some voltage which, if exceeded, could cause the diode to conduct when reverse biased, often resulting in excess current flow and destruction of the diode. A diode can literally explode if subjected to too high reverse voltage, and if there is sufficient current available when the diode's PN junction breaks down.

Another term used synonymously is PIV, for *peak inverse voltage*. Like $V_R$, PIV is a rating that should never be exceeded. For example, a diode with a $V_R$ or PIV rating of 100V should *never* be used to rectify 120VAC. It would be fine for use with a 5V, 12V, or 50V AC input.

### Forward current ($I_F$)

This is the amount of current the diode can handle when forward biased. Here we want a good margin of safety. If a circuit has a forward current of

400mA, a diode with $I_F = 0.5A$ would mean operating the diode near its current limit. A diode capable of handling 1A or more would be much better suited.

Exceeding the current rating can cause a diode to open, short, crack, or even explode.

Diode types are identified by an industry number, the most common having the form *1Nxxxx*, where *xxxx* can be two to four digits. For example, a common general-purpose germanium signal diode is the 1N34, and a popular silicon rectifier diode is the 1N4004.

The two main types of diodes with which you will be dealing in VCRs and most electronics are germanium signal diodes and silicon rectifier diodes. Germaniums will almost always have a cylindrical glass body with axial leads. Low-power Zeners and some high-speed switching diodes may be silicon, but may also be packaged in glass.

If a printed circuit board is marked *ZDxxx* next to a diode, this is most certainly a Zener. Be sure to replace with one having the same Zener voltage.

Always try to replace a diode with the exact same or equivalent type. Electronic parts supply houses have cross-references if you can't get an exact replacement type.

Glass diodes are usually quite small, about the size of a 1/6W or 1/4W carbon resistor. In general, you don't have to be concerned with forward current or PIV ratings for small signal or switching diodes.

For general-purpose silicon diodes, such as in a power supply, the main thing to be concerned about is that the replacement has *at least* the equivalent *forward current* and *reverse voltage* rating of the original. If either the current or PIV rating is higher than the original, no problem; go ahead. For example, a 1N5402 silicon diode rated at 3A, 200 PIV can very nicely replace a 1N4001 diode rated at 1A, 50 PIV.

Multiple diodes are often packaged together. The most common arrangement is four silicon diodes in a single unit with four terminals, called a *bridge rectifier*. A bridge rectifier is more convenient to work with than four individual diodes. Chapter 12 describes how a bridge rectifier works.

## Diode Failures and Checks

A diode can fail in one of three ways. It can become:

- Open
- Shorted
- Leaky.

An analog volt-ohm-milliammeter (VOM) or digital multimeter (DMM) can be used to check for all three conditions.

> CAUTION: Be certain that power is removed from any circuit before performing any of the following diode checks, otherwise meter or circuit damage could result.

There are variations in how different types of meters test diode and transistor junctions. Many multimeters have a function switch position for checking diodes that is marked with the diode symbol. Read the section in the instruction booklet for your specific meter to test diodes. The following is a general description of how to test them.

If there is no special diode test position, place the meter in one of the ohmmeter ranges, for example R × 1, R × 10, or R × 100. Any of these ranges should be acceptable for testing all silicon diodes. However, some analog meter instruction booklets may advise against using the R × 1 range for small germanium signal diodes, where ohmmeter current *may* be high enough to damage the diode. (Personally, I have *never* damaged any type diode when checking on the R × 1 range with several different analog meters, and I doubt you will. Nevertheless, follow your meter's instructions if you have any concerns about damaging diodes.)

As you test a diode, you are looking for it to have infinite, or very high, resistance when the meter leads are placed across it in one direction, and some much lower value of resistance when the meter leads are swapped around. In other words, when reverse biased by the battery in the ohmmeter, the diode should look like an open; when forward biased by the ohmmeter, the diode should read some value of resistance.

With most meters, the black test lead is connected to the negative side of the ohmmeter supply and the red lead to the positive. If the meter has a polarity reversing switch, this meter lead polarity holds true with the switch in the +DC position.

Connecting the black (−) lead to the cathode and red (+) lead to the anode should, therefore, forward bias the diode, because the cathode is negative with respect to the anode, and the meter should read some value of resistance. By reversing the leads, the diode is reverse biased and should look like an open. Test leads can be *electrically* swapped on meters with a polarity reversing switch by simply switching from +DC to −DC.

On an analog meter, the forward bias resistance value depends on the resistance range used, and is *usually not linear over different ranges*. For example, a good silicon diode may read 25Ω on R × 1, 180Ω on R × 10, and 1,400Ω on R × 100. All three readings are fine!

A digital meter usually displays the forward bias voltage drop ($V_F$) for the diode. For a germanium diode, this is about 0.4; for a silicon diode it is closer to 0.6. The meter is *not* set to a voltage range, but to a resistance range, or on some DMMs to a diode check position. It is just the nature of most DMMs that the forward resistance of a diode will indicate the same as the diode's forward voltage drop.

When the diode is reverse biased, the analog meter should read infinity, and a typical DMM indicates an overrange condition. Various model DMMs indicate overrange differently: for example, either **OL** or a blinking **1** leftmost digit may appear on the display.

Figure 10.13 shows forward and reverse checks of a diode using both an analog VOM and a DMM.

If you read zero ohms ($V_F$ = 0 on a DMM) in both directions, then the diode is shorted. A reading of ∞Ω, or DMM overrange, in both directions indicates the diode is open. A reading significantly less than ∞Ω in the reverse biased condition usually indicates a diode is leaky. You should not measure any leakage for a germanium diode and most silicon diodes; some silicons may show *some* amount of leakage in the reverse biased condition on the R × 1K or R × 10K ranges, but not in R × 1. When in doubt, check with a known good diode. Be suspicious of a diode showing leakage characteristics.

Most LEDs have a $V_F$ between 1.8V and 3.0V, depending on the color and specific diode junction, so in most cases you will not be able to test them in this manner with a multimeter. Many analog VOMs use a 1.5V battery on the lower ohms scales, which is under $V_F$ for most LEDs. A higher resistance range, such as R × 1K or R × 10K, on many VOMs uses a higher voltage battery (for example 9 to 22V), so you may be able to test an LED on one of these ranges.

Some DMMs may be able to check an LED if $V_F$ for that particular diode is under 1.999V, but the best way to check an LED is to see if it lights when connected correctly in a circuit. Remember, a current limiting resistor is needed in series with the LED!

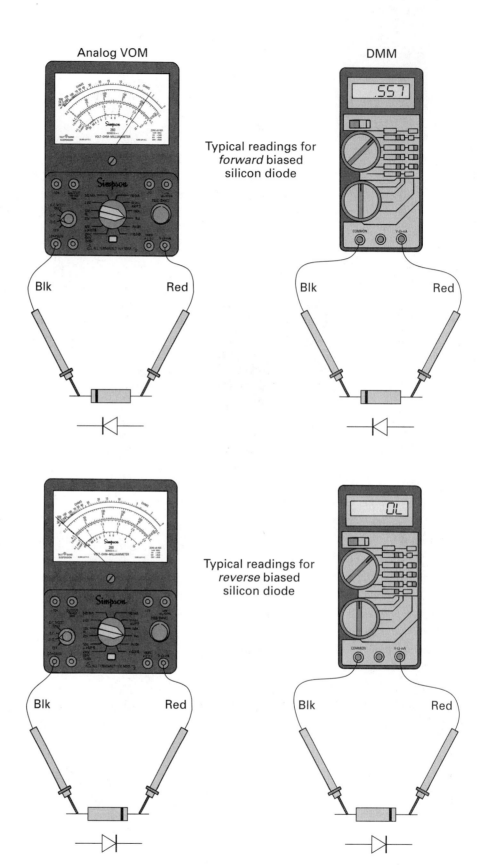

**FIGURE 10.13  Testing a diode.** *Courtesy of Simpson Electric Co.*

As with other components, you can't always be certain if a diode is good or not if you perform in-circuit tests, because of back circuits through other components. To be absolutely sure, you will need to lift, or disconnect, one diode lead from the circuit to avoid back circuits. However, this may *not* be necessary; a basic "good/bad" test can usually be made with the diode still connected in a circuit. Here's how:

- If you get a difference between the forward and reverse readings across a diode in a circuit, the diode is probably good
- If you do not read a low value resistance with analog VOM, or $V_F$ between 0.4 and 0.6 with DMM, in *either* direction, but some higher value, then the diode is probably open
- If you read near $0\Omega$, or near $0.0V_F$ with DMM, in *both* directions, the diode is probably shorted
- Because of back circuits, a diode leakage test cannot be reliably performed in most circuits. Most diodes fail by either opening up or shorting out, which you can usually determine with the diode connected in circuit.

A diode may also become noisy, with symptoms similar to those caused by a noisy resistor, as previously described. Spraying component cooler on a noisy diode may temporarily clear up the problem, thus identifying the culprit. Substituting with a known good diode is often the only way to prove that a noisy diode is causing a particular problem.

## Diode Summary

The following two diode checks with a multimeter *together* indicate a good diode:

- Check forward bias conduction
  — black (−) test lead to cathode, red (+) to anode
  — analog VOM, R × 1 or R × 10: some value, usually between $10\Omega$ to $500\Omega$, but can vary with meter
  — DMM, diode check position: $V_F$ for diode
    - 0.4 for germanium
    - 0.6 for silicon
- Check reverse bias condition
  — red (+) lead to cathode, black (−) to anode (or switch meter polarity to −DC)
  — analog VOM: $\infty\Omega$
  — DMM: overrange

When a diode is *forward biased,* electron current flow is:

- From cathode to anode
- *Against* the arrow of the schematic symbol
- From N-type to P-type silicon
  — think "negative to positive."

If you are comfortable with how a diode and its PN junction works, then go on to the next section about transistors. Otherwise, review this section. It is essential that you fully understand forward and reverse biasing of a PN junction and the effect on electron current flow before studying transistors.

## TRANSISTORS

A *transistor* is a semiconductor, similar to a diode in that it is made with both N- and P-type silicon. Some transistors are made of germanium; the vast majority are silicon. Unlike a diode, a transistor has three leads. Each lead connects to either an N or a P region in the transistor.

The main operational characteristic of a transistor is that a small voltage placed on one of the three leads can control a large amount of current flow through the other two leads. This enables a transistor to perform two basic functions:

- A transistor can act as an electronic switch, turning current flow On and Off
- A transistor can *amplify* a signal, making it larger in amplitude.

The three leads of a transistor are:

- Base **B**
- Emitter **E**
- Collector **C**.

Each lead connects to either an N or P layer inside the transistor. The base layer, sandwiched between the emitter and collector layers, is very thin compared to the other two.

Main current flow through a transistor is between emitter and collector. A small voltage, or signal, applied to the base of a transistor flows to the emitter, and controls current flow between emitter and collector, rather like a valve that controls water flow through a pipe. A small amount of base-emitter current controls a large amount of emitter-collector current.

This deserves repeating. *A small signal applied to the base of a transistor controls a large current flow between its emitter and collector.*

There are two configurations for a common transistor: *NPN* and *PNP,* as shown in Figure 10.14. Notice the different symbols for NPN and PNP transistors. The emitter arrow points away from the transistor body for an NPN, and toward the transistor body for a PNP. (Memory helper: NPN is **N**ot **P**ointing i**N**.) Either type transistor, NPN or PNP, can perform essentially the same function in an electronic circuit. The main difference between an NPN and a PNP transistor in a circuit is the direction in which electrons flow between emitter and collector.

| Transistor Type | Electron Flow |
|---|---|
| NPN | Emitter → Collector |
| PNP | Collector → Emitter |

Notice that, like a diode, electron flow is *against* the arrow in the schematic symbol. For an NPN transistor, the collector must be *positive* with respect to the emitter for current to flow. For a PNP transistor, the collector must be more *negative* than the emitter for electrons to flow.

Discussions in this chapter use NPN transistors as examples. Realize that a PNP transistor works essentially the same way as an NPN, except that the polarity of all applied voltages is reversed. In fact, NPN and PNP transistors are sometimes manufactured with the same characteristics, except they have op-

**FIGURE 10.14** NPN and PNP transistors.

posite polarity voltages and signals applied to them in the same circuit. Transistors like this are called *complementary;* each type works together with, or complements, the other type in a particular circuit, such as an audio amplifier power output stage. Complementary transistor circuits are rarely found in VCRs. However, a typical VCR *does* employ both NPN and PNP transistors, but not in complementary circuits.

Once again, a small signal applied to the base transistor lead controls current flow between emitter and collector. Note especially that when the base-emitter junction is *forward biased,* then current flows between emitter and collector. If the B-E junction is *reverse biased,* then no current flows between E and C. With this operating principle, a transistor can function as a switch: a small amount of B-E forward current turns the transistor On, allowing a much larger E-C current flow. Figure 10.15 illustrates a transistor switch. The resistor limits the amount of base current. A transistor can easily be destroyed if too much current flows through the very thin base region, so frequently you'll see a resistor in series with a transistor's base lead to restrict base current.

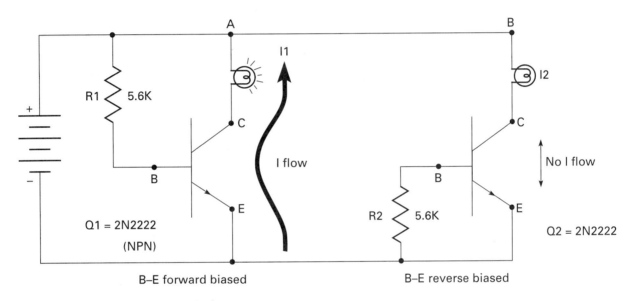

**FIGURE 10.15 Transistor switch.**

In Circuit A, the base-emitter junction is *forward biased:* B is positive with respect to E. Base current flows from the negative battery terminal to the emitter, out the base, then through R1 to the positive battery terminal. This turns the transistor On. Current flows from the negative battery terminal, against the emitter arrow, out the collector, through indicator lamp I1 to the positive battery terminal.

In Circuit B, the base-emitter junction is *reverse biased;* essentially, B and E are at the same potential. This turns the transistor Off: no current flows between emitter and collector, and so indicator I2 does *not* light.

In Figure 10.15, base resistors R1 and R2 connect to positive and negative supply voltages to turn the transistor On and Off, respectively. This is for illustrative purposes. In an actual circuit, the base input signal would come from a previous stage or some other electronic component, such as the system control microprocessor. Likewise, indicator lamps I1 and I2 represent a load device, such as a solenoid, relay coil, or DC motor in a VCR.

For example, a positive 5VDC output from a particular pin on the system control microprocessor IC could turn a transistor On. In place of the light bulb,

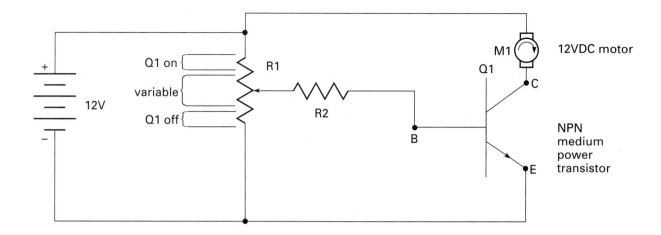

**FIGURE 10.16  Simple DC motor speed control.**

the transistor's load might be a 12V cam motor on the VCR transport. In this case the transistor functions as a *driver,* controlling current that runs the motor.

Figure 10.16 shows a simple DC motor speed control. Potentiometer R1 varies the voltage on the base of Q1. This varies the output current at transistor Q1's collector from 0 to the full amount, perhaps 0.5A, drawn by the motor.

Pot R1 controls how far Q1 is turned On, and therefore the voltage across M1. R2 limits Q1's base current to a safe value. Too much base current can destroy a transistor.

With R1's wiper toward the top, or positive side, of the supply, Q1's B-E junction is *forward biased:* transistor Q1 is fully turned On and maximum current drawn by motor M1 flows from emitter to collector. Q1 looks like a closed switch.

With R1's wiper toward the bottom, or negative side, of the supply, Q1's B-E junction is *reverse biased:* Q1 is completely Off and no current flows from emitter to collector. Q1 looks like an open switch.

As R1's wiper moves from negative (bottom) to positive (top), Q1's base-emitter junction at some point goes gradually from being reverse biased to being forward biased, allowing larger amounts of current to flow between Q1's emitter and collector. Thus the current through DC motor M1 varies, changing its speed.

Depending upon the particular characteristics of Q1, it may take a very small change in voltage applied to its base to go between totally Off and fully On, perhaps only 1 volt or less. Because a small change in input voltage to Q1's base produces a large change (0 to 12V) in the voltage at Q1's collector, Q1 is functioning as a DC amplifier.

One way to look at the circuit is to see Q1 as a variable resistor in series with motor M1. Q1's E-C resistance is controlled by the voltage applied to its base. When the B-E junction is reverse biased, Q1 looks like an open, $\infty\Omega$, and so no current flows through the motor. When the B-E junction is forward biased, Q1 looks almost like a short, with very low resistance; nearly the entire 12V supply is applied across the motor. A small voltage drop appears across the transistor when it is fully On, just like the forward voltage drop across a diode when it is conducting.

When pot R1 is somewhere in its midrange, transistor Q1 will appear to the circuit as some value of resistance between 0 and $\infty$ ohms. Suppose that Q1 acts like a 20-ohm resistor and the motor has a DC resistance of 20 ohms. Motor M1 and transistor Q1 therefore look like two $20\Omega$ resistors in series across the 12V supply. In this case, 6 volts is dropped across the motor and the other

6 volts appears across Q1. (Remember, the sum of voltage drops across series components equals the applied voltage.)

Now imagine that R1's wiper is even closer to the top, positive side, of the supply. Q1 turns On even more and has a lower resistance to current flow. Suppose Q1's effective emitter-to-collector resistance is now 10 ohms. What voltage appears across motor M1? What voltage appears between Q1's emitter and collector?

The total resistance of Q1 plus M1 is $10\Omega + 20\Omega = 30\Omega$. The resistance of the motor is therefore 20 ohms out of 30, or $20 \div 30 = 2/3 = 0.66$ of the total. This means that 2/3 or 0.66 of the applied 12 volts is across the motor, or 8V, leaving 4V between Q1's emitter and collector.

If you are having difficulty understanding how a transistor works so far, stop here. Go back and review the previous examples, applying Ohm's Law. With 12 volts applied across a $10\Omega$ resistor and a $20\Omega$ resistor in series, what is the voltage drop across each resistor?

So far we've seen how a transistor can be used as a DC switch or a DC amplifier. Next we see how a transistor can amplify an AC signal.

Figure 10.17 shows a simple AC amplifier. Resistor R1 has a value that keeps transistor Q1 biased about halfway between being totally Off and fully On, when there is *no input signal* applied to its base.

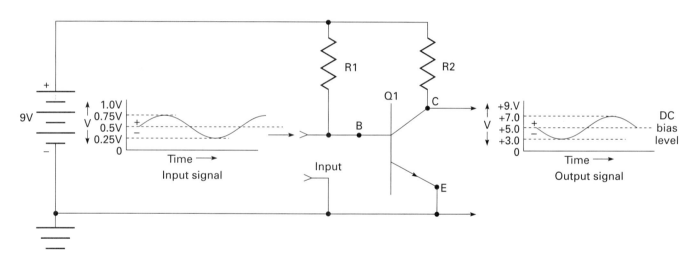

**FIGURE 10.17  Simple AC amplifier. Bias resistor R1 keeps Q1 partially turned On with no input signal. R2 is the output load resistor.**

An AC signal is now applied to Q1's Base. When the signal goes positive, Q1 turns On more, making its effective resistance go down, and therefore the voltage drop across Q1 goes down. Conversely, when the input signal goes negative, Q1 conducts less; its effective resistance goes up, and therefore the voltage drop across Q1 goes up. Note that both input and output signals are referenced to common ground, the negative side of the power supply.

Here is how the AC amplifier works. First, consider what happens with *no* input signal:

- Bias resistor R1 causes Q1 to conduct about midway between all the way Off and full On.
- Q1's collector voltage, with respect to ground, is +5VDC, identified as "DC Bias level" on the Output Signal graph.

Next, the *positive half cycle* of an AC signal is input to Q1's base:

- Q1's base-emitter junction becomes more heavily forward biased

- Q1 turns On further, so the effective emitter-to-collector resistance goes down
- Voltage drop across load resistor R2 goes up, and E-C voltage across Q1 goes down.

During the *negative half cycle* of the input signal:
- Q1's base-emitter junction becomes less heavily forward biased
- Q1 conducts less, so the effective emitter-to-collector resistance goes up
- Voltage drop across load resistor R2 goes down, and E-C voltage across Q1 goes up.

There are two important things to realize about the operation of this single NPN transistor AC amplifier:

1. Notice that the input signal applied to Q1's base varies between 0.25V above ground to 0.75V above ground, or a peak-to-peak input signal of 1/2 volt.

    The output signal resulting from this input varies between 3.0V above ground to 7.0V above ground, or a peak-to-peak output signal of 4 volts.

    This is amplification; in this case the input signal is amplified 8 times (0.5V × 8 = 4V).

2. Also notice that the output waveform is *inverted* with relation to the input signal. When the input goes more positive, the output goes more negative. (For this common emitter circuit, where the emitter is common to both input and output signals, the output is electrically 180 degrees out of phase with the input.)

The previous circuits show basic transistor operation. Variations of these circuits abound in electronic equipment, but the fundamental job performed by a transistor remains the same. A solid understanding of how a transistor functions can help you troubleshoot many more complicated circuits.

To summarize how a transistor behaves:

- Most current flows between emitter and collector, *against* the schematic symbol arrow.
- A small signal applied between base and emitter causes a larger change in current flow between emitter and collector.
- It is often helpful to think of a transistor as a variable resistor connected between E and C. As the voltage applied at the base forward biases the transistor, the E-C resistance goes down, and so there is less voltage drop between the emitter and collector.

## Types of Transistors

Transistors are designed for all sorts of applications. Some are in low-level DC, audio frequency (AF), and radio frequency (RF) amplifiers, handling small amounts of power. These transistors are physically small, produce little heat, and are generally called *signal* transistors. Others are *power* transistors, and are used in power supplies, radio transmitters, and audio power amplifiers. Figure 10.18 illustrates a few of the many styles of transistors available. Each style is manufactured as both NPN and PNP transistors.

Power transistors produce a lot of heat, and are therefore usually mounted to a piece of aluminum with fins, called a *heatsink*. The heatsink draws heat away from the transistor, allowing the transistor to handle more power than if there were no heatsink. Transistors with a TO-3, TO-220, or TO-126 package are power transistors, and so will be most likely mounted on a heatsink. Low-power signal transistors, like the TO-5, TO-72, and TO-92 packages, do not normally require heatsinking.

Like diodes, transistors have an industry type number, the most common having a *2Nxxxx* number, where *xxxx* is two to four digits; 2N3053 is an exam-

**FIGURE 10.18** Various types and styles of transistors. The numbers, such as TO-3 and TO-220, identify a particular size and shape of transistor package. (Note: TO-72 shown here has a fourth lead connected to the metal transistor case itself.)

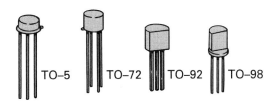

*Small signal-amplifiers, oscillators, switches, etc.*

*Power transistors*

ple, in this case a general-purpose, low power NPN silicon signal transistor with a TO-39 package. Many Japanese transistors have a *2Sxxxx* or *2Swxxxx* number, where *w* is a letter; 2SB641 is an example. Electronic supply houses have transistor cross-reference manuals that list transistor operating characteristics, lead identifications (E-B-C), and suitable replacement part numbers.

In many applications, several different transistors will work just fine. This is especially true for DC, audio, baseband video, and low-frequency RF applications. For high radio frequencies (for example, above 100MHz), it is best to use an exact replacement transistor if possible, but equivalents may work all right. Of course the transistor must be the same basic type, NPN or PNP, and be able to handle the amount of voltage and current in the circuit.

Again, consult a transistor cross-reference or substitution guide for suitable replacement types. The packaging or case style of a replacement may be different. For example, a listed replacement for a TO-92 transistor might have a TO-39 package. Lead identification, or *basing*—that is, which lead is emitter, collector, and base—varies between different style packages. Check a transistor manual for correct basing if a replacement transistor is not the same number as the original. Many transistors have their E, B, and C leads marked on the part itself. Usually a parts container or pegboard card has a graphic identifying the transistor leads, so you'll know how to connect the transistor in a circuit.

Transistors are usually designated with a Q or TR number on schematics and parts lists (for example, as Q207 or TR413). This number is just an identifier for the transistor in the particular piece of equipment, and has nothing to do with the transistor's type, style, or function. Reference these numbers to the electronic parts list in the manufacturer's service manual to find the transistor's industry type number.

## Transistor Failures and Checks

If you look back at Figure 10.14, you'll see that a transistor is quite similar to two back-to-back diodes. There's a PN junction between base and emitter and another PN junction between base and collector. It doesn't matter whether the transistor is an NPN or a PNP. Keep this in mind! Transistor failure modes and checks are quite similar to those of diodes.

Transistors have forward and reverse current and voltage ratings like diodes do. Exceeding either rating can destroy a transistor. For example, if the voltage between base and emitter exceeds the transistor's rating, the thin base layer may short out to the emitter layer, resulting in a base-emitter short. If there's too much collector current, the transistor can either open up, short, or even explode if it gets hot enough.

Following are ways a transistor can fail:

- B-E open
- B-E short
- B-C open
- B-C short
- Combinations of above
- A transistor junction (B-E or B-C) can become leaky or noisy, just as with diodes.

A multimeter can be used to perform a basic go/no-go test on a transistor. Test the transistor exactly as if you were checking two separate diodes, as explained in the earlier section on testing diodes. Select R × 1 on an analog VOM or the diode check function on a digital multimeter. An analog meter may have a different resistance range identified with the diode symbol; if so, use this instead of R × 1. Be sure that all power is removed from the VCR if performing in-circuit transistor checks.

Basically, you want to check for a good diode between base and emitter, *and* check for a good diode between base and collector. If either test fails, the transistor is bad.

To simplify things from here on out, consider a reading of $\infty\Omega$ on an analog meter or "overrange" on a DMM to be *High*. This is a good reading with a diode, or PN junction, reverse biased; it appears as an open circuit. Consider 10 to 500$\Omega$ on an analog VOM, or $V_F$ less than 0.8 on a DMM, to be *Low*. This is a good reading with a diode, or PN junction, forward biased; it appears as a low-value resistor.

It doesn't matter whether the transistor is an N-type (NPN) or a P-type (PNP). With test leads connected in one direction, B-to-E *and* B-to-C should read High. With test leads turned around (or meter polarity switch flipped), B-to-E *and* B-to-C should read Low.

Checking between emitter and collector is *not* a valid test. It will read High with test leads connected either way, for a good NPN *or* PNP transistor.

### Transistor Check

| Black (−) Test Lead | Red (+) Test Lead | Good Readings PNP | NPN |
|---|---|---|---|
| B | E | Low | High |
| B | C | Low | High |
| E | B | High | Low |
| C | B | High | Low |

Just as with a diode, checking a transistor while it is connected in a circuit may not give an absolute indication of its condition, but if there's a difference between forward and reverse readings for *both* base-emitter *and* base-collector junctions, chances are the transistor is OK.

Most circuit boards are marked E, B, and C to identify a transistor's leads. Some transistors have E, B, and C marked on the transistor body itself, but most do not. If you know the transistor type, you can look in a transistor manual to determine its basing.

You can determine whether a transistor is NPN or PNP with a multimeter; here's how. Refer to Figure 10.19:

**FIGURE 10.19** Identifying transistor type. First identify transistor base. If meter reads Low with negative lead on base, transistor is PNP. If meter reads Low with positive lead on base, transistor is NPN.

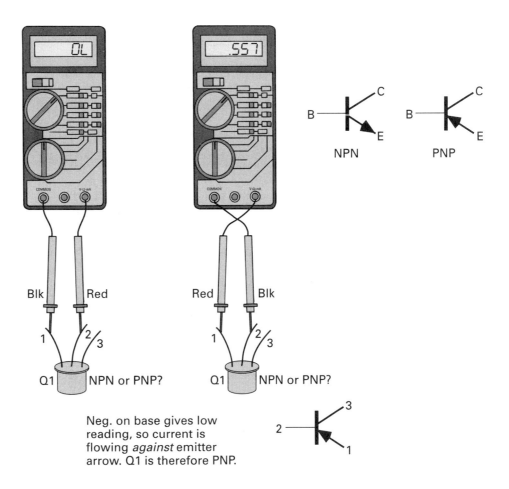

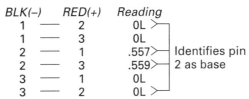

1. Experimentally, find the *one* transistor lead that reads Low to the other two leads in one direction, *but also* reads High to the other two leads with meter leads swapped. This one transistor lead is the base.
2a. If the test lead on the base is black (negative) and you read Low to the other two, this is a PNP transistor: current flows from negative test lead, *against* the emitter arrow to the positive test lead.
2b. Conversely, if you read High to the other two transistor leads with the negative meter lead on the base, this is an NPN transistor: current *cannot flow* from the negative test lead *in the same direction as* the emitter arrow.

For many of you, this probably seems like a lot of transistor information to absorb. In reality, this is only the surface of a fairly deep subject. Experimenting with a few transistors and a multimeter as you go back through the material is an excellent way to gain greater understanding of how to check transistors. You can purchase several types of general-purpose replacement transistors for less than $1.00 each from many electronics suppliers. Get a few NPNs and a couple PNPs, grab your multimeter, and see for yourself. With some actual hands-on activity, it won't take you long to become confident performing a basic transistor test.

## CAPACITORS

A *capacitor* is an electronic component that stores electrons. In its simplest form a capacitor consists of two metal plates separated by an insulating material. The insulating material between the two capacitor plates is called a *dielectric* (pronounced die-eh-LEK-trik). An electrical connection is made to each capacitor plate. Figure 10.20 shows the basic capacitor construction. Many capacitors use paper as the dielectric and metal foil comprising the two plates.

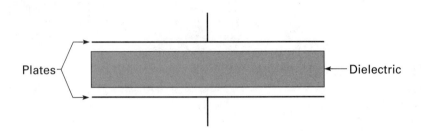

**FIGURE 10.20** Basic capacitor. A capacitor consists of two metal plates separated by a dielectric.

When the two capacitor plates are connected to an electric circuit, electrons are stripped from one plate and deposited on the other plate. This is termed *charging* a capacitor, shown in Figure 10.21. The battery supplies the energy to strip electrons away from the plate connected to the positive terminal and deposits them on the plate connected to the negative terminal.

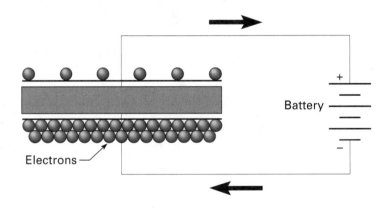

**FIGURE 10.21** Charging a capacitor. A capacitor is charged by removing electrons from one plate and depositing them on the other.

A capacitor is fully charged when the maximum number of electrons has been stripped from the positive plate and deposited on the negative plate, with a given applied voltage. The amount of charge or capacity of a capacitor is determined by the surface area of the two plates, the distance between them, and the type of dielectric material.

Once the capacitor in Figure 10.21 is charged, the battery can be disconnected and the capacitor will retain the charge. (In actuality, a capacitor won't retain its charge indefinitely. Depending on the composition of the dielectric, electrons will move across the dielectric at some rate, back to the positive plate.)

If we connect the leads of a highly sensitive voltmeter to the two plates of the charged capacitor, we would measure the same voltage as the battery that originally charged the capacitor. Electrons would flow from the negative capacitor plate, through the meter, back to the positive capacitor plate, as illustrated in Figure 10.22. This *discharges* the capacitor. As the capacitor discharges through the meter, we would see the voltage gradually drop down to zero. At this point, the capacitor is fully discharged. The voltmeter itself

**FIGURE 10.22** Discharging a capacitor.

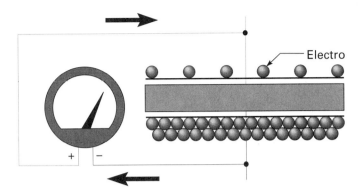

provides a path for electrons from the lower, negatively charged plate to the upper, positively charged capacitor plate.

As you already know, whenever electrons move, there is current flow, and so there is current flow whenever a capacitor is charging or discharging. Current flows during charging until the voltage across the capacitor plates equals the applied voltage. During discharge, current flows through an external circuit, like the meter, until the voltage differential between the two plates is zero.

Notice that current flows in one direction when the capacitor is charging (top to bottom plate, Figure 10.21); current flows in the opposite direction (bottom to top plate, Figure 10.22) when the capacitor discharges.

Now, if a resistor is placed in the charge path or discharge path of the capacitor, the rate of charge or discharge for a given value capacitor can be controlled by the value of the resistor. The higher the ohmic value, the more the resistor restricts current flow, and therefore the capacitor charges or discharges slower.

That is just one use for a capacitor: part of a simple resistor/capacitor (RC) timing circuit. That is, a resistor and capacitor together determine the rate of charge and discharge in a circuit. This rate is very precise for given values of capacitance and resistance, and can therefore be used to develop timing signals and pulses. Intermittent automobile windshield wipers work on the RC timing principle. A pot is usually used to control the amount of resistance in the RC circuit, and therefore the length of time between wiper blade swipes across the windshield.

A resistor and capacitor together can also change the shape of an electronic wave or signal. Electronic waveshaping is widely used in analog and digital circuits to provide precise switching times and to filter out unwanted frequencies or signals. Figure 10.23 illustrates schematic symbols for different types of capacitors. There are other uses for capacitors in electronic circuits, but first we need to talk a little more about capacitors themselves.

Capacitor Action

- Blocks DC voltages
- Allows passage of AC voltages

*The essential capacitor characteristic is that it blocks DC but allows an AC signal to go through.* For example, if we connect a capacitor in series with a 12V battery and a small 12V incandescent indicator lamp, the bulb will light up for perhaps a fraction of a second and then go out. Why? Because current flows from the battery, through the bulb's filament, to charge the capacitor. Once the capacitor is fully charged, no more current flows, and so the bulb goes out.

**FIGURE 10.23** Capacitor schematic symbols.

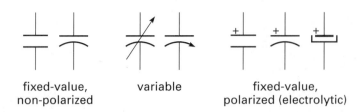

Now take the same capacitor out of the circuit and connect it across a similar 12V indicator. What happens? The charged capacitor discharges through the bulb's filament. The bulb lights for a fraction of a second and then goes out. In essence, the capacitor acts like a battery with a very short life!

Next, we'll connect the same capacitor and 12V indicator lamp in series with a 12V *AC* supply, like a 12-volt transformer. What happens? The bulb remains lit! Why? Because during one-half cycle, when the AC waveform is positive, the capacitor charges through the light bulb, just as with the 12V battery. Electrons pile up on one capacitor plate and are stripped from the other.

But then the AC waveform goes negative. The capacitor discharges through the bulb and then charges through the bulb in the opposite direction, or polarity. Electrons now pile up on the opposite plate of the capacitor, from when the AC waveform was positive.

This constant charge-discharge-reverse charge-discharge cycle repeats 60 times a second (for a 60Hz power source). This keeps current flowing back and forth through the light bulb, and thus it stays lit.

So a capacitor effectively allows an AC voltage to reach the light, but blocks a DC voltage after the very brief initial charging takes place. The term *coupling* describes a capacitor in a circuit that allows an AC signal to pass through it. The capacitor in the AC circuit couples the AC power source to the light bulb. In the DC circuit, the capacitor is a *blocking capacitor,* preventing DC from getting to the bulb.

Here's another principle about capacitor behavior: for a given size capacitor, as the AC frequency goes up, capacitive reactance goes down. *Capacitive reactance* is opposition to the flow of AC, and so is similar to resistance. You can think of capacitive reactance as a capacitor's resistance to the flow of an AC signal.

This means that as frequency increases, a capacitor acts like a lower value resistor to the flow of an AC signal. As the frequency goes lower and lower, the same size capacitor appears more like a high-value resistor. So a given size capacitor can look like a closed switch to high frequencies, but an open switch to low frequencies. Remember that a capacitor passes higher frequencies easier than lower frequencies.

It takes a much larger capacitor to transfer audio frequencies (20 to 20kHz) than a radio frequency, for example 54MHz. A capacitor that couples a 54MHz signal from one circuit stage to another in a VHF tuner looks like a very high-value resistor, almost $\infty\Omega$, to an audio signal.

## Capacitor Construction, Values, and Uses

As stated earlier, a capacitor is essentially two conductive metal plates separated by an insulating dielectric material. Common cylindrical paper capacitors are made of a strip of waxed paper sandwiched between two strips of thin metal foil. The three layers are then rolled into a cylinder with a wire lead connected to each foil strip sticking out an end of the roll. A paper capacitor is illustrated in Figure 10.24.

Capacitors are often identified by their dielectric material. In addition to paper, mylar, mica, oil, glass, ceramic, air, and even vacuum are used as dielectrics. These types of capacitors can be wired into a circuit in either direction; they are *nonpolarized*.

**FIGURE 10.24 Paper capacitors.**

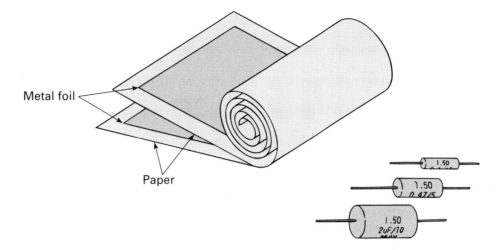

The number of electrons, or amount of charge, a capacitor can store is its *capacitance,* which is specified in *farads.* A 1-farad capacitor, however, would be huge. For all electronic work, capacitors are rated in much smaller units—the *microfarad* (MF or µF) and the *picofarad* (pF). The Greek letter µ means micro.

1 µF = 1 millionth of a farad, or 0.000,001 F ($10^{-6}$)
1 pF = 1 trillionth of a farad, or
0.000,000,000,001 F ($10^{-12}$)

Don't worry! You'll never have to calculate using these big numbers. Just remember that capacitors are rated in *microfarads* (MF or µF), and *picofarads* (pF), and that a capacitor rated in picofarads is much smaller than one rated in microfarads—one million times smaller, as far as its electrical capacitance is concerned.

Some capacitors are marked MFD: same thing as MF or µF, meaning microfarad. Some capacitors are marked MMF or MMFD, for *micro-microfarad,* which means the same as picofarad (pF).

> Microfarad = MF = MFD = µF
> Picofarad or Micro-microfarad = pF = MMF = MMFD

Capacitors rated in picofarads are found in RF and high-frequency circuits. Capacitors rated in microfarads are incorporated in low-frequency and DC circuits, like power supplies, audio amplifiers, and digital and timer circuits. From here on out, the abbreviations *cap* and *caps* are often used for capacitor and capacitors.

Capacitors also have a voltage rating, usually stated as WV for working voltage, or WVDC. Never wire a cap into a circuit with higher voltages than the cap is rated for. It's all right to replace a 0.15MF at 200WV cap with one rated 0.15µF, 400WVDC.

Following are a few typical nonpolarized capacitor values in VCR and other electronic circuitry:

4.7pF, 6.8pF, 18pF, 39pF, 110pF, 160pF, 220pF,
270pF, 390pF, 470pF
.001µF, .005µF, .01µF, .022µF, .047µF, .05µF,
.1µF, .22µF, .47µF, 1.0µF

These fixed-value, nonpolarized capacitors are used in many different types of circuits:

### As interstage signal coupling caps

A capacitor will not pass DC, but an AC signal will go through. Many times it is necessary to prevent the DC voltage at the collector output of one transistor from going to the base input of the following transistor stage. A capacitor serves the purpose; the DC is blocked, but the AC signal is coupled from one stage to the next.

### As part of a tuned circuit

When a capacitor and a coil are connected in parallel or in series, the two interact to form a *tuned circuit,* which is resonant at a particular frequency, depending on the values of the coil and capacitor. *Resonant* means that the natural frequency of the circuit is the same as an applied AC signal, and either maximum voltage or maximum current flow results.

As stated earlier, capacitive reactance goes up (looks like a higher value resistor) as frequency goes down. With a coil, just the opposite is true. Coils are described shortly, but for right now just learn that *when a cap and coil are connected together, they interact with each other.* How they interact depends on the signal frequency and how they are connected. For a particular value capacitor and a certain value coil there is one frequency or range of frequencies that produces:

- Maximum signal current flow *through* a cap and coil that are connected *in series*
- Maximum voltage drop *across* a cap and coil that are connected *in parallel*

### As part of a timing circuit

Connected with a resistor, the time it takes to charge and discharge a cap can be controlled. For a given size cap, the larger the value of resistance in series with it and the charging voltage, the longer it takes for the cap to reach the applied voltage. The resistance limits the amount of charging current, so it takes longer for the cap to become fully charged. Likewise, the larger the resistance in the cap's discharge path, the longer it takes the cap to reach zero volts. Figure 10.25 shows the charge and discharge curves of an RC combination.

### As waveshapers and filters

Essentially this is accomplished with one or more resistors and caps working together, just as previously described for a basic RC timer. An RC filter can be designed that will let some frequencies through, but block frequencies above (or below) them. A simple tone control on a table radio is an example, where a pot varies the filter frequency. As the control is turned one way, more and more of the high frequencies are sent to ground through a capacitor, rather than going on to the next amplifier stage. This reduces the loudness of the higher audio frequencies reproduced by the speaker. Figure 10.25 shows two examples of an RC combination changing the shape of a square wave. Wave shaping can provide a narrow output pulse from a broad input pulse.

- In the upper circuit, capacitor C1 charges through R1 when SPDT switch S1 is in the A position, and discharges through R1 with S1 in the B position.
- The lower circuit has a square wave input. DPDT switch S1 determines how the signal is applied to R1 and C1. In position A, the input goes through R1, then C1 to ground. In position B, the input goes through C1, then R1 to ground. Output is taken at the R1-C1 junction. In position A, R1-C1 forms an *integrator,* in position B, a *differentiator*.

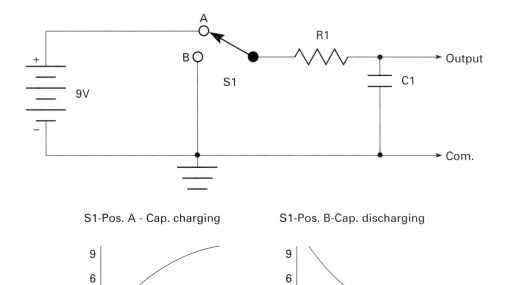

**FIGURE 10.25 Resistor-Capacitor (RC) circuits.**

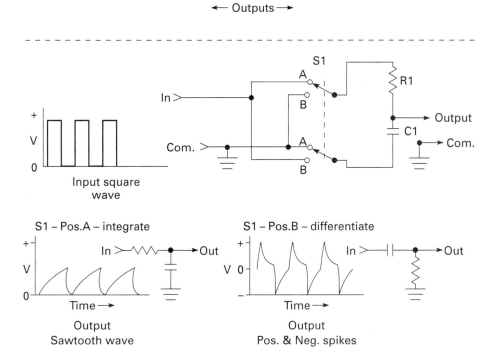

A particular type of cap, called an *electrolytic capacitor,* has a specially constructed dielectric, usually borax, aluminum, or tantalum oxide. An electrolytic cap *is polarity sensitive,* and is clearly marked with a plus ⊕ or minus ⊖ sign to indicate how it must be connected in a circuit. Connecting an electrolytic cap backward will destroy the capacitor. In some cases it will explode!

> CAUTION: Always observe correct polarity when installing an electrolytic capacitor. Also, make sure the voltage rating of the capacitor is equal to or greater than the applied voltage.

Following are circuits where you will most often find electrolytic caps:

- In a power supply (PS) to *filter* DC that has been rectified from AC

    Glance back at Figure 10.10, "Half-wave rectification." As you can see, the DC output is *not* steady; there is a series of rising and falling pulses, with gaps between them. This is termed *pulsating DC* or *raw DC*.

    Electronic equipment needs steady DC with no pulses. Enter the filter capacitor!

    In practice, a relatively large-value electrolytic capacitor is connected across the pulsating DC supply. The cap charges to the maximum, or peak, positive value of the half-wave DC pulses. Then, as the DC voltage drops back to zero, and in between pulses, the capacitor discharges, providing power to the connected circuits. The result is a steady, or *pure*, DC output voltage. You'll see power supply (PS) filter caps in Chapter 12.

- As *decoupling* caps on PS lines throughout a piece of electronic equipment

    This prevents any signal or electronic noise in one circuit from getting back into the power supply line and traveling to other circuits. For example, in a typical VCR, a common +5VDC power supply output goes to many different circuits. They might be audio, video, and digital control circuits. A decoupling cap to ground at the +5VDC input to each circuit effectively shorts to ground any AC signal riding on the DC power line, maintaining pure DC. In Figure 10.26, capacitor C3 decouples electronic noise and residual circuit signal on the common 9V power supply line.

- As *signal coupling capacitors* from one stage to the next, just as previously described for nonpolarized caps. Electrolytic caps are frequently used to couple audio stages. Coupling cap values are typically in the range from 10µF to 100µF. Figure 10.26 shows an *interstage* coupling capacitor (C2) between two transistor amplifiers.

Electrolytic caps have much larger capacitance values than nonelectrolytic caps of similar physical size. A VCR power supply might have an electrolytic cap as high as 10,000µF at 25WV as the main filter after AC is rectified to pulsating DC. There will be several electrolytics in a typical power supply,

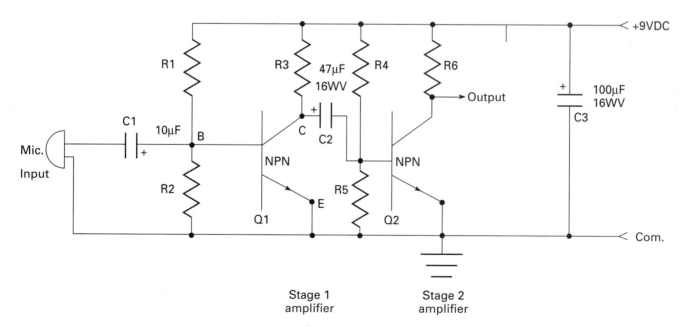

**FIGURE 10.26** Interstage coupling capacitors and decoupling cap. Capacitor C1 blocks any DC current from flowing into the microphone. C2 couples the output of transistor Q1 to the base of Q2. C3 is a power supply line decoupling capacitor. All three caps are electrolytics; correct polarity must be observed when they are connected.

with values ranging from perhaps 47µF to several hundred microfarads, and even several thousands.

Decoupling electrolytics, located at PS voltage input points on all circuit boards, and at voltage input pins ($V_{CC}$) to integrated circuit chips, typically have values between 10µF and 220µF. Often, a smaller nonelectrolytic cap, such as .01µF, is connected in parallel with an electrolytic decoupling cap. This small-value nonpolarized cap is more effective in shorting very high-frequency noise spikes to ground, while the larger value electrolytics smooth the DC and remove unwanted lower frequency noise.

The three caps on the left in Figure 10.27 illustrate two basic types of capacitors found in VCRs. Cylindrical caps with a wire coming out each end are *axial*-lead types; those with both leads coming out one end are *radial*-lead types. There is no electrical difference between axial-lead and radial-lead caps having the same capacitance and voltage ratings. Radial-lead components are designed primarily for PCB mounting. Axial lead components are also used on printed circuit boards, with their leads bent down at 90 degrees and then inserted through board holes and soldered.

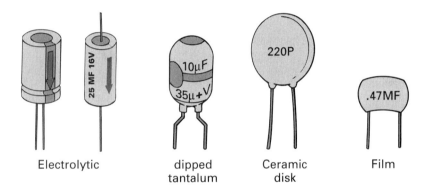

**FIGURE 10.27** Various styles and types of capacitors. The three caps on the left are polarized electrolytics; the two on the right are non-polarized, nonelectrolytics. Correct polarity must be observed when connecting an electrolytic cap into a circuit.

Many capacitors have their values printed on the cap body, and others are color-coded. See the Chapter 10 Appendix for this information plus additional illustrations of different style capacitors.

## Capacitor Failures and Checks

There are three primary ways in which a capacitor may fail. It can become:

- Open
- Shorted
- Leaky.

Failure of nonelectrolytic capacitors is fairly *in*frequent, especially with ceramic, mylar film, and polyester caps, which hold up better than paper types. Electrolytics are more prone to failure. With age, they can lose their capacitance, completely open, become excessively leaky, or short out. A high-voltage surge can also cause a capacitor to arc across the dielectric, leaving it shorted or highly leaky. The highest percentage of capacitor failures are probably electrolytics in power supplies.

You can use a multimeter to check for these conditions. For small-value nonelectrolytics, place the meter on its highest resistance scale, for example $R \times 10,000$ ohms. Watch the meter carefully as you connect the two test leads across the two capacitor leads. The meter pointer should give a slight kick upward, towards 0 ohms, and then return to ∞. This momentary meter deflection indicates the flow of charging current from the ohmmeter to the cap.

Then reverse the meter leads or flip the meter's polarity switch. The meter should kick again, this time to an even lower resistance reading, indicating discharge and then charge of the cap in the opposite direction. The meter should return to ∞ once the cap is charged in the opposite direction. Meter pointer deflection, or kick, may be very small, and might be nearly impossible to see for caps smaller than about 0.01µF.

On a digital ohmmeter, the display should show some value of resistance that climbs higher and higher until the overrange indication displays when the cap is first connected to meter leads. Reverse the meter leads, and the same thing should occur again. The reading may be very brief, depending on the particular DMM. In some cases, you may not see an indication at all, especially with low-value nonelectrolytic caps. An analog VOM is often better at displaying the brief charge and reverse charging of a cap than a DMM.

Indications vary depending upon the resistance range and individual meter. Switch meter polarity back and forth, or swap meter leads across the cap several times on the highest resistance scale, to see the effect.

The effect is more pronounced with an electrolytic; you may use a somewhat smaller resistance range, like R × 100, to speed up the charging and reverse charging cycles. With a large-value electrolytic, it is possible for the meter to read even less than zero ohms during the first part of the reverse charge, as the capacitor adds its charge voltage to the meter's ohmmeter battery. A DMM may actually indicate *minus* resistance, while an analog pointer may go to the peg at the right side of the meter momentarily.

Don't worry about reversing meter polarity across an electrolytic; the low voltage and current provided by the meter should not damage the cap, even when it is connected with reverse polarity.

If you don't get any meter deflection, even on the highest resistance range, the cap may be open. Again, the charge/reverse charge may be very difficult to see on some meters for caps smaller than about 0.01MF. If the meter reads near zero ohms in either direction, and stays at this reading, then the cap is shorted. If the meter doesn't return to infinity or overrange after the cap is connected for several seconds, but hangs at some resistance value, then the capacitor is leaky. A small amount of leakage, like one or two megohms, may be present and acceptable in an electrolytic, but be very suspicious of any leakage resistance under 100K ohms. It would be prudent to replace such a capacitor.

Over time, an electrolytic cap can decrease in value. It may pass the simple multimeter charge/discharge test, but have less than its rated capacity. This reduced capacitance is next to impossible to detect with the multimeter test. Comparing charge/discharge readings with a known good value cap of the same rating may show up reduced capacity in an old, dried out electrolytic. The meter needle won't kick as much, and charge and discharge will be noticeably faster than the old one.

Experimenting with several different values of known good non-polarized and electrolytic caps with your own meter can make you comfortable in interpreting this simple multimeter capacitor test.

As with other components, substituting a known good capacitor in the circuit is the surest way to confirm whether a particular cap is causing a problem. For nonpolarized caps, be sure the value is the same and the WV of the replacement is at least equal to the original. In many cases, the value of an electrolytic cap is not at all critical.

For example, if you determine that a 10µF, 16V coupling cap is open, replacing this with a 15µF unit at 20V should work just fine. Values are even less critical in a power supply. Replacing a 470µF, 50WV unit with a 750µF or even a 1,000µF cap at 50WV should cause no problem. Opt for a replacement with a higher value rather than one with a lower value. Never reduce the value of an electrolytic more than 10 percent; circuit problems such as hum or noise may show up.

For electrolytics, the operating voltage should not exceed by about 50 percent the working voltage of the original cap, or the circuit voltage. An electrolytic with a much higher voltage rating may, over time, decrease its amount of capacitance when operated on a considerably lower voltage. Technically, the capacitor cannot properly *form* itself at the lower voltage, and loses capacity.

For example, one could replace a 470µF cap rated at 16V with one rated at 35V, but a replacement with a 150V rating may not provide adequate capacitance down the road.

As with testing other components, be sure there is no power in the circuit being tested. To eliminate back circuits, one end of a cap must be isolated to get a true indication. Of course, if a cap is dead shorted, you can pick this off while it is in circuit.

## Capacitor Safety

A capacitor can store a charge for some time after equipment is powered Off. High-voltage electrolytic caps, and even large-value, low-voltage electrolytics, can pose a safety hazard. Usually these caps are in power supplies, and almost always have a resistor in parallel with the cap(s), called a *bleeder resistor,* to discharge the cap after power is switched Off.

If a bleeder resistor is open, or there is none, then the cap can retain a voltage charge after the unit is unplugged. A large 75V capacitor in a high power audio amplifier can easily melt the tip of a screwdriver, accompanied by flying metal sparks, if the tool shorts across the capacitor terminals. This could cause eye injury, burns, or secondary injury. The cathode ray tube itself in a computer monitor or TV set acts like a capacitor and can store thousands of volts! Although you won't find a CRT or caps storing anywhere near 1,000 volts and above in a VCR, it is good to know about the potential dangers.

Many VCRs have a switching power supply that rectifies the 120VAC line voltage prior to a step-down transformer. In these VCRs, an electrolytic filter capacitor could store as much as 300 volts. This can give you a nasty shock! Secondary involuntary muscle contraction can cause injury, such as cutting your arm on a corner of a VCR chassis.

When in doubt, take time to discharge power supply capacitors before working on a piece of equipment. Check across the cap with a DC voltmeter. The voltmeter will indicate whether the cap holds a charge. Leave the meter connected, and the voltmeter itself will discharge the cap in a few seconds to several minutes, depending on the meter's ohms-per-volt sensitivity and capacitor size. This is the safest way to discharge a cap.

Directly shorting a cap with a tool or piece of wire can be dangerous, and can also cause damage to the printed circuit board. Connecting a wire with a one-watt, one megohm resistor (brown-black-green-silver/gold) across a power supply filter cap is a much safer way to discharge a typical PS cap. Remember, keep one hand behind your back when working with voltages higher than about 42VDC. Check with your voltmeter that the cap is discharged before working on the equipment.

## COILS AND TRANSFORMERS

In general, a *coil* consists of many turns of wire wrapped around a common core. The core could be some ferrous material like iron or even air. When an electric current passes through the coil, a magnetic field is produced.

(Actually, a single loop of wire forms a coil, but most coils have anywhere from perhaps a half-dozen to several hundred turns.)

You've already learned about two coils in this book that produce magnetic fields: a solenoid coil pulls an iron plunger into a hollow core, and a relay coil attracts the armature to the coil, which transfers electrical contacts.

There are other uses for a coil, in addition to its ability to create a magnetic field. In an electronic circuit, a coil has certain electrical properties.

A coil in some respects acts just opposite a capacitor. A capacitor opposes a change of voltage *across* the cap; a coil opposes a change in current flow *through* the coil. A capacitor blocks DC while allowing AC to flow through it; a coil allows DC to flow through it while restricting AC current flow.

Another name for a coil is an *inductor*. The amount of opposition to AC current flow through an inductor is called *inductive reactance*. Recall that *capacitive reactance* may be thought of as some value of resistance in an AC circuit; you may also think of *inductive reactance* as some value of resistance to the flow of AC.

You just learned that with a capacitor, as the frequency goes up, capacitive reactance goes down; the higher the frequency, the more the cap acts like a straight wire.

With a coil, as the frequency goes higher, inductive reactance also goes up; the coil acts more like an open circuit. For high frequencies, the coil acts like an open switch, while allowing lower frequencies and DC to pass right on through.

Inductive reactance is measured in henries. For electronic circuits, a henry is too large an inductance to be of any use, so inductors are rated in smaller units: *microhenries* (μH) and *millihenries* (mH).

$1 \mu H = 1$ millionth of a henry, or $0.000,001H$ ($10^{-6}$)
$1 mH = 1$ thousandth of a henry, or $0.001H$ ($10^{-3}$)

Thus, a 330.0μH coil is the same as a 0.33mH coil (move decimal point three places to the left to convert μH to mH).

A coil's inductance is determined by its physical construction, including number of turns, spacing between turns, and the material around which the coil is wound, called the *core*. When a coil is wound around material, such as plastic, wood, or even air, that has no magnetic properties, it is called an *air core inductor*. A coil consisting of turns of wire wrapped around iron or steel is an *iron core inductor*. Two coils are illustrated in Figure 10.28.

Another popular core material, especially for high-frequency applications, is a hard, brittle material made of ferric oxide and other compounds, called *ferrite*. Rotating heads in a VCR consist of very fine wire wound around a ferrite core. Radio-frequency transformers, antenna coils (in AM receivers), and various types of noise filter coils have ferrite cores.

It is often desirable to prevent RF energy from going from one circuit to another in electronic equipment, while allowing DC to pass uninhibited. A coil

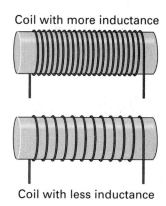

**FIGURE 10.28 Inductance is determined by the physical construction of a coil.**

**FIGURE 10.29 Types of coils and inductors. Some coils look similar to resistors or capacitors. Consult the schematic diagram to be sure. Most coils and inductors have an L designation on the schematic, and adjacent to the component on the circuit board (for example, L203).**

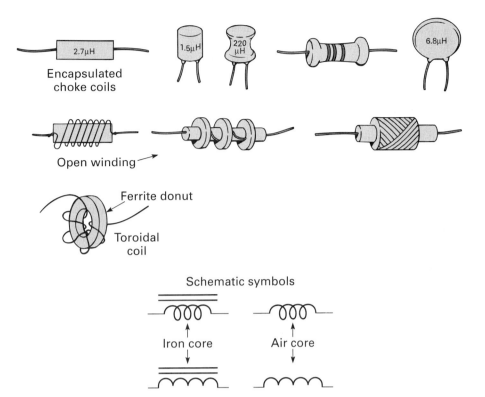

that performs this job is called a *choke*. Chokes are made with ferrite, iron, and air cores. Figure 10.29 illustrates different types of coils and chokes, along with their schematic symbols. Note the difference in the symbols for iron core and air core inductors.

Frequently, coils and transformer windings are wound around a doughnut-shaped core, called a *toroid*, giving them the names *toroidal coils* and *toroidal transformers*.

Recall how an electrolytic capacitor connected between a power supply line and ground decouples high-frequency signal components. The cap *in parallel* with the power supply voltage acts like a very low resistance, even a short, to ground for unwanted AC signals and noise on the voltage line. For a similar reason, a choke is sometimes connected *in series* between a power supply line and the circuit being powered. The choke lets DC voltage through to the circuit, but prevents signal components or noise generated in the circuit from getting back out on the power supply line, where it could interfere with other circuits connected to the supply.

To DC, a choke acts like a resistor of just a fraction to a few ohms. But for high frequencies, the choke appears like a very high value resistance ($\infty\Omega$).

A coil and capacitor together can resonate at one specific frequency, or over a range of frequencies. This coil-capacitor (LC) characteristic is used to select or tune to desired frequency signals and to reject or filter out unwanted frequencies. A coil and cap connected in series has a very low combined reactance at the resonant frequency, and therefore together act like a low-value resistor, allowing the selected signal to pass through. Wired in parallel, the same coil and cap together has a very high reactance at resonance, and a maximum signal voltage is developed across the LC combination.

## Transformers

When two coils are in close proximity to each other, a changing current in one coil produces a fluctuating magnetic field around the coil which produces an *induced* voltage in the other coil. A *transformer* consists of two or more coils wound on the same core. For a power transformer, the core material is usually iron. For a radio-frequency transformer, the core material is usually ferrite or air.

The basic property of a transformer is to change AC voltage. A transformer cannot change direct current voltage. A *step-down transformer* has a lower AC output voltage at its secondary winding than the AC input voltage to its primary winding. Conversely, a *step-up transformer* has a higher secondary than primary voltage. Figure 10.30 shows the basic construction of a power transformer.

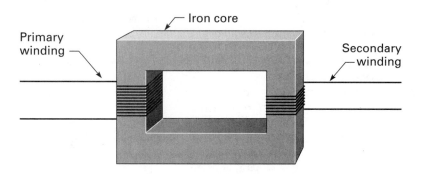

**FIGURE 10.30** Basic transformer construction.

The relationship between a transformer's primary and secondary voltage is determined by the ratio of the number of turns in the primary and secondary windings.

For example, if a transformer has 400 turns in its primary and 25 turns in its secondary winding, then the primary-to-secondary turns ratio is 400:25 or 16 to 1. The primary and secondary windings will then have the same voltage ratio. This means the voltage across the secondary will be one-sixteenth the voltage across the primary. If the primary is connected to a 120VAC outlet, the secondary voltage would be 7.5VAC (120 ÷ 16 = 7.5).

The same transformer can be used the other way around. If 7.5VAC is connected across the 25-turn winding, now acting as the primary, the 400-turn winding will produce 120 volts. As you can see, the terms *step-down* and *step-up,* and *primary* and *secondary,* really describe how a transformer is connected, not how it is constructed.

In many electronic devices, a power supply transformer has one 120VAC primary winding and several secondary windings, or a secondary winding with multiple taps. Figure 10.31 shows a transformer with multiple secondary winding taps.

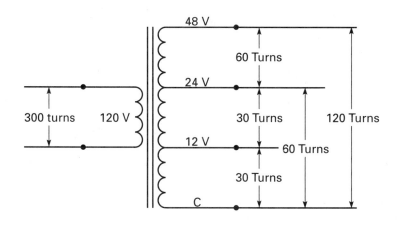

**FIGURE 10.31** Transformer secondary with multiple taps. Note that C is the common or ground connection.

Many VCR power supply transformers have several separate secondary windings. Figure 10.32 illustrates. Each secondary typically has a different AC voltage, such as 5V, 10.5V, and 32V. The output of each secondary winding goes to diode rectifiers to produce the several DC voltages needed to power the various VCR circuits and motors.

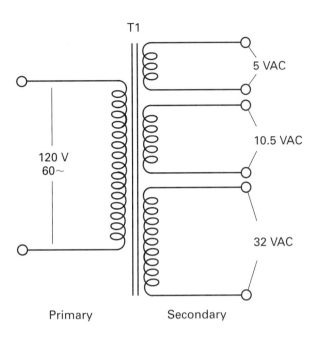

**FIGURE 10.32** Transformer with multiple secondary windings.

Transformers are usually designated with a T on schematics and circuit boards. For example, T304 could be an RF transformer in the tuner section and T101 might be the power transformer. These are the main types of transformers in a VCR, in addition to the important rotary transformer in the video drum.

The function of the rotary transformer is not necessarily to change voltages, but to couple the signal to and from the rapidly spinning heads and the VCR's record and playback circuits. A solid wire connection could not be used in the drum, because the heads are turning.

Here is one example where primary and secondary become relative terms. During record, the stationary portion of the transformer is the primary, coupling signals from the record circuits to the spinning secondary windings and then on to the heads. During playback, the spinning portion of the transformer becomes the primary, with the stationary portion the secondary.

Another transformer you may come across in electronic service work is an *isolation transformer*. An isolation transformer has a 1:1 turns ratio, and so the output voltage is the same as the input. So why do we need a transformer at all if the voltage out is the same as the voltage in? It is often desirable to isolate a piece of electronic equipment from the normal 120VAC power circuit, so that neither side of the powered equipment has ground as a reference. A 120V isolation transformer provides the needed isolation.

Some electronic equipment has what is called a *hot chassis,* where one side of the AC line connects directly to chassis ground, rather than being isolated with a power supply transformer. This can pose a shock and short-circuit hazard when connecting service bench test equipment to the unit under test. Powering the unit under test from an isolation transformer removes this danger. VCRs having a switching mode power supply (SMPS) should be powered by a 120V isolation transformer when on the test bench, for safety reasons. You will read more about this in Chapter 12.

## Coil and Transformer Failures and Checks

Coil and transformer failures are relatively rare in most electronic equipment, compared with other components. If too much current goes through a coil, the winding heats up and can either open up completely, or the insulation between turns of wire can break down, causing the coil to be partly or completely shorted. Many coils have just a thin coating of shellac or enamel over solid wire to insulate adjacent turns.

With small RF chokes and coils, you can perform a basic continuity check to see whether the coil is open or not. If too much current has gone through a coil (perhaps because another component failed, like a diode or transistor shorting out), you may see evidence of charring or cracking, or notice a burnt smell from the choke. There is no way to check for a partially shorted coil with just a multimeter, but these types of failures are exceedingly rare in VCR circuits. If a coil looks like it has gotten hot, replace it after determining and correcting the cause.

Just like a coil, you can check continuity of any primary or secondary transformer winding. The actual value of DC resistance you read is rather meaningless, but at least you will know whether a winding is open or completely shorted. For a power transformer, look for any evidence of overheating, such as darkened or blackened areas or a burnt smell. VCR power transformer failures are exceedingly rare.

With a power transformer, you can also measure the AC voltage at each secondary winding and compare this with values printed on the schematic diagram. Of course, a defective power supply component *after* the secondary, such as a shorted rectifier diode or filter capacitor, can severely load down the secondary, giving a much lower voltage reading than it would ordinarily have. In most cases, the mainline fuse in the primary winding circuit of the transformer will open under this overcurrent condition.

If the fuse has been replaced with one having a much higher current rating, the fuse may not blow, but the transformer could overheat and be damaged, usually by a winding opening up or partially shorting. If you suspect that something like this may have damaged a power supply transformer, disconnect one wire of each secondary winding from the power supply circuit and measure the AC output voltages at the transformer secondaries. Check the schematic; if a winding is tapped, you will need to disconnect all but one of the connections between the power supply board and the secondary to completely isolate the winding. Then you can measure the secondary output voltage. Schematic diagrams indicate what the correct AC voltages for secondary windings should be. A voltage within 10 percent of this value is probably all right.

A voltage much higher than expected probably indicates shorted turns in the primary, reducing the turns ratio and therefore raising all secondary voltages. The transformer will most likely run hot.

You may wonder how shorted turns in the primary can *increase* a secondary voltage. Here's how.

Suppose a good transformer has 360 primary turns and 24 secondary turns; the turns ratio is 360:24 or 15 to 1. If 120 volts is applied to the primary, the no-load secondary voltage will be $120 \div 15 = $ **8** volts. Now suppose wire insulation breaks down, shorting out 60 primary turns. This leaves 300 active turns in the primary. The pri-to-sec turns ratio is now 300:24 or 12.5 to 1. Now with 120 volts applied to the primary, the no-load secondary voltage would be $120 \div 12.5 = $ **9.6** volts.

A low secondary voltage probably indicates shorted turns in the secondary winding. Again, in most cases the transformer is the last item to suspect as the cause of a power supply problem.

## SUMMARY

This chapter described fundamental characteristics and uses of several electronic components, along with how to perform basic tests on them with a multimeter. Wires and individual conductors in a cable can be checked end-to-end for continuity. A poor crimp connection at a connector or a broken wire can be located in this way.

Short jumper wires on a printed circuit board are designated by J, such as J-21. These and test points, designated TP, are handy locations to attach test equipment probes.

Fuses and switches can also be tested with an ohmmeter. A good fuse has continuity, or zero ohms. Cylindrical glass fuses are the most common types in VCRs and most other electronic equipment. Never replace a fuse with one having a higher current rating. Fuses are designated with an F on schematic diagrams and circuit boards, such as F101.

Common switch types are single pole, single throw (SPST); SPDT; DPST; and DPDT. A DPDT switch is often used to change the polarity of a signal or voltage. On a schematic diagram, two or more operating levers of a multi-pole switch are connected by a dashed line, which means that they move or transfer together. Closed switch contacts should have zero resistance. Poor contacts show up as a higher value resistance. Clean switch contacts or replace the switch if you measure contact resistance. Switches are designated as S or SW on circuit boards and schematics: S207 or SW207.

A relay is basically an electromagnetically operated switch. Current flowing through a relay coil attracts an iron armature, which transfers normally open and normally closed relay contacts. Some relays have separate pick and hold coils. A kickback diode is often placed across the coil of a DC relay to swamp out, or short out, the high reverse polarity voltage that is developed when the coil is de-energized and its magnetic field collapses. Relays are usually designated by K, RY, or RLY on diagrams and circuits, like K7, RY3, or RLY801.

Resistors limit the flow of AC and DC current equally. Most resistors in VCRs are carbon or metal film types, 1/4 watt or less. The values of many axial lead resistors is determined by color-coded bands around the resistor body. First and second bands are significant digits, with band 3 being a multiplier. Resistors with a 10 percent tolerance have a silver fourth band, gold indicates 5 percent tolerance. A potentiometer, or pot, is a variable resistor having three terminals. A rotary tracking control is one place where a pot is found in many VCRs. Fixed resistors are designated by an R, such as R213 or R4. A pot may be designated with an R or VR, for variable resistor: R328, VR603.

A diode allows current to flow in one direction but not the other. A diode rectifies AC to produce pulsating DC. Electron flow through a diode is from cathode to anode, *against* the schematic symbol arrow. A diode is *forward biased* when the cathode is more negative than the anode; current flows through a forward biased diode. No current flows when a diode is reverse biased, that is, when the cathode is positive with respect to the anode. The cathode end of many diodes is marked with a single band or dot.

A Zener diode starts to conduct in the reverse biased direction at a specific voltage, for example 5V, 9V, or 12V. Once a Zener conducts, the voltage drop across the diode remains constant, even though the applied voltage increases. For this reason, Zeners are used as voltage regulators and voltage reference sources. The cathode part of a Zener diode symbol looks like the letter Z. When a Zener conducts in the reverse biased direction, current flow is *in the same direction* as the schematic symbol arrow.

Many diodes can be tested with a multimeter. With the negative ohmmeter lead connected to the cathode and positive meter lead connected to the anode, the diode is forward biased, and current flows through the meter. An analog meter in R × 1 or R × 10 will typically read between 10 and 500 ohms for a good diode. Digital multimeters usually have a separate diode check position. With a good diode, a DMM displays the diode's forward voltage drop, around 0.4V for a germanium and about 0.6V for a silicon diode. Either indication is a Low reading.

When meter leads are swapped so the negative lead is on the anode and positive on the cathode, a good diode will look like an open circuit, ∞Ω on an analog and "overrange" on a digital meter. Either indication is a High reading.

### Multimeter Diode Check

| VOM or DMM connection | Good Reading |
|---|---|
| Forward bias: cathode negative, anode positive | Low |
| Reverse bias: cathode positive, anode negative | High |

Diodes are designated as D on diagrams and PCBs, like D104. Zener diodes are usually ZD: ZD405. An LED may have a D or LED designation: D801, LED801. Often the diode symbol is printed on a circuit board; sometimes K is printed for the cathode end.

There are two basic types of transistors, NPN and PNP. Both have three leads: an emitter, a collector, and a base. A small current flow between base and emitter controls a much larger flow of current between emitter and collector. Thus, a transistor is a current amplifier.

An NPN transistor has P-type silicon for its thin base layer and N-type silicon for emitter and collector layers. An NPN transistor is forward biased with a positive voltage on the base with respect to the emitter. Electrons then flow from the negative emitter, *against* the emitter arrow of the NPN schematic symbol, to the positive collector.

A PNP transistor is just the opposite, having an N-type silicon base layer and P-type emitter and collector layers. A PNP transistor is forward biased when the base is negative with respect to the emitter. Electrons then flow from the negative collector to the positive emitter, *against* the emitter arrow of the PNP schematic symbol.

Transistors can be tested with a multimeter as if they were two separate diodes. The base-emitter junction forms one diode, the base-collector junction forms the other. For an NPN transistor, consider the base the anode of the two diodes, with emitter and collector acting like cathodes. For a PNP transistor, consider the base the cathode, with emitter and collector acting as anodes. A good transistor will test as two good diodes.

A transistor, sometimes abbreviated *Xistor*, is normally designated with Q or TR on schematics and boards, such as Q607 or TR607. Emitter, base, and collector leads are usually identified on a PCB with the letters E, B, and C.

Capacitors block the flow of DC current while allowing AC signals to pass through. As the frequency of an AC signal increases, a given size capacitor, or cap, acts like a smaller value series resistor, passing more of the signal. Another way of stating this is that the opposition to a *change* in voltage across a capacitor is *capacitive reactance*. Increase the frequency and capacitive reactance goes down. Caps are frequently used to couple AC signals from one stage to another, while blocking DC from flowing between the two stages.

### Multimeter Transistor Test

| Meter Connections | Good Reading |
|---|---|
| **NPN Xistor.** | |
| • B neg., E pos. | High |
| • B pos., E neg. | Low |
| • B neg., C pos. | High |
| • B pos., C neg. | Low |
| **PNP Xistor.** | |
| • B neg., E pos. | Low |
| • B pos., E neg. | High |
| • B neg., C pos. | Low |
| • B pos., C neg. | High |

Where: B = Base, E = Emitter, C = Collector
neg. = negative, pos. = positive
High = $\infty\Omega$ or "Overrange," Low = 10 to 500$\Omega$ or $V_F$

An electrolytic capacitor is polarity sensitive, and must be connected correctly, with its negative terminal wired to the more negative point in a circuit. These type caps are used as filters in power supplies, decoupling caps on voltage distribution busses, and as coupling caps for audio circuits.

A multimeter can perform a rough check on capacitors, usually 0.01µF and larger. With the meter set to its highest resistance range, the meter pointer kicks slightly as a cap charges from the meter's battery. Reversing the meter leads produces a somewhat larger meter movement, as the cap discharges and then charges in the opposite direction. An open capacitor results in no meter deflection. A shorted cap will read somewhere near zero ohms or a low resistance value. A leaky cap shows some value of resistance instead of infinity.

Capacitors are designated with a C number on schematics and wiring boards, such as C207.

A coil behaves just about the opposite from a capacitor. It allows the flow of DC current, but inhibits the passage of an AC signal. *Inductive reactance* goes up as frequency increases. That is, an inductor acts more and more like a high-value resistor as the frequency goes up.

When a coil and capacitor are connected together, they become resonant at one frequency, or over a range of frequencies. If coil and cap are *in series,* the combined reactance is *minimum* at resonance, and maximum current can flow through the components. When coil and cap are *in parallel,* they have *maximum reactance* at the resonant frequency, and so the voltage developed across them is at maximum.

A coil, also called a choke, is designated by L, such as L307, on a wiring diagram.

A transformer consists of two or more coils wound upon a common core. The main function of a transformer is to change AC voltage. The primary-to-secondary turns ratio of a transformer determines the secondary output voltage with a given primary input voltage. A transformer with a 4:1 turns ratio would produce 30 volts at its secondary with 120 volts across the primary winding (120 ÷ 4 = 30). Transformers are designated with a T on wiring diagrams and PCBs, such as T101. Transformer is sometimes abbreviated as Xfrmr.

Coils and transformers can be checked for an open winding with an ohmmeter, but unless the winding is completely shorted (0$\Omega$), a multimeter cannot detect a partially shorted winding. A power transformer can be tested by measuring the secondary voltages and comparing them with specifications. A difference of more than 10 percent normally indicates a partially shorted winding.

Whenever basic ohmmeter tests are performed on components when they are still wired into a circuit, the possible presence of one or more back circuits through other components must be considered. It may be necessary to lift, or isolate, one or more component leads from the circuit to more properly check the part.

Note that there is a Chapter 10 Appendix with additional information about some subjects covered in this chapter.

## SELF-CHECK QUESTIONS

1. Explain the relationship between the American Wire Gauge number and the amount of electrical current a wire safely can carry.
2. What is the most frequently encountered problem with wires and cables?
3. What is the main operational difference between a fast-acting and a slow-blow fuse? Give one example of where each type typically is used. How can you tell a glass slow-blow fuse from a fast-acting fuse?
4. Which type of switch allows a single electrical load to be turned On and Off independently from two locations? Draw a schematic diagram showing how this is done.
5. Describe a basic electrical relay. Include the names for the three basic relay parts.
6. What is the purpose of a diode connected in parallel with a DC relay coil?
7. Describe how to check a relay coil that has a diode soldered across it.
8. What are the three principal characteristics or specifications for describing a fixed resistor?
9. What specifically can you determine about a resistor that has the following color bands? Brown-Gray-Black-Gold
10. What should you do before adjusting a trim pot in a VCR so you can return it to its original resistance setting if you need to?
11. What is the fundamental characteristic of a diode?
12. Describe current flow through a forward-biased diode, and then through a reverse-biased diode. Correctly use the terms anode and cathode in your explanations.
13. Describe the electrical characteristics of a zener diode.
14. Describe how you would check a common silicon or germanium diode with an analog multimeter.
15. Describe in one or two sentences the basic operation of a transistor.
16. Describe how to perform a basic transistor test with a multimeter.
17. What two precautions must you take when replacing a defective electrolytic filter capacitor in a power supply?
18. Describe how you would check a 10µF coupling capacitor with an analog multimeter.
19. What is the purpose of a choke in series between a +5V DC power supply line and a printed circuit board containing the timer/operations microcomputer in a VCR?
20. What is the function of the power supply transformer in a VCR?

# CHAPTER 11

# HOW TO SOLDER

Almost every piece of electronic equipment relies on solder connections to combine components into a variety of circuits. Today the vast majority of electronic parts make electrical connection by being soldered to one or more printed circuit boards made of fiberglass and epoxy with copper wiring traces or runs. Not nearly as common, but also found in some equipment, is point-to-point wiring, where individual insulated copper wires and component leads are soldered together at terminal strips or tie points.

Frequently, electronic repair requires removing a faulty component and replacing it with a new one. In most cases this requires unsoldering the original and then soldering the replacement into the circuit. Similarly, wires and cables that are accidentally cut or separated from a connector can be repaired by making a soldered splice in the wire or by resoldering a wire end to a connector.

Therefore, it is important for an electronics technician to develop good skills in both unsoldering and soldering electronic parts, wires, and connectors. These skills will be introduced to you in this chapter. Correct procedures for unsoldering and soldering to make quality repairs are described. You will also learn how to determine whether a soldered connection is acceptable or not.

## CHAPTER OBJECTIVES

**Upon completing this chapter, you should be able to:**
1. Describe the steps to make a quality soldered connection between electronic components, including those mounted on printed circuit boards.
2. Describe two methods for removing solder from a printed circuit board when replacing a faulty component.

In addition to explaining correct procedures for electronics soldering and unsoldering—and describing the right tools to use for these tasks—this chapter also suggests practice exercises you should consider performing prior to working on a VCR or some other expensive piece of electronic equipment.

Also included in this chapter are plans for building a simple two-way tester that you can use for VCR troubleshooting. With this device, you'll be able to check out near-infrared LEDs and power small two-terminal DC motors, like those in many VCR front loaders and tape load assemblies. Constructing the tester lets you practice soldering skills as you build a useful piece of VCR test equipment.

Soldering is somewhat comparable to many other skills, like flipping pizza dough in the air: it looks simple until you actually try it yourself. Soldering is *not* difficult, but following a few precautions, practices, and techniques is essential to success.

## SOLDERING TOOLS AND SUPPLIES

Basic tools that you need for routine soldering and unsoldering were covered in Chapter 6, but here is a recap:

- Soldering iron or pencil: 20 to 30 watt. (Avoid soldering guns. They are usually too cumbersome for delicate electronic work.)
- Soldering tool holder or stand. (It is much better to get a cage-style stand than just a simple soldering iron rest, which leaves the hot tip exposed on the workbench.)
- Small gauge 60/40 *rosin core* solder
  — 60/40 means 60% tin, 40% lead
  — 63/37 rosin core solder is excellent, but may cost more than 60/40
  — Rosin helps prevent oxidation and cleans connection during soldering
  — *NEVER* use acid core solder or solder without a rosin core, such as used for sheetmetal work and plumbing
  — 0.040"- to 0.062"-diameter solder is good for general-purpose electronics work
- Desoldering braid: (often called *solder wick,* this is flat, braided copper wire on a small spool: sucks up old or excess solder)
- Optional tools and supplies
  — Solder helper/soldering aid tool: hand tool with small wire brush at one end and metal scraper tip at other end; assists in soldering and unsoldering
  — Clip-on heatsink: draws heat away from component body when soldering leads on transistors, diodes, and the like
  — Desoldering pump or desoldering bulb: creates vacuum to draw up unwanted molten solder through hollow Teflon tip
  — Rosin flux remover: often in spray can; cleans excess rosin from connection after soldering
  — Tip cleaner sponge: cleans soldering tip of oxidation, burned rosin.

Take a look through the tool section of electronics parts suppliers' catalogs and you will see many different types of soldering pencils, solders, tools, and ancillary items. You needn't spend a lot of money. These suppliers usually offer several low-cost soldering pencils and accessories that are adequate. Figure 11.1 illustrates a few soldering tools and related items.

Once you purchase a soldering iron or pencil, read any instruction booklet that comes with it. This can give you suggestions on caring for and using the tool.

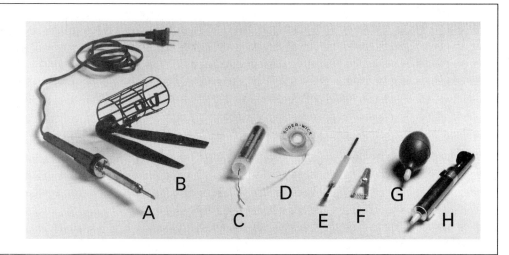

**FIGURE 11.1** Soldering tools and accessories. (A) 30-watt pencil; (B) soldering iron stand; (C) rosin core radio solder; (D) solder wick or desoldering braid; (E) soldering helper; (F) clip-on heatsink; (G) desoldering bulb; (H) desoldering vacuum pump.

## SUCCESSFUL SOLDERING TECHNIQUES

As stated earlier, soldering looks easy when you see someone else doing it, but there is a certain amount of skill involved. Like most skills, you gain proficiency through practice and more practice. Later in this chapter are some suggestions and a simple project that will let you try your hand at soldering before working on a "real" piece of equipment.

You can purchase a soldering iron or pencil and a roll of solder at an electronics supply house or through several electronic parts mail order companies. Radio Shack is probably the most convenient source for most people. You'll do well with a pencil-type soldering iron with a wattage rating somewhere around 25–35 watts. The exact wattage is not too critical. You *don't* want a 50–125 watt iron; it puts out too much heat. A 15–20 watt unit is a little on the light side for all but very fine work, like soldering an integrated circuit with small, closely spaced component leads to a printed circuit board. A pencil or small chisel tip will do fine for most applications. Solder guns are awkward to use on delicate electronic equipment; stay away from these.

A holder is an important accessory for your soldering iron. Place the iron in a holder when it's not actually in your hands. This prevents many problems that can occur with a hot iron just lying about the work surface. Soldering iron holders, or stands, are inexpensive and well worth their cost. They usually employ some sort of cage, which often looks like a coiled spring, mounted on a small platform. The cage protects the hot tip of the iron from touching things it shouldn't. Simple soldering iron rests are better than nothing, but a hot iron can easily roll out of many types of rests, and you can easily touch the hot iron with your arm if you're not careful. The cage-style soldering stand is *the* way to go.

For solder, you'll want *rosin core* "radio" or electronics solder with 60% tin and 40% lead. This is marked on the roll of solder as "60/40." You definitely do *not* want solder for plumbing work, or acid core solder. That solder doesn't have a rosin core and it is usually too large in diameter for convenient use on electronics. *Never use acid core solder,* which is corrosive to soldered electrical connections and components. Solder between .050" and .062" in diameter is just fine for most electronics work.

The rosin, called *flux,* in the center core of radio solder is a paste that helps clean any thin layer of oxidation from component leads, wires, and metal being soldered together. It also aids in preventing oxidation when parts are heated for soldering. Once again, *be sure you get rosin core solder!*

Another very useful soldering accessory is a small sponge. Keep it damp and frequently clean the tip of your iron as you work. Burned rosin flux and oxidation on the tip of a soldering iron interfere with a nice, clean solder connection.

When you purchase a new soldering pencil, the first thing you should do, after reading any booklet that comes with it, is to *tin* the tip of the iron. Tinning prepares the iron by filling open pores in the tip and forming a thin coat or layer of solder over its working surface. This is necessary for solder to adhere to the tip. If the tip isn't tinned, hot solder will just drop off onto the work when you apply solder to make a connection. The thin coat of solder also prevents the tip from oxidizing.

Your soldering iron may come with a pre-tinned tip. If it looks silver, it probably is. Most soldering iron tips are made of copper. If the surface of the tip is copper-colored, you definitely need to tin the tip. Follow any instructions accompanying the soldering tool for preparing the tip. It won't hurt at all to tin a pre-tinned tip. Before using your soldering iron, read over the following section.

## Soldering Safety

Molten solder and a working soldering iron are hot stuff! They can cause nasty burns. Be especially careful not to accidentally brush away molten solder so that it comes anywhere near your eyes or gets on your face. Wearing glasses while soldering is strongly recommended.

Work on a surface that won't be harmed by molten solder, or at least a surface that you won't be concerned about if it does get burned. A smooth, hardwood workbench surface is ideal; it is a good electrical insulator and won't easily burn. Molten solder dropped on such a surface can be easily pried off after it has hardened. Simply use a fingernail or an old credit card.

Use a soldering tool holder or stand, or at least something like an old ceramic dish on which to rest the hot iron. There is really not much equivalent to a cage-type holder, which shields the hot iron when it's not in use.

Provide adequate ventilation. Rosin flux burns. Although it won't produce an open flame, it will produce some smoke. Avoid breathing this. A small fan at one end of the workbench is an asset.

## Tinning the Tip

Following are general instructions on how to tin a soldering iron tip:

1. Plug in the soldering iron.
2. As soon as the tool comes up to temperature, generously apply solder to the entire working surface of the tip.

    It is important that you not wait! Have the solder ready to apply as soon as the tip gets hot enough to melt it. For an untinned copper tip, you'll notice the tip start to darken, with streaks of dark bluish-purple appearing on the surface of the copper as it reaches temperature.

    With solder on the tip, take a damp sponge or cotton cloth and quickly wipe the solder around the entire surface of the tip, working the solder into all areas.

3. Generously reapply solder to the tip and allow it to soak in or cook for a few minutes.
4. Again wipe the tip with a damp sponge or cloth.

    Examine the tip. You should see a thin layer of silver (the solder) over the entire working surface of the tip.

    If there is a void somewhere, reapply solder and attempt to work it into the void with a damp sponge or cloth. If solder refuses to adhere, take a small wire brush or metal file and wipe over that area of the tip while the iron is still hot. This removes any oxidation on the tip that may be preventing the solder from sticking.

> CAUTION: When you use a wire brush, like that on one end of a solder aid tool, be certain to brush *away* from yourself so hot solder doesn't fly into your face and clothing.

5. Reapply solder to the entire working surface of the tip and allow to cook for a few minutes.
6. Wipe the tip again with a damp sponge or cloth.

    At this time there should be a shiny, silver-colored coating over the entire working area of the tip, with no voids. If this is not so, repeat steps 4 through 6.

**Never leave the soldering iron plugged in when you are not actually using it!** Besides being a safety hazard, this increases the rate of oxidation of the tip. Get in the habit of unplugging your iron as soon as you are finished using it.

Service shops that must keep a soldering iron ready to go at all times often use a controlled-temperature soldering station, which automatically backs off the heat of the iron when it is placed in its holder. This prevents premature tip oxidation. These handy devices can cost $100 or more.

In time, the tips of all soldering irons will wear, mainly due to oxidation. Keeping the tip well tinned and clean while you are using the soldering tool, and unplugging the iron at all other times, will give you maximum tip life.

When the tip *does* get a bit ragged, you can "dress" it with a metal file. Use the file to restore the tip to its original shape and retin it. After a while, the tip will get worn down so much that you will have to replace it. All good-quality soldering irons have replaceable tips. Now that you know how to tin the tip of a soldering iron, and how to keep it in good condition, here are some simple procedures that usually work well when soldering:

> TIP: If your particular soldering iron has screw-on tips, consider buying a small tube of anti-seize lubricant. Apply this to the threads when installing a new tip. The lubricant helps prevent metal operating at high temperatures from seizing, making it easier to unscrew a tip for exchange or replacement.

- Start with a hot iron
  — Test the iron before using it the first time after plugging it in. It normally takes three to five minutes for the tip to reach full operating temperature. Touch the end of the roll of solder to the tip; solder should melt immediately.
- Keep the tip of the iron clean at all times
  — Get in the habit of wiping the tip with a damp sponge or cloth every time you pick it up and before you set it back down. This helps ensure good, clean solder connections.
- Start with a clean connection
  — Components to be soldered *must be clean metal* to get a really good solder connection. Grease, oil, paint, dirt, melted insulation, corrosion, or oxidation on metal parts prevent solder from adhering properly.
  — Older wires and components can develop a thin layer of oxidation on their surfaces. Even individual strands of an insulated stranded wire can have a film on them after several years. You'll have to remove this before soldering. Apply paste flux, available in small tins, directly to the wire or connection surface to help clean off oxidation prior to soldering. If the metal has a dull gray appearance, rather than the sheen of new metal, you'll need to clean the surfaces before soldering.
  — Fine sandpaper or an emery stick is useful in cleaning connection surfaces of heavy oxidation or other contaminants before soldering. Use the small wire brush end of a solder helper to remove oxidation on printed circuit board traces or other components. Do not use steel wool! In fact, keep steel wool far away from the electronics service bench. It is too easy for a fine steel fiber to make its way to someplace it shouldn't be and short something out.
  — The rosin flux in the solder helps clean a connection and prevent oxidation while you are soldering, but you have to start clean.
  — Metal can oxidize very quickly when it is heated. If you don't get a good soldered connection right at first for some reason, you may have to re-clean the connection surfaces before you try again.

- Make a good mechanical connection
  - For example, two wire ends should be twisted together so that they hold together before being soldered. For components on pin-in-hole printed circuit boards, the leads are normally bent slightly to one side, or *clinched,* on the wiring side of the board. This holds the component in place while soldering.
  - Wires or component leads to be soldered to chassis terminal strips, tie points, switch terminals, etc., should be hooked around the terminal post and lightly crimped or squeezed so the wire or lead is mechanically secure before soldering.
- Apply the heat to the work
  - This means to hold the tip of the iron against the component leads or wires you're soldering together. Don't just heat the solder; it won't flow properly unless the work itself is at or near the same temperature.
  - A small dab of solder on the tip of your soldering pencil will help transfer heat to the junction to be soldered. For most soldered connections, applying heat to the surfaces to be soldered for about 3 to 5 seconds is usually enough.
  - It's important that *both* metals to be soldered together are heated to high enough temperature to melt the solder.
  - When the junction is at temperature, apply just enough solder so that the entire connection area is evenly coated. Solder should flow quickly and evenly around the connection. Capillary action will pull molten solder into small cracks and crevices. There should be no excess solder blobs around the connection.
- Use just enough heat
  - Most people tend to use too much heat when they first begin soldering. Solder should be applied as soon as the connection is hot enough for the solder to melt and flow freely.
  - Too much heat causes all sorts of problems: wiring traces can detach from a circuit board; plastic insulation on wires melts, leaving wire exposed that shouldn't be; plastic-bodied connectors get soft or melt while soldering to a pin or contact.
  - Excessive heat can also damage some components. It's a good idea to attach a clip-on heatsink to the component lead being soldered (or unsoldered), between the solder joint and the component body. Extra heat that could possibly damage the component is absorbed by the heatsink. This is more important with small *germanium* diodes and transistors. Most components are *not* easily damaged by soldering.
  - Solder should melt nearly instantaneously when you touch it to the connection. If you find yourself "force feeding" the solder, or the solder sort of balls up at the end of the roll instead of flowing freely, the connection is not hot enough.
  - If the connection is not hot enough when you apply the solder, you'll get what's called a *cold solder joint.* Not good! A cold solder joint has a dull gray, grainy or pitted appearance, rather than a smooth, shiny coating of solder. Reheat the connection so solder flows smoothly throughout the entire connection.
- Apply solder *to the connection*
  - Be prepared to apply solder to the heated joint so that it flows as soon as the work is at temperature.
  - Touch the end of the roll of solder to the area where the two or more component leads, wires, terminals, etc. come together, where the solder

tip rests. *Do not* apply solder to the soldering iron tip itself and then transfer it to the joint.
— After the solder flows in and around the entire connection, remove the solder. Hold the tip of the iron on the work for about one more second, then take it away.
— Most novice solderers have a tendency to use too much solder. Hot solder should be sucked into the connection by capillary action. As soon as this occurs, and the connection has a smooth, shiny coating, remove the solder.
— Excess solder can drip and burn whatever it touches on the way down. On a circuit board, too much solder can easily flow across adjacent wiring traces, or runs, shorting them out. It's easy to have a mess on your hands. Go easy with the solder!

- Allow the connection to cool before moving it
  — If a connection moves while the solder is hardening, the solder will crack or become grainy. The result is a cold solder joint. If this happens, reheat the connection. You may not need to add more solder.
- Practice, practice, practice
  — As we said, it's not as easy as it looks. You can read and understand all that's been said here, but until you get some hands-on experience, you won't gain any actual soldering skills. Only by *doing* will you learn just how much heat is enough, how much solder to apply, and when to take the heat away.
  — Take the time to solder some components to scrap circuit boards before you tackle the real thing. Make some wire splices with solid and stranded wire. Solder some wires and component leads to a terminal strip or tie point. You'll gain confidence! Soldering should become so natural that you hardly even think about what you're doing to get perfect solder connections every time.
  — You can inexpensively pick up some components like 1/2W carbon resistors, chassis terminal strips or tie points, small spools of solid and stranded 20- or 22-gauge hookup wire, and a small pre-drilled PCB with copper wiring grids at electronics parts suppliers. (This type of PCB is sometimes called a *project board*.)
  — Try your hand at soldering wires and components to the tie point lugs and PCB solder pads. Make some soldered wire splices. Keep at it until you're satisfied that the soldered connections look professional.
  — Another possibility is to take a tour of the neighborhood the evening before trash pickup. You may very well find discarded VCRs, clock radios, and other goodies sitting at curbside. These are great for practicing soldering and unsoldering skills. You could also ask a local TV repair shop to save you a "clunker," if they don't already have some unrepairables lying around that are yours for the asking.
  — Don't be discouraged if your first attempts at soldering are disasters. They probably will be. Keep at it and review the basic procedures listed here. You'll know when you've soldered correctly: there'll be a nice, shiny, clean connection with no excess solder, and wire insulation or anything else in the immediate vicinity won't be burned to a crisp.

Figure 11.2 illustrates a few mechanical connections, how to heat a connection, and how to apply solder.

**FIGURE 11.2 Examples of solder connections.**

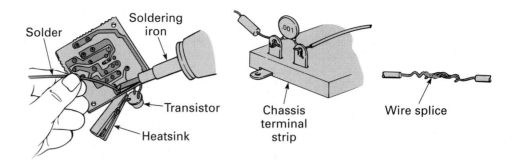

> **Soldering Summary**
> - Start with hot iron
> - Keep iron tip clean
> - Clean connection surfaces
> - Make solid mechanical connection
> - Apply heat *to the work*
> - Use just enough heat
> - Apply solder *to the connection*
> - Allow connection to cool
> - Practice

## Surface Mount Devices

Many newer circuit boards incorporate *surface mount devices* (*SMDs*). Instead of the pin-in-hole connections used by axial and radial lead through hole mount (THM) components, an SMD has flat, conductive pads that are soldered directly to wiring pads on the surface of a printed circuit board. SMDs include resistors, diodes, capacitors, and integrated circuits. An SMD IC is also called a *leadless chip carrier* (LCC).

A big advantage of SMDs is that they take up considerably less surface area or real estate on a circuit board than an equivalent THM component, making smaller boards possible. SMDs are being used more and more, especially where compactness is important, such as in camcorders, notebook computers, and smaller tabletop VCRs.

Besides requiring less board real estate, surface mount devices are better suited to automated manufacturing, where robots place and solder the components to PCBs. (No pin-in-hole alignment!) This reduces the cost of an assembled board. SMD and THM components are frequently both present on the same printed circuit board.

With great patience and care (and luck), you *can* solder and unsolder a surface mount device (SMD) similarly to conventional components; a small tweezers is pretty much mandatory for holding a device, which can be quite small. A typical SMD resistor is flat and about the size of a paper match head.

For soldering and unsoldering SMDs, a 15-watt iron is recommended. Small-diameter solder is a must here, such as .020", .025", or .031". You need to be *very cautious* not to apply excessive heat or use too much solder. Solder pads on components and on printed circuit boards are quite small, very delicate, and closely spaced. Too much heat can cause solder pads to detach from circuit boards and excessive solder can easily bridge closely-spaced circuit board traces.

Electronic parts and tool catalogs from larger supply houses often list soldering aids designed for SMD rework and repair. These tools usually cost in excess of $50.

## INSULATING CONNECTIONS

Although insulating is not directly related to soldering, this is a good place to briefly discuss insulating different types of connections. Most electronic connections are within some enclosure, and it's not necessary to insulate the majority of them. But perhaps a wire has been spliced, or a bare component lead soldered to a tie point, switch terminal, or jack is close to another bare lead or chassis ground. It's a good idea to insulate a splice or bare lead that could otherwise short to a nearby component or metal frame, or be touched.

Two types of insulating material are ideal for insulating component leads and wire splices: spaghetti and heat shrink tubing. *Spaghetti* is just hollow plastic or fiber tubing that can be slipped over a bare component lead before making a connection. It's really like insulated wire, *without the conductor*.

You can easily make your own short lengths of spaghetti by carefully stripping insulation from a wire or cable. For example, a 3/4-inch length of insulation removed from a piece of 20-gauge hookup wire is ideal for insulating resistor, diode, and transistor leads that may otherwise come into contact with something else. Most electronic parts suppliers have variety packs of spaghetti, with several different diameters and colors of tubular insulation in one-foot lengths. This is handy material to have, especially for do-it-yourself electronics projects.

*Heat shrink tubing* is similar to spaghetti, except its diameter shrinks when you apply heat, making a snug, insulated covering for the connection. Heat shrink is especially good for splices in insulated wire. Place a length of heat shrink over a wire before making the splice. Choose a diameter that is slightly larger than the existing wire insulation so it will slip on easily.

Solder the splice, being careful that the heat shrink tubing isn't close enough to the splice to get hot. After the splice has completely cooled, move the tubing over the splice. It should cover the splice, extending over the wire insulation on each side of the splice 1/4 to 1/2 inch.

Now, heat the shrink tubing. For occasional use, a cigarette lighter works fine. Rotate the wire and move the lighter flame back and forth beneath the heat shrink so that the entire surface and length of the heat shrink tubing are uniformly heated. Keep the flame tip at least an inch or so away. The tubing will shrink and form a nice tight covering for the splice. A hair dryer on high heat can also work as a heat source.

Heat shrink tubing is available in a wide variety of diameters, usually in three- or four-foot lengths, or by the foot, at electronic parts outlets.

While you practice your soldering techniques, you might insulate a wire splice or component lead with spaghetti, heat shrink, or both. I say "or both," because for a splice in a wire that will get handled a lot, a heftier insulation can be achieved by first putting regular spaghetti over the splice, and then a slightly longer piece of heat shrink over the spaghetti and splice.

Time spent practicing soldering, unsoldering, and insulating techniques will pay off if you decide to construct the two-way tester described later in this chapter. These skills are part of what it takes to be successful at VCR repair.

## UNSOLDERING

There will be many times you'll need to unsolder an entire component to remove it, or perhaps you have to unsolder just one lead of a part from a printed circuit board. This could be to replace a shorted electrolytic capacitor on a VCR power supply board or to lift the base lead of a transistor so it can be tested with a multimeter without possible back circuits. To do this, solder usually must be removed from the connection.

At other times, excess solder must be removed; perhaps you accidentally applied too much, and a solder bridge formed between two adjacent wiring traces on a printed circuit board. This short circuit must be eliminated.

Unsoldering, especially when a component is mounted on a printed circuit board, must be done *very carefully!* Inattention or haste while desoldering can lead to disaster. It is very easy to apply too much heat, or pressure, or both and have a section of wiring trace detach from the board's surface. Molten solder can also get where it's not supposed to be, possibly causing a short circuit. As with soldering, desoldering should be practiced on scrap materials first.

To remove unwanted solder, it's hard to beat what most technicians call *solder wick,* which is flat, braided copper wire that sucks up molten solder from a connection. Solder wick is low-cost and easy to use. Also known as *desoldering braid,* it is sold under several brand names, including Soder-Wick®. This product comes on a small plastic spool in diameters from .030- to .110-inch. A diameter of .030-inch is suited for small surface mount device solder pads, whereas .060- or .080-inch solder wick is better suited for general-purpose work. Solder wick is available at all electronic supply outlets. Here's how to use solder wick or desoldering braid:

1. Pull a few inches of the flat, braided wire from its supply spool.
2. At the free end, fluff up the braid a bit so the strands are loose, with small air spaces between them. This increases the braid's surface area and makes it easier to absorb molten solder.
   — Hold the braid between thumb and forefinger, about an inch or so from the free end.
   — With thumb and forefinger of your other hand, push on the free end, back towards the supply roll. The braid will fatten up, with small gaps visible between individual copper strands.
3. Place the braid end between the soldering iron tip and the solder you want to remove. As the solder melts, capillary action will draw it up into the copper wick. You may have to move the braid to fresh copper until all excess solder is removed from the connection.
4. Remove wick and soldering iron as soon as all solder flows into the wick. Avoid excess heat which can cause a printed wiring trace to lift from the board.
5. After the braid cools, cut off and discard the section that has sucked up solder to expose fresh braid.

Figure 11.3 illustrates how to remove solder from a printed circuit board connection with solder wick.

In addition to desoldering braid, there are a few other relatively inexpensive methods for solder removal. You might prefer one of these instead. They each rely on vacuum to withdraw molten solder from a connection or circuit board, and are referred to by many technicians as *solder suckers.* The simplest of these is a rubber squeeze bulb with a hollow Teflon tip that can withstand the temperature of liquid solder. A desoldering bulb is pictured in Figure 11.1.

To use a desoldering bulb, first heat the solder to be removed with an iron. Squeeze the bulb closed, place the hollow tip in direct contact with the hot solder, and release the bulb all at once. The resulting vacuum when the bulb is released vacuums the solder up into the bulb. The Teflon tip can be removed to empty the solidified solder pellets from the bulb after many uses.

It is important to keep the Teflon tip clean and trimmed. Use a sponge or rag to wipe the tip after each use. If the tip gets ragged, it can be carefully trimmed with a small utility knife. Replacement tips are available.

**FIGURE 11.3
Removing solder with desoldering braid.**

Another solder removal tool is the *desoldering pump,* also pictured in Figure 11.1. Like the desoldering bulb, the pump works by drawing hot solder up through a hollow Teflon tip into a chamber with vacuum action.

A pump has a spring-loaded plunger in a cylinder. The pump is cocked by pushing the plunger from the rear of the tool toward the tip end, where it latches. After the hollow tip is in contact with molten solder, pressing a trigger on the side of the pump releases the spring-loaded plunger. The plunger flies toward the back of the cylinder, creating a vacuum and pulling the solder up into the chamber.

After several uses, the pump is easily disassembled to discard solder beads in the cylinder. Each time the pump is cleaned, the rubber O-rings on the plunger should be wiped with a cloth or sponge and then lubricated with a thin grease coating. Heatsink compound or silicone grease used between power transistor bodies and heatsinks to assist in heat transfer is fine for this purpose. Maintain the Teflon tip as described earlier for the desoldering bulb.

The next step up is the dedicated solder sucker or desoldering iron. This tool is a cross between a soldering iron and the desoldering bulb described earlier. It is a specially designed soldering iron that has a hollow metal tip directly heated by the iron. A hollow metal tube runs from the top of the tip back to a rubber squeeze bulb adjacent to the soldering iron handle. To operate, simply squeeze the bulb, place the hot tip against solder to be removed, and release the bulb when solder becomes molten.

Desoldering irons cost more than the other tools just described, but are easier to use, especially for more than occasional solder removal. They can be easily operated with a single hand. As with other irons, it is important to maintain a clean, well-tinned tip on a desoldering iron. Because the hollow tip is much more delicate than a solid soldering tip, it is especially important not to let a desoldering iron stay hot during periods of nonuse; the tip will oxidize and become ragged. Replacement tips are available.

Beyond this equipment, there are many specialized desoldering tools for removing an entire integrated circuit (IC) easily, and even motor-driven vacuum desoldering tools. These are generally too expensive for occasional use, but you may want to check out special soldering iron tips that simultaneously heat all pins of an IC for easier component removal. Mail order catalogs and larger electronic supply houses generally have a variety of desoldering tips designed for specific IC pin arrangements.

Whatever desoldering method you decide on, practice and experiment on scrap PCBs before tackling a "live" repair job. It looks simple—and looks can be deceiving!

## BUILD A TWO-WAY TESTER

This project is fairly easy to construct. You'll get some additional experience soldering, *and* you will end up with a convenient tool to help you troubleshoot VCRs.

The two-way tester (TWT) detects near-infrared light and serves as a reversible power supply for two-terminal DC motors. These motors are typically found in VCR front loader assemblies and as the cam or tape load motor in many VCR models.

All parts for the TWT are low cost and easily obtainable. The "RS" number in parentheses in the parts list is Radio Shack's stock number; similar parts are available at nearly all electronic parts suppliers.

### TWT Circuit Description

Refer to Figure 11.4 during the following discussion.

Near-infrared photo transistor Q1 and visible light-emitting diode LED1 together form the IR detector portion of the TWT. When near-infrared light energy strikes Q1, it conducts. Current then flows from the negative terminal of

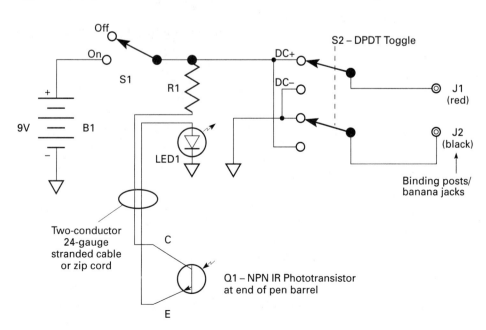

**FIGURE 11.4** Two-way tester schematic diagram. Refer to parts list for component values.

9V battery B1, through red LED1, Q1, R1, and S1, to B1's positive terminal. The red LED lights up, indicating the presence of IR light striking Q1. R1 limits forward current through Q1 and LED1 to about 15mA with Q1 fully conducting. SPST switch S1 functions as an On/Off switch for the TWT.

In actual construction, phototransistor Q1 is on a wand, or IR probe, connected to about 2-1/2 feet of cable. This allows Q1 to be conveniently positioned close to an IR light source, such as the light post that sticks up on a VHS VCR reel table and protrudes into the middle of a loaded cassette.

The second portion of the two-way tester consists of battery B1, jacks J1 and J2, and DPDT switch S2. With power switch S1 On and S2 in its DC+ position, the negative side of B1 is connected to J1 (black) and B1's positive terminal is connected to J2 (red). Flipping S2 to its DC- position reverses the polarity of the 9 volts at J1 and J2, so that J2 (black) is now positive and J1 (red) is negative.

Jacks J1 and J2 can accept banana plugs, spade lugs, or bare wire. A pair of test leads with banana plugs at one end and miniature spring-loaded hook clips at the other end are ideal for connecting the TWT's reversible 9V battery to a VCR motor. This is explained further in Chapter 13.

## TWT Parts List

Following is the parts list for the two-way tester. Substitutions may be made using other available parts. The value of R1 is not critical. Any resistor between 330 and 560 ohms will work satisfactorily, although the sensitivity of the IR detector is slightly greater with lower R1 values. R1 may also be a 1/4W resistor.

**Two-Way Tester Parts List**

B1:   9V rectangular "transistor-radio" battery
J1, J2:   Banana jack/binding post (RS 274-661)
LED1:   Red light-emitting diode (RS 276-041)
Q1:   Infrared Phototransistor (NPN) (RS 276-145)
R1:   390Ω, 1/2W carbon resistor* (RS 271-018)
S1:   SPST toggle switch (RS 275-612)
S2:   DPDT toggle switch (RS 275-614)

**Miscellaneous Parts**

Small project box (RS 270-221 or 270-231*)
9V battery holder (RS 270-326)
9V battery connector (RS 270-325)
Holder for LED1 (optional*) (RS 276-079 or 276-080)
Empty felt-tip pen barrel
Hookup wire, spaghetti, heat shrink tubing
Misc. hardware, labels
Epoxy cement or silicone sealer
* = see text

## Two-Way Tester Construction

Building the TWT is very straightforward, with no critical areas. All components except IR phototransistor Q1 are mounted in a small project box. Should you decide to purchase a project box from Radio Shack, 270-221 is all plastic, whereas 270-231 is plastic with an aluminum top cover.

If you do not have a small hand drill and assorted bits for making the round holes in aluminum, an all-plastic box may be easier to work with. Mounting holes for switches and combination jacks can be made by melting a small hole in the plastic with a hot scribe or screwdriver tip. Then enlarge the holes to size with a rat-tail file, tapered reamer, or thin, sharp knife. (Be sure to always *cut away* from yourself!)

Make appropriate size holes in the top cover for mounting S1, S2, J1, J2, and the holder for LED1. Also make a small feedthrough hole for the two-conductor wire from the IR probe. Top cover component layout is not critical, and is left up to you. Refer to photos later in this chapter for a possible layout. Make sure that the cover will fit back on the box after the parts are mounted and that there is no interference with the 9V battery. Top cover parts should be securely mounted before soldering to them.

The LED holder is optional. You can carefully make a hole so that the LED fits very snugly from the inside of the box cover. A generous dab of silicone sealer or epoxy cement on the back side of the LED will hold it in place. Let the cement cure before continuing.

For intercomponent connections, 20- to 22-gauge insulated hookup wire, either solid or stranded, is adequate. Solid is easier to work with in some respects, but it is not as flexible as stranded.

A holder for the 9V battery is optional, but it's nice not to have the battery just bouncing around inside the tester. You can mount the holder almost anywhere on the project box, but make sure that the location doesn't interfere with components on the cover when it is in place. One good place is along one long side, close to the bottom of the box. This leaves the outside bottom of the box free of a protruding screwhead.

You may need to splice additional wire onto the short pigtail leads of the 9V battery connector. Be sure to insulate the splice(s) with spaghetti or heat shrink tubing.

When connecting LED1 and Q1, be sure they are oriented correctly. The shorter of the two leads marks the LED cathode and the phototransistor collector. Both of these leads are also identified by a small flat spot on the circumference of the base of the device.

Leave the leads on LED1 and Q1 full length when you solder connecting wire to them. Wrap the wires around the leads about 3/4-inch away from the body of the component. Place spaghetti or shrink tube around the wire so you can slide it up toward the component and insulate the two leads after the connecting wires are soldered. This is necessary for Q1, optional for LED1.

As an extra precaution, put a clip-on heatsink or alligator clip on the lead between the solder connection and component body while soldering.

To cut the excess component lead after the connecting wire has been soldered, first grasp the lead *firmly* with long-nose pliers between the component body and solder connection. Then cut the lead just beyond the connection with small diagonal cutters. This prevents mechanical shock waves produced when you cut the lead from traveling up into the component and possibly shattering a PN junction.

## IR Probe

The IR probe consists of phototransistor Q1 mounted to the business end of an empty pen barrel. A two-conductor cable soldered to Q1's leads runs up the pen barrel, out a hole in the top of the barrel, and on to the two-way tester box. A felt-tip pen barrel is usually larger in diameter than that of a ballpoint pen, making it easier to get the two-conductor cable through the barrel.

Here are some suggestions for constructing the IR probe as illustrated in Figure 11.5.

First solder one end of a length of two-conductor cable to Q1. Then mount this to an empty pen barrel. Finally, connect the other end of the cable inside the TWT.

Select three feet or so of small-diameter two-conductor cable. Two-conductor, 24-gauge stranded zipcord, often sold as cheap speaker wire or intercom cable, will work fine. Stranded wire cable is better than solid, because it is more flexible, with less chance of a wire breaking in normal use.

Notice that there is some characteristic that identifies one of the two zipcord conductors. You must be able to tell which wire is which to make correct polarity connections to LED1 and Q1. Often the outside edge of the insulation is smooth on one conductor, but ribbed or serrated on the other. Or one side may be flattened, whereas insulation on the other is rounded. On some cables, the strands of one wire are silver-colored, while the other wire is copper-colored. Some two-conductor cables have different-colored insulation for each wire, or one wire has a white tracer along its length.

At one end of the zipcord, separate the two conductors for about 3 inches, and place about 1-1/4 inches of spaghetti or heat shrink over one conductor. It doesn't matter which; the idea is to prevent the two leads of Q1 from touching each other. Strip about 1/2- to 3/4-inch of insulation from each wire. Be careful not to cut into the wire itself.

Now, wrap the bared end of each wire tightly about each lead of Q1, heatsink the leads, and solder. Whichever distinguishes the two conductors in the cable you are using, make a note as to which wire you solder to which lead of Q1. For example:

*ribbed = collector*

Cut off any excess lead, as described earlier, and slide the spaghetti or shrink tube down over the solder connection after it cools. Q1's leads should now be insulated from each other. If you are using spaghetti and the fit is very loose, so the tubing won't stay over the splice, put a dab of silicone glue or caulk on it to hold it in place. If using heat shrink, apply heat to the insulating tubing.

Remove and discard the innards from a plastic felt-tipped or ballpoint pen. You may need to use a utility knife to cut into the barrel, just above the writing tip. Make a hole at the upper end of the pen barrel large enough for the two-conductor cable to pass through. You now have a hollow plastic tube.

Thread the free end of the probe cable through the opening at the pen tip, through the barrel, and out the hole at the top end of the barrel. Glue Q1 to the

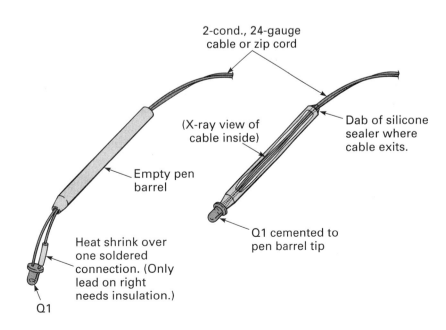

**FIGURE 11.5  IR probe assembly.**

bottom of the pen barrel with epoxy cement or silicone sealer. You might also put some cement or sealer at the other end of the barrel where the wire exits.

The entire body of Q1 should be exposed at the tip of the pen barrel. Wipe away any sealer, glue, or cement that may have gotten on the surface of Q1. Set the probe assembly aside to cure for 24 hours.

Figure 11.5 shows what the IR probe should look like before and after affixing Q1 to the end of the pen barrel. Note that one of Q1's leads is completely insulated from the other with spaghetti or heat shrink tubing.

After sufficient cure time, pass the free end of the probe cable through the feedthrough hole in the tester box cover. Make a simple overhand knot a few inches away from the end of the cable on the inside of the box to prevent it from being pulled out.

Optionally, you can feed the cable through a rubber grommet in the cover hole, and also put a generous dab of silicone sealer around the knotted probe cable with the knot against the inside of the top cover. This prevents any possible chafing of the cable against the cover, especially the aluminum one, and stops the probe cable from twisting, which could pull on component leads inside the box. Allow sealer to cure before proceeding.

## TWT Final Assembly

With all components mounted to the top cover, make all connections to these parts with hookup wire and one end of R1. Realize that the three downward-pointing arrows in the Figure 11.4 schematic are common points, and must be interconnected. Remember to make a clean mechanical connection, and doublecheck your wiring with the schematic prior to soldering.

Strip about 1/2-inch of insulation from each wire of the two-conductor probe cable. Slip a 1-inch length of spaghetti or heat shrink over each conductor. Refer to the note you made when constructing the probe assembly. Solder the wire connected to Q1's collector to the free end of R1, and the other wire to LED1's anode. Cover each connection with the insulated tubing. Figure 11.6 shows the completed two-way tester before the top panel is installed on the project box.

On the underside of the aluminum cover panel, white silicone sealant has been applied around the IR probe cable and LED1. A soldered splice between one end of R1 and the probe wire from Q1's collector has been insulated with heat shrink tubing. S1 is at bottom left and S2 is near bottom center of the panel. Binding post/banana jacks J1 and J2 are at the right end of the cover panel. The 9V battery and its U-shaped holder can be seen inside the plastic project box.

## Final Testing of TWT

Before screwing on the top cover, test the assembled two-way tester. First, recheck all wiring connections. Make sure there are no shorts and that all connections are neatly soldered. If you used an aluminum panel, check with your ohmmeter that Jacks J1 and J2 are insulated from the panel. Once you are satisfied that everything appears OK, turn S1 to its Off position and install a fresh 9V battery *(alkaline preferred)*.

Turn S1 On, and move the probe tip toward an *incandescent* light bulb. (Fluorescent light has very little infrared light output.) As the probe gets nearer the light, the red LED should light more brilliantly. If the LED fails to light, turn S1 Off and recheck your wiring. Be sure that the battery, Q1, and LED1 are wired with the correct polarity. With S1 On, shorting the anode of LED1 to the end of R1 connected to Q1 should cause it to light.

Next, set your multimeter to its 10VDC range, or whatever range can safely measure 9 volts. Connect the negative (black) meter lead to J2 (black) and the

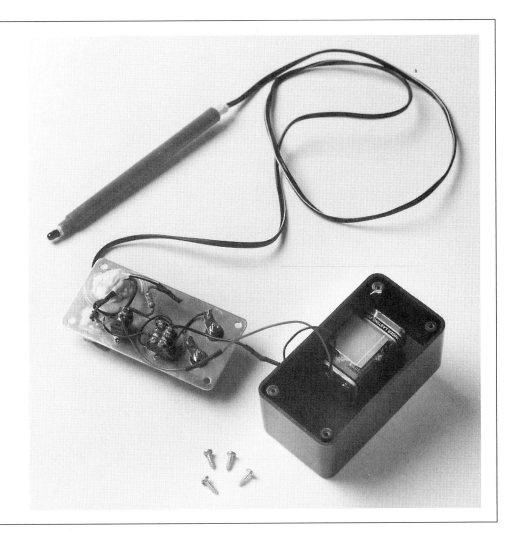

**FIGURE 11.6**
Completed two-way tester prior to securing aluminum cover panel to plastic project box.

positive (red) meter lead to J1 (red). With S2 on the TWT in its +DC position, the meter should measure about 9.3 volts with a fresh alkaline battery. Remove the meter leads, flip S2 to its -DC position and connect the meter leads in the opposite direction to J1 and J2 (or change the meter's polarity switch to -DC). The meter should again register about 9.3 volts.

When everything checks out OK, turn S1 Off, install the top cover onto the TWT and repeat the tests, just to make sure nothing shorted out or broke loose when the cover was put on.

Label the completed tester as you wish. Embossed tape labels or transfer lettering are possibilities. There's nothing wrong with hand-lettered stick-on labels, trimmed to size.

## USING THE TWO-WAY TESTER

Your newly assembled two-way tester can be a help in diagnosing a few VCR failures. One of the frustrations with infrared is that it can't be directly seen by the unaided human eye—and infrared light is incorporated in nearly all present-day VCRs.

Both supply and take-up end-of-tape sensors are phototransistors that detect IR light shining through the transparent leader at each end of a VHS tape.

Reel sensors that determine whether supply and take-up reels are turning also employ IR LEDs and phototransistors. In many VCRs, the mode switch consists of two IR LED and phototransistor pairs that sense the position of the cam and tape load mechanisms.

The most common application for IR in consumer electronics is undoubtedly in handheld remote controls. These send out digital infrared pulses that are picked up by an IR receiver in the VCR, TV, CD player, and stereo gear, or cable TV box. Handheld remotes have one or more IR LEDs that produce the IR pulses when a button on the unit is pressed.

When an IR sensor or remote control unit malfunctions, or is suspect, it is now easy to determine whether the IR LEDs involved are working. Simply turn on your two-way tester and place the IR probe tip close to the IR LED in question. If IR light is present, then red LED1 on the TWT will glow. For remote units, place the probe directly in front of the window of the unit and press a control button. A rapidly blinking red LED on the TWT confirms that the unit is sending out IR pulses.

As an IR detector, the TWT works best when ambient IR light is low. Avoid using it in the presence of strong infrared light sources, such as direct sunlight or bright incandescent lighting. Fluorescent light has little IR content, and provides ideal background illumination when detecting IR light with the TWT.

With a little experimentation, you will quickly become familiar with how the TWT responds to near-infrared light and the best position for the probe tip. Best pickup is usually with the probe tip straight on; LED1 gets brighter the closer the probe is to the IR source. Also, the more similar the near-IR frequency of the source to the IR response curve of Q1, the brighter LED1 will glow. Figure 11.7 shows the completed TWT detecting IR output from a handheld remote control unit while one of its buttons is held depressed.

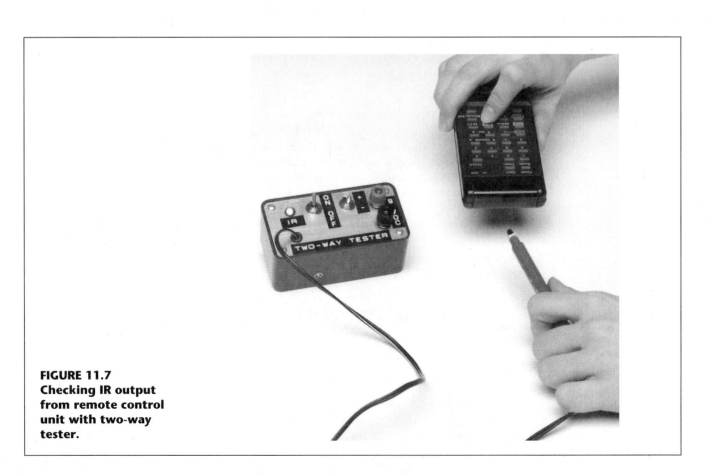

**FIGURE 11.7
Checking IR output from remote control unit with two-way tester.**

The second function of the TWT is to provide a convenient means of powering small 9V to 12V two-terminal DC motors. Most small 12V motors found in VCRs will work quite well from a fresh 9V battery. Simply connect test leads from J1 and J2 on the TWT directly to the two motor terminals (for instance, on a VHS front loader motor). S1 then starts and stops the motor, while flipping S2 changes the polarity of the voltage going to the motor and so reverses the direction in which the motor spins.

> CAUTION: To prevent possible damage to electronic drive components, at least one side of the motor should be disconnected from VCR circuitry when powering it with the 9V battery. Usually this can be accomplished by simply unplugging a connector near the motor. The VCR should also be unplugged from the AC line.

Chapter 13 provides additional information about checking motors with the TWT. Be sure to read this material before checking motors.

## SUMMARY

In this chapter you learned the proper techniques for successful soldering and desoldering. A 25–35-watt pencil is adequate for most VCR and other electronics soldering jobs. Properly tinning and maintaining the tip on a soldering iron is essential to achieving a good solder connection. Only rosin core solder should be used for electronics work. Solder that is 60 percent tin and 40 percent lead is the most common type for general-purpose electronics work. Solder packs or rolls are normally marked "60/40."

Most novice solderers apply too much heat and too much solder to a connection. Practicing with different gauges of solid and stranded wire, scrap PCBs, and inexpensive components like a few 1/2W carbon resistors is essential before tackling a real repair. A good solder joint is clean and shiny, with solder around and through the entire connection, but no excess solder. A soldered joint that looks dull gray, grainy, or pitted is a cold solder joint. This is usually caused by insufficient heat or movement of the connection before the solder solidified.

One of the easiest ways to remove unwanted solder is with solder wick, flat copper braid that absorbs molten solder from a heated connection. A desoldering bulb, pump, or special desoldering iron may also be used to remove unwanted solder. These devices work with a vacuum that draws molten solder away from a connection.

The chapter concluded by describing how to construct a simple diagnostic tool called the two-way tester (TWT). The TWT can determine whether infrared LEDs are working and can also power small 9- to 12-volt two-terminal DC motors in either direction. Building the TWT gives you valuable experience in soldering and following a schematic wiring diagram, as well as providing a useful VCR diagnostic tool.

## SELF-CHECK QUESTIONS

1. What wattage soldering iron is adequate for most VCR electronics work? What type solder should be used?

2. Solder wick also is called _____ (non-tradename). What is this product used for?
3. List two safety precautions to take when working with solder and soldering iron.
4. Describe how to tin a soldering iron.
5. What should you do each time you pick up a hot iron, just before using it, and before putting it back into its holder?
6. What three conditions should exist before applying solder when making a connection?
7. Describe a cold solder joint, including its appearance.
8. What are two common causes for a cold solder connection?
9. What could happen if heat is applied for too long when desoldering a component from a printed circuit board?
10. How can you identify the leads of most discrete LEDs and of most discrete phototransistors—like those in the two-way tester project in textbook Chapter 11—without using an external reference, such as pegboard product card or semiconductor manual, or an ohmmeter?

CHAPTER  **VCR POWER SUPPLIES**

VCR electronic circuitry is generally quite reliable, providing years of trouble-free operation. If there is one electronic area that seems to break down and cause problems more than others, it is the power supply section. The reasons for this are readily explained.

Much of the power supply is powered On whenever the VCR is plugged into a wall outlet, so parts of it are working all the time. Voltages that operate the entire VCR are produced by the power supply, so all power goes through this section. For this reason, more heat is produced by the power supply than other circuits. The power supply also takes the punishment of voltage spikes, transients, and surges on the power line, like when there's lightning nearby. These voltage disturbances can cause power supply components to fail before other VCR circuitry.

Because a VCR power supply may be more prone to failure than almost any other single VCR circuit, this chapter describes typical power supply (PS) operation, failures, and basic repairs. First you will learn about two types of full-wave rectifiers. Next, voltage regulators and basic PS control circuits are described. These concepts are taught for a *transformer-operated* or conventional power supply, one having a step-down power transformer connected to the 120-volt AC line. Finally, you will learn the fundamental operation of a *switching mode power supply* (SMPS), a type of supply that is found increasingly in today's VCRs.

# CHAPTER OBJECTIVES

**Upon completing this chapter, you should be able to:**
1. Describe two methods of power supply full-wave rectification and how to test each for correct operation.
2. Describe basically how a solid state voltage regulator functions and how to test one for proper operation.
3. Describe the fundamental operation of a typical VCR power supply control circuit.
4. Describe the basic operation of a switching mode power supply (SMPS), and precautions to take when working on a VCR with this type supply.

Before beginning this chapter, it is extremely important that you generally understand the material presented in previous chapters. You should be familiar with an AC sine wave and know the operating principles of a power transformer and a diode, in particular how a half-wave rectifier changes AC to pulsating DC. AC sine waves and direct current are discussed in Chapter 8. Half-wave rectification and how an electrolytic capacitor functions as a power supply filter are described in Chapter 10.

Take time to review before going on in this chapter if you aren't comfortable with AC, DC, and half-wave rectification. This will make the subject matter in this chapter much easier to comprehend.

## POWER SUPPLY OVERVIEW

Here is a big-picture view of what goes on inside a typical VCR *conventional-style* power supply (PS). Don't be concerned about the details right now. Just try to understand these basic power supply functions that change power from the wall outlet to the various types of power used in VCR circuits.

- Power transformer: changes 120VAC to several lower AC secondary voltages
- Diodes: rectify AC secondary voltages to pulsating or raw DC
- Filter capacitor(s): smooth pulsating DC into steady, pure DC
- Transistor switches: turn some DC output voltages On when Power button is pressed. Other PS output voltages are always present.
- Voltage regulators: solid state devices that keep some of the DC output voltages very constant or steady.

You know already that VCR power supplies provide several different output voltages for operating various electronic circuits and DC motors. Some voltages are available whenever the VCR is plugged into a wall outlet, whereas others come On only after the VCR Power switch is pressed. Some outputs are regulated, providing highly stable voltages, whereas other voltages are unregulated.

In a *conventional* transformer-operated power supply, as opposed to a switching mode power supply (SMPS), 120VAC power goes through a mainline fuse to the primary winding of the power transformer. Two, three, or even more secondary windings provide different AC voltages that are all lower than the 120-volt input, for example, 16, 10, and 6 volts AC. Each secondary AC voltage is then rectified with one or more diodes to produce pulsating direct current. Electrolytic capacitors filter the pulsating DC to provide smooth DC output voltages.

Following are PS output voltages for one model VCR. Those in boldface are always present whenever the VCR is plugged in; the others turn On electronically when the Power button is pressed.

- **5.2 VAC**
- **5.2 VAC**       Both these lines are for the fluorescent display. There is no direct ground reference.
- **−30 VDC**       to timer circuit
- **+5.6 VDC**       to timer circuit
- **+5 VDC**       to system control (syscon) circuits
- **+50 VDC**       to RF tuner circuit
- +12 VDC (reg)       to video and audio circuits
- +12 VDC (sys)       to control and motor circuits
- +16 VDC       to servo circuit
- Ground       Common for *all* DC voltages

How are voltages turned On electronically when the Power button is pressed?

A signal from system control, which is always powered up, forward biases transistors and voltage regulators in the power supply. The transistors then act like a closed switch and feed the DC voltage out from the supply. Voltage regulators turn On in a similar manner.

Figure 12.1 is a *simplified* block diagram of a portion of a conventional VCR power supply, with 120V power transformer. Notice that the outputs at

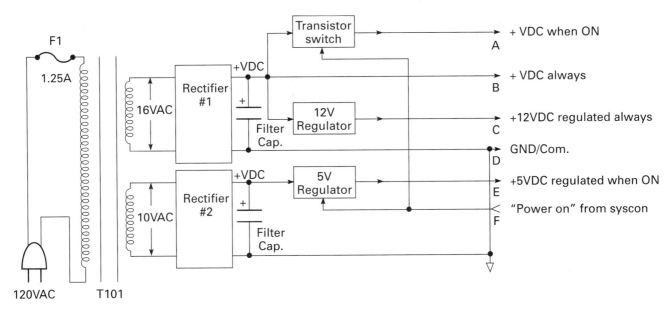

**FIGURE 12.1** Simplified power supply block diagram. "Power On" signal from system control turns On transistor switch connected to Rectifier #1 and also turns On +5V regulator connected to Rectifier #2.

"B" and "C" are always On, but that outputs at "A" and "E" are turned On with a "Power On" signal from system control. Note: The arrows in Figure 12.1 are representative of signal flow for instructional purposes, *without* respect to either electron or conventional current flow.

Some DC output voltages are developed with half-wave rectifiers; others employ full-wave rectifiers. You learned all about half-wave rectifiers in Chapter 10. A single diode allows one half of the AC sine wave to pass through, while holding back the other half cycle. For electronic circuits that require more current, or very pure DC, or both, the power supply frequently incorporates *full-wave rectification*. In full-wave rectification, *both halves* of the AC sine wave are rectified to form DC; the entire, or full, alternating current wave is converted to direct current.

## FULL-WAVE RECTIFICATION

There are two basic configurations for a full-wave rectifier. One employs a center-tapped secondary transformer winding and has two diodes. The other consists of four diodes, together called a *bridge rectifier,* which is used when there is no transformer center tap.

First, the full-wave bridge rectifier having four diodes is described. We'll trace the flow of current from the transformer secondary winding through the diodes and a load resistor for the positive half of the AC cycle, and then for the negative half of the AC waveform. Refer to Figure 12.2. For the first portion of this description, consider that SPST switch S1 is in its Open or Off position, so that C1 is not connected in the circuit.

To understand how a full-wave bridge rectifier works, first recall that a diode allows current to flow in only one direction, *against* the arrow of the diode symbol. During the first half of the AC cycle, Point A on the transformer secondary is positive with respect to Point B. Current flows from negative to positive, so we shall trace current flow in the circuit from B back to A:

- Current flows from negative B to the junction of diodes D1 and D4. Since current can only flow *against the arrow*, it flows through D1.

348 ■ Practical VCR Repair

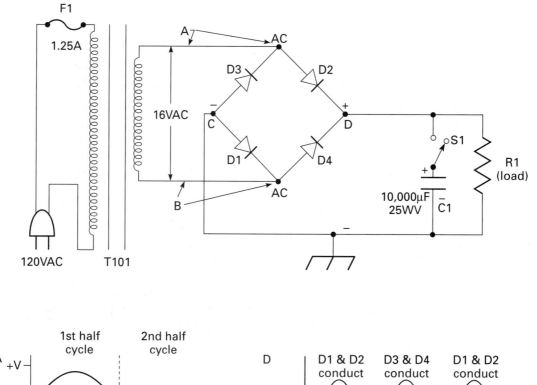

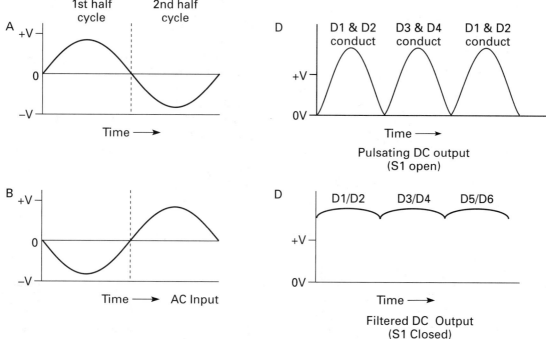

**FIGURE 12.2** Full-wave bridge rectifier. Diodes D1-D4 form a bridge rectifier. C1 is the electrolytic filter capacitor. R1 represents the power supply load.

- At the junction of D1 and D3, current cannot flow through D3.
- Current flows out C to the bottom of R1, and then through R1 to Point D.
- At junction of D2 and D4, current goes through D2 to A, which is positive.
  (Current does *not* flow through D4, because that goes back to B where we started, and there is no potential or voltage difference across D4, so it doesn't conduct.)
- The output at D is the first "hill" of the pulsating DC output, with D1 and D2 conducting.

Next is the negative half cycle, in which Point A goes negative while Point B goes positive. Current flows from negative to positive, so we shall trace current flow in the circuit from A back to B:

- Current flows from negative A to the junction of Diodes D2 and D3. Since current can only flow *against the arrow,* it flows through D3.
- At the junction of D3 and D1, current cannot flow through D1.
- Current flows out C to the bottom of R1, and then through R1 to Point D.
- At junction of D2 and D4, current goes through D4 to B, which is positive.

    (Current does *not* flow through D2, because that goes back to A where we started, and there is no potential or voltage difference across D2, so it doesn't conduct.)
- The output at D is the second "hill" of the pulsating DC output, with D3 and D4 conducting.

Notice that the pulsating positive DC output at D is the rectified, or converted, halves of the entire alternating current cycle. But, even though the voltage never goes negative, it rises and falls from zero volts to some positive value, back to zero, and then back to a positive voltage. This is *pulsating DC*.

Rising and falling, or pulsating, DC voltage (also called *raw DC*) is no good for powering an electronic circuit! Steady, nonpulsating DC like that from a battery is required.

This is where electrolytic filter capacitor C1 comes into play. By closing S1, C1 is connected across the pulsating DC output. C1 charges to the peak voltage, at the tops of the pulsating DC waveform. When the rectified AC goes back to zero volts each half cycle, capacitor C1 discharges. This puts voltage *back into the circuit,* across load R1. C1 "fills in the valleys between the hills." In this way, C1 holds the voltage at nearly the peak value, so there is only a very slight voltage dip between half cycles.

Of course, in a real power supply there would not be a switch to connect or disconnect a filter capacitor; C1 would be connected at all times. S1 is in Figure 12.2 just to illustrate what the DC output waveform would look like with and without C1 in the circuit.

A full-wave bridge can consist of four separate, discrete diodes, but often a bridge rectifier is a single component, with the four diodes packaged together. Electrically, there is no difference between four discrete diodes and a unified bridge rectifier. There are just fewer connections to make with a single component.

A single-component, or unitized, bridge rectifier has four leads or terminals, marked as follows:

- **AC** or ~ one side of AC input
- **AC** or ~ other side of AC input
- **+** positive DC output
- **−** negative DC output.

Just as with discrete diodes, a bridge has forward current ($I_F$) and peak inverse voltage (PIV) ratings that must not be exceeded. As long as these ratings are within circuit parameters, four separate diodes may be replaced with a single bridge or vice versa. Normally it is good practice to replace all four individual diodes if one of the four in a bridge configuration fails. One failed diode often puts a strain on the remaining three, possibly weakening one of them.

Now, if you had difficulty following current flow in the full-wave bridge rectifier, take time now to go back over the diagram and explanations before proceeding. It might be helpful to review how a diode performs half-wave rectification, back in Chapter 10.

Notice in the full-wave bridge rectifier that neither minus ⊖ nor plus ⊕ DC outputs are themselves directly connected to a power transformer secondary winding. Contrast this with the next type of full-wave rectifier.

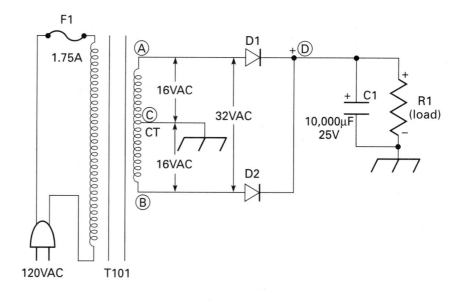

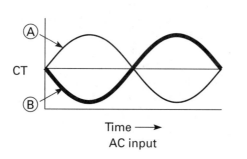

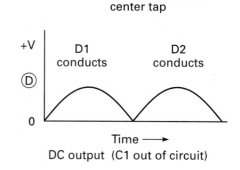

**FIGURE 12.3** Full-wave rectifier with center-tapped transformer secondary. Diodes D1 and D2 form a full-wave rectifier. C1 is the electrolytic filter capacitor. R1 represents the power supply load.

The other full-wave rectifier configuration frequently found in power supplies has two diodes and a *center-tapped secondary winding* on the power transformer, as in Figure 12.3. We shall trace electron current flow similarly to the bridge rectifier, from negative to positive.

Notice in Figure 12.3 that the transformer secondary is tapped halfway between each end. This center tap (CT) is connected to common or chassis ground, which becomes the negative side of the DC supply. When tracing current flow, it may be helpful to think of the secondary as two *separate* windings, one from Point A to the center tap, and another from Point B to the CT.

During the first half of the AC input cycle to transformer T101, Point A will be positive with respect to the CT (and also with respect to Point B). During the second half cycle, Point B will be positive with respect to the CT (and also with respect to Point A).

We first trace electron current flow (I) for the first half cycle (A pos, CT neg):

- I flows from the transformer center tap to ground, and then into the bottom of R1.
- I flows through R1 to the cathodes of D1 and D2.
- I flows against the arrow of D1 to Point A, which is positive with respect to the negative CT.

    (It does *not* flow through D2, because Point B is negative with respect to the CT during this half cycle, and so D2 is reverse biased and does not conduct.)

For the second half cycle (B pos, CT neg):

- I flows from the transformer center tap to ground, and then into the bottom of R1.
- I flows through R1 to the cathodes of D1 and D2.
- I flows against the arrow of D2 to Point B, which is positive with respect to the negative CT.

  (It does *not* flow through D1, because Point A is negative with respect to the CT during this half cycle, and so D1 is reverse biased and does not conduct.)

The pulsating DC output in Figure 12.3 is the raw full-wave DC output of the rectified AC voltage, *without filter action* by electrolytic capacitor C1. With C1 connected into the circuit, filtered DC output becomes a nearly steady positive voltage. C1 charges to the peak voltage of the rectified AC sine wave, and then discharges during the valleys between half cycle peaks, delivering power to the load. In this way the load has a fairly constant voltage across it.

Both bridge and center-tapped transformer full-wave rectifiers are used extensively in electronic power supplies. They can provide higher current DC output with a given AC input than a half-wave rectifier, because *both halves* of the AC input cycle are converted to DC, not just one.

Depending on the particular model VCR, you may come across power supplies with one or both types of full-wave rectification, along with one or more half-wave rectifiers.

## RECTIFIER AND FILTER FAILURES

Before looking at solid state voltage regulators and Power-On control transistors in a typical VCR power supply, this is an appropriate time to discuss possible failure symptoms of rectifiers and filter capacitors. As you learned in Chapter 10, both diodes and electrolytic capacitors can open up or short out. As previously stated, it is rare for a power transformer problem to develop.

The dielectric material in an electrolytic cap can dry out with age, causing the capacitor to decrease in value, rather than completely opening. A power supply filter cap that was originally 4,700μF may have only 1,000 microfarads capacitance several years later.

A shorted rectifier diode or filter cap will cause excessive secondary current to be drawn from the transformer. This will almost always blow a fuse. Some power supply designs have just a single fuse wired in series with the transformer primary, like F1 in Figures 12.2 and 12.3.

Other designs have a primary fuse plus one or more fuses in the low-voltage secondary circuits. Just because you find one fuse that is good doesn't mean there isn't another fuse somewhere that is open. Some power supplies have one or more fusible resistors, resistors with low ohmic values that open with excessive current flow. Values of 0.47Ω to 500Ω are typical. It is always best to consult the schematic wiring diagram for the particular VCR with which you are working before trying to diagnose a power supply problem.

Give the entire power supply a close visual examination if you suspect power problems. A shorted diode or electrolytic cap will frequently show signs of overheating: darkened or burned, bulged, or surface cracked. In some cases the component may have ruptured or even exploded. Look especially around the base of radial-lead electrolytic caps and at the two ends of axial-lead electrolytics for any sign of oozing dielectric material. This normally shows up as a dry white or light gray, powdery material. Replace any cap with these signs.

An open diode in a *half-wave* rectifier will result in a complete absence of DC output voltage. However, in a *full-wave* rectifier, a single open diode will not cause the DC output voltage to drop to zero. Instead, the rectifier becomes a half-wave rectifier. DC output voltage will be less than what it should be. Depending on the circuit, the voltage could be about half, or just a small percentage lower than specification.

An open filter capacitor, or one that has decreased substantially in value, will also cause the DC output voltage to be less than it should be. If there is insufficient capacitance to supply power to the circuit between rectified peaks, the average DC output voltage goes down.

The same capacitor defects also cause power supply *ripple,* which is pulsations on the DC output instead of a pure DC voltage. Again, the cap no longer can fill in between rectified voltage peaks, but allows the voltage to substantially decrease between them. In audio circuits, this causes hum (and often a buzz) to be heard. The hum will be 120Hz with a full-wave power supply and 60Hz with a half-wave supply.

In video, ripple can be seen as "hum bars," one or two wide, dark, horizontal bars on the screen. Inadequate power supply filtering can also cause the picture to roll vertically, because the ripple interferes with the vertical synchronization circuits. Almost any malfunction you can think of can be caused by low power supply output voltage, power supply ripple, or both. In fact, both conditions are often present together, because inadequate filtering causes ripple and allows the overall DC voltage to go down.

Again, one easy way to detect power supply ripple without an oscilloscope is to listen for hum or buzz in the audio. Correcting this may very well cure other problems as well.

There is another reasonably effective method for detecting power supply ripple. Simply connect a fairly large nonelectrolytic capacitor in series with an AC voltmeter probe and check for the presence of AC voltage (ripple) on the DC supply line. Because a capacitor blocks DC but allows AC to pass, the meter will measure only the ripple voltage content.

For example, suppose you measure a +12 VDC power supply output with your multimeter set to read DC voltages. It measures 11.1 volts. This is just a little low, but not alarmingly so. You wonder if there is any ripple on the supply line; there shouldn't be. Perhaps the lower-than-expected DC voltage is due to a filter cap that has decreased in value. To find out, connect a 0.1μF nonpolarized capacitor *in series* with your meter probe and the voltage test point. The cap should be rated at 100WV or more. Measure the voltage again, but this time put your meter on an *AC* scale.

Ideally the meter will read zero volts; we don't want AC riding along on a DC supply line. The series capacitor blocks any DC from reaching the meter, but will couple any AC or ripple signal to the meter circuit. Suppose you measure 1.5 VAC. This says that there is 1.5 volts of AC ripple riding on the 11.1-volt DC line. Not good! The DC supply line has more than 10 percent ripple. In this case, suspect a bad filter cap first. Substituting with a known good capacitor of the same or higher value in microfarads (μF), and the same or slightly higher voltage rating (WV or WVDC), is the best way to prove it. Observe capacitor polarity!

You do, and now the 12-volt output reads 12.2VDC. Checking ripple content with the 0.1μF cap in series, with the multimeter set to *AC,* shows about 0.18 volts of AC ripple. This is more like it, less than 1 percent ripple voltage.

Also note that a leaky or open diode in a full-wave rectifier could cause similar symptoms. An open diode would change the full-wave rectifier to a half-wave rectifier. Only half of each AC input cycle would be converted to DC, and even a perfectly good filter cap would be unable to fill in for the missing half cycles. How to check diodes and capacitors is covered in Chapter 10.

Checking the power supply is often the first thing you should do when troubleshooting electronic equipment. Incorrect power supply voltages and high levels of ripple can cause all sorts of problems. We've already mentioned hum in the audio, horizontal hum bars on the screen, and vertical sync problems. Here are some other indications that may point to the power supply as a problem source:

- Intermittent shutdown for no apparent reason, especially when either the front loader or cam motor is operating.
- Sluggish operation of the front loader or cam motor. Motor may also hum or buzz rather than turn, or may turn slowly.
- Dim or flickering LEDs or front panel fluorescent display.
- Panel display or LEDs dim during cassette load or tape load operations. This could also be caused by a partially shorted loader or tape load motor.
- Erratic timer operation.
- VCR changes channels all by itself.
- Highly intermittent and bizarre problems during multiple machine operations.

Perhaps a VCR doesn't do anything when you plug it in; the time-of-day display even remains blank. Start at the power supply.

Maybe a VCR won't turn On when the Power button is pressed, but the time-of-day *does* display. Again, start at the power supply. First, check all voltages that should be present when the unit is plugged in, but *not* powered On. These must be available to power the system control microprocessor so it, in turn, can send a power control signal back to the supply to turn On the remaining outputs when the Power button is pressed.

If these voltages are all OK, then check the power control line from syscon. Does the voltage on this line change when the power button is pressed? If so, the cause of the VCR failing to turn On is probably in the power supply itself.

If there is *no* indication on the power control line from syscon when the Power button is pressed, *and Power Off voltages are present,* then the problem is most likely elsewhere, not in the power supply. Perhaps the syscon chip or wiring between it and the power supply is defective. Check for correct power supply voltages at the syscon chip. Check continuity of the power control line between syscon and the power supply.

There could be a problem with the timer/operations microprocessor to which front panel buttons and IR remote detector are connected, or maybe there's an open connection between this microcomputer and the system control microcomputer. Perhaps there's a defective front panel Power pushbutton or a broken wire between the IR detector and operations microcomputer. Either of these could prevent the operations microprocessor from signaling power On to the system control microprocessor. You'll definitely need a schematic to check voltages at the pins of these two ICs.

These are some basic diagnostic approaches for possible power supply problems. It is nearly impossible to do much more than replace a blown fuse or a component that is visually defective without a schematic diagram. In some cases, you can't even replace a part that is obviously burned out without a schematic.

For example, perhaps a power supply transistor has shorted, and as a result has "cooked" a resistor connected to its Emitter lead. There are no visible numbers on the transistor, so you don't even know whether it's an NPN or PNP. And the resistor is so charred that all color-coding is long gone. Without a schematic, it's a pretty helpless situation.

In fact, you'll need a schematic even to check power supply voltages. How else will you know what voltage should be present at which power supply connector?

Depending upon how a particular VCR's power supply is packaged, you *might* be able to swap out the main power supply circuit board for a new one. But, among the risks of doing so is the possibility that there is more to the entire power supply circuit than what's contained on the main PS board alone. You might be ordering a $50 part to fix a $2.00 problem that is located elsewhere, such as a shorted voltage regulator mounted to a heatsink separate from the main supply board. You could replace the power supply board only to have the new one not fix the problem, or what's worse, a problem elsewhere might damage the new supply. Without a schematic, it's a guessing game. That's why it's important to learn how to read a schematic and troubleshoot based on information it provides. We'll be doing quite a bit of that from here on out.

We next look at power supply diagrams for one VCR model. It is typical of many VCR power supplies. The unit has a *conventional* style supply, with a step-down transformer connected to the 120-volt AC input. As you'll see later in this chapter, switching mode power supplies do *not* have a transformer connected to the AC line.

This particular model was chosen because of the completeness of the service literature, which includes second-level block diagrams, or simplified schematics, in addition to the actual power supply schematic. Block diagrams and simplified schematic wiring diagrams are very helpful when learning electronic circuit operation.

## Working with a PS Schematic

Some of the rest of the material on VCR power supplies in this chapter may at first seem highly technical, difficult to follow, or even confusing. Don't be overly concerned about this. If you are new to electronics, it would be rather surprising if you *did* understand everything during your first read; you are not really expected to immediately grasp all that's going on.

Read through the discussions that follow more than once, of course. Attempt to get just an overview of the subject, without getting hung up on any details. Try to follow the *general* procedures for troubleshooting a typical VCR power supply. Then read through it a second and even a third time. Later, when you've had some actual hands-on experience with VCRs, you can come back and go through the material again. It will undoubtedly make more sense as you become more familiar with power supply repairs!

For those of you who *are* familiar with electronics, the following discussions and troubleshooting scenarios should help you learn some of the characteristics of VCR power supplies. These supplies are generally much more complicated than those in many pieces of electronic gear, with which you may have experience. They are certainly more elaborate and more difficult to understand and service than power supplies in most home stereos, radios, and even some TVs.

We will look at the same VCR power supply with the manufacturer's service diagrams in three ways. First, supply outputs *without pressing the manual Power button* are described using a simplified block diagram schematic. Then, power supply voltages controlled by the Power switch are depicted on a similar simplified schematic. Finally, most of the actual schematic wiring diagram is presented, showing all relevant components and circuit operating voltages. Possible power supply failures and suggested troubleshooting steps are discussed for each of the three power supply diagrams.

The simplified schematics that follow are for instructional purposes, and to get an overview of power supply operation. Some details are omitted. Do not try to trace actual current flow using this diagram. Complete current paths and return paths may not always be shown.

## Always, Unswitched, and Ever Output Voltages

Figure 12.4 is a simplified schematic diagram of a typical conventional style VCR power supply. Bold lines trace those supply outputs that are active *at all times,* as long as the VCR is plugged into a 120-volt outlet. Depending on the manufacturer, power supply outputs that are On whenever the VCR is plugged in are labeled "UNSW" (unswitched), "ALWAYS," or "EVER."

Before discussing the circuits in Figure 12.4, it is helpful to know what some of the designations and abbreviations mean.

- The dashed rectangle around the power transformer at the left of the diagram means it is a separate component, not part of another assembly.
- Notice the large dashed rectangle around most of the circuitry. This is a printed circuit board, which the manufacturer calls "REGULATOR CIRCUIT."
- CN means a multi-pin connector
  — Locate CN1, CN2, CN3, CN4, and CN91 on the diagram
- IC means integrated circuit
  — Locate IC1 and IC2
  — Notice that there is a dashed box around IC1. This indicates that the IC is mounted on a separate heatsink rather than on the REGULATOR CIRCUIT PCB.
- TR means transistor
  — Locate TR1 through TR11
- REG means regulator or regulated
- UNREG means unregulated
- POWER CONT. (CN2 pin 6) is the power control line from syscon to turn On portions of the power supply
- CAP. DRIV (CN2 pin 8) is capstan motor drive voltage
- CAP. CONT. (CN2 pin 9) is capstan motor control signal
- GND (various pins—CN1, CN2, CN3, CN4) is ground or common. The symbol with a short horizontal line and three slanted legs means *chassis ground.* All DC voltages on the schematic are with respect to ground.

Now that you have your bearings, take a few moments to trace the circuits that provide power supply output voltages when the VCR is plugged in, *without pressing the Power pushbutton:*

- +50V ... at CN1, pin 2
  — derived from half-wave rectifier diode D6
- −30V ... at CN3, pin 4
  — derived from half-wave rectifier diode D5
- 5.2VAC ... between CN3 pins 1 and 2
  — direct output from bottom secondary winding of the power transformer
- +5V ... at CN2, pin 3
  — derived from half-wave rectifier diode D8, then to voltage regulator IC1
  — notice that the output of 5V regulator IC1 also feeds three other portions of the power supply that are not yet turned On: SYS 12V, INVERTER, and REG 12V.
- +5.6V ... at CN3, pin 5
  — derived from half-wave rectifier diode D8, then to voltage regulator IC2.

## Troubleshooting Unswitched Outputs

Now let's suppose we have a VCR that for all intents and purposes is dead. We plug it in, and all we see are some random segments lighted on the fluorescent display, not the usual blinking **12:00** that appears before the time-of-day clock is set. Nothing appears to happen or change when the Power button is pressed.

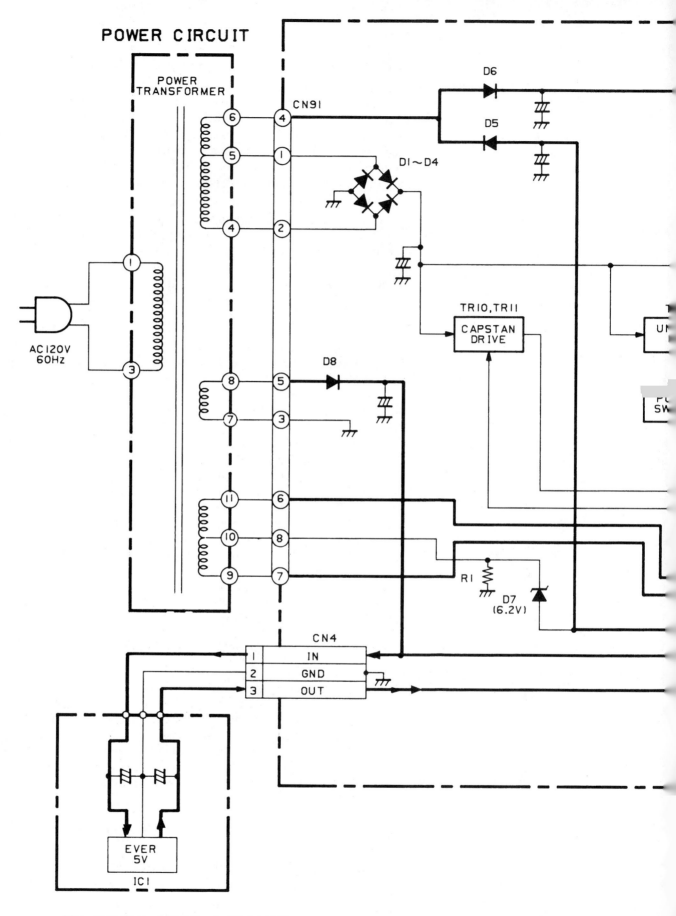

**FIGURE 12.4** Power supply without manually turning On the Power switch. Courtesy of JCPenney Co., Inc.

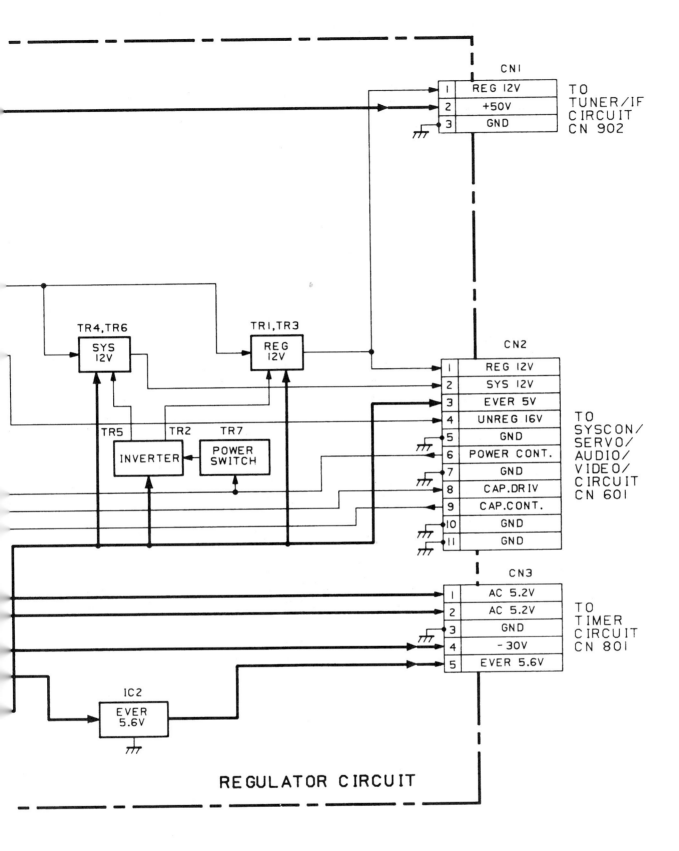

Because at least something (even if it is "garbage") displays, we assume that at least part of the power supply must be working. We decide to check the power supply outputs that *should* be present when the VCR is plugged in. With the simplified schematic wiring diagram of the power supply before us, we check output voltages at power supply Connectors CN1, CN2, and CN3.

Here's *where* we probe and *what* we measure:

- Black lead to CN1-3 (chassis ground/common)
  Red lead to CN1-2 (+50V)
  VOM reading: 58V

  This is probably OK! The +50V output is *not* regulated, and with the VCR not fully powered On we would expect that unregulated outputs might read a little on the high side, since the power transformer and other circuits are not loaded down at all.

- Black lead to CN2-5, 7, 10, or 11 (gnd/com)
  Red lead to CN2-3 (EVER 5V)
  VOM reading: 5.03V
  This certainly seems reasonable.

- Black or Red lead to CN3-1 (AC 5.2V)
  Other lead to CN3-2 (AC 5.2V)
  VOM reading: 5.4V (on AC scale)

  Looks okay! This is an AC voltage, so the multimeter must be switched to its AC voltmeter function. For an AC voltage, it doesn't matter which meter lead probes which of the two connections. Notice that neither meter lead goes to ground. This AC voltage has no direct ground reference, but is *balanced* with respect to ground.

- Red lead to CN3-3 (gnd)
  Black lead to CN3-4 (-30V)
  VOM reading: 33.6VDC

  This looks fine! Just like the +50V line, it is a little high, but this is to be expected for an unregulated voltage, especially when the VCR is not fully powered up.

  Notice that because we were measuring a negative voltage, we placed the *Red* meter lead on ground and probed the voltage pin on connector 3 with the Black meter lead. Of course, if the meter had a polarity reversal switch, we could have flipped this switch to its −DC position and put the black lead on ground. For a DMM it wouldn't matter. With Black on ground, a minus sign (−) would appear before **33.6** on the display.

- Black lead to CN3-3 (gnd)
  Red lead to CN3-5 (EVER 5.6)
  VOM reading: 0.8V

  Oops! Something doesn't seem right here; the voltage *should* be around 5.6 volts. We doublecheck. Yes, the meter *is* set up to measure DC voltage, and we confirm that the meter leads are on the correct pins. We even try another ground pin, with the same results. Looks like we're narrowing down the problem: there is no 5.6V output when and where there should be.

Looking back at the schematic in Figure 12.4, observe that the DC voltage produced by half-wave rectifier diode D8 goes to *two different places:* 5V REGULATOR CIRCUIT IC1 and EVER 5.6V IC2. IC2 is actually another voltage regulator. Because the output of IC1 is all right at CN2-3 (EVER 5V), D8 and its associated transformer winding and filter capacitor must be working correctly. This says that the problem almost *has* to be IC2.

Just to make sure there is no broken PCB wiring pattern or bad connection, perform the following checks:

- With the VCR still plugged in, place black meter lead to a ground point and probe IC2's input lead. The meter reads +11.5VDC. This is all right! The DC input voltage to a solid state regulator must always be higher than its output voltage. A voltage regulator of this type can only reduce voltage, never raise it.
- With the VCR *unplugged,* check for continuity between a ground point and IC2's ground lead. The measurement is 0 ohms. Great! No problem here.
- With the VCR still unplugged, check for continuity between IC2's output lead and CN3-5. The measurement is 0 ohms. Great! No problem here.

With these checks, it sure looks as if IC2 must be open. But before just going ahead and replacing IC2, it would be prudent to make one more check.

Perhaps the output of IC2 is shorted on the TIMER CIRCUIT board, or somewhere else. If this is so, a replacement for IC2 *might* be destroyed right away! To check for this possibility, we want to measure the resistance between IC2's output pin and ground. Here's the setup and reading:

- VCR unplugged!
- Red ohmmeter lead to CN3-5 (5.6V output of IC2)
  Black lead to CN3-3 (gnd)
  VOM reading: 2,500 ohms

    Well, this certainly isn't a short, which would be close to zero ohms. It looks safe to go ahead and replace IC2.

In this power supply, IC2 is a three-lead component that looks like the TO-92 transistor in Figure 10.18, in Chapter 10. It is soldered to the REGULATOR CIRCUIT printed circuit board very near connector CN3.

Replacing IC2 fixes the VCR. Everything works great!

In actuality, we were probably overcautious here, but that never hurts. Once it was determined that there was no +5.6V output at CN3-5, the additional checks were most likely unnecessary. Most solid state voltage regulators are internally protected against a short across their outputs, and there was no particular reason to suspect a broken printed circuit wiring pattern going to IC2. Of course, a hairline PCB crack between IC2's output and CN3-5 would have given the same symptoms.

Suppose you have the same symptom: a dead VCR. You decide to first check the power supply output voltages. This time, when you check the output voltages on CN1, CN2, and CN3, all readings are acceptable *except for CN3-5 (EVER 5.6V), which measures +10.4V*. What might you suspect as the cause of this problem? Don't read the next paragraph until you refer back to the schematic in Figure 12.4 and try to answer this question, okay?

What's your answer? If you suspect that IC2, the 5.6V voltage regulator, is probably shorted, you're doing great! Instead of regulating the DC input voltage from rectifier diode D8, IC2 is acting like a straight wire between its input and output. The excessively high output voltage prevents the timer microcomputer from functioning correctly. In this particular VCR model, as in many, front panel pushbuttons connect to the timer/operations microcomputer, so the Power button doesn't even function.

Replacing IC2 restores the VCR to proper operation.

For a third time, suppose you have the same symptom: a dead VCR. This time, when you check the output voltages on CN1, CN2, and CN3, all readings are acceptable, *except for CN3-4 (−30V), which measures only −3.8V*. What might you suspect as the cause of this problem? Study Figure 12.4 and think it through.

If you suspect that diode D5 might be open, you've got the right idea! Since the +50V output at CN1-2 is all right, you can assume that the upper transformer secondary winding is all right. If D5 was good and the electrolytic filter capacitor (unlabeled in this simplified schematic) connected to D5's anode was open, or had decreased in value, you would probably read less than −30V, but more than −3.8V.

Now it could be that the filter capacitor is shorted. This could cause D5 to open. It might also blow a fuse. In a similar manner, a shorted diode can destroy the associated filter capacitor. Whenever a power supply diode or capacitor is open or shorted, it is best also to check related components before just replacing one.

Those are some examples of how you might troubleshoot the unswitched, always, or ever outputs of a VCR power supply. Next, examine the outputs that become active when the VCR is turned On. These are often called *switched* outputs.

## Switched, or Power-On, Outputs

Refer to Figure 12.5. This is the same simplified schematic of the VCR power supply in Figure 12.4, except circuits that are controlled by the front panel Power button are printed with bold lines.

Switched power supply outputs that become active when the Power button is pressed are:

- REG 12V (CN1-1 and CN2-1): audio and video circuits
- SYS 12V (CN2-2): motor drive circuits
- UNREG 16V (CN2-4): servo circuit.

Basically what happens when the Power button is pressed is that system control sends a power control voltage to the power supply. This signal turns On some switching transistors. And the switching transistors then forward bias other transistors sufficiently for either 12V or 16V to pass from a 20V bridge rectifier to the three power supply outputs. It's a process similar to powering the single transistor DC motor speed control circuit described in Chapter 10. You might want to look back there before we continue!

Here are further details on how the three switched output voltages are developed and controlled.

All three outputs are derived from the full-wave bridge rectifier consisting of diodes D1–D4, the output of which is about 20VDC.

TR9 is the driver, or *pass transistor,* for the 16V output. The terms *driver, driver transistor,* and *pass transistor* are sometimes used synonymously. The term *driver transistor* usually relates to a transistor that directly powers a motor, relay coil, solenoid, light bulb, or LED. *Pass transistor* normally describes the output transistor of a voltage regulator or a transistor output switch in a power supply. Some service literature also calls these *driver transistors.*

The two 12V power supply outputs, REG 12V and SYS 12V, have nearly identical circuitry. TR6 is an amplifier for pass transistor TR4 on the SYS 12V output, and TR3 is an amplifier for pass transistor TR1 on the REG 12V output. Both of these are regulated output voltages. Regulated 5V from the output of IC1 (EVER 5V) is supplied as a stable *reference voltage* to amplifiers TR3 and TR6.

You might be wondering why there are *two* separate, yet nearly identical, 12-volt power supply outputs. The main reason is to provide increased isolation between two different types of circuits. The SYS 12V supplies the VCR's motors, a brake solenoid, and some system control circuitry, whereas REG 12V powers audio and video circuits. By feeding these separately, electrical noise from motors, switching circuits, and digital microcomputer circuitry is much less likely to interfere with the audio and video. There is also tighter voltage regulation with two separate pass transistors.

So three PS outputs are turned On and Off by the system control microcomputer. Notice the POWER CONT. line coming into the power supply REGULATOR CIRCUIT board at CN2-6. This line comes from the Collector of a switching transistor on the SYSCON/SERVO/VIDEO printed wiring board (PWB). The system control microcomputer, in turn, supplies the Base signal to turn this transistor On and Off.

When Power is Off, the POWER CONT. line at CN2-6 is at a 0-volt level. It goes to +5.0 volts when syscon turns VCR power On. Looking at Figure 12.5, trace the power control line to two switching transistors, TR7 and TR8.

With the POWER CONT. line at +5 volts, transistor TR8 forward biases TR9 so that 16V is passed to the PS output. TR7 turns On two secondary switching transistors, TR2 for the REG 12V output and TR5 for the SYS 12V output.

### Troubleshooting Switched PS Outputs

Suppose the VCR doesn't turn On when the Power button is pressed. You've already checked all the unswitched, or always, voltages and they are within specification. What would you do next?

Probably the first thing is to see if the POWER CONT. line at CN2-6 ever goes to +5V when the Power button is pressed. Set up the multimeter to measure 5 volts DC. With the VCR plugged in, proceed as follows:

- Check POWER CONT. line:
  — Black lead to ground
  — Red lead to CN2-6
  — VOM reading: 0 volts
     That is what you would expect to measure at this time.
  — Press and release VCR Power button
  — VOM reading: goes to 4.97V for a second or so, then drops back to 0 volts

This confirms that syscon *is* telling the supply to turn On. If power supply outputs don't come up within a few seconds, syscon drops the power control line back to 0 volts. If this VCR were operating all right, the reading at CN2-6 would remain at nearly 5 volts after the Power button was released.

Next you want to see if the three switched outputs are turning On. Set up the multimeter to measure DC volts.

- Check REG 12V output:
  — Black lead to ground
  — Red lead to CN1-1 or CN2-1 (REG 12V output)
  — Press and release Power button
  — VOM reading: goes from 0 to 12.3 volts, then drops back to 0
     The REG 12V output seems to be all right.
- Check SYS 12V output:
  — Black lead to ground
     Red lead to CN2-2 (SYS 12V output)
  — Press and release Power button
  — VOM reading: goes from 0 to 0.4 volts, then drops back to 0
     Aha! The SYS 12V output is *not* turning On! It is wise not to stop here, but to continue checking the switched DC output voltages. This may give us some help in identifying the cause of the problem.
- Check UNREG 16V output:
  — Black lead to ground
  — Red lead to CN2-4 (UNREG 16V output)
  — Press and release Power button
  — VOM reading: goes from 0 to 18.2 volts, then drops back to 0
     The UNREG 16V output appears all right. Yes, it is a little high, but this *is* an *un*regulated voltage. It would also tend to be somewhat high because there is no current drain from the SYS 12V output, which otherwise would lower the input voltage to TR9 slightly.

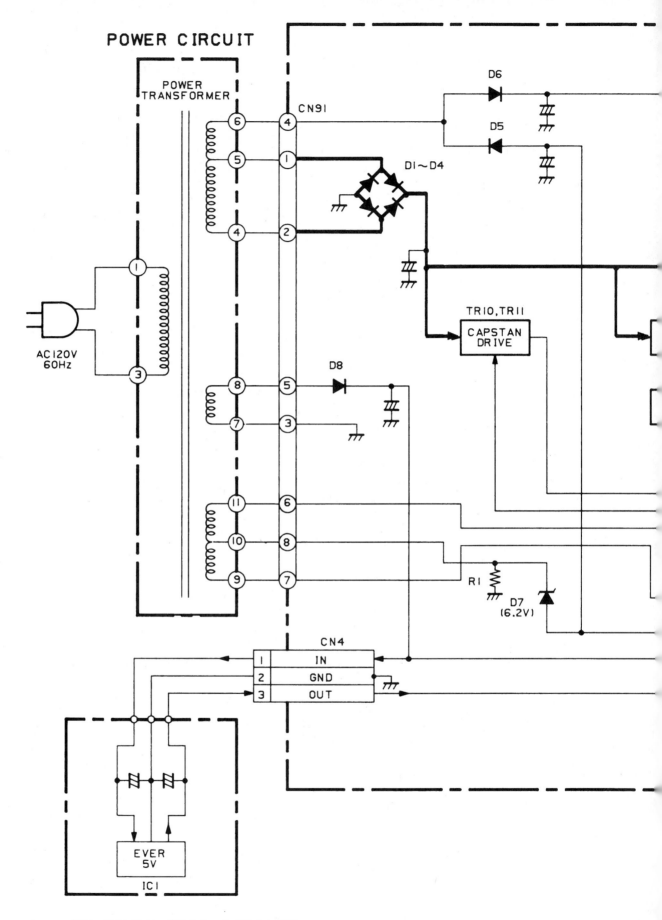

**FIGURE 12.5** Power supply controlled by the Power switch. Courtesy of JC Penney Co., Inc.

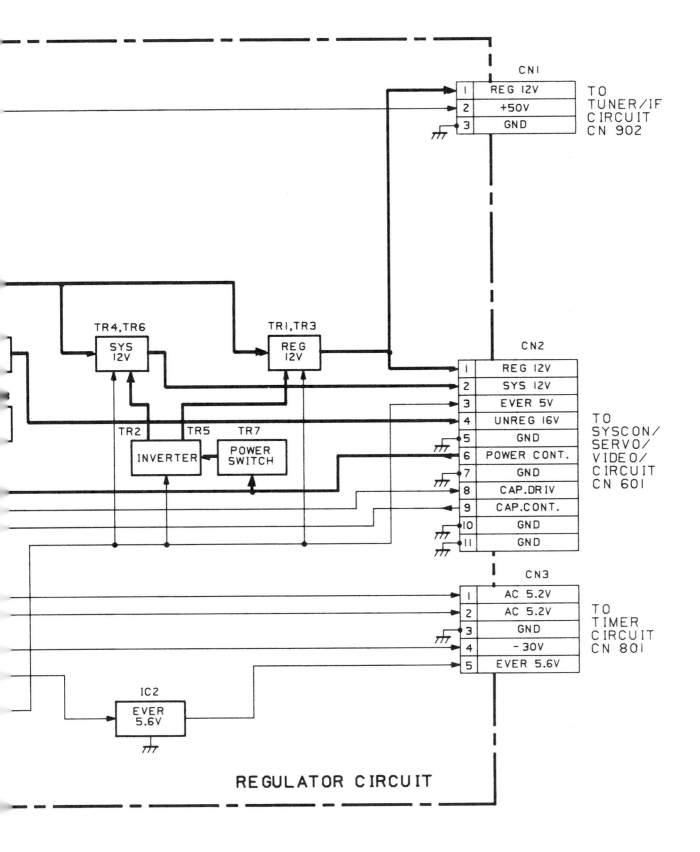

This narrows things down quite a bit. Since the unregulated 16V and REG 12V outputs are all right, this says the bridge rectifier and filter are working properly, as are power switches TR7 and TR8. It would appear that there is something wrong with transistors TR5, TR4, or TR6, or associated circuitry. These components control and regulate the SYS 12V output.

Of course, a full schematic is required to identify the exact problem. Suppose that pass transistor TR4 is open between Base and Emitter. Again, it would be prudent to check that there is not a short to ground at CN2-2 before replacing the transistor.

Suppose you are working on a VCR with the same problem: it won't power up. You check the power supply with a multimeter and get the following readings:

- Unswitched or always outputs:
  - CN1-2 (+50V): +56.4V
  - CN2-3 (EVER 5V): +5.1V
  - CN3-4 (-30V): −32.8V
  - CN3-5 (EVER 5.6V): +5.6V
- Power control input:
  - CN2-6 (POWER CONT.): 0V, goes to +5V when Power button is pressed and released, then returns to 0V after a second or so.
- Switched outputs:
  - CN2-1 (REG 12V): 0.0V
  - CN2-2 (SYS 12V): 0.0V
  - CN2-4 (UNREG 16V): +18.8V

What do you think could be the cause for the VCR not powering On? With these readings you know that:

- All *unswitched* outputs are OK
- Correct turn-On signal is received from system control
- Bridge rectifier (D1–D4) and filter cap probably OK, because the unregulated 16V output seems all right
- *Both* 12V regulators fail to turn On and produce an output.

What would be a likely cause for *both* 12V outputs failing? Refer to the simplified power circuit in Figure 12.5.

The *one* component these two outputs have in common is POWER SWITCH TR7. That would be a prime suspect! It is unlikely that two different components, one in each of the two 12V regulator circuits, would fail simultaneously.

Without knowing the exact wiring scheme, the simplified block diagram schematic *seems to indicate* that TR2 and TR5 work together to control the two 12V regulators. So if you thought one of these components or associated circuitry was failing, you've got the right idea. In reality, TR2 and TR5 work independently: TR2 turns On the REG 12V regulator, consisting of TR1 and TR3; TR5 turns On the SYS 12V regulator, made up of TR4 and TR6.

If you have followed along with little difficulty so far, and have understood the troubleshooting procedures using the simplified block diagram schematics, that's great! On the other hand, if you got totally lost, take time now to go back over those areas that gave you trouble. The "Aha! Light" often turns On the second time through!

When you're comfortable with what's been covered so far, continue. Next we'll look at the complete power supply schematic diagram for this particular VCR, rather than simplified block diagrams.

## CONVENTIONAL POWER SUPPLY SCHEMATIC

Figure 12.6 is the complete power supply schematic, represented by simplified diagrams in Figures 12.4 and 12.5. Take a few moments to compare this diagram with the earlier diagrams. Locate components on this diagram that are depicted in the earlier block diagrams.

Note the following in Figure 12.6:

- The wide prong on the AC power plug (lower left) is the "cold" or *neutral* conductor.
- The rectangle above the power plug, just below the Underwriter's Laboratory (UL) symbol, is a two-prong 120V convenience outlet on the rear apron of the VCR.
- The large, somewhat irregular rectangle above the power plug is the power transformer assembly, which includes a circuit board. Mounted to the circuit board are:
  — F91  main line fuse (1-1/4 amp.); spring clips on board hold tubular glass fuse
  — C91  AC line noise suppression capacitor
  — L91  AC line noise filter choke, composed of two separate 0.8mH coils
- Notice the fuse labeled T.F. inside the transformer itself. This is a high-temperature fuse that opens only if the transformer gets extremely hot. The fuse is embedded within the transformer and is not normally replaceable. It opens to prevent the transformer from reaching a temperature that could cause a fire. As long as main line fuse F91 is the correct value, it should blow well before the transformer gets hot enough for T.F. to open.
- Capacitor C1 across the AC input to the bridge rectifier (D1 ~ D4) is for noise suppression. Many bridge rectifiers do not have this cap. It is not essential to full-wave rectification. Sometimes you will see a small value capacitor, like 0.001µF, in parallel with a rectifier diode. This capacitor helps suppress diode switching noise, produced at the points in an AC input cycle when the diode turns On and turns Off.
- Notice secondary fuse F1 (4 amp.) at the output of the bridge rectifier. If this fuse were open, there would be no REG 12V, SYS 12V, or UNREG 16V outputs. Also, there would be no voltage at the collector of capstan driver transistor TR10.
- Locate the following electrolytic filter capacitors and notice the polarity of each, indicated by a + mark:
  — C2, 10,000µF filter for bridge rectifier
  — C4, 100µF filter for +50V unswitched supply
  — C3, 100µF filter for −30V supply
  — C15, 2,200µF filter for half-wave rectifier diode D8.
- Locate R16 (lower center). This 0.47-ohm, 1-watt resistor is a *fusible resistor*. It opens if too much current flows through capstan motor driver transistor TR10.
- Trace the POWER CONT. line from CN2-6 to the base of each switching transistor, TR7 and TR8.
- Trace the positive output of the bridge rectifier at the cathode junction of D3 and D4 to:
  — Collector of TR1, NPN driver transistor for REG 12V output
  — Collector of TR4, NPN driver transistor for SYS 12V output
  — Emitter of TR9, PNP driver transistor for UNREG 16V output
  — Collector of TR10, NPN capstan motor driver transistor
- Notice the dashed boxes enclosing transistors TR1, TR4, and TR10. This indicates they are mounted on an aluminum heatsink on the power supply board.

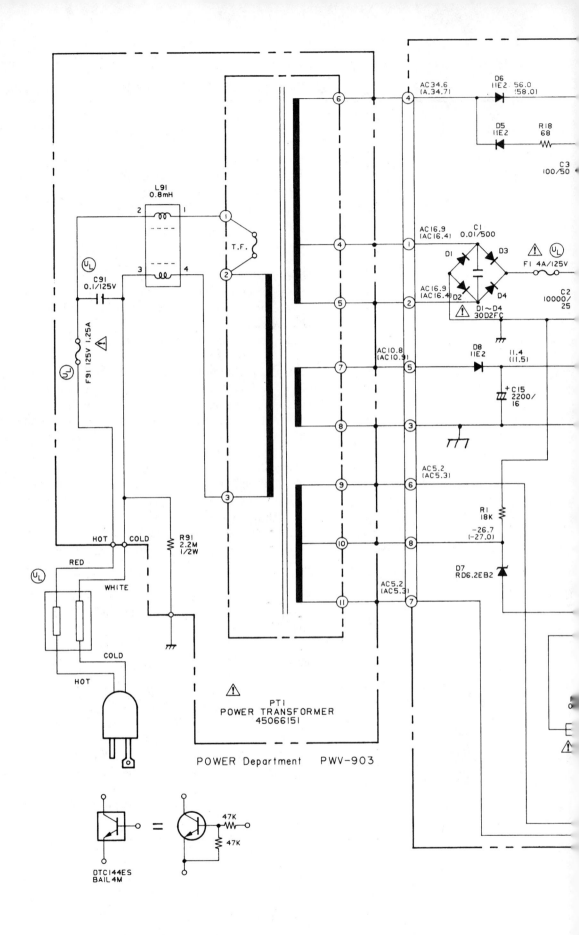

**FIGURE 12.6** Conventional power supply schematic. Courtesy of JC Penney Co., Inc.

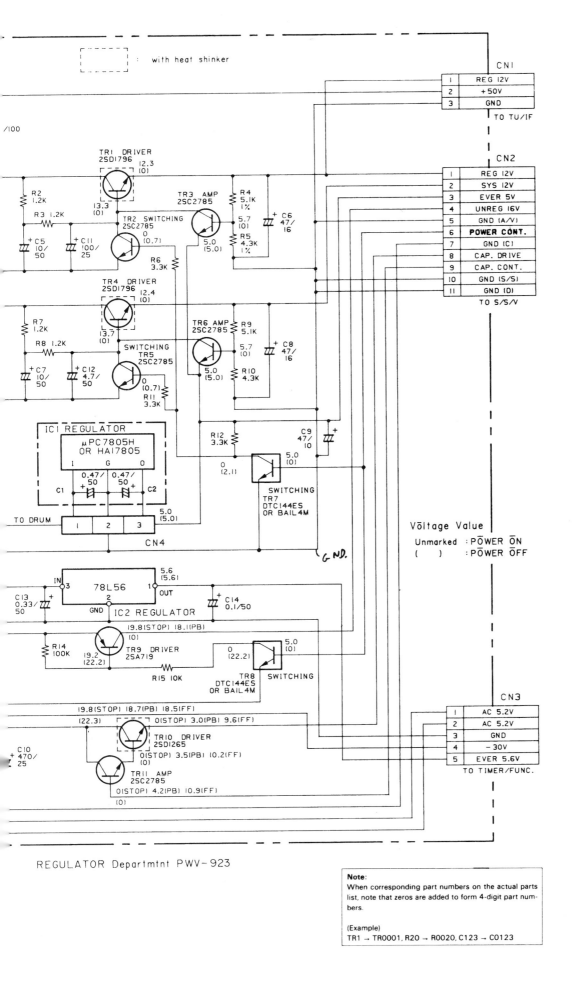

REGULATOR Department PWV-923

**Note:**
When corresponding part numbers on the actual parts list, note that zeros are added to form 4-digit part numbers.

(Example)
TR1 → TR0001, R20 → R0020, C123 → C0123

- Locate IC1, the EVER 5V regulator.
  — On this particular VCR, this component, along with two small 0.47µF electrolytic caps, is mounted on a separate metal heatsink bracket on the back side of the lower, stationary part of the video drum. A three-conductor cable connects the IC to the power supply board at CN4.
  — I, G, and O on the IC symbol identify the **I**nput, **G**round, and **O**utput leads, respectively.
- Trace the output of IC1 to:
  — CN2-3, PS output connector
  — Emitter of TR6, SYS 12V regulator amplifier transistor
  — Emitter of TR3, REG 12V regulator amplifier transistor.
- Locate IC2, the EVER 5.6V regulator.
- Notice that in many places on the schematic there are two sets of numbers, one above the other, with the lower number enclosed in parentheses. These are actual voltage readings taken at various points on a working VCR. The upper value is with the VCR powered On, the lower number in parentheses is the voltage with power Off.
- What is the input voltage to IC1 and IC2?
  — Tracing the input lines back to the left and up, you should arrive at the cathode of D8. Right above that line are two voltage measurements, 11.4 {VCR On} and (11.5) {VCR Off}.
  — There is very little difference between power On and Off voltages, because IC1 and IC2 provide *un*switched regulated voltages.
- What is the voltage at the base of TR6, the SYS 12V regulator amplifier, when:
  — power is On?
  — power is Off?
      Locate the voltage measurements just to the right of TR6's base lead. The base voltage should be about +5.7 volts when the VCR is On, and 0 volts when it is Off.
- Compare the REG 12V circuit, consisting of transistors TR1, TR2, and TR3, with the SYS 12V circuit, comprised of TR4, TR5, and TR6. Except for the very minor difference in the values of electrolytic caps C11 and C12, the two circuits are identical. Both are turned On by switching transistor TR7.
- Notice the voltage readings on the leads of TR10 and TR11, capstan motor amplifier and driver transistors, respectively. Separate measurements are given with the VCR in Stop, Playback (PB), and Fast Forward (FF) modes.

When an electronic circuit fails, you can often locate the failing component by comparing actual voltage readings to those printed on the schematic diagram.

### PS Troubleshooting with Schematic Diagram

Most manufacturers' schematics have voltage levels printed at various circuit points. These are helpful in diagnosing circuit malfunctions. If you take a voltage measurement that is over 20% different, you can be reasonably sure there is a problem affecting that point in the circuit. Regulated voltages should normally be within 5 percent of the values printed on the schematic.

For example, suppose the VCR won't power On. You first decide to check the POWER CONT. line at CN2-6, but you have no idea what voltage *should* be there. You trace the line down to the Base of TR7 or TR8. Here you see that the line should be 0 volts with the VCR Off, and 5 volts with it On.

You check at CN2-6 and sure enough, when you press and release the Power switch the meter swings from 0 to 5 volts. A second or so later it drops back to zero, because syscon shuts down the line when power doesn't successfully come up.

Next you elect to check the voltage at the Collector of TR7:

- Transistor TR7 collector voltage measurements:
  — Black meter lead to ground
  — Red lead to TR7 collector
  — VOM reading:   4.8V
  — Press Power On button
  — VOM reading:   4.8V

"Something's not right here!" you think to yourself. And you're correct. TR7's collector should go to 0 volts as the POWER CONT. line goes to 5V, when you press the Power button. Also, the Power Off collector voltage is more than twice the 2.1-volt level specified on the schematic.

It appears as though TR7 is not working. Before going further, though, it would be prudent to check that the Emitter of TR7 *is* at ground potential. An open wiring pattern to the Emitter would give the same indications! In this particular instance, it might be difficult to check TR7 with the basic ohmmeter test, described in Chapter 10.

Why? Look at the very lower left of the schematic. Here we see that TR7 is actually a transistor *plus two 47KΩ resistors,* in the same package. You can't get directly to the base of TR7 with an ohmmeter lead, because there's a 47KΩ resistor between the actual transistor base and the base lead on the component.

If the B-E junction was *shorted,* you would *not* measure 0 ohms. Instead, you would read 47K ohms, the horizontal base lead resistor. If the B-E junction was *open,* you would read the values of both resistors in series, or about 94K ohms. In other words, the basic High/Low tests would not apply. A forward/backward ohmmeter check of TR7's B-E junction might not be conclusive.

But, according to the voltage readings at TR7's collector, it sure looks like the transistor is never turning On, even though there's the proper signal at its Base. You decide to replace TR7, and now the VCR powers On as it should.

Although there are many differences in conventional VCR power supplies, they all work similarly. Some power supplies in older VCRs have one or more relays that switch voltages On and Off, rather than accomplishing this with transistor switches. Armed with a schematic diagram that has printed voltage measurements, you can probably repair many defective conventional power supplies with what you've learned so far.

Next we'll look at a power supply that is significantly different from what you've just studied.

## SWITCHING MODE POWER SUPPLY

Up until now, this chapter has dealt with transformer-operated, or what are sometimes called conventional type or *linear* power supplies. A very different type of power supply is coming into increasing use for electronic equipment, including VCRs: the *switching mode power supply* (SMPS). One of the first items to employ a SMPS on a wide scale was the personal computer, back in the early 1980s. PCs today almost universally have switching mode power supplies.

A switching mode power supply is sometimes called a *transistor-switching supply* or just a *switching power supply.* Don't confuse this with a power supply's *switched* outputs, which are DC outputs that turn On when the Power pushbutton is pressed, in both conventional *and* switching mode power supplies.

One major advantage that a switching supply has over a conventional power supply with a step-down transformer is that it can be built much smaller and substantially lighter, and still produce the same amount of output power.

Parts cost is also less. All of these factors contribute to more compact, lighter, and less-expensive VCRs. Switching power supplies are also more efficient than conventional supplies, where heat produced in the power transformer wastes energy.

Decreased size, weight, and cost, and better efficiency are mainly attributable to the lack of a heavy, iron step-down transformer and the use of much smaller electrolytic filter capacitors. Capacitance values in microfarads are smaller, which means the caps are smaller physically, too. You'll understand more about this as switching supplies are described.

The following is a *generalized description* of switching mode power supplies. We are not going to explain their operation in as much detail as we did conventional, transformer-operated supplies earlier in this chapter. Just as with power supplies having a 120V step-down transformer, there are many different designs of switching power supplies. Refer to the block diagram of a typical switching power supply in Figure 12.7 while reading the following basic description of circuit operation.

In a switching supply, the 120-volt, 60Hz AC line goes *directly* to a full-wave bridge rectifier and filter capacitor. The DC output voltage after rectification and filtering is higher than 120, usually 170 volts or even more. Recall that a filter capacitor charges to the peak value of an AC rectified wave, which is higher than the RMS value (refer to discussion of AC and DC in Chapter 8, especially Figure 8.15).

This DC voltage powers an oscillator, which produces high-frequency pulses. These high-frequency pulses, typically in the range of 40kHz to as high as 100kHz, are input to a power transistor. The power transistor then turns the flow of DC current through the primary of a transformer On and Off at the high-frequency rate.

You can think of the pulses as a high-frequency AC signal applied to the primary winding of the transformer. Instead of 60Hz power going into the transformer primary, as with a conventional transformer-operated supply, it is *many thousands* of hertz.

A separate feedback winding on the transformer in some designs is part of the oscillator circuit.

On the secondary side of the transformer, things work pretty much as they do in a more conventional supply. Secondary windings produce several low AC voltages, which are rectified and filtered to produce several DC voltages. Some DC voltages go to voltage regulators, whereas others go unregulated. Some DC voltages are always active, or unswitched (UNSW); others are switched On and Off by system control (SW).

Most switching power supplies incorporate a DC *feedback* circuit, or electronic servo, from one of the secondary DC outputs back to the oscillator circuit. This feedback regulates the frequency or the pulse width of the oscillator, or both. All this results in overall output voltage regulation plus reduction of wasted power when the supply doesn't need to put out as much power, as when the VCR is turned On but the capstan and drum motors *aren't* energized.

For example, if the monitored secondary voltage decreases, like when a load motor runs, the oscillator might increase in frequency or produce wider pulses. This increases current flow through the transformer primary and therefore brings the secondary voltage back up. Likewise, with the VCR powered down, an unswitched output would tend to increase in voltage due to the reduced load on the supply; the feedback reduces the width of pulses going to the transformer primary, reducing primary current flow, and therefore regulating secondary transformer voltages.

It is assumed in the design that if one secondary voltage goes down, the rest do also. Changing supply loads, such as when the drum and capstan motors are On or Off, thus controls the oscillator, keeping secondary voltages more constant.

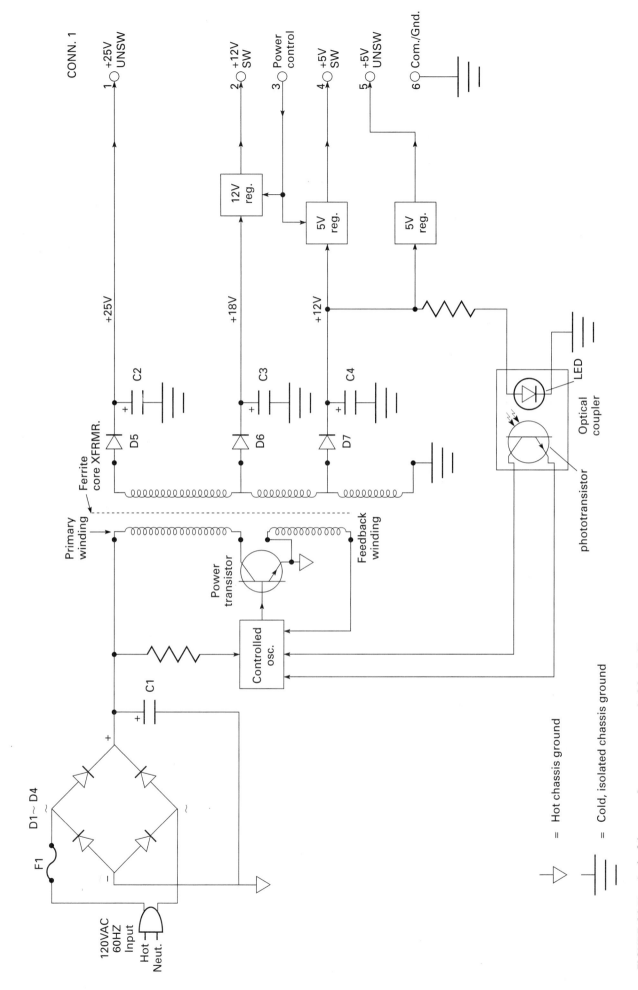

**FIGURE 12.7** Switching mode power supply block diagram.

371

To keep the secondary and primary voltages totally isolated from each other, an optical coupler, or *optoisolator,* is generally used to send this feedback information from the monitored secondary voltage back to the oscillator. An opto coupler, or photo coupler, consists of an LED and phototransistor packaged together in a single module.

Now let's return to the question of why switching supplies are so much smaller, lighter, and more efficient than their transformer-operated relatives. The chief reason is the big difference in the frequency applied to the transformer primary winding. At high frequencies, a transformer can be about *one-tenth the size* of one operating at 60 hertz, where both transformers supply the same amount of secondary power. Also, at high frequencies, lightweight ferrite material can be used for the transformer core, instead of the heavy iron laminations used in a 60Hz transformer.

It doesn't stop there. Without getting technical, a transformer operating at high frequencies wastes much less power in the core material than one working at 60Hz. Less heat is produced by the transformer itself.

Furthermore, when secondary voltages are rectified, instead of having just 60 (1/2-wave) or 120 (full-wave) rectified AC peaks each second, there are now *thousands,* depending on the frequency applied to the transformer primary. This means that a filter capacitor can charge to full peak voltage hundreds to thousands of times more often in a second. It also has much smaller intervals in which it has to fill in between the dips or valleys of the rectified AC to provide pure DC to the load.

What this all boils down to is that a much smaller capacitor is needed for adequate power supply ripple filtering at high frequencies than with 60Hz input to the transformer. For the same reasons, a half-wave rectifier can sometimes suffice where a full-wave rectifier would be required at 60Hz.

## SWITCHING MODE POWER SUPPLY SAFETY

Be aware that there's a downside to the switching power supply marvel: *they can be very dangerous to work on!* Notice in Figure 12.7 that the hot side of the AC line essentially goes to all power supply components on the primary side of the transformer. Now if you were to touch anything in this circuit and ground at the same time, such as a grounded piece of test equipment, there would be a path for electricity to flow through your body back to the neutral side of the 120-volt AC line. Neutral and ground are the same back at the service entrance panel. You *could* receive a severe or lethal shock.

Equipment of this type, where one side of the AC line connects to a circuit ground, is referred to as having a *"hot"* chassis.

Never connect any test equipment probes to anything in the primary circuit of a switching power supply with the unit plugged into a regular AC outlet! Don't even think of working on a switching power supply when the equipment is plugged into a standard AC outlet. Here's why. In a more conventional supply, a *very limited portion* is connected to the hot AC line: just the mainline fuse and transformer primary connections, and perhaps a convenience outlet and filter choke. But in a switching supply, a great deal more is hot.

For service work, a 120V *isolation transformer* is a very necessary accessory when working on a switching mode power supply. An isolation transformer has a 120-volt primary winding that plugs into a regular AC outlet and a 120-volt secondary winding that powers a separate AC outlet, into which you plug equipment being serviced. There is no direct electrical connection between the primary and secondary windings. By plugging a VCR into the secondary outlet, the 120-volt hot lead is no longer present in the power supply.

That is, neither side of the AC now powering the VCR is 120 volts above ground; there is no ground reference.

Isolation transformers for powering 120V, 60Hz equipment are available from electronic supply houses and through mail-order catalogs. The power rating of an isolation transformer is usually rated in volt-ampere (VA). You'll want one rated at 100VA or larger for powering VCRs. Many 120V isolation transformers have an AC power cord and output outlet included.

> CAUTION: *Do not* service a switching-type power supply without powering it from an isolation transformer!

In a pinch, you can construct a "poor person's" isolation transformer using two *heavy-duty,* low-voltage transformers, each having the same secondary voltage (for example, 12.6 volts). Connect the secondary winding of one transformer to the secondary of the other. Plug one transformer into an AC outlet and power the VCR from the 120V winding of the second transformer.

The current capacity of this back-to-back transformer lash up must be high enough to power a VCR. Measure the AC voltage going into the VCR *with it powered On* and playing a tape. If it is more than 10 percent lower than the AC line voltage at the standard 120-volt wall outlet, the transformers are too small or inefficient for this application.

Do not confuse an isolation transformer with a variable AC transformer, often called by the brand name Vari-AC (pronounced VARY-ack). Many transformers of this type *do not* provide power line isolation!

## SWITCHING MODE POWER SUPPLY SCHEMATIC

Figure 12.8 is the schematic wiring diagram for a switching mode VCR power supply, included in this chapter just so you can get an idea of what one looks like. We will not be providing any detailed circuit description, but will point out some items of interest.

Following are some items of interest in the switching supply schematic:

- Notice that the schematic symbol for the full-wave bridge rectifier is a single diode inside the diamond shape normally associated with a bridge. This symbol sometimes appears on schematics rather than four diodes, especially where a single bridge rectifier module is used instead of four discrete diodes, as is the case here.
- C105 is the input filter capacitor for the full-wave rectifier.
- The oscillator and power transistor functions of the previous block diagram are contained in integrated circuit IC101.
- An exclamation mark within a triangle indicates that these components are critical for safety. Only the manufacturer's part numbers should be used for replacement.
- There are several lines on the schematic marked **B+** and a couple that are labeled **B−**. This is a throwback to the very early days of radio when the "A battery" powered tube filaments and the "B battery" supplied positive plate voltage to the tubes. Today, B+ simply means any positive DC supply voltage, and B- indicates a negative voltage.
- Note the opto coupler, IC201, in the lower left of the schematic. This feeds back information about secondary output voltage to the oscillator circuit, while keeping the primary and secondary circuits electrically isolated from each other.

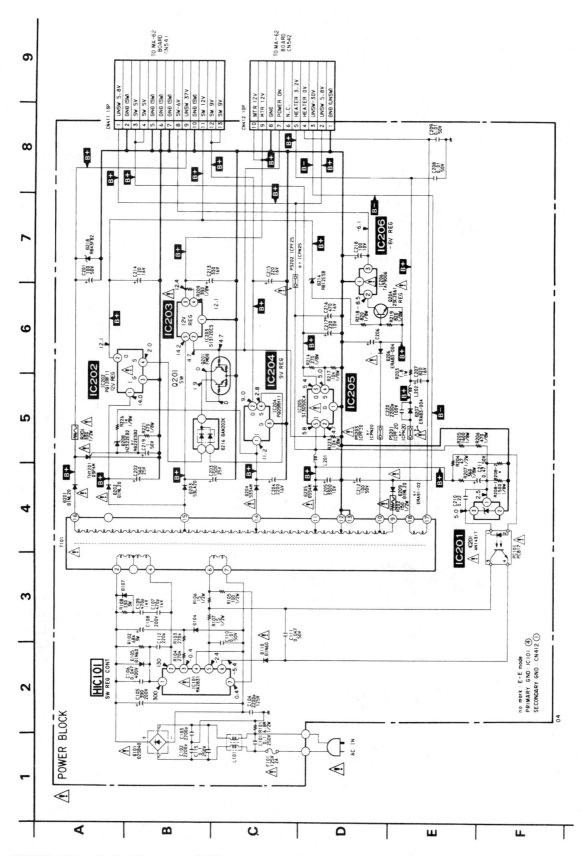

**FIGURE 12.8** Switching mode VCR power supply. Courtesy of Sony Electronics, Inc.

- Read the notes in the lower left of the schematic. Yes, there are *two* ground reference points on this power supply. It is important to use the correct ground reference when measuring voltages within a switching mode supply.
    a. Ground reference for the primary circuit—basically everything on the left side of transformer T101—is at IC101, Pin 4. This is also the negative output of bridge D101, which rectifies the 120-volt AC line.
    b. Ground reference for the secondary circuits is connector CN412, Pin 1. The heavy black lines on the right side of T101 are ground for all switched and unswitched DC outputs.
- Locate the 330Ω, 1/2W fusible resistor, R201, at the top center on the unswitched 37V line.
- Locate fusible resistor R202.
- Find the following components and trace the output voltage line from each to power supply connector CN411 or CN412.
    — Switched 12V regulator IC202
    — Motor 12V regulator IC203
    — Switched 9V regulator IC204
    — Switched 5V regulator IC205
    — Switched −6V regulator IC206.
- Notice the POWER ON line at connector CN412, Pin 7. This is the power control signal from syscon that turns On the switched power supply voltages.
    — Trace the POWER ON line to switching transistor Q201.
- Besides 2-amp. mainline fuse F101 and fusible resistors R201 and R202, this particular power supply also has three *fusible links* soldered to the printed circuit board. See if you can locate them:
    — PS201 (ICPF20 or ICPN20)
    — PS202 (ICPF25 or ICPN25)
    — PS203 (ICPF20 or ICPN20)
    A fusible link functions as a fuse, opening when excessive current flows.

As you can see, the secondary rectifiers, turn-On circuit, and voltage regulators are similar to those in the conventional power supply, described earlier. With a schematic diagram you can check both the *switched* and *unswitched* outputs of a switching power supply.

In general, switching mode power supplies are more difficult to troubleshoot than transformer-operated supplies. This is especially true if none of the outputs are On, usually caused because the oscillator circuit is dead. Unless the oscillator is working properly, there will be no transformer secondary voltages. The transformer in a switching power supply is much more likely to fail than the power transformer in a conventional power supply, where the transformer primary is connected to the 120V, 60Hz power from the line cord.

A failure of almost any component in the primary circuit can prevent oscillation. Sometimes the failure of one component in the primary circuit destroys others, especially if a part develops a short. It is best to check all parts as thoroughly as possible, rather than just replacing the first component that you find defective.

Once more, do not even consider troubleshooting a switching power supply with the VCR plugged into a regular AC outlet. *Always use a 120-volt isolation transformer when working on the supply while it is powered.*

Oftentimes, when it comes to a switching type power supply, it is more economical to replace the entire supply, rather than spending a lot of time (and frustration) diagnosing and repairing it. It's not a bad idea to check on availability and cost of the replacement power supply before digging in too deeply.

## SUMMARY

Of the relatively small percentage of VCR problems that are electronic rather than mechanical in nature, the power supply is probably responsible for the most frequent failures. Higher currents and operating temperatures compared with other circuits and the susceptibility to power line spikes and surges accounts for this.

This chapter described two basic types of low voltage supplies used in VCRs. One type has a step-down transformer whose primary winding is connected to the 120-volt AC line. This is a conventional style supply, sometimes called a linear power supply. However, 120V-powered electronic devices increasingly are employing switching mode power supplies. In this type of supply, the 120-volt AC line does *not* go to a transformer primary, but is directly rectified to DC. The DC then runs a high-frequency oscillator that powers the transformer primary. Instead of 60Hz AC power, primary transformer input is DC pulses in the 40kHz to 100kHz range. This high frequency enables a much smaller and lighter step-down transformer to be used in the power supply. Also, smaller value filter caps (in microfarads) can be employed than in a linear or conventional supply.

For both types of supplies, transformer secondary windings produce several low AC voltages. These are rectified and filtered to provide various DC outputs. Some power supply outputs are unswitched, meaning they are On whenever the VCR is plugged in. Other outputs are switched, turning On when the Power button is pressed. A power control line from the system control microprocessor turns the power supply switched outputs On and Off.

For many VCR failures, it is wise to first check power supply output voltages. In some designs, a missing voltage causes system control to shut down the supply. A DC voltage that is lower than it should be, or one with high ripple content, can cause intermittent, hard-to-diagnose problems. PS ripple can be measured by connecting a 0.1µF capacitor in series with an AC voltmeter when probing PS DC outputs.

A schematic wiring diagram is needed to troubleshoot and repair most power supply failures. Some manufacturers' service literature has second-level or simplified block diagram schematics, which make understanding the complete schematic easier.

Servicing a switching power supply while it is plugged into a regular AC outlet is *extremely dangerous*, because circuitry to develop the high frequency pulses that drive the primary winding of the power transformer is not isolated from the AC line. Working in this area with grounded test equipment, or while contacting ground, could cause a severe electrical shock. A 120VAC isolation transformer should always supply 120V power to a VCR when working on a switching power supply.

## SELF-CHECK QUESTIONS

1. In a single phrase, describe the function of each of the following power supply components.
   a. Power transformer:
   b. Rectifier diode:
   c. Filter capacitor:
   d. Voltage regulator:
2. Refer to Figure 12.1. Explain where the signal at point F comes from and what it accomplishes.

3. Complete each of the following diagrams so the circuit has full-wave rectification and one electrolytic capacitor to apply filtered direct current to load resistors R1 and R2. Apply the positive DC voltage to the top of each resistor, with respect to the bottom. Mark the polarity of the capacitor.

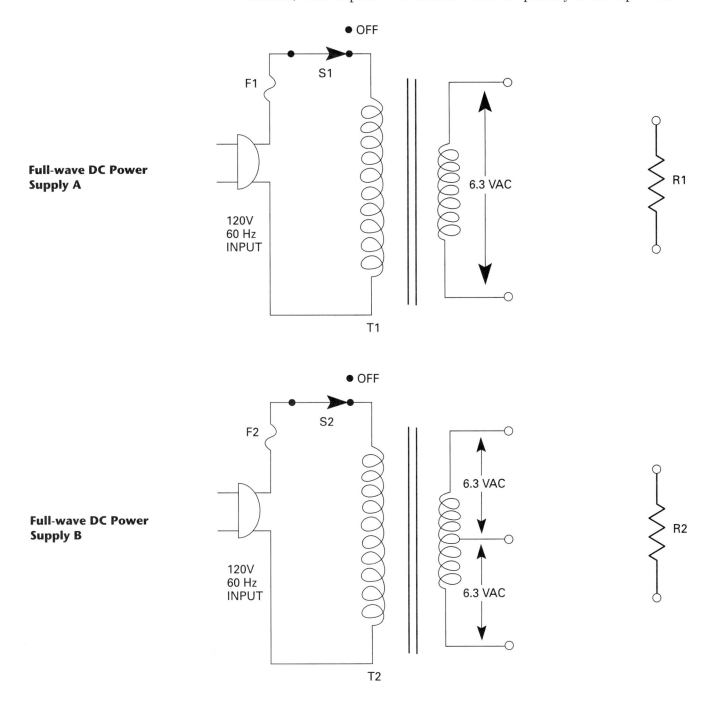

**Full-wave DC Power Supply A**

**Full-wave DC Power Supply B**

4. In most VCR power supplies, what would be the result if a diode in a full-wave rectifier shorted out? What might be the result if a diode in a full-wave rectifier opened up?
5. List one or two VCR symptoms that might lead you to suspect a power supply electrolytic capacitor has dried out, resulting in reduced filter capacitance? How could you check for power supply ripple on a +5V DC line using a multimeter?

6. What is the best way to see if a dried out power supply electrolytic filter capacitor is causing a particular VCR problem, without using a meter?
7. Refer to Figure 12.6. What power supply component(s) would you suspect might be defective if you measured 0V at CN4–1, with respect to ground? (CN4 is a three-pin connector near the center of the schematic.) Approximately what voltage would you expect to read at CN4–1 with the power supply operating normally?
8. Refer to Figure 12.6. What power supply component(s) would you suspect might be defective if you measured +10.5V at CN3–5 and +5.0V at CN2–3, with respect to chassis ground? Approximately what voltage would you expect to read at CN3–5?
9. Refer to Figure 12.6. The VCR is powered On. What power supply component(s) would you suspect might be defective if you measured about 0V at CN2–4, with respect to GND? All other voltages at CN1, CN2, and CN3 are correct. What further checks would you make to determine the cause of the problem?
10. How can you determine that a VCR has a switching-type rather than a conventional power supply by just looking at the supply, without using a multimeter or referring to a schematic diagram or service literature?

# CHAPTER 13 CHECKING MOTORS, OPTICAL SENSORS, AND REMOTES

VCRs rely on numerous motors and several electronic sensors. Motors load a cassette, load tape, move tape through the transport, and spin the rotary tape heads. Sensors check several areas of the VCR and report their findings back to the system control microcomputer.

This chapter describes how to diagnose common electrical problems in simple two-terminal DC motors, such as the cassette load and cam motors. You'll also learn how optical sensors and infrared remote control units work and how to service them. Details about what motors do in a VCR and how sensors are employed to monitor mechanical operations are covered earlier in this textbook, especially in Chapters 3 and 6.

In most VCRs the capstan and drum motors are actually multi-phase AC motors, so they cannot be thoroughly tested in the manner described here for the simpler two-terminal DC motors. However, this chapter provides some basic electrical tests that *can* be made on these motors in some cases.

## CHAPTER OBJECTIVES

**Upon completing this chapter, you should be able to:**

1. Describe how to operate a typical two-terminal VCR DC motor and associated mechanics using a 9-volt battery.
2. Describe how to determine whether a two-terminal DC motor or its drive circuitry is defective.
3. Describe how to check the operation of infrared (IR) optical sensor components.
4. Describe how to disassemble and make common repairs to a handheld remote control unit.
5. Describe how to check the operation of a handheld IR remote control unit and the companion IR photodetector receiver in the VCR.

### INDEPENDENTLY POWERING A TWO-TERMINAL DC MOTOR

When troubleshooting or performing maintenance on a VCR, it can be helpful to operate one of the DC control motors independently from the VCR itself. Following is an example of why you might want to do this and how to go about it.

Suppose you discover a broken gear in a front loader assembly, probably caused by someone forcing a cassette into the machine while it was powered Off or even unplugged. (Many newer VCRs automatically power On when a cassette is inserted, to help prevent user abuse.)

You have removed the loader assembly from the machine and replaced the broken gear. Although you have hand cycled the mechanism back and forth through its cassette load and unload cycles, you want to make sure the assembly works mechanically several times. You also want to be sure it works under power before completely reinstalling the assembly. Incidentally, hand cycling can become terribly tedious! This loader, like many, has a two-terminal DC drive motor on the side of the assembly. The motor may be called the *cassette motor, cassette load motor,* or *front loader motor.*

Rather than reconnecting the loader back in the VCR, operate the motor from another power source. You'll gain several benefits by doing so. Because of the short multi-conductor VCR cable that connects the loader on some VCRs, you can't easily see all areas of the loader when it's back in place in the machine. If you lay the loader on top of a circuit board so you *can* see, you risk accidentally damaging a part or causing a short. (If you do this, place a piece of cardboard beneath the loader to insulate and protect the parts under it.)

Similarly, there may be times when you'd like to operate the tape load motor without powering On the VCR to do so. This motor is also called the *cam* or *mode motor.* Perhaps you have replaced a broken cam follower arm or some other undercarriage component and have also had to remove the mode switch to install the new part. Now you want to run the mechanism back and forth to make sure everything works before replacing and timing the mode switch. But without the mode switch installed and timed, system control can't really operate the cam motor; it hasn't a clue as to where the mechanism is positioned without the mode switch connected. You need some way to run the motor without relying on the VCR itself to do so.

A fresh 9-volt "transistor-radio" battery will operate most small DC motors quite well, even if it is a 12-volt or a 6-volt motor. A 12-volt motor will turn just a bit slower, but in most cases you won't even notice. Because voltage is applied for such a brief period of time, usually only a few seconds at most, a motor rated at less than 9 volts won't be damaged. Twelve-volt DC motors seem to be the norm for VCR front loaders and cam mechanisms.

Here's how to run the DC motor:

First, *unplug the VCR!*

Although the next step may not be necessary in all cases, it is a precaution to prevent damaging a VCR motor drive integrated circuit (IC) or transistor. This is very unlikely, but it doesn't hurt to take the precaution.

Disconnect the motor from the VCR. This simply requires unplugging a connector. Trace the wires from the two terminals on the motor to the connector. Some DC motors are mounted on a small PC board with a connector right adjacent to the motor.

Next, connect the two terminals on the DC motor to the two terminals of a fresh 9V battery with two alligator clip test leads. Terminals are usually at the rear or side of the motor, and at the shaft end for some motors mounted to a PC board. Should the motor have more than two terminals, do not use this procedure. Note that a motor may have a ground terminal, clearly connected to the metal motor frame. *Do not count this as a third terminal.* Figure 13.1 illustrates powering a cam motor from a 9V battery.

Depending on how the motor terminals are connected to the positive and negative battery terminals, the motor should rotate either clockwise or counterclockwise. If the motor turns in the wrong direction, reverse the clip leads at the battery terminals.

Be cautious not to run the motor past its limits in either direction, which could *possibly* break a part; stand by to disconnect one of the test leads.

**FIGURE 13.1** Powering a cam motor from a 9V battery.

Remember, *you* are taking the place of system control; *you* must stop the motor at the end of travel for the mechanism. Fortunately, part breakage is rather unlikely, as most mechanisms have some sort of slip clutch arrangement that prevents overtravel damage. Also, the 9V power source limits the torque capabilities of the motor, which will likely stall before a part is damaged. But it pays to take precautions.

Simply reverse the battery connections to move the mechanism in the opposite direction. If you constructed the two-way tester described in Chapter 11, it is easier to start and stop the motor and change its direction with the On/Off and polarity reversal switches on this unit.

## TESTING TWO-TERMINAL DC MOTORS

This same procedure is also ideal for *testing* a small two-terminal DC motor. Suppose you are working on a VCR where the front loader or cam mechanism just won't operate at all. To diagnose the problem, a good approach is first to see if there are any binds in the mechanism caused by a broken or damaged part. Do this by hand cycling.

If hand cycling shows the mechanism to be in working order (that is, you can easily turn the motor shaft), the next logical step is to determine if the motor itself is working. In this case you absolutely want to disconnect the VCR cable going to the motor. A shorted motor driver IC or transistor could give you a false indication, otherwise: By disconnecting the motor cable, you will be testing just the motor, without driver electronics, which could be shorted, leading you to believe the motor is defective.

Simply connect the suspect motor to the 9V battery, as described earlier, and see if it works.

A common failure of small, inexpensive DC motors is that they develop a dead spot somewhere within their 360 degrees of rotation. This is usually caused by a high-resistance contact between one of the brushes and the commutator inside the motor. (Don't worry if you don't know what commutators and motor brushes are. Next time you get a dead DC motor, take it apart and you will see. Note, though, that usually these motors are not meant to be disassembled, and once apart they're like Humpty Dumpty—*very* hard to reassemble! In most instances, these motors are probably not worth trying to repair.)

A dead spot on the commutator will not normally show up unless the motor is *under load,* or just happens to stop on the dead spot. If it did happen to stop on the dead spot, manually rotating the motor shaft just a little bit will usually get the motor going. This, however, is *not* a fix; the motor should be replaced.

A motor with a dead spot will sometimes fail intermittently. Sometimes a slight jar to the VCR or a vibration may get it going. If you suspect a dead spot, operate the mechanism from the battery and put an additional load or drag on the mechanism with your fingers. This slows the motor down so it can't easily coast over a dead spot on the commutator. In this manner you may locate a bad motor.

There may be a small capacitor soldered to the two motor terminals. Its purpose is to suppress commutator noise. Of course, if this cap should become shorted, then the motor won't work. You can check this with an ohmmeter. First, disconnect the VCR cable to the motor. If the reading across the motor terminals is less than 12 ohms, suspect a shorted cap or shorted motor windings. You can simply unsolder or clip one lead on the capacitor if you suspect it might be shorted. Its absence will not affect motor operation.

Under some conditions, a VCR *may* keep a small motor energized even though it is stalled. It might be that a cassette jams during eject, and system control fails to shut down the motor. (A well-programmed syscon microprocessor will turn the motor Off if the switch or sensor in the loader doesn't detect that the loader is in the full eject position within a few seconds. Not all of them do!)

If a stalled motor remains energized, it can heat up and even develop a partial or complete short. In some cases, where the motor is mounted to a plastic bracket, it gets hot enough to soften the plastic and warp the bracket. If you see evidence of this, replace the motor. It very likely has developed a partial short, even though it may still operate.

A partially shorted motor may even *appear* to work fine when powered from the 9V battery (after the mechanism is unbound), but it is drawing too much current and will cause problems down the road. Another indication of a partially shorted motor is dimming of the VCR display or front panel LEDs when the motor operates. It is drawing too much current from the power supply. A shorted or partially shorted motor can also destroy its driver IC or transistors, because of the lower motor resistance and increased current demand on its driver.

Now suppose you have a front loader that just won't load a cassette. You've checked the motor with the battery and there doesn't seem to be any problem; the loader works great. Next, check the switch that is supposed to transfer when a cassette is placed into the loader. In fact, check all switches on the front loader for proper mechanical operation and electrical contact transfer. Check continuity of switch contacts and any connectors between them and the system control microcomputer. Also check continuity and connectors of the cable powering the motor.

If everything seems in order, the DC motor's driver integrated circuit or driver transistors may be defective. You will need the schematic wiring diagrams to further diagnose the problem. Most VCRs have a single driver IC for both the cam motor and the cassette load motor. A VCR may have separate driver transistors. If discrete transistors are used, there will be at least two, one for CW and the other for CCW rotation.

Inputs to the loader and cam motor driver IC are from the system control microprocessor chip, which determines *which motor* to run and *in what direction*.

Drive circuitry to turn the motor in one direction may be defective, even though circuitry to operate it in the opposite direction is OK. You can test for this by hand cycling the mechanism to the opposite end of travel from where it is currently stopped, and then see if the VCR will operate the motor. Defective driver circuitry for one direction may, however, prevent the motor from operating in either direction.

With the schematic, you can check for proper input and output voltages at the motor driver integrated circuit. This IC may also be labeled "motor switching" or just "switching." Input lines are from the syscon chip, and outputs go directly to the motors.

Figure 13.2 is the schematic for both cassette and mode motors in a typical VCR. On an actual schematic, there would be many more input and output pins on the system control IC, as well as other circuitry unrelated to the two motors. These items have been omitted to more clearly show motor drive circuitry.

Notice the three output lines labeled Motor1, Motor2, Motor3, from the system control microprocessor going to the motor driver IC. The condition, or state, of each of these three lines determines which motor is energized, and in what direction. The syscon output line truth table in Figure 13.2 tells the story.

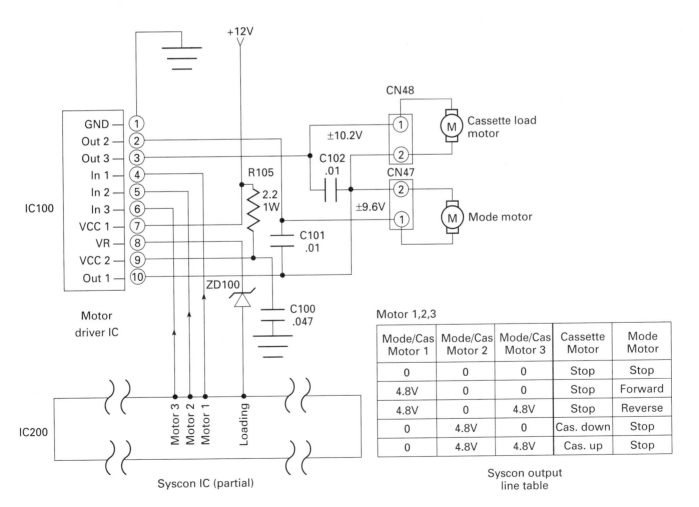

**FIGURE 13.2 Cassette and mode motor circuit.**

For example, if all three outputs are at 0 volts, neither motor runs. With Motor1 at 4.8V and the other two at 0V, the mode motor runs forward. If Motor2 and Motor3 are both 4.8V, then the cassette motor ejects a cassette (Cas. Up).

Input power to the motor driver IC is +12VDC. Depending on the voltages on the three input pins (IN1, IN2, IN3), IC100 sends about 9-1/2 volts, polarized one way or the other, to one of the two motors. Note that OUT1 is common to both motors, OUT2 drives the mode motor, and OUT3 drives the cassette motor. Also notice that neither terminal on either motor is directly connected to ground or common.

IC 100 puts out between +9 and +11V on its VR pin as feedback to the microprocessor when either motor is being energized, in either direction. Zener diode ZD100 reduces this voltage to 5V before going to the LOADING input pin of IC200. The three capacitors are for electrical noise suppression.

A shorted driver IC or transistor *could* keep a motor energized when it shouldn't be. If a motor runs continuously, this is a possibility. But first, suspect a problem with one of the limit switches on the front loader if the loader motor won't shut Off, or with the mode switch if the cam motor continues to run.

Again, a good syscon *should* shut down the signal to the motor driver if it doesn't receive input from a switch or sensor indicating that the mechanism has moved appropriately. However, a shorted driver transistor or IC will continue to pass voltage to the motor. Because it is shorted, it can't shut Off when directed to by syscon.

> Note: Some VCRs have just a single DC motor for both cassette loading and tape loading operations.

## Checks on Other VCR Motors

VCR motors with more than just two terminals cannot be tested or operated with a battery, as previously described. These are not simple DC motors, but are multi-phase AC motors. Often there are three motor windings connected in a Y configuration, where one end of each of the three windings are connected together, and the other end of each winding is a voltage input pin on the motor. These motors contain no brushes or commutators, as two-terminal DC motors do. Integrated circuits convert DC from the power supply to pulses of varying widths to drive the three motor windings.

Both the capstan and drum motors typically have three motor windings, and frequently use an IC and one or more Hall devices that work together to properly energize the motor coils. The motor driver IC may be an integral part of the motor itself, or may be on one of the VCR PC boards.

Even though you can't run this type motor by simply applying a battery voltage, there are still *some* checks you can make.

First, determine the mechanical condition of the motors. Drum and capstan motors are much more critical than any other VCR motors. They both have to rotate precisely and smoothly to produce a stable picture. If they don't rotate freely, they won't be able to rapidly make minor speed variations to keep videotape tracks and rotating heads synchronized. Drum, capstan, and tape synchronization, performed by the VCR's servo circuitry, are explained further in Chapter 14.

With no power applied, check that the drum and capstan motors spin freely. Most newer VCRs have a direct drive capstan, where the motor shaft itself is the capstan and the flywheel is an integral part of the motor. If there is

Chapter 13 Checking Motors, Optical Sensors, and Remotes ■ 385

a belt around the flywheel pulley, temporarily remove it. For indirect drive capstans, remove any belts around the motor pulley(s). Virtually all VHS video drums are direct drive. Do not touch the side of the cylinder that contacts videotape; turn it from the top.

Slowly rotate each motor by hand, while feeling for any grittiness, rough spots, or drag caused by a worn bearing. The motors should rotate smoothly. Give each a spin. Listen for any bearing noise. On some models, there is a brake pad that rests against the rim of the capstan motor flywheel. Manually disengage the brake while performing these tests. If there is evidence of a bad bearing, in most cases the entire motor will have to be replaced, as component parts are rarely available.

Electrically, there is not a whole lot you can do to test these type motors. If they don't work, most likely the motor itself or a driver IC chip is failing. Of course this is assuming that system control has the right conditions to turn the motor On in the first place. You can consult the schematic diagram and make *some* voltage and continuity checks.

There are many variations on how drum and capstan motors are operated. Figure 13.3 is a portion of the schematic diagram for one VCR model, just to give you an idea of how they may be connected, rather than to analyze this particular circuit.

Notice the following in the partial schematic for drum and capstan motor circuitry:

- Each motor has three windings (V, U, W) connected in a **Y** configuration. These three coils produce a rotating magnetic field that spins the motor.

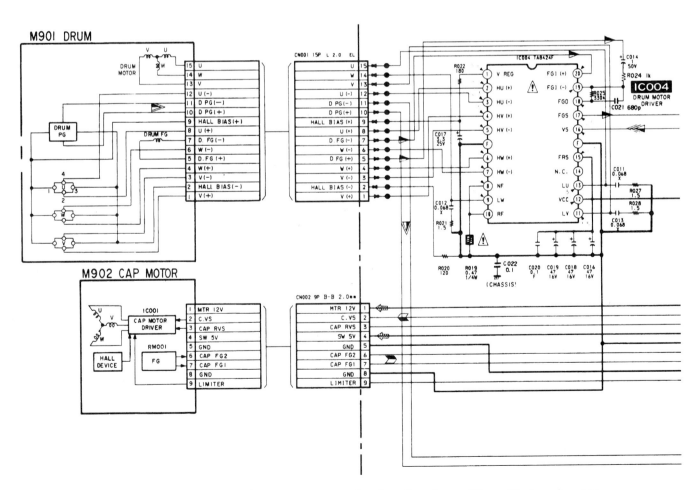

**FIGURE 13.3** Partial drum motor and capstan motor circuits. **Courtesy of Sony Electronics, Inc.**

- Drum motor is driven by separate integrated circuit IC004, whereas capstan motor has a built-in driver chip, IC001.
- Within the drum motor schematic, the four rectangles (U, W, V) are Hall devices that work with IC004 to produce a rotating magnetic field in the three motor windings.
- Within the motors, FG (frequency generator) and PG (pulse generator) are either magnetic induction coils or Hall devices that send signals to the VCR servo circuitry. They report capstan speed, and drum speed and angle of rotation, or position, within each 360° revolution.
- DC power input to IC004 is +12V on Pins 12 and 15. Although not shown in this partial schematic, the source is the MTR 12V power supply output.
- Notice the 0.47Ω, 1/4W fusible resistor R019 connected between ground and pins 8 and 10 of IC004.
- There are two DC inputs for the capstan motor, +12V at CN002-1 (MTR 12V) and +5V at CN002-4 (SW 5V).

Consult the manufacturer's service literature and schematic wiring diagrams for motor connections and circuit voltages. Although you may not be able to thoroughly check the circuit and motor with a multimeter, at least you can see if power supply voltages are present where they should be, and that there is continuity on the motor cables.

Accurate checking of pulse generator (PG) and frequency generator (FG) signals developed by the rotary videodrum and FG signal produced by the capstan motor cannot be done with just a multimeter. This requires an oscilloscope, which displays signal waveforms on a cathode ray tube. This chapter therefore does not troubleshoot these sensors. However, an explanation of how they are used in a VCR servo circuit is given in Chapter 14.

## CHECKING OPTICAL SENSORS

The vast majority of VCR optical sensors consist of a light source and a phototransistor working together. When light falls on the phototransistor, it is similar to placing a forward bias on the Base of a regular transistor: The phototransistor conducts current between its Emitter and Collector.

Optical, or photo, sensors are found in many places in a modern VCR. Besides providing VCR operation from a handheld remote control, they can report the status of the front loader, end of tape, cassette reel rotation, and position of the tape load and cam mechanics. Increasingly, optical sensors are replacing mechanical switches as feedback devices to system control. They are more reliable.

Failure of a mechanism to operate may be due to a defective photo sensor. A bad mode sensor, for example, may prevent tape from loading, or may cause system shutdown at the completion of tape load, instead of playing the tape.

All photo sensors have two fundamental parts: a light source and a light detector. Nearly all VCRs employ infrared (IR) LEDs as light sources and IR-sensitive phototransistors as detectors. This section describes how to check both.

### Checking Optical Light Sources

Early VCRs employ small incandescent lamps as optical light sources. With these, it is often easy to tell whether they are working by merely observing if the lamp is lit—but not always. Some older top loaders do not energize the incandescent light that sticks up into the middle of the cassette until the cassette-down or cassette-in switch transfers at the completion of cassette load. It may be hard to see whether it's lit with a cassette covering it. Solution:

"dummy up" cassette load as previously explained with a videocassette test jig. Manually transferring the cassette-in switch may be all that is needed to turn the tiny lamp On.

Infrared LEDs have replaced incandescent lamps as light sources. The trouble is, how can you tell whether they are lit or not? There are two ways. Use an infrared sensor card or an electronic IR detector.

Many electronic parts outlets sell a plastic card that has a small patch of IR-sensitive material on its surface. Hold the card near an IR source, and the patch glows orange. This can easily show whether an IR LED is putting out.

Another method for detecting IR radiation is to use an IR phototransistor in series with an ohmmeter or a battery and visible light LED. Chapter 11 contains construction plans for an IR detector called the two-way tester (TWT). When IR energy strikes the IR-sensitive phototransistor at the tip of a handheld wand, the red LED on the tester illuminates.

You can also use a multimeter and an IR phototransistor. When IR light strikes the transistor, it conducts, and the meter indicates a lower value of resistance. The transistor must be connected across the meter leads with the right polarity; otherwise it won't conduct at all.

It is important, when using a phototransistor IR detector, like the TWT, that the ambient lighting *not* contain much IR energy. An incandescent lamp has quite a bit of IR output and will cause the phototransistor to conduct. Fluorescent light is better suited to lighting your work area when you're checking for IR output. Hold the phototransistor as close as possible to the IR LED you are testing. Observe the ohmmeter or red LED on the TWT. You should readily see if the LED is working.

Experiment with either the infrared sensor card or electronic IR detector on known good IR sources. You can then readily determine whether an IR LED is functioning as it should.

## Checking Optical Detectors

The essential fact to remember about an IR phototransistor is that when infrared light strikes it, its resistance goes down, and therefore current can flow through the device. Think of the phototransistor as a SPST switch: when light strikes, the switch is On; when it is dark, the switch is Off.

In most VCRs, the various photo sensors send signals to the system control microprocessor. This signal indicates whether or not the phototransistor is exposed to light. Nearly all VCR sensor circuits operate from a +5-volt power supply line.

Depending on the VCR model, an illuminated phototransistor can be indicated by an input to syscon that is nearly 5 volts (*High*), or nearly 0 volts (*Low*). How the particular system control microprocessor chip is designed determines whether it recognizes *High* or *Low* as an indication that light is falling on the IR detector. The end result is the same: syscon "knows" whether the detector is exposed or dark.

Figure 13.4 shows both commonly used optical sensor circuits. For each circuit, think of the IR LED as the light post sticking up into the center of the cassette, and the phototransistor as the supply end-of-tape detector. When tape plays to the end, light shines through the transparent leader, striking the transistor and turning it On.

This is how the two similar photo sensor circuits work:

- Circuit A:
  — With no light hitting Q1, it is like an open switch; it might as well not be there. Five volts then appear at the input pin of IC101 via Q1 load resistor R2 and current limiting resistor R3.

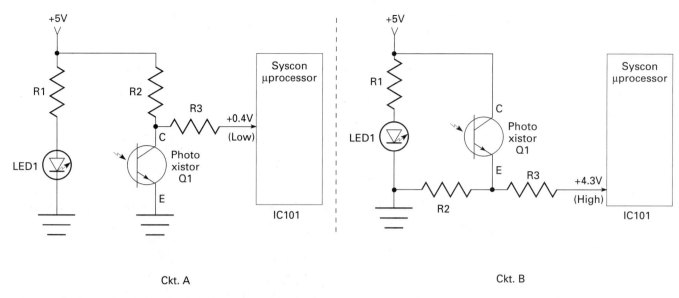

**FIGURE 13.4** IR photo sensors. Circuit A: Input to syscon is Low, about 0 volts, with phototransistor illuminated. Circuit B: Input to syscon is High, about +5 volts, with phototransistor illuminated.

— When light strikes Q1, it conducts. Q1 acts like a closed switch, placing Q1's collector at nearly the same potential as its emitter, or at ground. This is coupled through R3 to IC101. (With Q1 conducting, the voltage drop across load resistor R2 is nearly 5V.)
- Circuit B:
  — With no light striking Q1, it acts like a switch that is Off; there is no path for +5 volts to reach IC101. Syscon input is thus at ground potential, applied through emitter resistor R2 and current-limiting resistor R3.
  — When light strikes Q1, it conducts, acting like a switch that is On. Five volts then has a path via Q1 and R3 to the input pin of IC101. (When Q1 conducts, the voltage across load resistor R2 is nearly 5V.)

For either circuit, R1 limits current to the IR LED. Typical values are in the range of 270 to 470 ohms. Q1's load resistor is usually around 50KΩ, while the value for R3 is in the neighborhood of 1K to 3.3K ohms.

In actuality, you usually don't need to know which of these basic photo sensor circuits is being used. What you *do* care about is that there is a shift in the output of the photo transistor between light and dark conditions. You may not even need a schematic in most cases.

For example, suppose you suspect that the take-up (TU) reel sensor may not be working. The VCR goes into shutdown a few seconds after tape starts moving in play or record. You can see that the TU reel *is* turning, but if syscon "thinks" it isn't, it will unload tape and power down.

You look in the undercarriage and see a very small printed circuit board (~1/2" x 3/4") underneath the TU spindle. There are three wires going to the board; you can see and probe where they are connected. Most likely there is an IR LED/phototransistor pair on the other side of the PCB. Light from the LED bounces off reflective segments on the underside of the TU spindle as it rotates, alternately turning the phototransistor On and Off.

But what are the three wires? Look back at Figure 13.4. It could be that one is ground, one is +5V, and the third is the output of the phototransistor. Resistor R1 may or may not be on the little PCB. R1 may also be connected to the ground or cathode side of the LED.

It really doesn't matter whether the configuration is like Circuit A or B. The main thing we are concerned about is that one of the wires should show a definite voltage shift when the TU spindle rotates. Voltage should swing back and forth between about 0 volts and close to 5 volts. So how to proceed?

- Power up the VCR.
- Set up the multimeter to measure 5 volts, and connect the black lead to chassis ground.
- One by one, touch the red lead to the three connections on the sensor PCB while manually rotating the TU spindle *very slowly*.
- Suppose the three wires are red, orange, and brown. If all is working OK, this is what you might measure.
  — Red wire: 5.1V, no change when spindle rotates
  — Orange wire: 3.8V, no change when spindle rotates
  — Brown wire: 0.6V 4.9V 0.6V 4.9V ... as the spindle rotates

Aha! The brown wire is showing sharply defined transitions as the spindle rotates. This is a good indication that the TU reel sensor is working. In a similar manner, you can test other photo sensors within the VCR. Look for a sharp transition on the output of the phototransistor between when it is dark and light.

Of course, if you have a schematic, you could make the measurement at the input pin to the system control microprocessor, or at either end of R3. R3 is normally located very close to the syscon input pin, and is usually easier to probe than the very closely spaced IC leads. A spring-loaded hook-style miniprobe is ideal for hooking onto discrete component leads on a PCB. Checking at the input pin of the syscon IC ensures that it is really receiving the sensor's output.

Don't be fooled! Ambient light, especially direct sunlight or nearby incandescent lamps, can make a phototransistor appear to be leaky. Instead of its output going between nearly *zero* volts and 4.8 volts, perhaps the output swings between only *3.5* and 4.8 volts. The 3.5V output might give unpredictable results in some circuits. It is in the no man's land or ambiguous area between High (5V) and Low (0V); that is, it is not clearly On *or* Off. Not taking into account the effect of ambient IR energy when servicing a VCR may lead you to an incorrect diagnosis, or totally confuse your troubleshooting efforts.

The phototransistors most vulnerable to ambient light while a VCR is being serviced are the two tape-end detectors. Most others are much less exposed to room light, even with the covers off. When checking each of these two detectors for its Off state, completely cover the aperture, or small hole, where light enters.

## REMOTE CONTROL PROBLEMS

A VCR's handheld remote is often subject to more rough handling than the VCR itself. It gets dropped, it takes a bath in soda and beer spills, and it gets clogged with pizza crust crumbs. Although the remote unit is a fairly simple device, as compared with the rest of the VCR, it can cause many problems.

A remote consists basically of several pushbuttons, usually one IC, a quartz crystal, one or more infrared LEDs, an LED driver transistor, a few miscellaneous electronic components, and one or more batteries. When a button is pressed, the integrated circuit develops coded pulses that energize the IR LED.

Infrared pulses from the remote are picked up by an IR receiver on the front panel of the VCR. Receiver output is sent to an IC which decodes the pulses. This IC is normally the same chip that the VCR front panel pushbuttons are connected to, usually the timer/operation microprocessor. Figure 13.5 is the schematic diagram of a typical remote control unit.

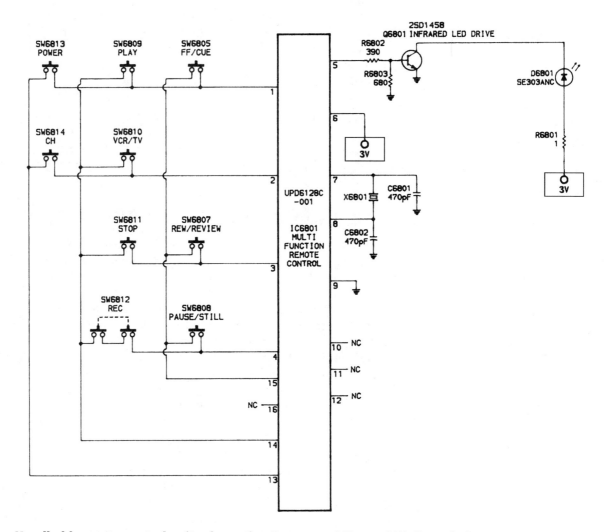

**FIGURE 13.5** Handheld remote control unit schematic. Courtesy of Howard W. Sams & Co.

Notice that Q6801 is the driver transistor for IR LED D6801. Component X6801 connected between Pins 7 and 8 of the IC is a *quartz crystal,* used as the basis for a highly stable oscillator to produce coded pulses. The NC beside pins 10, 11, 12, and 16 means "No Connection"; don't confuse that with N.C. for *normally closed.*

A typical remote control unit produces IR output of approximately 40kHz. The 40kHz carrier is then pulsed On and Off in different patterns, depending on which button is pressed. That is, a digital bit stream from the IC turns the 40kHz IR carrier On and Off in bursts. This is a form of pulse modulation.

Following are common remote control problems:

- Weak batteries
  — This can cause the remote to not work at all or to work intermittently. The answer, of course, is to simply replace the batteries.

— Checking a battery's voltage with a multimeter is totally *invalid!* There is no current being drawn from the battery, and a weak battery may show acceptable voltage with no load; when actually powering a device, though, the voltage dips way down.
— Most battery testers have built-in resistors to place a load across the battery when its voltage is being measured. Even some of these testers do not place an adequate load across a battery. It may read good with the needle "in the green," but be unable to reliably operate a remote control unit or other device. (With your multimeter, measure the voltage of a 9V battery from a smoke detector sometime after the unit starts beeping its weak battery signal. You may be surprised to see that the battery still reads close to 9 volts, and will even read in the Good zone on *some* battery testers. But its internal resistance has gone up with age, and no longer can supply the current to reliably operate the detector and sound a smoke alarm.)
— Rechargeable nickel-cadmium (nicad or Ni-Cd) batteries have a lower voltage per cell than a conventional carbon or alkaline battery. A single-cell nicad (AAA, AA, C, D), for example, delivers 1.25 volts when fully charged rather than about 1.5 volts for the other types. A fully charged "9-volt" nicad actually supplies only about 7.2 volts! Some remotes may not work satisfactorily with nicads because of these batteries' lower voltage.

- Distorted, weak battery contacts
  — On some remotes the battery contacts can lose tension, especially when the unit has been dropped. Units with AA or AAA batteries are especially vulnerable. A jeweler's screwdriver or needle-nose pliers can usually form the contacts to restore contact pressure. Often, a small piece of paper or cardboard wedged between the back side of a battery contact and the end wall of the battery compartment works well to increase contact tension against battery terminals.
- Dirty or corroded battery contacts
  — High contact resistance can result from slight oxidation or corrosion on battery terminals. This is more likely to happen when bargain carbon batteries are used. These can leak gases that quickly corrode battery contacts.
  
  Worse, is that corrosive gases can easily eat away tiny printed circuit wiring foils inside the remote. Whenever you find corrosion inside a battery compartment, always take the unit apart and inspect for this type of damage to the PCB or other components. If you're fortunate, you may be able to clean any corrosion off wiring patterns before they are eaten through. Clean battery contacts, terminals, and other affected areas using the procedures described in Chapter 6.
  
  Because PCB wiring traces are usually so small and delicate, it is nearly impossible to repair a remote that has open wiring patterns. You can *try* to bridge an open wiring trace with 30-gauge wire. Good luck!

All this also holds true for personal, portable stereos and other small, battery-operated equipment.

- Broken, cracked PCB
  — Despite the fact that remotes get dropped quite a bit, cracked PCBs are not all that common. Remotes are light enough to bounce rather than break. However, a board can crack, perhaps only a hairline. An 8X or 10X loupe or a magnifying glass and a high-intensity light are helpful in finding hairline cracks.
  
  You *may* be able to flow solder across a cracked wiring trace or bridge it with 30-gauge wire. It isn't always necessary to bridge the open

trace right at the point of the break. Follow the trace on both sides of the break and you may find more convenient points for soldering the ends of a small jumper wire.

Depending on how badly the board is cracked, you might want to repair it with epoxy cement either before or after you have made electrical repairs and checked it out. If a piece has totally broken off, cement first. If there's just a small crack, bridge open wiring traces, then shore things up with epoxy. Be very careful to not get epoxy on printed circuit switch contacts, described shortly.

- Cracked quartz crystal
  — An electronic crystal is a very thin sliver of quartz that vibrates at a particular frequency, depending on its size and thickness, when excited by an electrical pulse. A quartz watch has a precise oscillator whose frequency is determined by a quartz crystal.
  
  The IC in the remote contains the oscillator circuit for the crystal. Its precise frequency is then used to produce the 40kHz infrared carrier and digital pulse stream when a button is depressed.
  — Most quartz crystals are packaged in a flat, rectangular, silver-colored metal container, with two leads coming out one end. However, some come in a brightly colored plastic package, such as yellow or orange. A crystal is usually designated with an X number on schematics and circuit boards, such as X801. Xtal. and Xtl. are abbreviations for crystal.
  — If subjected to hard mechanical impact, such as dropping the remote, the fragile crystal can crack, rendering the remote inoperable. There is no simple multimeter check you can make to check a crystal. When the rest of the remote appears OK, suspect a bad crystal if the unit was dropped just before it stopped working.
  — A replacement crystal *may* be available from the manufacturer; check the parts list section of the service manual. However, replacing a crystal could prove to be very difficult because of the tiny wiring patterns on the remote's circuit board. Use a 15W pencil, a clamp to hold the board, and a lot of patience.

- Stuck pushbuttons, dirty pushbutton contacts
  — There are basically two types of remote control pushbuttons. Some remotes have discrete microswitches or membrane pushbuttons. These are relatively rare, but you may come across them. Either type of pushbutton makes a very soft click sound and has a positive tactile feel when pressed, like buttons on the front panel of a VCR. In most cases, if one of these types of pushbuttons fails, the entire remote control unit must be replaced, but it won't hurt to take the unit apart and have a look.
  
  Much more common are printed circuit pushbutton contacts. These are silent and have a very soft, even feel when pressed, no click or tactile feedback. A round conductive pad on the back of each rubber button completes a circuit between two interlaced or zigzag wiring traces on the top surface of the circuit board when the button is pressed. (See Figure 13.6.)
  — Some remotes have individual plastic buttons for each function, which stick through the remote's top cover; most have a single rubber button pad that incorporates all individual buttons. A conductive area on the back of each button touches the PCB wiring traces when it is pressed.
  — A common cause for buttons sticking down is that a beverage or something else has been spilled onto the remote.
  — Usually when a button fails to work, or works intermittently, it is caused by a dirty printed circuit board or button contact. That is, the zigzag

Chapter 13  Checking Motors, Optical Sensors, and Remotes ■ 393

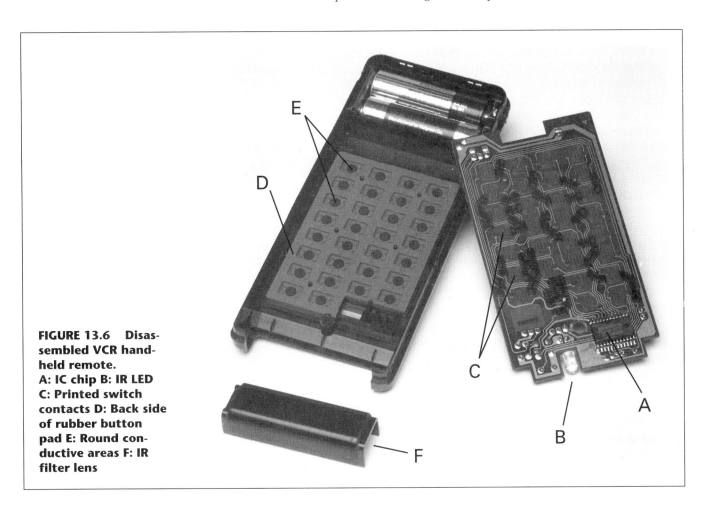

**FIGURE 13.6** Disassembled VCR handheld remote.
A: IC chip B: IR LED
C: Printed switch contacts D: Back side of rubber button pad E: Round conductive areas F: IR filter lens

wiring traces on the board and the conductive area on the back of a button have a light film of residue, and so create a high-resistance contact.

For either of the preceding two problems, take the unit apart, then clean around the button openings in the top cover. Clean the individual buttons or the entire unitized rubber button pad, paying particular attention to the conductive areas on the back of each button. Clean the zigzag printed wiring patterns on the circuit board. Mild dishwashing liquid, warm water, and cotton swabs are fine for cleaning all these parts. Alcohol or degreaser will also do, but these *may* eradicate lettering on the top cover or dull the finish; stick with mild dishwashing liquid.

Be careful when cleaning the delicate circuit traces that form the contacts. Use a swab moistened with alcohol, degreaser, contact cleaner, or mild dishwashing liquid. Apply *very light pressure* here; rub too hard and you may cause the wiring trace to detach from the board. If using water and dishwashing liquid, rinse the buttons or one-piece pad and printed circuit board thoroughly under running water. Make certain there is no residue on the conductive areas on the backside of the buttons or on the printed circuit switch contacts. Avoid touching contact surfaces with your fingers. Shake off excess moisture and allow the button pad and circuit board to dry *thoroughly* before reassembling the unit. Do not wash a circuit board with membrane or individual microswitches in water.

- Defective IR LED
  — This is *not* a very common problem. You can easily determine whether a remote is transmitting IR. Hold an IR sensor card, or electronic tester like the TWT, up close to the remote's window. Then hold a button depressed. You should see about 8 to 12 On/Off pulses per second. This is also a great way to test all the individual buttons on a remote.
  — Some remotes have only one IR LED; others have two or even three. Take the unit apart to determine this. You can then test each LED with one of the IR detectors.

Figure 13.6 shows a typical VCR remote control unit disassembled. The quartz crystal is on the backside of the circuit board in the photo.

## Disassembling a Remote Control Unit

If you like puzzles, you'll enjoy figuring out how to take a remote apart. It's not all that difficult, but sometimes it presents a bit of a challenge. First, look for and remove any visible screws. Take the batteries out and inspect the compartment; there may be one or more screws inside.

Some remotes hide screws underneath rubber feet on the back. Pry out the feet with a small flat-blade screwdriver. If there is a stick-on label on the back of the unit, there may be one or more screws beneath it. You can usually feel an indentation where a screw cup is with your fingertip. Then use a small utility knife to cut through the label, all around the screw cup. Remove the piece of label within the cut to get at the screwhead.

Inspect the front of the remote for any decorative plastic strip or manufacturer's nameplate that might easily pry out, revealing a hidden screw.

Many remotes have only one, or perhaps no screws holding the two halves together. Plastic latches, perhaps not easily seen, are common. Look all around the edges of the remote for small slots or plastic latches. Take a small flat-blade screwdriver and very gently push in or twist to release these latches as you gently pry the top and bottom pieces apart.

Some remotes come apart by removing screws or releasing plastic latches, or both, and then sliding the bottom cover toward the back end of the unit a short distance. This releases latches, at which point the two halves can be separated.

If the remote doesn't use any of these assembly schemes, take a thin, flat-blade screwdriver and force it into the crack along one side, between top and bottom halves, and gently twist to pry the two sections apart. Try this at various places along one side. At some point the two pieces should begin to separate. Work along the entire length of the side like this. You may have to repeat the procedure on the other side, or the halves may easily hinge open and separate once you've released all the hidden latches on one side.

It is important to examine the unit all over, and then *be patient,* especially if you need to pry the sections apart. Don't use too much muscle! It is easy to gouge or crack the plastic case. When reassembling, gently *squeeze* the two halves back together to engage plastic latches, if of this type. Just snug any screws; don't overtighten, as this can easily strip the plastic threads in the cover pieces.

## Checking VCR IR Receiver

On most VCRs, the detector that receives input from the IR remote consists of a *photodiode.* The photodiode is usually mounted inside a small metal enclosure with an aperture hole. This is the *infrared detector module.* It is soldered to a front panel circuit board, normally the same PCB on which front panel pushbuttons, controls, and LEDs are mounted. The detector module usually contains components to amplify and shape the IR signal received from the remote unit.

A photodiode works like a phototransistor: its resistance changes when light strikes it. Photodiodes have extremely fast response times. They are well suited for IR communications with rapid data streams. The change in current through the IR receiver photodiode is coupled to the timer/operation microprocessor, after being amplified, shaped, and conditioned by an IC. The timer/operation microprocessor decodes the pulses to perform VCR functions, just like the front panel pushbuttons that are directly connected to the timer/operation microcomputer.

Figure 13.7 is the schematic of a VCR IR receiver. Diode D1 connected to pin 7 of IC1 is the IR receiving photodiode. At the right side, Jumper 1 connects the output of IC1 pin 1 to the timer/operation microprocessor IC. IC1 amplifies the output of the photodiode and performs wave shaping and detection of the received 40kHz IR pulses.

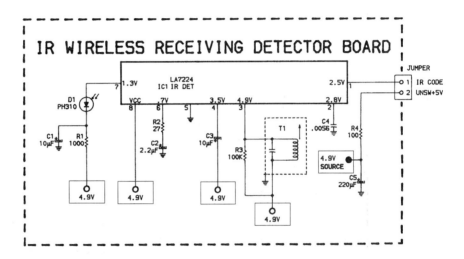

**FIGURE 13.7 VCR IR receiver/infrared detector module. Courtesy of Howard W. Sams & Co.**

Testing a VCR's IR receiver with just a multimeter is rather inconclusive, but you can usually tell whether it is functioning at all. In the circuit in Figure 13.7, and most similar VCR circuits, measure the voltage at the *output* of the IR detector module (Jumper 1 in this schematic). Normally you will read somewhere around 4 volts.

Then press and hold down a button on a functioning remote control unit aimed at the receiver. You should see some dip in the meter reading if the diode is receiving pulses. On an analog meter, you will likely see the meter quivering rapidly back and forth a few tenths of a volt about some voltage, typically around 2.5 to 3.5 volts. On a DMM, you should see the voltage drop, perhaps .5 volt or less, when the remote button is pressed, and while the button is pressed, the reading won't settle down or stabilize. Instead, the right most one or two digits will continually change, due to the constantly changing voltage caused by the pulses.

Place your meter lead on the remote control input pin on the timer/operation microcomputer IC. You may have to trace the output of the IR detector to the timer chip on the schematic to locate this lead. Here you should see a similar indication when the IR detector is receiving light pulses from the remote.

To perform this basic test, you don't need the specific remote for the VCR on which you are working. Almost any functioning *IR* remote control unit may be used. The VCR probably won't respond to whatever commands are being sent, because the coding of the IR pulses is different, but the IR detector module should still receive and detect them.

# SUMMARY

This chapter described how to check VCR motors and optical sensors. Most VCR motors having just two terminals operate on 12 volts DC. They are normally the cassette load motor in the front loader assembly and the cam, or mode, motor. These can be operated and checked independently from the VCR with a 9V battery. It is best to disconnect these motors from VCR circuitry when testing or operating them with an external power source.

Capstan and drum motors cannot be checked in this manner, because they have more than two terminals and work with precisely controlled DC pulses. Often these motors have three coils and rely on a driver IC to rotate the magnetic field so the motor can spin. The driver IC may be on a PCB or incorporated into the motor itself. With a schematic, a multimeter may be used to check for DC input voltages to the motor and driver IC, and to check continuity of the cable going to the motor.

Many VCR sensors consist of an infrared (IR) LED light source and an IR phototransistor detector. When light strikes the transistor, it conducts between Emitter and Collector, much like a transistor that has its Base forward biased. In effect, light is the input to the transistor Base, turning the transistor On.

IR LED output can be detected with an infrared sensor card. A small IR-sensitive patch on the card glows orange in the presence of IR energy. IR light can also be detected with an IR phototransistor connected to an ohmmeter or in series with a battery and a visible LED. In the presence of IR, the ohmmeter reads a low value of resistance and the visible LED glows. A make-it-yourself electronic IR detector is described in Chapter 11.

In most VCR circuits, the On/Off output of an IR phototransistor goes to an input pin on the system control microprocessor IC. Depending on the particular circuit configuration, the input pin goes to nearly 0 volts (*Low*) or to nearly 5 volts (*High*) when light strikes the phototransistor. The main thing is that there is a definite change at the transistor output, and syscon input, between when the detector is light and when it is dark.

IR phototransistor output can be checked with a multimeter. A good Low output is usually in the range of 0 to about 1 volt, whereas an acceptable High is above 2 volts, but will normally lie between 4 and 5 volts.

VCR remote control units produce coded infrared pulses that are received by a photodiode detector in the VCR. Common remote control problems are weak batteries, poor battery connections, and dirty printed circuit switch contacts. A cracked quartz crystal, resulting from a handheld remote being dropped, can also render a remote inoperative.

Output from a remote can be checked with an infrared sensor card or by an electronic device like the two-way tester described in Chapter 11. Hold the card or tester wand close to the remote's IR window. You should see about 8 to 12 On/Off pulses per second when a button on the remote is held depressed.

Basic operation of the photodiode IR detector module in a VCR can be detected with a multimeter. Set the meter to read 5 volts DC and probe the *output* pin of the IR detector module, normally a small metal box soldered to the front panel PCB. There will usually be one or more ground pins and a +5V pin as well as the output pin connecting the module to the circuit board. While probing the output connection, the pointer of an analog meter will normally drop from about 4 volts to the neighborhood of 2.5 to 3.5 volts when a button on the remote unit is held depressed. The pointer will rapidly quiver back and forth a few tenths of a volt about this lower voltage.

A DMM will normally decrease the voltage reading to about 3.5, and show instability in the low-order digits on the right of the display when pulses are received. That is, the last digit or two will jump from number to number as the meter reads the ever-changing voltage of the pulses being received and detected.

## SELF-CHECK QUESTIONS

1. Briefly describe how to power a typical front loader motor when a VCR is unplugged from the AC line.
2. Suppose a particular VCR has a 6V DC motor in the front loader. Discuss the effect of using a 9V transistor radio battery to exercise the loader a few times.
3. What is the most common voltage for VCR cassette load motors?
4. Explain why it is a good idea to disconnect a two-terminal DC motor from the VCR before powering it from an external power source.
5. What should you do while powering a DC motor which you suspect may have a dead spot to confirm that it does?
6. What are two possible causes for a cam motor being continuously energized?
7. Refer to Figure 13.3. What two DC voltage checks would you make if the capstan motor does not run at all?
8. Describe the relative resistance of an IR phototransistor when no light strikes it. Describe when infrared light does strike it.
9. Refer to Figure 13.4, Circuit A. Approximately what voltage would you expect to read at the input pin of the syscon microprocessor if Q1 were shorted between emitter and collector?
10. Describe how you would check whether or not a particular button on a handheld IR remote control unit was working, without using the associated VCR, TV, or other base device.

# CHAPTER 14

# VCR MICROPROCESSORS AND SERVOS

Many VCR electronic operations are handled by integrated circuits and microprocessors. These are complex components that have many individual circuits combined in a single package. An integrated circuit (IC) may consist of hundreds or even thousands of transistors, diodes, and resistors all manufactured on a single slab of silicon, called a *chip*. In VCRs, a *microprocessor* is a complex integrated circuit that has built-in programming instructions. The programming determines how the microprocessor handles a variety of inputs to produce some output.

Earlier VCRs have perhaps only two or three ICs and microprocessors to handle basic system operations, with most circuits made up of discrete transistors and related components. Increasingly, VCRs use many more ICs rather than separate parts in various circuits. This helps reduce production costs while increasing reliability, because a single IC replaces so many individual parts. (Incidentally, sometimes microprocessors are referred to as *microcomputers*.)

This chapter describes how ICs and microprocessors are used in VCRs and how to locate pins on an IC chip. Two different digital technology logic levels are also covered. You will learn what a digital logic probe is and how to use one for diagnosing digital circuits.

Servo systems synchronize videotape movement with the tracks laid down on tape by the rotary heads. This chapter explains the fundamental operation of VCR drum and capstan servos.

## CHAPTER OBJECTIVES

Upon completing this chapter, you should be able to:

1. Describe how to locate any pin or lead of an integrated circuit.
2. Define the acceptable logic levels for both TTL (transistor/transistor logic) and CMOS (complementary metal-oxide semiconductor) circuits.
3. Describe a digital logic probe and how it may be used to troubleshoot digital circuits.
4. Describe the overall operation of VCR drum and capstan servo circuits.

## INTEGRATED CIRCUITS AND PIN IDENTIFICATION

Integrated circuit chips come in a wide variety of packages, with as few as 4 pins to as many as 100 or more. Chips with 8, 14, 16, 18, 22, 28, 42, and 64 pins are common in VCRs. Most ICs having 64 pins or less and have 2 parallel rows of pins. An IC of this type is called a DIP, for *dual inline package*.

Chips with more than 64 leads frequently have pins on all four sides, but ICs with fewer pins may also have their pins arranged on all four sides of the chip.

It is necessary to know how to count the pins on an IC when troubleshooting. You can then check to see if a particular signal exists on a pin, as indicated by the schematic diagram.

All ICs have some identifying mark so they can be oriented correctly when soldered to a printed circuit board, and to identify Pin number 1. Most ICs have a small U-shaped notch or marking at one end. Looking straight down at the chip, Pin 1 is just to the left of this marking, going counter clockwise from the identifying mark. Many ICs identify Pin 1 with a small circle or dot marked adjacent to Pin 1. Yet other ICs have the corner nearest Pin 1 cut at an angle.

Beginning at Pin 1, other IC pins are numbered going *counterclockwise* (CCW) around the chip, looking down at the top of the component. For a DIP, the highest numbered pin is thus directly opposite Pin 1. Figure 14.1 illustrates a few ICs and their pin numbering schemes. Frequently, several IC pin numbers are marked on the printed circuit board. For example, a 64-pin DIP might have Pins 1, 5, 10, 15, 20, 25, 30, and 32 identified along one side of the chip and pins 33, 35, 40, 45, 50, 55, 60, and 64 marked on the other side.

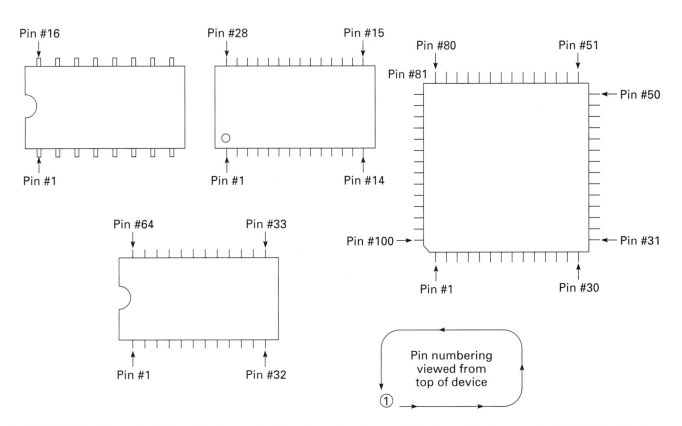

**FIGURE 14.1 Integrated circuit pin numbering. Looking down at the top of a chip, going CCW, Pin 1 is to the left of: A) a U-shaped notch or mark, B) a circle or dot, C) a corner cut. Pins are numbered counterclockwise around the chip.**

Most VCR ICs are soldered to PCBs; you will rarely, if ever, find a socketed chip. Should you ever replace an IC, be sure to orient it correctly before soldering it to the board. Installing an IC in the wrong direction may destroy the IC as well as other components in the VCR. Most PCBs are marked with the outline of each IC, including the U-shaped notch or a cut corner, or Pin 1 is designated.

While we're on the subject of IC replacement, consider soldering an IC socket onto the PCB first. Then, should there be something else in the circuit causing the IC to fry, you won't have to unsolder another IC. Printed circuit boards do not hold up well to repeated soldering and unsoldering! An IC socket also has a U-shaped cutout or mark on one end, or a cut corner, or Pin 1 is designated with a circle or dot; be sure to orient the socket correctly when placing it on the circuit board.

There are two general categories of ICs: analog and digital. VCRs use both. Analog ICs process audio and video signals during record and playback. Digital ICs control VCR operating modes and automatic timer functions. Digital ICs are also used as character generators to form on-screen programming menus and prompts, and for producing special effects such as picture-in-picture (PIP).

## DIGITAL LOGIC LEVELS

The chief difference between analog and digital ICs is their input and output voltage levels. Analog ICs that handle audio, video, and RF signals operate over a range of constantly changing input and output voltages, whereas digital circuitry operates on *only two voltage levels*. One voltage means logical On and a different voltage means logical Off. (You may want to review *"Analog and Digital Circuits"* in Chapter 8.) The On or Off logic state of a digital signal can also be expressed by 1 or 0, High or Low, Up or Down, Yes or No, and by True or False.

**Digital Logic Levels**

On = 1 = High = Up = Yes = True
Off = 0 = Low = Down = No = False

There are several "families" of technologies of digital circuits. The main characteristic that distinguishes one family of digital ICs from another is the voltage levels that represent logical On and logical Off. Two common digital technologies are TTL and CMOS (pronounced "See-moss"). *TTL* stands for transistor/transistor logic, which works at one set of voltage levels; *CMOS* means complementary metal-oxide semiconductor, a type of low-power chip that works at a different set of logic levels. LS (linear series) technology is the same as TTL; MOS (metal-oxide semiconductor) technology is the same as CMOS.

An important fact to remember is that both TTL and CMOS circuits work correctly *only* when digital pulses or signals go from a logical low voltage level to a logical High level or vice versa. Signals or pulses that are somewhere between logical High and Low voltage levels can cause erratic operation. Figure 14.2 shows TTL and CMOS logic levels.

For TTL circuits, voltages between 0 and +0.8V are considered logic Low, voltages between +2.8 and +5V are logic High, and voltages between +.8 and +2.8V are questionable. Questionable means the circuit could interpret voltage levels in this area as High or Low *or* neither High nor Low, which is an invalid level.

CMOS devices do not necessarily have a specific operating voltage level, but operate over a *range* of the power supply voltage. The questionable, or ambiguous, area for CMOS logic is therefore a percentage of the power supply volt-

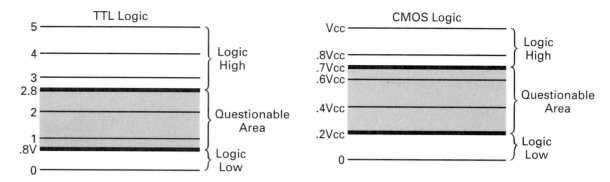

**FIGURE 14.2  TTL and CMOS logic levels.  Courtesy of *Sencore News*.**

age, or $V_{CC}$, applied to the IC. On schematics the input power supply voltage pin is marked $V_{CC}$. This pin may also be printed on the circuit board adjacent to the DC voltage input pin. Ground pins may be identified as Com, Gnd, C, etc., but not always. Look back at Figure 13.3 and you will see that for IC004, ground connects to two F-pins on this particular IC. Pin 12 is the power supply input, or $V_{CC}$ pin.

As can be seen in Figure 14.2, a questionable CMOS voltage lies between 20 percent and 70 percent of $V_{CC}$. For example, for $V_{CC}$ = 10 volts, voltage levels between 2 and 7 volts would be questionable. Logic Low would be 2 volts and less, while logic High would be 7 volts or higher.

Suppose you are working on a VCR that intermittently unloads tape and rewinds while playing a tape. Perhaps the supply end-of-tape sensor is sending an ambiguous, or questionable, voltage level to system control (syscon). VCR syscon microprocessor chips usually work with TTL levels. Depending on the particular VCR, when tape blocks the supply end-of-tape sensor, either logical High ($\approx$+5V) or logical Low ($\approx$0V) is sent to the syscon IC (the symbol $\approx$ means "approximately"). On this particular VCR, input to syscon should be Low with the sensor blocked.

Now if the IR phototransistor is somewhat leaky (or incandescent light hits the sensor while the VCR is being serviced), it may output around 2.8V. Referring back to Figure 14.2, this is right at the edge of the questionable, or ambiguous, voltage range for TTL devices. It is possible that at some time syscon might interpret this voltage level as logical High, when it is really supposed to be Low. As a result, syscon thinks the end of the tape has been reached, so it unloads and rewinds the tape.

*All* digital devices within a technology family, such as TTL, don't necessarily have the same ambiguous voltage ranges as depicted in Figure 14.2. For example, a given TTL device may not recognize a signal as being High unless it is higher than approximately +4 volts. Just remember that High is close to +5 and Low is close to 0 volts. By referring to the manufacturer's service literature and schematic diagrams, you can determine the operating voltage levels at various pins of an IC.

## VCR MICROPROCESSOR INTEGRATED CIRCUITS

Most VCRs have two separate microprocessors that work together to control the machine: the Timer/Operations and System Control microcomputers. Next, we'll look at block diagrams of these two circuits for a typical VCR, starting with the 64-pin TTL timer/operations IC shown in Figure 14.3. The timer/operations IC may also be called just the *timer* or the *function microcomputer*.

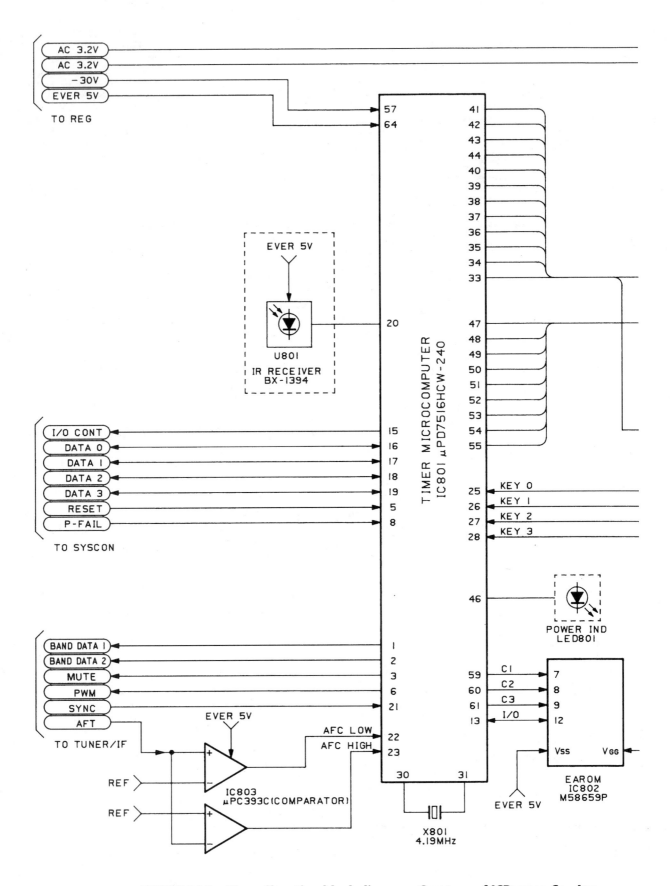

**FIGURE 14.3** Timer/function block diagram. Courtesy of JCPenney Co., Inc.

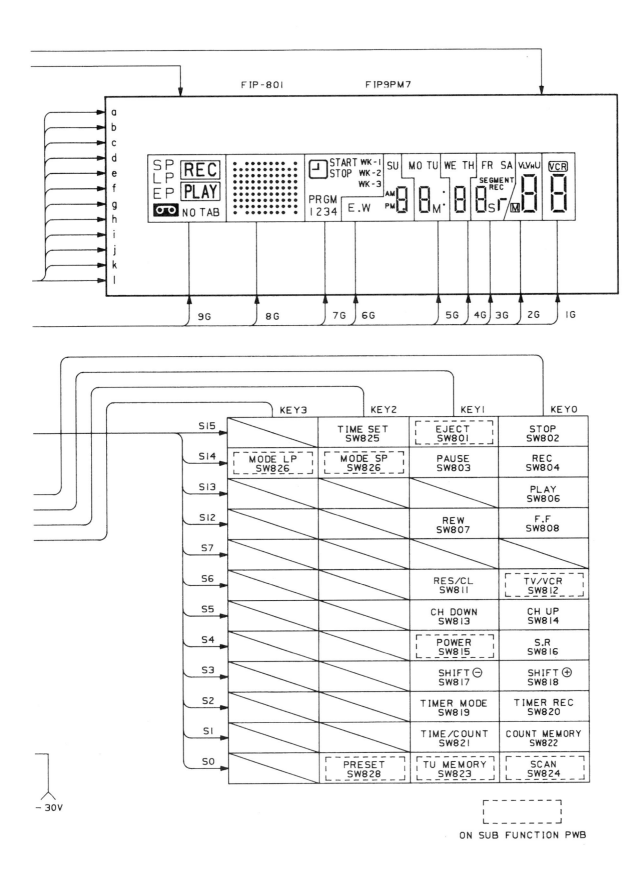

Notice the following items on the timer/function diagram:

- At upper left, four power supply voltage lines, labeled TO REG, power the IC and front panel display.
  — REG stands for power supply regulator board.
  — primary chip power is +5V to Pin 64 from the EVER 5V power supply line, which is active as long as the VCR is plugged in.
  — ground lines are not normally shown on block diagrams.
- IR detector module, consisting of a photodiode and an IC housed in a small metal enclosure (in dashed rectangle), receives and detects 40kHz coded pulses from remote control unit.
- At center left, seven lines interconnect timer/operations and syscon microprocessors.
  — I/O CONT is Input/Output Control.
  — DATA 0–DATA 3 are bidirectional signal lines between timer/operations and syscon microcomputers.
  — P-FAIL signals failure of the power supply to turn On or to remain On.
- At lower left, six lines connect the timer/operations IC to the VCR tuner and intermediate frequency (IF) section.
- At bottom of the IC is the 4.19MHz quartz crystal, X801.
- At upper right is the front panel fluorescent display, driven by the Timer/Operations or Function IC.
- At lower right is the manual pushbutton and control *matrix*.
  — This is a 4 × 12 matrix: Key 0 through Key 3 across the top and S0 through S7 and S12 through S15 along the left side.
  — With the matrix, essentially two microprocessor lines are connected to each other when a button is pressed. For example, Rewind pushbutton SW807 connects the S12 and KEY1 lines when it is depressed.

Figure 14.4 is the block diagram of a System Control microprocessor. Like the timer/operations IC, it also happens to be a 64-pin TTL chip. By now, most of the items in the syscon diagram should be familiar to you. The following is a quick rundown of some of the major features, starting at the upper left.

- *Mode Control Motor* loads tape around drum, moves pinch roller against capstan (also called tape load or cam motor).
- *Sensor LED* is the IR LED that sticks up into center of VHS cassette to illuminate end-of-tape sensors.
- *End Sensor* is part of front loader assembly, indicates end-of-tape at the supply reel. IR light from the sensor LED shines through transparent leader at the end of a tape and strikes the end sensor.
- *Cassette Motor* loads and ejects a cassette.
- *Cassette Lamp*—This VCR has a small 12V lamp inside the loader that turns On when a cassette is loaded. The amount of tape on supply and take-up reels can then be seen on a small flip-down mirror inside the loader, through a transparent cassette compartment door.
- *Insert Switch*—A microswitch that transfers when a cassette is gently pushed into the loader. Starts the cassette motor to lower the cassette onto the reel table.
- *Up End Switch*—Stops the cassette motor when the loader has fully ejected a cassette.
- *Start Sensor*—Part of front loader assembly, indicates end-of-tape at the take-up reel. IR light from the Sensor LED shines through transparent leader at the beginning of a tape and strikes the start sensor.
- *Brake Solenoid*—This VCR has a solenoid that controls supply and take-up spindle brakes when Stop is pressed, or between FF and Rewind, etc. Actually, the brakes are *released* with the solenoid coil energized; coil voltage drops to apply the brakes. This design prevents tape from spilling if

there is a power failure while tape is being rewound, or moved; when the power goes Off, the spring-loaded plunger pulls back out of the solenoid core and applies the spindle brakes.
- *Record Safety Switch*—Detects when record safety tab on cassette is missing to prevent VCR from entering Record mode.
- *Cassette-in Switch*—Transfers when a cassette is loaded onto the reel table.
- *Take-up Sensor*—IR LED/phototransistor pair. Phototransistor emits pulses when take-up spindle rotates.
- *Mode Sensor*—Twin IR LED/phototransistor pairs detect position of cam and mode assembly. Note the two small triangles pointing to the left. These are amplifiers that increase the output of the two phototransistors.
- Finally, at the lower left, notice the seven lines interconnecting the system control and timer/operations microprocessors.

Did you also notice that the syscon microprocessor has its own 4.19MHz quartz crystal, X101? EVER 5V at Pin 32 keeps the microprocessor active whenever the VCR is plugged in. IC102 is the driver for the mode and cassette motors. On the right side of the diagram, syscon outputs feed servo, audio, and video circuits.

Remember, Figures 14.3 and 14.4 are just *block diagrams* for Timer/Operations and System Control circuits. Many IC pins, signal lines, discrete components, and voltage levels that are on the full schematic diagrams have been eliminated to more clearly show the overall operation of these circuits.

## TROUBLESHOOTING DIGITAL CIRCUITS

On rare occasions, a microprocessor or digital IC will *latch up* and appear to be malfunctioning. This could just affect one of its operations. For example, a VCR may fail to go into Record mode at the time for which it has been programmed, or perhaps it refuses to load a cassette. It may be that the microcomputer or digital chip has become "confused," and has conflicting internal indications that its built-in programming cannot resolve.

This type of lock-up or other malfunction can be caused by voltage spikes on the power line, electrostatic discharge near or upon the VCR, or high-power RF transmission in the vicinity, like from a commercial or ham radio transmitter. In rare cases, latch-up can even be caused by hitting two or more control pushbuttons in rapid succession, when your fingers sort of stumble across several buttons at once.

It is necessary to reset the microcomputer(s), so that their internal programs start all over afresh. Simply unplug the VCR for several minutes and then plug it back in to reset the microprocessor(s) and restore correct operation if the system locked up under one of these conditions. It is necessary to *unplug* the VCR, not just power it Off, because the timer/operations and system control microcomputers remain powered as long as the line cord is plugged in. This may explain a no-trouble-found (NTF) service call, where you can't duplicate what the customer says is wrong. Unfortunately, not all microcomputer and digital IC problems are so easily resolved.

There are times when you can use a multimeter to determine whether a microprocessor pin is at the right level, or is changing levels when it should. For example, suppose you have a VCR that will not load a cassette when one is gently pushed into the loading compartment. You have already checked the cassette motor with a 9V battery, as described in Chapter 13, and the motor and loader mechanics work fine.

Next you decide to see if syscon is receiving a signal from the front loader Insert Switch on Pin 27. We will look at just a small portion of the full system

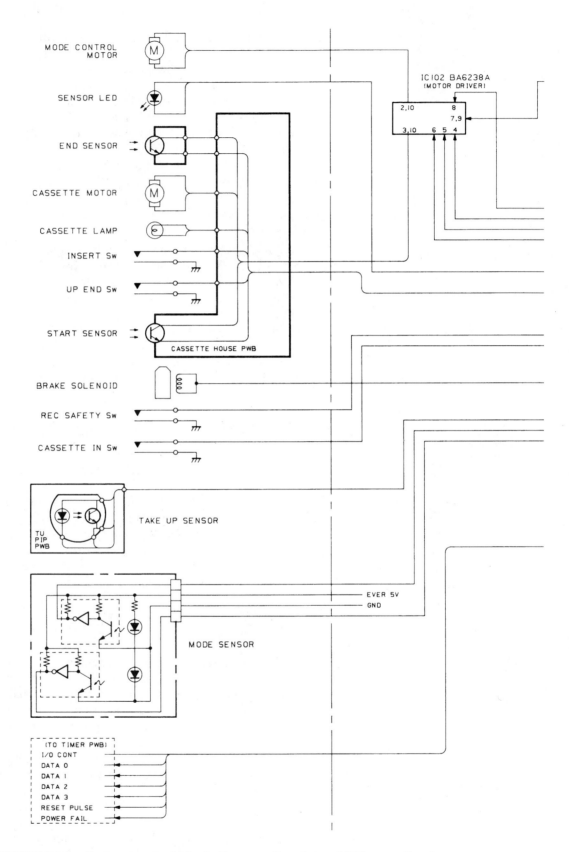

**FIGURE 14.4** System control block diagram. Courtesy of JCPenney Co., Inc.

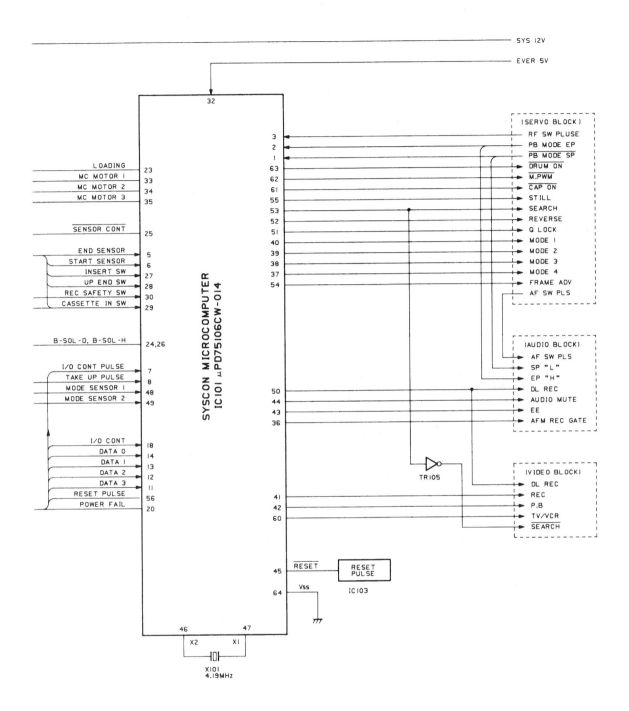

control microcomputer schematic to see how to do this, and what voltage levels to expect. Figure 14.5 is the partial schematic.

Take a minute to locate the following on the schematic:

- Cassette-insert switch, labeled IN
  — Located on cassette PWB (printed wiring board) on the front loader.
  — Notice it is shown as a normally open (N.O.) push button.
- Ground connection to one side of cassette-insert switch at CN102-1 (GND).
- +5V (EVER 5V) input at S20 (upper left).
- Pin 27 (INSERT SW) on the left side of the microcomputer chip.

Now, notice that there is a solid line *over* the words INSERT SW at Pin 27. This line means the same as the word *NOT*, so the line is really labeled "NOT INSERT SW." The voltage at Pin 27 is given as 4.9. Together, this means that when the Insert Switch is *NOT* transferred, the voltage at Pin 27 should be ≈+5V.

Follow this path to see how Pin 27 becomes about +5V:

1. Start at 5V input at S20 (upper left).
2. Drop straight down to the fourth connection point, just to the right of CN103-5.
3. Now go the right to the third connection point you come to (*don't* count the dot to the right of CN103-5 as the first one!)
4. Go up to the fourth connection point. You should be at the left end of 10KΩ Resistor R130.
5. Go to the right through R130 to the junction of R130 and 3.3KΩ Resistor R126.
6. Go to the right through R126 to Pin 27.

And so +5V is placed on Pin 27 through R130 and R126.

Now, trace the circuit when the cassette-insert switch on the front loader is transferred:

1. From ground at CN102-1, go through the closed cassette-insert switch and out CN102-2.
2. Follow line to the right, down, then right again to the junction of R130 and R126.
3. At this time, the right end of R130 is at ground while the left end of R130 is at +5V. All 5 volts are therefore dropped across R130. So ground, or 0 volts, goes to Pin 27 via R126.

Cassette-insert switch open:     Pin 27 = +5V
Cassette-insert switch closed:   Pin 27 = 0V

And now it's a simple matter to check with a voltmeter to see if Pin 27 switches from +5 to 0 volts when the cassette-insert switch closes. If it doesn't, but remains at +5 volts, then perhaps the switch is defective. Or there might be an open connection somewhere between ground and one side of the switch, or between the other side of the switch and R126.

In a similar manner, you can trace schematic wiring and check the voltage at a syscon pin for all the other switches and sensors. This is a good way to know for sure that syscon is getting appropriate input signals.

This diagnostic procedure works well for signals that change state from High to Low or vice versa when a switch closes or opens, or when a sensor changes from light to dark or from dark to light. However, some TTL signals at syscon only pulse *very briefly* when they are activated, or there might be a stream of pulses, like a string of data bits for communications between two processors. You would not be able to tell if a pulse occurred or not with a meter. A meter is no good for signals that rapidly turn On and Off or pulse—it just can't respond fast enough.

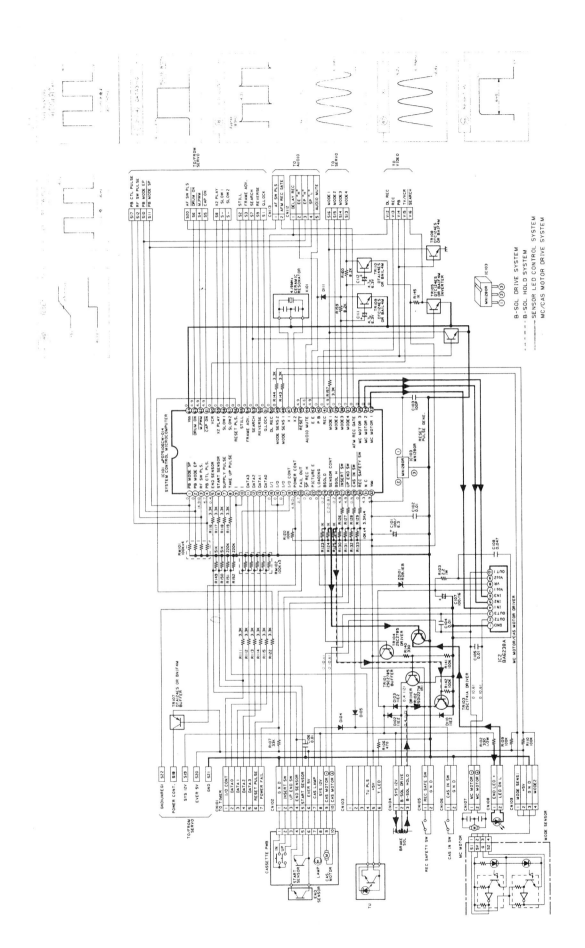

**FIGURE 14.5** Partial system control schematic diagram. Courtesy of JC Penney Co., Inc.

**NOTE: A larger fold-out version of this schematic can be found at the end of the book.**

For example, the four lines DATA0 through DATA3 that interconnect the Syscon and Timer/Operations microcomputers swing from 0V to somewhere between 4.1V and 4.9V when active. This is a good Low to High TTL transition. But the signal is a short pulse, or several pulse bursts, lasting somewhere between 0.7 and 1.9 milliseconds! One millisecond is 1/1,000 of a second, so you can appreciate that a conventional multimeter could not measure such a brief pulse. If you did probe Pin 11, 12, 13, or 14 with a meter, the meter would just remain at 0 volts, even though one or even several of these pulses were present. On a steady stream of On/Off pulses, the meter might read *some* value of voltage, but generally this would be a questionable logic level, between a true logical 1 and logical 0 for the particular logic family. (*Data stream*, *pulse train*, and *stream of pulses* all mean the same thing—a series of pulses.) Fortunately, there *is* a way to disclose the state of TTL and CMOS family logic levels and a means to detect brief pulses and pulse trains, without the expense of purchasing an oscilloscope.

## DIGITAL LOGIC PROBE

A *digital logic probe* is designed to display the presence of very brief pulses with two LEDs. One LED, usually labeled High or Up, indicates logical High or 1; the other, labeled Low or Down, indicates logical Low or 0. Some probes also have a two-tone audible signal to indicate High and Low in addition to the two LEDs. Circuitry in the probe keeps either LED lit longer than the actual width of the pulse, up to about a second. Without this feature, you would not be able to detect an LED coming On for a thousandth of a second or less.

A logic probe normally has a switch to select TTL or CMOS, and perhaps other logic technology levels. Most probes also have a Hold, Latch, or Memory switch that keeps an LED lit once a probed line has gone to that logic level, even for a single, short pulse. This feature is handy for "babysitting" a line. You don't have to keep an eye on the probe all the time.

Suppose you want to know if a line *ever* goes High, even briefly, as you work on another area of the unit being serviced. Connect the logic probe to the point you want to monitor and turn On the Hold, Latch, or Mem switch. Then, if the line *does* go High, the Up LED will remain lit until you reset it by turning the Mem switch Off.

A typical logic probe is illustrated in Figure 14.6. Some are quite small, about the size of a large pen. A logic probe may be battery-powered, or powered from an external source, such as the equipment being serviced. A battery-powered probe is easier to use, but most probes on the market are designed to be powered from the equipment being tested. *For either type, the probe's ground wire must be connected to circuit ground for the lines being probed!*

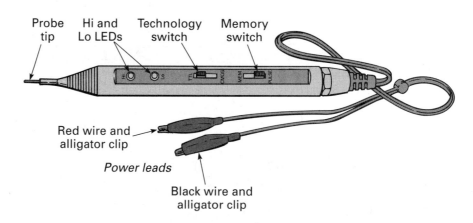

**FIGURE 14.6** Typical digital logic probe.

For the logic probe pictured, you would connect the black alligator clip to ground on the equipment you are servicing and the red clip to some positive voltage. Most probes will work on any DC voltage between +5 volts and +15 volts. On a VCR, you could connect the probe's black lead to chassis ground or a ground pin and the probe's red power lead to a +5V or +12V power supply line.

You can also power the probe from a 9V battery or from the two-way tester (TWT) described in Chapter 11. If using the TWT, be sure the DC polarity switch and connections to the probe are correct before turning the tester On.

If you power the probe in this manner, the probe's black ground lead *absolutely must* make electrical connection with ground for the circuit you are probing. One way is to clip the probe's black ground lead to ground in the VCR, then connect another test lead between VCR ground and the negative battery terminal or DC minus jack on the TWT.

Now suppose you have the probe set for TTL and the Memory or Hold switch turned Off. Here are possible indications when probing a TTL signal line, such as Pin 18 on IC101 in Figure 14.5:

- Hi LED **On**  Steady TTL High or 1 level (+5V)
  Lo LED Off
- Hi LED Off
  Lo LED **On**  Steady TTL Low or 0 level (0V)
- Hi LED **On**  TTL line is pulsing,
  Lo LED **On**  (0 ~ +5V ~ 0 ~ +5)
- Hi LED Off  Invalid or ambiguous TTL level,
  Lo LED Off  or no connection

So if both LEDs on the logic probe light and stay lit, this means there is a continuous stream of pulses on the line. You won't know the pulse rate, width, or shape, but you will know that the line is pulsing between valid TTL High and valid TTL Low voltage levels.

Now imagine you are working on a VCR. The customer says that every once in a while, for no apparent reason, it just stops when it is playing a tape. So you set it up and play a tape. Sure enough, about 15 minutes into the tape the machine unloads tape and stops. It's off with the covers! You poke around a little bit, but can't find anything wrong. The next time you play a tape, it goes for nearly an hour before stopping.

Well, you can't spend all day watching this one, so why not let a digital logic probe babysit for you? You wonder if maybe the supply side end-of-tape sensor is erroneously telling syscon that tape has run out. *This* would cause a VCR to stop (and to automatically rewind on some models). Referring to the schematic, Figure 14.5, you connect a digital logic probe to Pin 5 of IC101 (END SENSOR).

With a cassette loaded and playing, the end sensor line indicates Lo on the probe. Then turn On the probe's Mem or Hold switch. The Lo LED should remain lit and the Hi LED should remain Off. Now you can let the VCR play and let the probe babysit the line to see if the end sensor line ever goes High.

After replacing the pinch roller and a couple of belts in another VCR, you come back to the self-stopping VCR. Sure enough, it stopped playing sometime in the last hour or so, *before reaching the end of tape.*

Aha! Looking at the probe, you see that the Hi LED is lit! Switching the Mem or Hold switch Off, just the Lo LED lights. This means that sometime between when you started playing the cassette and now, the end sensor line *did* go High, but that right now it's Low, indicating tape *is* blocking IR light from hitting the end sensor phototransistor.

It seems like something is falsely causing the end sensor line at Pin 5 to go High, even though videotape is still blocking the end-of-tape sensor. The line may have gone High for just a fraction of a second, but the probe has latched the

High LED On. And now you know you're on the right track! Upon checking around the little end sensor printed circuit board on the left side of the front loader, you find a loose strand of wire on one of the two wires soldered to the board. This tiny strand looks like it's touching the other solder connection, which is only about 1/8-inch away.

That would do it! If occasionally the wire strand *did* make electrical contact with the other soldered connection, it would short the phototransistor's emitter and collector leads. This would have the same effect as the transistor conducting when IR light strikes it. (Perhaps the technician who replaced the phototransistor the last time the VCR was serviced wasn't too careful when soldering the two wires to the new circuit board.)

There are a variety of digital logic probes on the market, many in the $15 to $30 price range. Some probes also contain a *logic pulser,* which can directly inject a signal into a logic circuit. A logic pulser produces a steady stream of short duration, logic High pulses at the correct voltage level for either TTL or CMOS technology. Typical pulse width is 10μsec (microseconds). (A microsecond is one-millionth ($10^{-6}$) of a second.) Some pulsers can produce different pulse rates, for example 1 pulse every 2 seconds or 400 pulses per second. In some service work it is helpful to inject a pulse into a circuit and see if the circuit behaves as it should from that point on. A pulser is not used as frequently as a logic probe in most digital service work.

Separate logic pulsers are also available, rather than being built into a logic probe. Sometimes it's handy to have both instruments at once: the pulser to pulse a line and the probe to see the effect of the pulses elsewhere in the circuit. You can't do this with most combination logic probe/pulsers.

A logic probe and pulser are not necessities for most VCR service work. But, for their (relatively) low cost, you might consider purchasing these pieces of test equipment after you are experienced in using a multimeter to its fullest capabilities. As with any piece of test equipment, thoroughly read the operator's manual to use a logic probe or pulser safely and effectively.

## VCR SERVO CIRCUITS

It is absolutely essential, when playing a videotape, that rotary video heads A and B precisely pass over the exact same magnetic video tracks that were originally recorded by heads A and B. Recording heads A and B may be on different machines, as with a commercially pre-recorded tape, but it makes no difference. If during playback head A scans what head B recorded, or doesn't scan all of what head A recorded, there will not be a stable, clear picture, if any at all.

Heads A and B are constructed differently. If head A scans tracks laid down by head B and head B reads tracks recorded by head A, there will be no picture at all.

Thus it is necessary during playback to precisely control video drum rotation *and* videotape forward motion. By controlling capstan and drum rotation, the spinning heads align with and correctly read the tracks recorded on tape. This is accomplished by the VCR drum and capstan *servomechanism,* called *servo* for short.

What is a servomechanism? A *servo* is an automatic control system where some output is constantly compared with some input, through feedback, so that any error or difference between input and output is used to control the output. For example, an automobile cruise control is one form of servo mechanism. As a car starts up a hill, cruise circuitry senses a slight reduction in driveshaft rpm. The control then increases throttle for more fuel flow to the cylinders, which delivers more engine rpm. The end result is that driveshaft rpm remain constant, and the car maintains speed while climbing the hill.

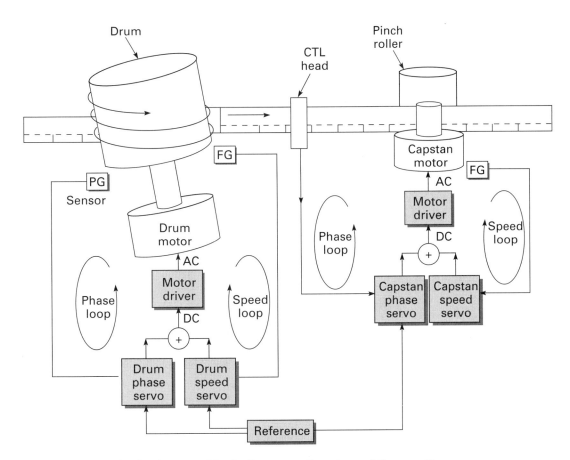

**FIGURE 14.7** The basic servo block diagram. Courtesy of *Sencore News.*

The output (driveshaft rpm) to input (amount of throttle) signal is called a *servo loop* or *feedback loop*.

A household furnace and room thermostat make up a form of servo. When the room cools below the thermostat setting, the thermostat signals the furnace to turn On (input). When the furnace has produced enough heat (output) so the room temperature equals the thermostat setting, the thermostat (input) shuts down the furnace (output). Again, there is a feedback loop between output (heat) and input (thermostat turning furnace On or Off).

The following description of VCR servos is reprinted with permission of *Sencore News,* a publication for electronic equipment servicers by Sencore Electronics Inc., Sioux Falls, South Dakota, a manufacturer of electronic service test equipment.

### The Basics of Servos

VCR servos control movement of both videotape and rotary videoheads. Servos are a combination of mechanical devices and electronic circuits. The capstan servo controls the movement of the videotape through the VCR, while the drum servo controls the positioning of videoheads. Together they ensure that the correct videohead is positioned exactly over the corresponding video track on the magnetic tape. Since videoheads spin close to 1800 RPM, and the recorded track is only a few thousandths of an inch wide, servo operation must be exact.

The capstan servo uses a motor and pinch roller to pull videotape through the VCR. An electronic motor driver supplies the current to run the motor. In order for the motor to run at the correct speed, a servo control "loop" monitors the motor rotation and another loop monitors the tape position. They supply

414 ■ *Practical VCR Repair*

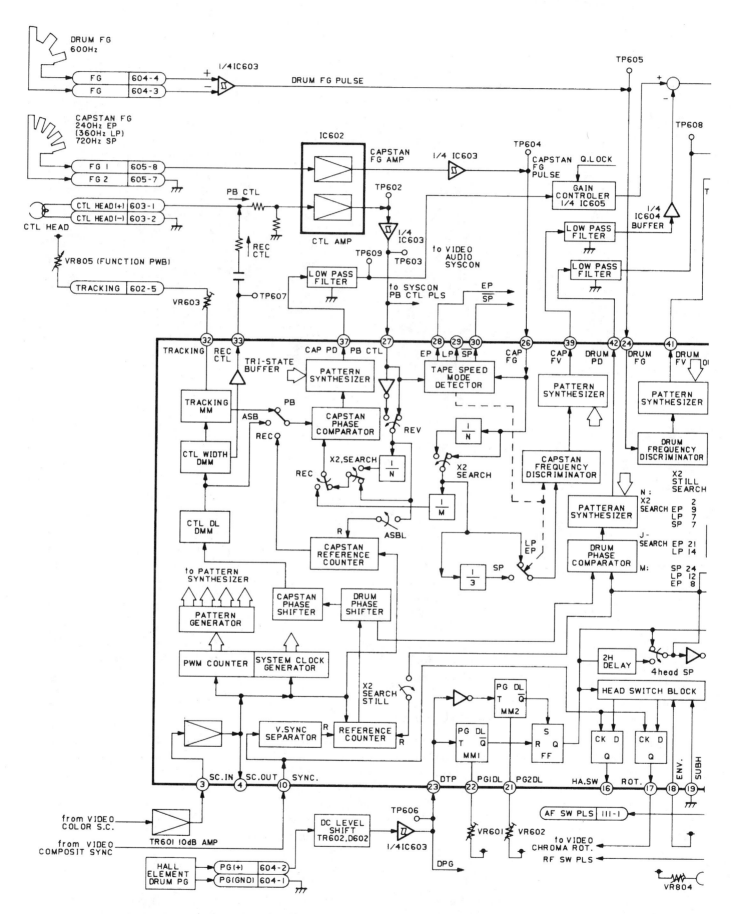

**FIGURE 14.8** Servo control block diagram. Courtesy of JCPenney Co., Inc.

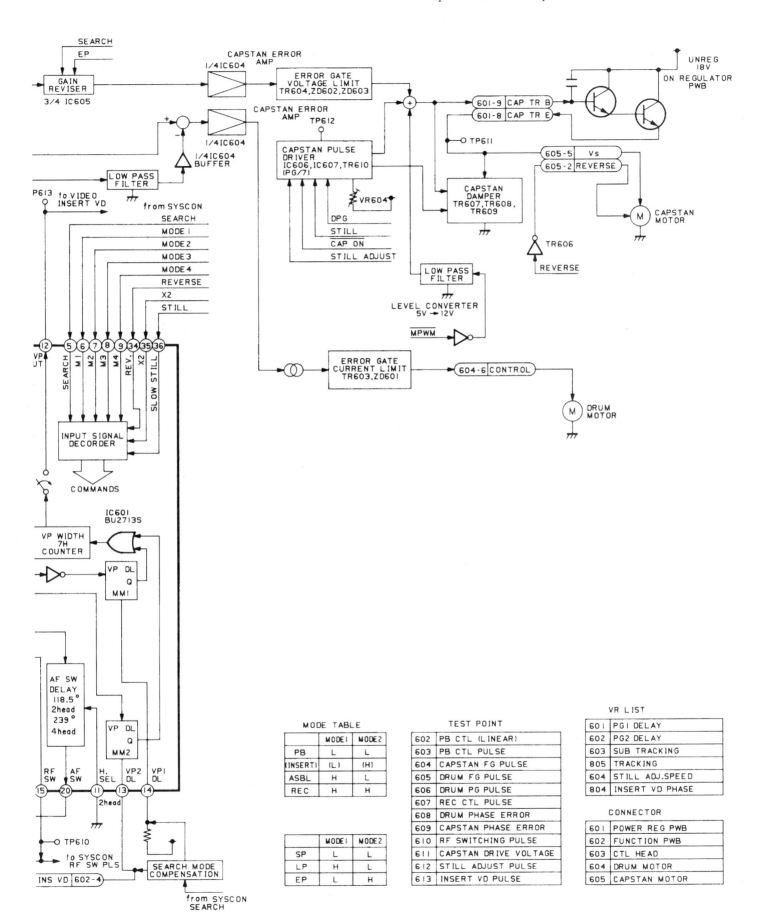

correction signals to the motor driver to correctly position videoheads to pick up a signal.

A frequency generator (FG) sensor, located next to the motor, develops a signal for the servo speed control loop to tell it how fast the motor is turning. The speed servo loop compares this signal to a reference signal and sends a correction voltage to the motor driver to correct for any motor speed variations.

A second signal, called the control track logic (CTL) signal, is obtained from videotape using a CTL head. It tells the capstan phase loop where the tape is at any instant in time. The CTL signal is compared to the reference signal and a correction signal is sent to the motor driver to speed up or slow down the motor to get the tape in the correct position at the correct time.

The drum servo controls the speed of the drum motor which rotates videoheads at approximately 1800 RPM. A similar electronic motor driver supplies the current to run the drum motor. In order for the drum motor to run at the correct speed, two control "loops" are again used to monitor the speed of the spinning heads and their position.

Like the capstan servo, the drum servo uses an FG sensor to create a signal that indicates how fast the drum is turning. This signal is monitored by the drum speed loop and a correction voltage is created to speed up or slow down the motor.

The drum servo uses a pulse generator (PG) signal to tell the drum servo phase loop *where* videoheads are in their 360° of rotation. The PG signal is compared to the same reference signal used by the capstan circuit to create a correction voltage that places videoheads at the correct position at any instant of time.

Since both the drum and capstan servo use the same reference signal to lock in their signals, they are also locked to each other. (End of *Sencore News* excerpt.)

Note that in the previous explanation, "CTL head" is the control track head in the Audio/Control (A/C) head assembly.

In many VCRs, the reference which capstan and drum servos lock onto is a 30Hz signal. This signal is derived from a much higher frequency, crystal-controlled 3.58MHz oscillator in the VCR's color circuits. That is, the stable 3.58MHz (3.579MHz) signal is electronically divided or counted down thousands of times to create a stable 30Hz (29.97Hz) reference signal for the servo circuitry. (Some VCRs use other methods for producing a servo reference signal. Consult the manufacturer's service literature.)

Servo problems can be difficult to diagnose, even with an oscilloscope and manufacturer's schematics showing proper waveforms at various test points in the circuitry. Sencore's *VC93 All Format VCR Analyzer* checks for proper servo operation and simplifies servicing servo problems.

Without an oscilloscope or other sophisticated test equipment, there is a limit to how much of a VCR's servo circuits you can diagnose. In many VCRs, however, a majority of the servo circuits are contained on a single integrated circuit chip. You can check some DC voltages and continuity of lines to the chip, and perhaps even use a logic probe to check for the presence of various pulses. (Further servo system diagnosis is beyond the scope of this textbook.)

Just to give you an idea of the circuitry involved in a VCR servo system, Figure 14.8 is a servo control block diagram for a typical VCR. Note that the majority of the circuitry is contained within a single IC. The following are some items you may want to locate on the servo block diagram:

- Inputs:
  — Drum FG signal (upper left), amplified by 1/4 of IC603, to servo IC Pin 24.
  — Drum PG signal (lower left), shaped by transistor TR602 and diode D602, amplified by 1/4 of IC603, to servo IC Pin 23.
  — Capstan FG signal (upper left, beneath drum FG), amplified by 1/2 of IC602 and 1/4 of IC603, to servo IC Pin 26.

- CTL pulses from control head (upper left, beneath capstan FG), amplified by 1/2 of IC602 and 1/4 of IC603, to servo IC Pin 27.
- 3.58 MHz oscillator signal enters Pin 3 of servo IC after being amplified by transistor TR601 (lower left).

- Outputs:
  - For capstan motor: servo IC Pins 37 and 39. Trace each line through low pass filter and other circuitry to the capstan motor, upper right. Follow arrows on block diagram lines.
  - For drum motor: servo IC Pins 41 and 42. Trace each line through low pass filter and other circuitry to drum motor, upper right.

One additional item deserves mention before concluding the discussion of servos. In playback, a signal derived from the drum pulse generator (PG) signal switches between the outputs of Videoheads A and B. The PG signal indicates the *position* of the drum and therefore the position of the two heads. To reduce noise pickup, it is desirable to take the playback video signal only from the head actually contacting videotape. You will see how this head-switching signal is used in the next chapter, during the discussion of the VHS playback block diagram.

Again, the intent in describing VCR servos and their block diagram here is just so you have an overview of these circuits. As you gain electronic servicing experience, *and* obtain access to an oscilloscope and other test equipment, you may be able to economically diagnose and repair some servo failures. Without this equipment, or a specialized VCR servo analyzer, like that which is included in Sencore's VC93, even the most experienced electronic technician would be hard pressed to identify and correct many problems in this area of a VCR.

## SUMMARY

This chapter began by explaining what an integrated circuit (IC) is and how to identify pin or lead numbering on an IC. An integrated circuit can have just a few or hundreds of thousands of transistors, diodes, and resistors manufactured on a single slab of silicon, called a chip. It thus replaces many discrete components. An IC may have as few as 4 leads or as many as 100 or more.

IC pins are numbered counterclockwise (CCW) around the edge of an IC when looking down on the top of the component. Pin number 1 is often identified by a dot or circle on the top of the IC. Pin 1 is just to the left of a U-shaped notch or mark at one end of an IC, or just to the left of a cut-off corner, going CCW.

An IC may be designed to work in an analog circuit with constantly changing voltage and frequency, like audio and video circuits, or in a digital circuit. Digital ICs have signals of only two voltage levels, one signifying On or 1, the other level meaning 0 or Off. Digital ICs are used as timer/operations (or function) and system control microprocessors in a VCR. Digital ICs also form character generators for on-screen menus and prompts and provide special effects.

There are two fundamental families or technologies of digital ICs: TTL and CMOS. Each operates with specific voltage levels to indicate On or logical High, and Off or logical Low. TTL stands for transistor-transistor logic, which is sometimes called LS technology, meaning linear series (LS). (Linear series ICs can also be analog chips, such as an operational amplifier (Op Amp).) CMOS is complementary metal-oxide semiconductor, often called just MOS technology. Logic High and Low voltage levels for these two technologies are determined differently. For TTL, 0 volts is Low and 5 volts is High. TTL levels are ambiguous between +.8V and +2.8V. For CMOS, $V_{CC}$ is the power supply voltage to the device; logic Low is from 0 volts to 20 percent of $V_{CC}$ and logic High is from 70 percent to 100 percent of $V_{CC}$. Ambiguous CMOS levels are between 20 percent and 70 percent of $V_{CC}$. Depending on the particular component, digital signal levels in the ambiguous or questionable area may produce erratic or unpredictable operation.

### Technology Logic Levels

| Technology | Off/Low/0 | On/High/1 | Ambiguous |
|---|---|---|---|
| TTL/LS | 0 ~ +.8V | +2.8 ~ 5V | +.8 ~ 2.8V |
| CMOS/MOS | 0 ~ .2$V_{CC}$ | .7$V_{CC}$ ~ $V_{CC}$ | .2$V_{CC}$ ~ .7$V_{CC}$ |

where $V_{CC}$ is power supply voltage to the particular component.

A digital logic probe is a handheld piece of test equipment for checking TTL and CMOS circuits. A probe has two LEDs, one labeled Hi, the other Lo. These LEDs indicate whether the probed point is at a logical High or logical Low level. Both LEDs light if there is a continuous pulse stream on the line.

Many probes have a Hold, Latch, or Memory switch that keeps the Hi or Lo LED lit after a logic High or Low pulse is sensed on the line. This is a handy feature for tracking a circuit that may have only an occasional logic transition or pulse. Some digital logic probes also have a built-in logic pulser, which can inject logical High pulses into the circuit under test. Logic pulsers are also available as a separate piece of equipment. Pulsers are sometimes helpful when troubleshooting digital circuits. Although not essential for most VCR repairs, a digital logic probe and digital logic pulser may be worthwhile pieces of equipment to have on hand, once the servicer is experienced using a multimeter to isolate problems.

VCR drum and capstan servos precisely control the position of rotary videoheads and the speed of the tape through the transport. This is essential if the two videoheads are to be properly positioned over the recorded tracks on tape during playback. Capstan speed control is obtained by comparing a frequency generator (FG) signal that monitors capstan speed with a stable reference signal. Fine tape positioning control of the capstan motor is obtained by comparing the 30Hz control track logic (CTL) signal recorded near the bottom edge of the tape with the reference signal.

In a similar manner, drum speed is acquired by comparing an FG signal that monitors drum rpm with the reference signal. Head positioning control is acquired by comparing a pulse generator (PG) signal that monitors drum position with the reference signal.

Many VCRs count down the 3.579MHz oscillator in the color circuits to 29.97Hz, which is then used as a reference signal in the drum and capstan servo circuits.

In later model VCRs, a majority of servo circuits are contained in a single IC. In general, an oscilloscope or specialized piece of test equipment is required to efficiently troubleshoot servo problems. However, a multimeter and a logic probe may be used in some cases to identify a missing or incorrect DC voltage, open wiring, or missing pulse in a servo circuit.

## SELF-CHECK QUESTIONS

1. Describe the various methods for orienting an IC and locating lead number 1.
2. Describe how to locate pin number 37 on a DIP IC having 64 pins.
3. Give three other ways in which a digital logical 1 level, or state, can be written or expressed. What are three other ways of expressing a logical 0 level?
   a. Logical 1 =
   b. Logical 0 =
4. What voltage range is a valid logic-1 level in a TTL circuit. What voltage range is valid for representing a logic-0 level?
   a. TTL logical 1 =
   b. TTL logical 0 =
5. Refer to Figure 14.3. Which timer/operation microcomputer lines are connected when the VCR front panel Play button is pressed?
6. Refer to Figure 14.4. Which syscon pin would you probe with your multimeter to see whether the Cassette-in switch on the reel table was transferring correctly? What voltage, with respect to ground, would you expect to see on this pin with the cassette-in switch transferred?
7. How do you reset a microcontroller that you suspect may be latched up, for example, when the timer/operations microcomputer doesn't turn the VCR on at the required time for unattended recording or the VCR seems to be locked up, refusing to do anything?
8. Refer to Figure 14.5. Which component directly keeps or holds the brake solenoid energized during Play, Fast Forward, Rewind, and other modes when tape is in motion?
9. Refer to Figure 14.5. How could you detect the presence of data bits on the DATA2 interface line between the timer microcomputer and the syscon microcomputer on pin 12 of IC 101? Notice the waveform in the second block from the top at the right of Figure 14.5, which represents pulses on data lines 0 through 3.
10. What tells the drum servo phase loop where the video heads are in their 360 degrees of rotation?

# CHAPTER 15
# HOW A TV PICTURE IS MADE

In this chapter you will study the fundamental operations that produce a picture on a TV screen. Understanding how a TV picture is made becomes important when diagnosing many VCR problems. After all, a VCR has to perform many of the same basic operations to create a picture as a TV camera or videotape deck at a television broadcasting studio.

First you will learn how a black-and-white image is produced on a TV screen, and then the basics of creating a color picture. During these discussions, you will discover how a cathode ray tube (CRT), or TV picture tube, works. Electronic signals that control the creation of a TV image are described in general, fundamental terms. It is not the intent of this chapter to delve into details of the various video signals and their wave shapes, but instead to give an overview of what these video signals accomplish.

## CHAPTER OBJECTIVES

**Upon completing this chapter, you should be able to:**
1. Describe basically how a black-and-white image is produced on a TV screen, including the terms:
   - Pixel
   - Raster
   - Horizontal and Vertical beam deflection
   - Retrace and blanking
   - Video field, video frame, and interlaced scanning.
2. Describe the fundamental method used to produce a color TV picture, including the terms:
   - Luminance or Y signal
   - Chrominance or C signal
   - Color burst signal.
3. Describe the basic construction of a color cathode ray tube and how it works.

## PRODUCING A BLACK-AND-WHITE TV IMAGE

It probably surprises many people to learn that the entire image on a TV screen is made by just one very small spot of light, but that's how it works.

Imagine yourself in a dark room, holding a flashlight pointed at a wall about four feet away. If you move the flashlight very rapidly left and right, and up and down, instead of seeing just a single circle of light on the wall, you'll see what appears to be solidly lit streaks or lines or geometric shapes, depending on how you move and point the flashlight's beam.

It is the human eye's *persistence-of-vision* characteristic that makes the light appear to remain in one area even though the actual beam has moved. The visual impression is that the flashlight beam is illuminating many places on the wall at the same instant, when in fact it is really lighting just one spot at a time.

Persistence of vision makes movies possible. Images on a movie theater screen are created by projecting about 24 separate frames of film each second; the screen actually goes black between each frame. But because of persistence of vision, we see smooth motion on screen, with no interruptions.

Television is also possible because of persistence of vision, but instead of an entire frame or picture being projected on the screen at one time, as with film, only a single dot, moving rapidly around the screen surface, produces the image we see. So in TV, a single dot has to sweep over the entire surface of the screen, *and* the dot's intensity has to be varied as it moves to produce black, gray, and white areas of the image.

## BASIC CRT OPERATION

In a TV picture tube, or *cathode ray tube* (CRT), an *electron gun* at the rear of the tube's neck fires a narrow beam of electrons toward the face of the screen. The inside surface of the screen is covered with phosphor material that lights up white when struck by these electrons.

Electromagnetic *yoke* coils around the outside of the picture tube neck move the narrow electron beam back and forth (horizontally), and up and down (vertically) to cover the entire surface of the screen. A magnetic field *deflects* an electron beam, either toward the field or away from it, depending on whether the field is north or south.

Inside the CRT, a control element called a *grid* changes the intensity of the electron beam that is swept over the tube's face by the yoke coils. This control grid can vary the intensity of the electron beam from completely Off, producing black on the screen, to full On, producing bright white. At intensities between full Off and On, the rapidly moving dot can produce many shades of gray.

Figure 15.1 is a simplified drawing of a CRT. A filament heats the cathode, which emits electrons. The control grid varies the intensity of the electron beam, while the horizontal and vertical coils in the yoke deflect the beam to cover the entire tube face.

The very small dot or area lighted on the tube face by the electron beam at any instant in time is called a *pixel*, which is short for *picture element*. A pixel is the smallest part of a picture produced on a CRT or display screen. (*Pel* is an equivalent term for pixel.)

Figure 15.1 shows two pixels, A and B. Realize that both pixels could never be lit at the exact same time. Pixel A is illuminated with no horizontal or vertical deflection of the beam by the coils in the yoke, and so the pixel appears in the very center of the screen.

Pixel B is above and to the right of Pixel A. The electron beam was moved here by the horizontal and vertical deflection coils in the yoke. The horizontal coils deflected the beam to the right of center; at the same time, the vertical coils deflected the beam upward. By varying the sweep signal to the horizontal and vertical coils in the yoke, the beam can be moved so it strikes anywhere on the face of the CRT.

The many thousands of pixels covering the entire tube face are called a *raster*. A raster is the illuminated and scanned area on the face of a picture tube. How is this term used? You've probably seen a TV screen where some problem prevented the image from filling the entire screen; instead, there was a black border at top and bottom, or perhaps all the way around the image. This would be described as a *shrunken raster*.

**FIGURE 15.1   Basic CRT operation.**

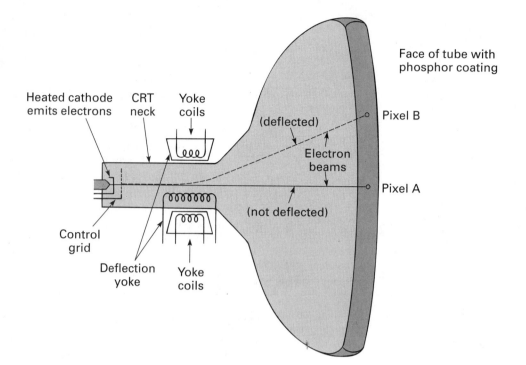

Next is a description of how a TV raster is produced in the United States, with standards defined by the National Television System Committee (NTSC). European TV standards are different, and although the fundamental principles of operation are the same, a TV set or VCR designed to NTSC standards will not work in parts of the world using other standards, such as PAL (phase alternation line) or SECAM (sequential couleur á memoire). *NTSC and PAL/SECAM are incompatible!*

## VIDEO FIELDS AND FRAMES

In the United States, a TV raster consists of 525 horizontal lines of pixels on the screen, from top to bottom. All 525 lines of the raster are called a *frame*. Thirty video frames are produced each second.

Each of the 30 frames produced each second in turn consists of two halves, called *fields*. Field 1 is made up of "even" horizontal lines, whereas Field 2 is made up of "odd" lines. That is, Field 1 makes 262-1/2 lines and Field 2 makes the other 262-1/2 lines for the total of 525 horizontal lines in a frame.

At the beginning of Field 1, the beam strikes the upper left-hand corner of the CRT face. The horizontal deflection coils then move, or *trace*, the beam to the right side of the screen at a rate of 15,750Hz. At this rate, it takes 63.4 microseconds for the dot on the face of the tube to go from left to right:

Time = 1 ÷ Frequency = 1 ÷ 15,750 = 0.000,0634 seconds

(A microsecond is one-millionth ($10^{-6}$) of a second.)

At the same time the beam is tracing from left to right under control of the horizontal deflection coils, the vertical deflection coils pull the beam downward at a much slower rate of 60Hz. The end result is a downward sloping scanning line, going from left to right.

Once the beam finishes scanning or tracing one line from left to right, the beam is electronically turned Off. Then the horizontal deflection coils very rapidly return the beam to the left side of the screen. This is called *horizontal retrace*. Turning the beam completely Off during retrace is called *blanking*.

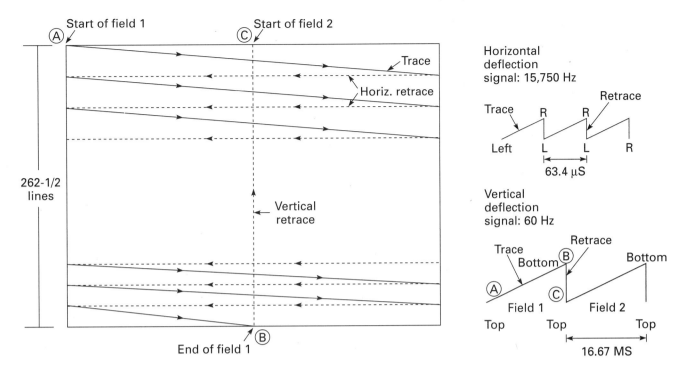

**FIGURE 15.2** Video Field 1 scanning. The horizontal and vertical deflection signals show the voltage amplitude produced by the TV's sweep circuits. These sweep signals are sent to the two sets of deflection coils in the CRT yoke. The X-axis (horizontal) is time. The Y-axis (vertical) is voltage.

As soon as the beam is back to the left side, it is turned back On and the next scan line begins, tracing from left to right, sloping downward at a slight angle. This trace/retrace sequence is repeated until all 262-1/2 lines of Field 1 are scanned. At the end of Field 1, the electron beam is at the very bottom, center of the raster. Because the vertical scanning rate is 60Hz, it takes 1/60 of a second, or 16.67 milliseconds, to scan Field 1. Figure 15.2 illustrates Field 1 scanning. For simplicity, only the first few and last few of the 262-1/2 scan lines are shown.

At the completion of Field 1, the beam is turned Off, or blanked, during *vertical retrace,* which positions the beam at the top center of the raster, where Field 2 begins.

Of course, during Field 1, the CRT control grid is constantly varying the intensity of the beam as each line is traced to produce a black-and-white image, with shades of gray also, on the face of the tube.

Video Field 2 is produced similarly to Field 1. The beam starts at top center. Under control of the horizontal and vertical deflection coils, it traces left to right, working down the screen until it completes 262-1/2 lines at the bottom right corner of the screen. At this point Frame 2 is complete and the beam is blanked. Vertical *and* horizontal retraces together return the beam to the upper left corner to begin Field 1 of the next video frame. Figure 15.3 illustrates video Field 2. For clarity, only the first few and last few of the 262-1/2 scan lines are shown.

This method of producing an image is called *interlaced scanning,* because of the "even" numbered and "odd" numbered lines in Fields 1 and 2, respectively, that together make up a full video frame. Because Field 1 takes 1/60 of a second and Field 2 also takes 1/60 of a second, the entire frame takes 2/60 or 1/30 of a second. So, in 1 second, 30 video frames are produced.

The phosphor on the screen surface has some amount of persistence; that is, it tends to remain aglow a short period of time after the exciting electron beam has moved on. This characteristic, along with the human eye's persistence of vision, works to produce an apparent smoothly changing picture.

**FIGURE 15.3** Video Field 2 scanning.

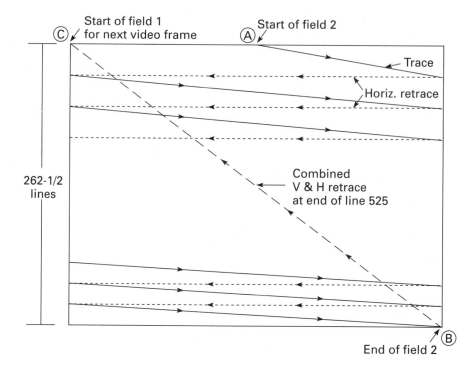

To produce a black-and-white picture, a TV station or cable channel produces a complex signal. This signal must do two primary things. It must vary the brightness of the pixels as the electron beam sweeps the face of the CRT; this creates the image. It must also synchronize (sync) the scanning rate of the electron beam in the CRT with the TV camera that produced the picture in the first place. This ensures that each pixel picked up by the TV camera is positioned at the correct location on the TV screen.

A composite black-and-white video signal contains:

- Brightness information, also called the *luminance* or *Y signal,* which varies the intensity of the electron beam as it traces the face of the CRT.
- Horizontal sync pulses
  — a sample of the horizontal scan signal at the TV camera
  — sweep circuit in TV produces drive signals for horizontal deflection coils that are synced, or locked, to the broadcast horizontal sync pulses.
- Vertical sync pulses
  — a sample of the vertical sweep signal at the TV camera
  — sweep circuit in TV produces drive signals for vertical deflection coils that are synced to the broadcast vertical sync pulses.
- Horizontal and vertical blanking pulses turn the CRT electron beam Off during horizontal and vertical retrace intervals.

The words *scan, sweep,* and *trace* are used interchangeably, so don't get confused by these different terms. You've no doubt seen a TV screen rolling vertically. This is what happens when the TV's vertical sweep circuits are not locked onto the vertical sync pulses being broadcast. Horizontal rolling or instability is loss of horizontal synchronization. In either case, the lines making up a video frame do not scan or trace in step with the TV camera.

## PRODUCING A COLOR TV IMAGE

It may come as a surprise that a color TV picture tube does not actually produce all the colors you see. It really creates different intensities of just three colors: red, green, and blue. A color CRT has three different types of phosphor coatings on the face of the tube, arranged in either a dot-triad pattern or in vertical stripes, depending on the manufacturer. One phosphor glows red, another green, and a third blue when struck by an electron beam.

A *dot triad* consists of one tiny dot of each color phosphor (red, green, blue) arranged in a tight triangle. Each dot triad forms a pixel on the screen. There are thousands of these dot triads on the CRT face. A vertical-stripe screen has extremely narrow strips of the three phosphor colors across the CRT face.

The three color dots in a triad, or three adjacent vertical color stripes, are so close together that unless your face is right up to the screen the individual red, green, and blue colors blend together. By mixing different amounts of these three primary colors, all possible colors, as well as white, can be created.

White is produced with a mixture of 30 percent red, 59 percent green, and 11 percent blue. Different percentages of these three colors produce flesh tones, yellow, purple, pink, aqua—you name it. For example, red and green make yellow.

Within the CRT, three separate electron guns produce three distinct electron beams. The intensity of each of the three beams is independently controlled. Each beam is precisely aimed electronically at only one color phosphor. The electron guns are therefore referred to as the red gun, green gun, and blue gun. Because each gun can independently vary the intensity of its beam, and therefore the brightness of the color phosphor excited by that gun, all colors at all brightness levels can be produced.

Again, only different intensities of red, green, and blue are actually produced on the face of the CRT, but because the dots of a triad, or thin vertical stripes, are so close together, the human eye blends the three colors into one overall color. Close inspection of the face of an operating color tube with a magnifying glass will reveal the individual color dots or vertical stripes.

All three electron beams from the three guns are deflected horizontally and vertically to sweep the face of the CRT at the same time, just as in a black-and-white CRT. There are very small but precise differences in where the three beams are aimed. This ensures that electrons from the red gun hit only red phosphor, green gun electrons strike only green phosphor, and blue gun electrons excite only blue phosphor. Inside the CRT, a *shadow mask* or *aperture grill* blocks each electron beam from impacting any color phosphor but its own as the beams are swept horizontally and vertically.

Figure 15.4 is a very simplified view of how a color CRT works. At any instant in time, the beams from the three electron guns in the neck of the CRT pass through the same opening in a shadow mask inside the tube, and then strike their respective colored phosphors. *Convergence* is the process of precisely controlling the three beams magnetically and electrostatically so each gun's electrons impact only one color phosphor over the entire face of the CRT.

Dot-triad screen CRTs are called "delta system" because the three colored dots form a triangle, like the Greek letter delta, Δ. A Sony Trinitron system, technically speaking, has only a single gun, but three separate electron beams are produced.

Color TV broadcasts contain all the signals of a black-and-white transmission, including the overall luminance Y signal, plus a *chrominance* signal, called *chroma* for short. The chroma signal conveys color information. In a TV set, the chroma signal determines the intensity of each of the three color guns at any instant of time as the screen is being scanned.

The following is a broad overview of how the chroma signal is produced from a color TV camera. It may be a bit technical for you at this point. Don't be concerned if you don't fully understand these explanations. As long as you get the overall idea of how color transmissions are made, that is all that is necessary. As you become more experienced in VCR repair, the following information should help you follow schematic diagrams and troubleshoot a VCR's color circuits.

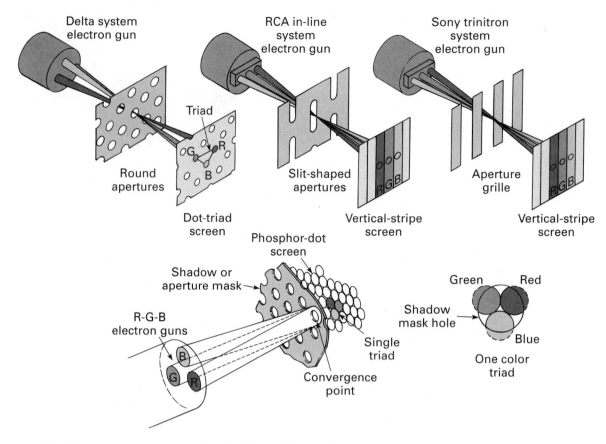

**FIGURE 15.4** Color CRT operation. Adapted from *Sencore News*.

- A color TV camera has three color filters that separate the image its lens "sees" into red, green, and blue. Each color produces a separate signal in the camera that varies in amplitude with the intensity of the three primary colors.
- The three color signals are combined in the proportion 30 percent red, 59 percent green, and 11 percent blue to produce the luminance (Y) signal, which is transmitted. This enables a black-and-white TV to receive a color broadcast.
- The blue output of the camera is combined with the Y signal to produce an I signal.
- The red output of the camera is combined with the Y signal to produce a Q signal.
- Both I and Q signals separately *modulate,* or change the characteristics of, a 3.58MHz subcarrier that is transmitted along with all the regular signals for a black-and-white broadcast.
  — The I signal is shifted 90 degrees from the 3.58MHz subcarrier, whereas the Q signal is shifted 57 degrees. This phase difference is detected in the TV set to separate the I and Q signals.
  — The 3.58MHz subcarrier is suppressed, which means the carrier itself is not transmitted, but only sidebands produced by the I and Q signals. But a small sample of the subcarrier is transmitted during each horizontal blanking interval, when the electron beams are turned Off and moved from the right side of the screen back to the left. This 3.58MHz *color burst* synchronizes a 3.58MHz oscillator in the TV or VCR. This 3.58MHz signal is then used to *demodulate,* or recover, the chroma, or *C signal,* consisting of the separate I and Q signals.

Notice that no specific information is transmitted for green. The I signal relates to the amount of red; the Q signal relates to the amount of blue; the Y signal is the overall intensity of red, green and blue together. By essentially subtracting the intensities of red and blue—determined by chroma signal components I and Q—from the Y signal in the TV set, the amount of green is determined.

A VCR demodulates the C signal and records it on videotape. There is no need for a VCR to separate I and Q signals from the chroma signal. After all, it has no separate electron guns! The luminance (Y) signal, plus horizontal and vertical sync and blanking signals, are also recorded on videotape. How the luminance and chroma signals are recorded is covered in Chapter 16.

During VCR playback, both Y and C signals are combined and sent to the TV as a composite line-level signal (RCA Video Out jack), or come from the VCR's modulator, which converts these signals to VHF Channel 3 or 4. The TV set then separates the chroma signal from the combined Y and C signals, using the 3.58MHz color burst that occurs during horizontal blanking as a reference. The TV further separates the C signal into I and Q signals. The intensities of the red, green, and blue signals as originally produced by the color TV camera are redeveloped and control the three CRT beams.

Yes, this is a lot of information. We hope that much of this material will become clearer in Chapter 16, where you'll see some block diagrams of VCR record and playback circuits.

Right now, you should know that the overall color TV video signal, whether it is broadcast on the air, comes through a cable, is recorded on a VCR, or is played back on a VCR, contains two main video signals:

- Luminance, or Y signal
  — Primarily conveys black-and-white intensity.
  — Also contains:
    • Horizontal and vertical sync signals, to synchronize horizontal and vertical sweep generators in TV set.
    • Horizontal and vertical blanking signals, to turn Off electron beams during horizontal and vertical retrace.
    • Color burst; sample of 3.58MHz chroma subcarrier, synchronizes 3.58MHz oscillator in TV or VCR. Burst occurs during each horizontal retrace interval.
      • Black-and-white TV ignores color burst.
      • Insufficient level of color burst signal causes color TV set or VCR to produce black-and-white on a color program.

- Chroma, or C signal
  — Conveys red and blue intensity
  — Demodulated from received signal (broadcast or cable) using 3.58MHz oscillator, synced by color burst.
  — VCR does not need to separate chroma signal any further.
  — TV separates C signal into I signal, related to red, and Q signal, related to blue. With Y, I, and Q signals, TV recreates intensity of red, green, and blue originally captured by color TV camera lens.

These are the main video signals that make up a TV transmission. You may be wondering about the audio. Audio is sent out by an entirely different transmitter and occupies a different portion of the TV channel bandwidth. Video is *amplitude modulated,* whereas the audio is *frequency modulated,* similarly to a standard FM radio station signal.

In fact, if there is a Channel 6 TV station on the air in your area, you can most likely tune in the audio portion on an analog tuning FM radio or receiver at the low end of the dial, just below 88MHz. (Most digital receivers won't tune below 88.1MHz.)

# SUMMARY

This chapter began by describing how a black-and-white picture is produced on a TV picture tube or cathode ray tube. An electron gun in the base of the tube produces a beam of electrons, which is fired onto the back of the phosphor-coated screen surface. The phosphor glows when struck by electrons. A control grid in the tube varies the strength of the electron beam, and therefore the brightness of the impacted phosphor.

The electron beam inside a CRT is swept or scanned across the face of the tube by horizontal and vertical deflection coils in the yoke around the neck of the tube. In the United States, the horizontal coils operate at a sweep frequency of 15,750Hz, while the vertical coils operate at 60Hz. The area of phosphor that can be excited by the swept beam is called a raster.

In the United States, a TV picture consists of 30 frames each second. Each frame is 525 horizontal lines, consisting of two interlaced video fields of 262-1/2 lines each, called Field 1 and Field 2. Within each field, the electron beam starts at the top of the screen and is swept left to right to make one horizontal line. At the end of each horizontal line, the beam is turned Off or blanked during retrace, when the beam is rapidly moved back to the left side of the screen. At the same time the beam moves left to right under control of the horizontal deflection coils, it also moves downward under control of vertical deflection coils. At the completion of each field, the beam is at the bottom edge of the screen. It is blanked during vertical retrace, which relocates the beam at the top edge of the CRT.

Black-and-white intensity is conveyed by the luminance or Y signal. Also present in a black-and-white signal, as well as in color broadcast or cable signals, are horizontal and vertical sync and blanking pulses.

Color TV pictures are produced by combining different intensities of red, green, and blue on the face of a picture tube. Within a color CRT, three electron beams are aimed so that each impacts only its own color phosphor on the screen. A shadow mask or aperture grill behind the CRT face blocks electrons from striking the two colors other than the one to which it is assigned. All three beams in a color CRT are swept to create video Fields 1 and 2, just as in a black-and-white CRT.

Color information is broadcast as a chroma or C signal. The chroma signal consists of an I and a Q signal. The I signal relates to the intensity of red in the image, and Q relates to the blue. The chroma signal is transmitted on a 3.58MHz suppressed carrier subcarrier along with the black-and-white signals on the main carrier. A sampling of the 3.58MHz suppressed carrier is transmitted during each horizontal blanking pulse. This color burst signal synchronizes a 3.58MHz oscillator in the VCR or TV set so that the chroma signal can be demodulated or recovered.

Although a VCR separately demodulates the luminance (Y) and chrominance (C) signals, it does not need to separate the I and Q signals from the chroma signal.

The luminance, horizontal and vertical sync and blanking, and color burst signals are recorded and played back together on a VCR. The chroma signal is separated from the composite received signal before being recorded on a VCR.

# SELF-CHECK QUESTIONS

1. Describe the basic operation of a black-and-white TV picture tube.
2. For NTSC TV, as used in the United States, describe the relationship between video *fields* and video *frames*. How many of each are produced

each second? How many lines of horizontal video information are in a full *raster*?

3. What is the term that describes the CRT electron beam moving from the right end of a horizontal scan line back to the left side of the screen? What term describes turning the electron beam Off during this action?
4. What is the name of the video signal that varies the intensity of a black-and-white TV picture? This signal is also known by a single-letter name; what is it?
5. Describe the purpose of horizontal and vertical sync signals.
6. How many different color phosphors are there in the coating on the internal screen surface of a color TV picture tube? What are the colors? Describe how the phosphor colors can be arranged on the CRT face.
7. What is the name of the video signal that conveys color information? This signal is also known by a single-letter name; what is it?
8. What type of modulation is used for TV *video?* What type of modulation is used for TV *audio?*

# CHAPTER 16 RECORDING ON VHS VIDEOTAPE

This chapter first explains the fundamental principles of magnetic tape recording, and then describes how VHS videotape is recorded. Basic magnetic tape recording is essentially the same for videotape as it is for reel-to-reel and cassette audio tape. If you already know how sound is recorded and played back, the first part of this chapter will be a review.

Because video material contains much higher frequencies to record and play back than audio, it is not possible to directly record the baseband signals on tape, as with audio. Instead, several departures from and enhancements of audio recording technology are employed in a VHS VCR to record the wide frequency range present in a video baseband composite signal.

You will learn that both the luminance (Y) and the chrominance (C) signals that make up a color TV picture are first separated from the composite baseband signal. Each signal is then changed or modified before being recombined and sent to the two rotary video tape heads for recording. Luminance and chroma signals were described in Chapter 15.

Understanding the overall operation of several VCR video circuits is very helpful in diagnosing VCR record and playback problems. This chapter describes record and playback video circuit functions at the block diagram level for a typical VHS machine.

Although some of the discussions are somewhat technical, it is not essential that you fully comprehend all the detail. The primary intent of this chapter is for you, the student, to get an overview of how VHS video recording and playback take place.

## CHAPTER OBJECTIVES

Upon completing this chapter, you should be able to:
1. Describe the basic principles of magnetic tape recording and playback.
2. Describe fundamentally how the luminance (Y) and chroma (C) video signals are recorded and played back on a standard VHS VCR.
3. Describe overall operation of standard VHS VCR record and playback video circuits, using a block diagram as reference.

It is *not necessary* to know all the information in this chapter to diagnose and repair the majority of VCR malfunctions, but some readers may be interested in the theory and concepts presented. Many readers may wish only to skim this material. That is fine! You may appreciate why rapidly rotating tape heads are needed to record and play back video, but don't get bogged down in the details of head gap width, azimuth, wavelength, tape velocity, and frequencies. The main thing is to understand the fundamental way in which VHS videotracks are recorded on magnetic tape and then played back.

As you become more experienced in electronics and VCR servicing, a basic understanding of video record and playback circuit operations will enable you to diagnose and repair problems in these areas, using manufacturers' service literature. As stated previously, *most VCR repairs do not involve these circuits. Rather, most repairs are made to mechanical systems, power supplies, switches, motors, and sensors*, which have been covered in previous chapters. What is more, troubleshooting video circuits is nearly impossible without an oscilloscope, and perhaps a video pattern generator, as minimum test equipment.

Before beginning, the following definitions of terms may be helpful:

- The Greek letter lambda (λ) stands for *wavelength*. As used in the discussion to come, wavelength is just the *distance* on tape, in meters, that it takes to record a complete AC cycle, such as from the beginning of one positive-going half cycle to the beginning of the next positive half cycle.
- G stands for the width of the gap in a magnetic tapehead, measured in meters.
  Short wavelengths and small head gaps can be expressed in micrometers (μm). One μm is one millionth (1/1,000,000 or $10^{-6}$) of a meter.
- V stands for the velocity, or speed, of magnetic tape passing a point, measured in meters per second.
- f stands for frequency, measured in Hertz (Hz) (or cycles per second).
- The symbol ≤ is a combination of < and =, and means "less than or equal to."
- Each octave represents a doubling of the frequency. For example, the span from 440Hz (A above middle C on a musical instrument) to 1,760Hz is two octaves. (440 to 880 = first octave; 880 to 1,760 = second octave.)
- dB stands for decibel, which is a measure of signal strength. Six dB represents a doubling (or halving) of the voltage level of a signal.

The following is reprinted with permission from J C Penney Co., Inc. This well-presented material is printed in *Technical* Information *Supplement A* of the Technical Service Data for J C Penney Model No. 686-6076 VHS VCR.

## BASIC VIDEOTAPE RECORDING

To understand the VHS format, it is wise to first review the basic principles of videotape recording.

Like audio tape recording, video information is stored on magnetic tape by means of a small electromagnet, or head. The two poles of the head are brought very close together, but they do not touch. This creates magnetic flux to extend across the separation (gap) [as shown in Figure 16.1].

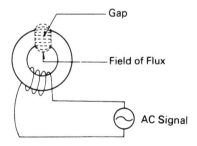

**FIGURE 16.1 Field of flux.**

If an AC signal is applied to the coil of the head, the field of flux will expand and collapse according to the rise and fall of the AC signal. When the AC signal reverses polarity, the field of flux will be oriented in the opposite direction and will also expand and collapse. This changing field of flux is what accomplishes the magnetic recording. If this flux is brought near a magnetic material, it will become magnetized according to the intensity and orientation of the field of flux. The magnetic material used is oxide coated (magnetic) tape.

Using audio tape recording as an example, if the tape is not moved across the head, just one spot on the tape will be magnetized and will be continually remagnetized. If the tape is moved across the head, specific areas of the tape will be magnetized according to the field of flux at any specific moment. A length of recorded tape will therefore have on it areas of magnetization representing the direction and intensity of the field of flux. For instance, the tape [in Figure 16.2] will have differently magnetized regions, which can be called North (N) and South (S), according to the AC signal. When the polarity of the AC signal changes, so does the direction of magnetization on the tape, as shown by one cycle on the AC signal. If the recorded tape is then moved past a head whose coil is connected to an amplifier, the regions of magnetization on the tape will set up flux across the head gap, which will in turn induce a voltage in the coil to be amplified. The output of the amplifier, then, is the same as the original AC signal. This is essentially what is done in audio recording, with other methods for improvement like bias and equalization.

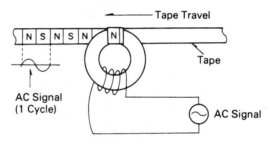

**FIGURE 16.2 Magnetic Recording.**

There are some inherent limitations in the tape recording process which *do* effect videotape recording, so they will be examined now.

[As shown in Figure 16.2], the tape has North and South magnetic fields which change according to the polarity of the AC signal. What if the frequency of the AC signal were to greatly increase?

If the speed of the tape past the head (head to tape speed) is kept the same, the changing polarity of the high frequency AC signal would not be faithfully recorded on the tape [as shown in Figure 16.3].

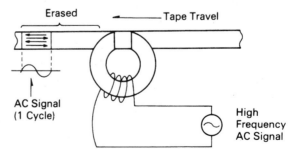

**FIGURE 16.3 High-frequency signal recording.**

As the high-frequency AC signal starts to go positive, the tape will start to be magnetized in one direction. But the AC signal will very quickly change its polarity, and this will be recorded on much of **the same portion** of the tape, so North magnetic regions will be covered by South magnetic regions and vice versa. This results in zero signal on the tape, or self-erasing. To keep the North and South regions separate, the head to tape speed must be increased.

When recording video, frequencies in excess of 4MHz may be encountered. Through experience, it is found that the head to tape speed must be in the region of 10 meters per second in order to record video signals.

The figure of 10 meters per second was also influenced by the size of the head gap. Clearly, the lower the head to tape speed, the easier it is to control that speed. If changes in head gap size were not made, the necessary head to tape speed would have been considerably higher. [How the gap size influences this can be explained by Figure 16.4.]

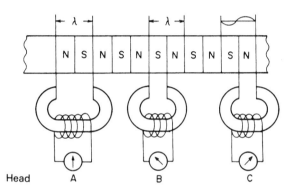

**FIGURE 16.4 Gap size and magnetization.**

Assume a signal is already recorded on the tape. The distance on the tape required to record one full AC signal cycle is called the **recorded wavelength** or $\lambda$. Head A has a gap width equal to $\lambda$. Here, there is both North and South oriented magnetization across the gap. This produces a net output of zero, since North and South cancel. Heads B and C have maximum output because there is just one magnetic orientation across their gaps.

Maximum output occurs in heads B and C therefore, because their gap width is $1/2\lambda$. (Heads B and C would also work if their gap width is *less* than $1/2\lambda$.) The same is also true for recording. The maximum useable (no self-erasing) transfer of magnetic energy to the tape occurs when the gap width, G, can be expressed as:

$G \leq \lambda/2$

The **recording wavelength** can be expressed as:

$\lambda = V/f$

where V is the head-to-tape speed and f is the frequencies to be recorded.

So,

$G \leq V/2f$

as V increases, G is also allowed to increase for the same **maximum** frequency. Conversely if G is made very small, V is allowed to be reduced.

In practice, G can be made as small as (and smaller than) 1μm (1 × $10^{-6}$ meters) and this puts V in the area of 10 meters per second. A head to tape speed of 10 meters per second is a very high speed, too high in fact to be handled accurately by a reel to reel tape machine of reasonable size. Also, tape consumption on a high speed reel machine is tremendous.

The method employed in video recording is to move the video heads as well as the tape. If the heads are made to move fast, across the tape, the linear tape speed can be kept very low.

In 2-head helical video recording (the only format which will be discussed here) the video heads are mounted in a rotating drum or cylinder, and the tape is wrapped around the cylinder. This way, the heads can scan the tape as it moves. When a head scans the tape, it is said to have made a TRACK. [This can be seen in Figure 16.5.]

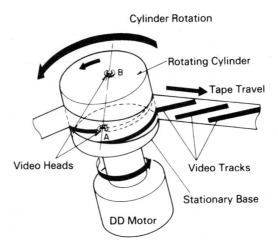

**FIGURE 16.5** Direct drive head cylinder.

In 2-head helical format, each head, as it scans across the tape, will record one TV field, or 262.5 horizontal lines. Therefore, each head must scan the tape 30 times per second to give a field rate of 60 fields per second.

The tape is shown as a screen wrapped around the head cylinder to make it easy to see the video head. There is a second video head 180° from the head shown in front. Because the tape wraps around the cylinder in the shape of a helix (helical), the videotracks are made as a series of slanted lines. Of course, the tracks are invisible, but it is easier to visualize them as lines. The two heads "A" and "B" make alternate scans of the tape.

An enlarged view of the video tracks on the tape can be shown [as in Figure 16.6]. The video tracks are the areas of the tape where video recording actually takes place. The guard bands are blank areas between tracks,

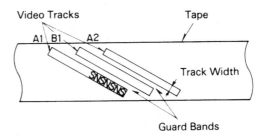

**FIGURE 16.6** Video tracks.

preventing the adjacent track's crosstalk from appearing on the track where the video head is tracking.

There is one more point about video recording which will be discussed here. Magnetic heads have the characteristic of increased output level as the frequency increases. Then, as determined by the gap width, the maximum output occurs at approximately

$$G = V/2f.$$

In practice, the lower frequency output of the heads is boosted in level to equal the level of the higher frequencies. This process, as also used in audio applications, is called *equalization*.

Video frequencies span from DC to about 4MHz. This represents a frequency range of about 18 octaves. Eighteen octaves is too far a spread to be handled in one system (one machine). For instance, heads designed for operation at a maximum frequency of 4MHz will have very low output at low frequencies.

Since there is 6dB/octave attenuation, $18 \times 6 = 108$dB difference appears. In practice this difference is too great to be adequately equalized. To get around this, the video signal is applied to an FM modulator during recording. This modulator will change its frequency according to the instantaneous level of the video signal.

The energy of the FM signal lies chiefly in the area from about 1MHz to 8MHz, just three octaves. Heads designed for use at 8MHz can still be used at 1MHz, because the output signal can be equalized. Actually speaking, heads are designed for use up to about 5MHz. Therefore, some FM energy is lacking but it does not affect the playback video signal, because it is resumed in the playback process. Upon playback, the recovered FM signal must be equalized then demodulated to obtain the video signal.

## Converted Subcarrier Direct Recording Method

The one method of color video recording that will be discussed here is the converted subcarrier method. In order to avoid visible beats in the picture caused by the interaction of the color (chrominance) and brightness (luminance) signals, the first step in the converted subcarrier method is to separate the chrominance and luminance portions of the video signal to be recorded. The luminance signal, containing frequencies from DC to about 4MHz, is then FM recorded, as previously described.

The chrominance portion, containing frequencies in the area of 3.58MHz, is down-converted in frequency in the area of 629kHz. Since there is not a large shift from the center frequency of 629kHz, this converted chrominance signal is able to be recorded directly on the tape. Also note that the frequencies in the area of 629kHz are still high enough to allow equalized playback. In practice, the converted chrominance signal and the FM signals are mixed and then simultaneously applied to the tape.

Upon playback, the FM and converted chrominance signals are separated. The FM is demodulated into a luminance signal again. The converted chrominance signal is reconverted back up in frequency to the area of 3.58MHz. The chrominance and luminance signals are combined, which reproduces the original video signal.

## Video Head Azimuth

*Azimuth* is the term used to define the left-to-right tilt of the gap if the head could be viewed straight on. In VHS, the video heads have a gap azimuth other than 0°. And more, one head has a different azimuth from the other. The 2 values used in VHS are azimuth of $+6°$ and $-6°$. [Refer to Figure 16.7.]

These heads make the VHS format different from most other VTR formats. Exactly how the azimuths of $\pm 6°$ help to keep out adjacent track interference is explained next.

**436** ■ *Practical VCR Repair*

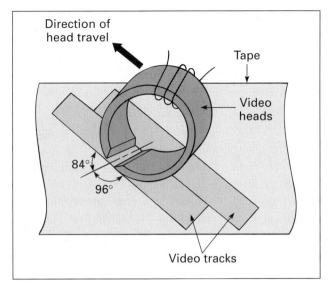

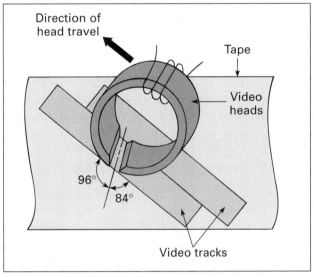

**FIGURE 16.7** −6° azimuth and +6° azimuth.

## Azimuth Recording

Azimuth recording is used in VHS to eliminate the interference or crosstalk picked up by a video head. Again, because adjacent tracks touch, or "cross talk," a video head when scanning a track will pick up some information from the adjacent track. The azimuths of the head gaps assure that video head "A" will only give an output when scanning across a track made by head "A". Head "B", therefore, only gives an output when scanning across a track made by head "B". Because of the azimuth effect, a particular video head will not pick up any crosstalk from an adjacent track. (End of reprinted material from J C Penney *Technical Information, Supplement A*.)

The important thing to realize here is that the two video heads in the spinning upper cylinder are not the same; each is manufactured with a different azimuth, or gap angle. That is why it is absolutely essential, when you replace an upper cylinder, to connect the heads correctly. If Head A is connected to the Head B terminals on the upper portion of the rotary transformer, the VCR will not play back previously recorded tapes.

## VHS FORMAT

You may not have followed all of the previous discussion, but don't be too concerned. The main thing is that an NTSC composite TV signal is really divided into two separate parts before being recorded on a VCR:

- Luminance (Y) signal, plus horizontal and vertical sync and blanking, and color burst signals are frequency modulated (FM) in the VCR to higher frequencies.
- Chrominance (C) signal is down-converted to a lower frequency. This is called *color-under format.*

Both the FM luminance (Y) signal and color-under chroma signal (C) are then recorded together on the videotape by the two spinning heads. Head A records Field 1 and Head B records Field 2 of each video frame. Figure 16.8 shows the relationship between a composite NTSC signal and the signal recorded by a VHS VCR.

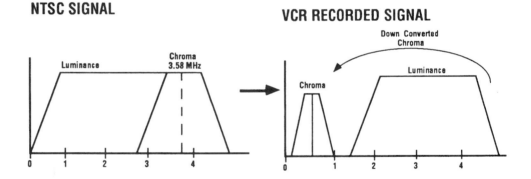

**FIGURE 16.8 NTSC and VCR recorded signals. The frequency of the chroma signal is down-converted in the VCR color-under format. Courtesy of *Sencore Tech Tips*.**

## RECORD CIRCUITRY BLOCK DIAGRAM

Figures 16.9 and 16.10 are block diagrams of VHS luminance and chroma *Record* and *Playback* circuits, respectively. We will not be discussing these in detail, but merely pointing out major items of interest. In descriptions that follow, numbers enclosed in brackets [ ] are the "balloons" or numbers in black circles in the diagrams. There may well be differences in how a particular model VCR accomplishes video recording and playback, so it is necessary to consult the service literature for the model on which you are working for specific information.

Notice the following items on the video *record* block diagram.

- Two major sections of video record circuits:
  — Luminance
  — Chroma.
- Video In [200] is NTSC composite video. This could be from the Video-In RCA jack, or a demodulated signal from antenna or cable, as selected by VCR tuner.
- Filter [400] separates out chroma signals, having 3.58MHz as a center frequency, from composite video and sends signals to chroma section.
- Baseband luminance signals are frequency modulated [206].

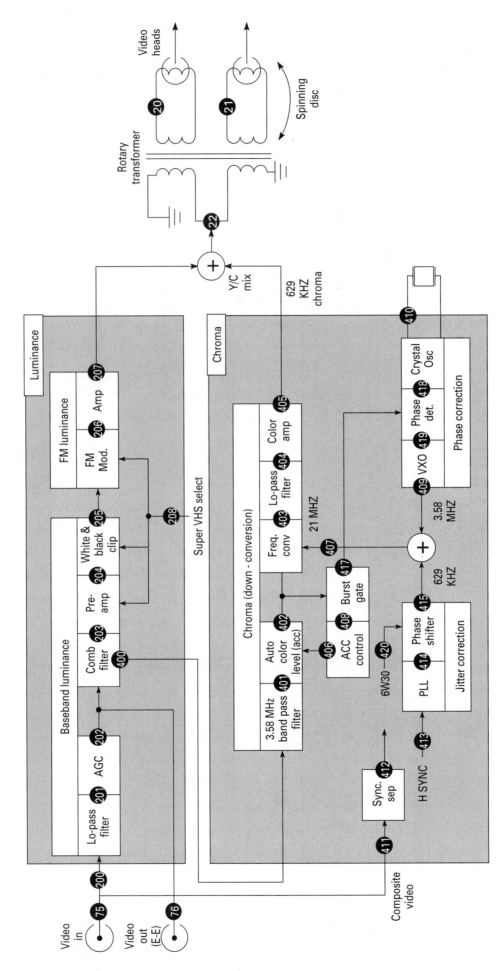

**FIGURE 16.9** Universal VHS video record block diagram. *Courtesy of Sencore Electronics Inc.*

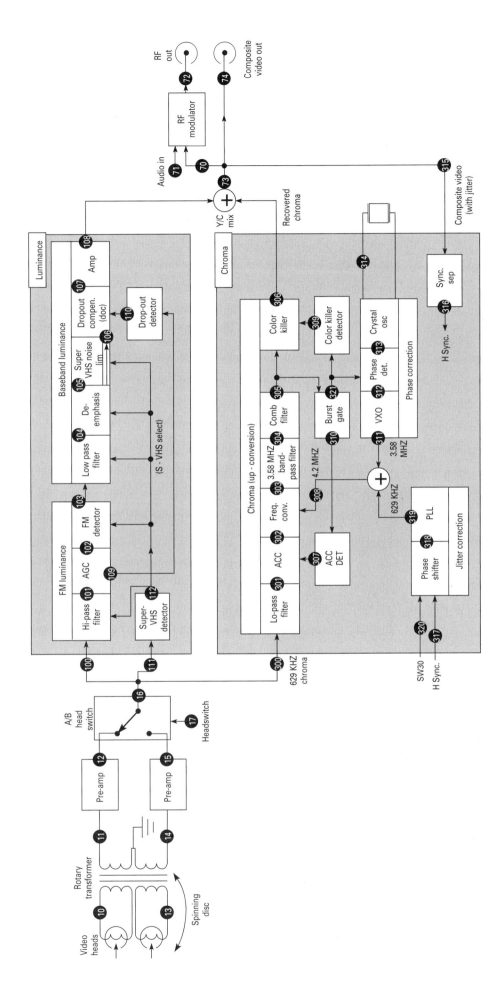

**FIGURE 16.10** Universal VHS video playback block diagram. Courtesy of *Sencore Electronics Inc.*

- Amplified luminance output (Y) [207] also contains:
  — Horizontal and vertical sync signals
  — Horizontal and Vertical blanking signals
  — 3.58MHz color burst signal.
- Note 3.58MHz chroma crystal oscillator [418] and quartz crystal [410]
  — Color burst signal that occurs during horizontal sync and blanking synchronizes this oscillator.
- Amplified color-under chroma output (C) [405]:
  — Centered at 629kHz for VHS (688kHz for Beta).
- Y and C signals mixed before going to both video heads [20 and 21]
  — Note that the combined signal goes to both heads simultaneously, but because only one head is in contact with the tape at a time, tracks on tape alternate between heads A and B. Head A records video field 1; head B records field 2.

Although the other video record circuit blocks shown are not discussed in detail, here are what some of the abbreviations stand for and the basic functions performed:

| | |
|---|---|
| Lo-Pass Filter: | Allows lower frequencies to go through, blocks higher frequencies. |
| AGC: | Automatic gain control; smooths out large differences in signal levels. |
| Comb Filter: | Sharply separates different frequencies. |
| Pre-Emp: | Pre-emphasis; equalizer that boosts high frequencies. |
| Band pass filter: | Allows frequencies within a certain range to go through, blocks higher and lower frequencies. |
| ACC: | Automatic chroma control, similar to AGC for color levels. |
| Burst Gate: | Recognizes 3.58MHz color burst signal riding on horizontal blanking pulse and allows color burst to go on to synchronize 3.58MHz chroma oscillator. |
| Freq. Conv.: | Frequency converter (or conversion); adds or subtracts two frequencies. Here, 3.58MHz is subtracted from 4.21MHz to obtain 629kHz color-under frequency. |
| Sync. Sep.: | Sync separator; recovers synchronization signals from composite input. |
| PLL: | Phase locked loop; an electronic servo system. |
| VXO: | Voltage-controlled oscillator; a correction voltage changes the frequency or phase, or both, at which the oscillator operates. |
| Phase Det.: | Phase detector; recognizes minor differences in the timing of two signals having the same frequency. |

Again, the intent of this chapter is to understand the big picture of how video is recorded and played back in a VCR, so don't be concerned about the operation of all these circuit blocks.

## PLAYBACK CIRCUITRY BLOCK DIAGRAM

Next is the block diagram for VHS video *playback* circuits. Note the following in Figure 16.10:

- Video input is from the two Videoheads A and B [10 and 13]
  — Each video head has its own preamplifier [12 and 15] to increase signal strength read from magnetic tape.

— Notice that the output of the two head pre-amps is switched back and forth by the A/B Head Switch [16]. As explained in Chapter 14, it is desirable to receive output only from the video head that is actually in contact with the tape. This eliminates random video noise that might be picked up by the other head.
— The 30Hz head switching signal [17] is derived from the drum pulse generator (PG) signal, which tells the drum servo circuit the position of the drum as it rotates.
- Video head output goes to both luminance and chroma circuits [100 and 300].
- Luminance circuit demodulates FM signal recorded on tape [103] to produce baseband Y signal, which also contains sync, blanking, and color burst signals.
- Chroma circuit up-converts 629kHz color under signal [303] to original frequencies in composite video.
- Luminance (Y) and chroma (C) signals are mixed to produce composite video signal [73].
— Composite video goes to Line-Out RCA jack [74] and Channel 3/4 RF modulator [70] in VCR.

Although the other video playback circuit blocks shown are not discussed in detail, here are what some of the abbreviations stand for and the basic functions performed:

| | |
|---|---|
| De-emphasis: | Reduces level of frequencies that were originally emphasized during recording. Pre-emphasis and de-emphasis work together to flatten out the frequency response characteristics of the video heads and magnetic tape. |
| Dropout detector: | Detects very short-term reduction in signal strength of FM luminance carrier, below that which would produce a good picture. For example, if scratch in tape coating prevented a head from reading one horizontal scan line in a video field, dropout detector would recognize this signal loss. |
| Dropout Compen. (DOC): | Electronically replaces missing video information when a dropout occurs. Luminance information from preceding horizontal line is stored temporarily in a *delay line*. When dropout occurs, previous horizontal line is output from delay line, so that one scan line actually appears twice. This is a very small part of the picture, and so is usually unnoticeable. |
| Color Killer: | Detector looks for presence of 3.58MHz color burst from burst gate. If burst is absent, color killer shuts down output of chroma signal, leaving just luminance signal to go to RF modulator and composite output jack. This produces a "clean" black-and-white picture, rather than an unfaithful (read gaudy!) color picture, for example during a black-and-white broadcast. Insufficient color burst level on a color program also causes color killer to shut down the chroma circuits, leaving a black-and-white. This circuit is often called ACK, for "automatic color killer." |

This chapter concentrated on how standard VHS *video* is recorded and played back, without mentioning the audio portion of the program. Linear audio is recorded the same as a conventional reel-to-reel or cassette audio tape recorder. As you learned earlier, an audio track (two for stereo) is laid down on a linear track near the top edge of the videotape. The audio head in the A/C (Audio/Control/Audio erase) head assembly both records and plays back. This is described in more detail in Chapter 17, as well as Hi-Fi Stereo recording, which uses separate rotary heads.

## SUMMARY

In this chapter you read about how a VCR records and plays back video. Composite video from a video Line-In jack or demodulated RF signal from antenna or cable service consists of two main components, a *luminance* (Y) signal and a *chroma* (C) signal. These two signals are separated, processed, and then recombined during record.

The luminance signal is frequency-modulated to a higher range of frequencies. This reduces the number of octaves that the video heads need to record and play back from tape. Horizontal and vertical sync and blanking pulses, and color bursts, are also contained in the frequency-modulated luminance, or Y signal.

Chroma information is separated from the composite video signal and then down-converted to frequencies with 629kHz as a center frequency. This is referred to as color-under format, because the chroma signal is now at frequencies that are less than, or under, the frequency-modulated luminance signal.

The frequency-modulated luminance signal (Y) and down-converted chroma signal (C) are mixed together and sent to two video heads, A and B.

Videotape wraps around the spinning video cylinder at an angle, or in the shape of a helix. Because of this, both Heads A and B record diagonal tracks on the tape. The combination of forward tape motion and rapidly spinning heads produces a head-to-tape velocity that is high enough to record the FM luminance signal.

Each of the videoheads, A and B, has a different azimuth, which is the angle of the head gap with respect to the videotape. Head A has a −6° azimuth, head B a +6° azimuth. This prevents either head from picking up crosstalk from adjacent video tracks during playback.

Head A records 262.5 horizontal lines of one video field; head B lays down 262.5 lines of the second field. Each field is contained in a single slanted video track on tape. Both fields together produce one video frame.

During playback, an electronic A/B head switch selects the output of either head A or B, depending on the position of the video drum, as determined by the drum PG signal. Head output then goes to both luminance and chroma circuitry. The luminance circuit demodulates the FM signal to recover the original Y component of the composite video. The chroma circuit up-converts the 629kHz color-under signal to its original 3.58MHz range.

Both the demodulated luminance and up-converted chroma signals are mixed to form the original composite video. This is sent to the Channel 3/4 RF modulator and composite video Line-Out jack on the VCR.

## SELF-CHECK QUESTIONS

1. Briefly describe how a tape head works with magnetic tape to record and play back audio or video information.
2. What are the two main factors related to magnetic tape and tape heads that determine the highest frequencies that can be recorded and played back?
3. Briefly describe 2-head helical video recording.

4. Describe how black-and-white video information is recorded in the 2-head helical scan VHS format.
5. Describe how color video information is recorded in the 2-head helical scan VHS format.
6. Explain why Videohead A cannot play back information recorded by Head B and vice versa.
7. What synchronizes the color circuits in a VCR or TV so that reproduced colors are the same as the video source?
8. What is the purpose of the dropout compensator (DOC) in a VCR?

# CHAPTER 17
# BEYOND STANDARD VHS

The vast majority of VCRs sold in the United States are standard or basic VHS models. They reproduce picture and sound quality, on a TV set, that is totally acceptable to most people. Most hookups probably use the modulated Channel 3 or 4 output from the VCR to play back a videotape. As previously discussed, this method of connecting VCR output to TV gives poorer playback results than using baseband connections. It is likely, but sadly true, that even when people own TVs with monitor capabilities, the RF connection is frequently utilized rather than the superior A/V connections.

However, better quality video and audio *are* desired by an increasing number of people. Owners of high-quality, large-screen video monitors with five-channel Dolby Pro Logic surround sound home theater systems demand more than standard VHS. The answer for better audio is a Hi-Fi Stereo VCR. Greater picture resolution, or detail, is available on any TV or video monitor with the Super VHS (S-VHS) format; however, a TV or monitor with an S-Video connector is required to realize the maximum benefit of S-VHS.

So far you have learned how *standard* VHS video is reproduced, and a little about how linear audio tracks are recorded. This chapter first describes how Hi-Fi Stereo, as well as linear track audio, is recorded and played back. Next, VHS HQ is described, followed by an explanation of the differences between standard format VHS and Super VHS VCRs. There are several enhanced features that are increasingly being included on VCR models, such as VCR Plus+, some of which are also discussed briefly in this chapter.

## CHAPTER OBJECTIVES

**Upon completing this chapter, you should be able to:**
1. Describe how linear audio and Hi-Fi Stereo are reproduced by a VHS VCR, using record and playback circuit block diagrams.
2. Describe the fundamental differences between standard and S-VHS formats.
3. Describe S-VHS user controls and the interconnection between an S-VHS VCR and a monitor or TV having S-Video capability.
4. Describe basic functions of enhanced VCR features, such as a flying erase head and freeze frame head.

It is important that you, as a potential service technician, understand the principles of operation and terms used for enhanced VCR functions and features. Most of this won't have an immediate, direct effect on the actual repair of a malfunctioning unit, but without this knowledge you could easily make an incorrect diagnosis. Attempting to resolve an audio problem by replacing a linear head won't do much good if it is the Hi-Fi audio section that is failing!

As stated at the beginning of Chapter 16, as you gain electronic knowledge and service experience, you can use the information you gain here to troubleshoot enhanced feature circuits. In most cases the manufacturer's service literature and schematic diagrams will be needed, along with advanced test equipment, such as an oscilloscope, NTSC pattern generator, and frequency counter.

# HI-FI STEREO VCRS

One of the most common features beyond standard VHS is Hi-Fi Stereo. Video reproduction is the same, but audio is recorded in an entirely different way, providing higher quality sound. A standard VCR with linear audio has a frequency response from about 50Hz to around 8kHz, whereas a 20Hz to 20kHz response is available in a Hi-Fi Stereo model. The sheen, openness, and sparkle is just not there when audio is limited to roughly 8kHz.

*Dynamic range,* the difference between softest and loudest reproduced sounds, is about 45 dB on linear audio tracks, but 80 dB when recorded in Hi-Fi. On a standard VCR, a whisper may be somewhat lost in background audio noise, and the crescendo of a full orchestra seems terribly restrained. The wider dynamic range of audio CDs and Hi-Fi Stereo on videotape more faithfully recreates the full range of sound intensity.

Wow and flutter are also reduced using Hi-Fi over linear audio. *Wow* is slow changes in tape speed, and therefore audio pitch; *flutter* describes rapid changes in tape speed. Listen to long, sustained piano chords played back on a tape machine. Wow is present if the sound wavers up and down in pitch. Flutter is more like a warbling effect in the tone, instead of a steady note.

Hi-Fi audio is also lower in distortion (both harmonic and intermodulation) and has a lower noise floor than linear audio reproduction.

The following is reprinted with permission of Sencore Electronics Inc. from *Tech Tips* #190.

## Basics of VCR Hi-Fi Stereo

Linearly recorded audio suffers in quality due to the low head-to-tape speed which limits the audio frequency range that can be recorded. Conventional recording limits the frequency response, signal-to-noise ratio, and produces high amounts of wow and flutter.

By increasing the head-to-tape speed and recording the signal using an FM carrier, all of the deficiencies inherent to linearly recorded audio are eliminated. Since the speed of the video tape is fixed by the tape format, the only way to increase the head-to-tape speed is to use rotary heads for audio as well as for the luminance and chroma information. By modulating an FM carrier with the audio signal, the signal-to-noise ratio, frequency response, and distortion are greatly improved. This is precisely what is done in VCR Hi-Fi Stereo. In fact, this method produces results that compare closely with audio CDs.

In the VHS format, separate audio heads are contained in the rotating drum assembly. In order for this system to work, however, the audio information cannot be recorded at the same frequencies as the video or chroma information. Instead, the FM audio information is recorded in the frequency spectrum between the down-converted chroma and the luminance information. The audio information is shifted to the intended position in the frequency spectrum by FM modulating a carrier at the desired frequency.

## VHS Hi-Fi Stereo

VHS Hi-Fi Stereo uses two separate FM carriers to record the stereo information onto the video tape. The audio is recorded linearly as well so that non-Hi-Fi Stereo VCRs can play back the tape. The left channel audio information modulates a 1.3 MHz carrier while the right channel audio information modulates a 1.7 MHz carrier [as shown in Figure 17.1]. Since the left and right audio information is kept separate and is not matrixed together, there are few stereo separation problems with this system.

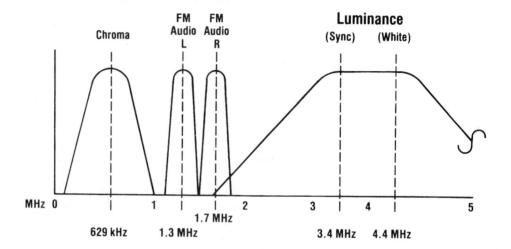

**FIGURE 17.1 VHS Hi-Fi Stereo uses a separate carrier for the left and right audio information. Courtesy of Sencore.**

VHS Hi-Fi is the only format that uses separate audio and videoheads on the rotating drum. The audio head first records the audio information. Then the videohead records the luminance and chroma information over the top of the audio.

At first this might appear to be a problem since the videohead would, theoretically, erase the audio information. This is overcome by recording the audio information with heads containing wider gaps and by increasing the record current so that the signal records deep into the tape. The video information is then recorded only on the surface, leaving some audio information still embedded deep in the tape. The process of deep recording necessitates a high quality tape with sufficient oxide coating. Some problems in VHS Hi-Fi VCRs can be traced to the consumer not using a high quality tape that has a deep enough oxide layer.

Interference is avoided by using separate pairs of heads to record the audio and video information. The audio heads have a different azimuth than the video heads. The FM audio information is deep recorded into the tape and the video information is shallow recorded over it. [See Figure 17.2.]

Since the audio and video heads are not at the same location on the head drum, headswitching is more complex. In the case of videoheads, a headswitching signal is used to select the appropriate head. This is also done for the audio heads, but since the audio heads cross the tape before the videoheads, a different audio head switching signal is needed. An adjustable headswitch signal is provided to switch the audio heads on at the correct time. The timing of this audio headswitching signal is critical for proper operation of the Hi-Fi Stereo signal.

Because the recorded audio and video signals are close in frequency and are laid down on top of each other, a method is needed to prevent audio and video signal crosstalk. This is done by using a 30° azimuth angle between the audio and video heads.

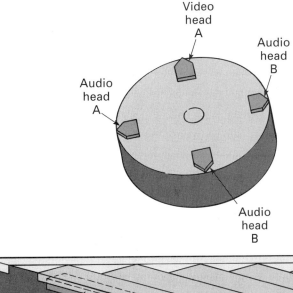

**FIGURE 17.2 VHS Hi-Fi Stereo recording.**

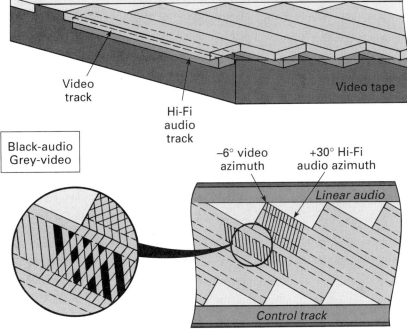

There are several factors that can cause the audio signal to be totally or intermittently missing. A dirty or bad audio head, incorrect headswitching, or a poor quality tape can cause total or intermittent loss of the audio signal. Whenever this happens, a popping noise or static is heard in the audio. This would be objectionable for most people if no preventative action were taken. For this reason, an audio muting circuit is used in the playback circuits. The mute circuit monitors the output of the rotating Hi-Fi audio heads for the presence of FM audio carriers. If the carriers drop out, for even an instant, the muting circuitry switches from Hi-Fi Stereo audio to linear mono audio. (End of reprinted material from Sencore *Tech Tips* #190.)

Figure 17.3 is a block diagram of audio *record* circuits, both linear stereo and Hi-Fi Stereo. In the descriptions that follow, items in the diagram with a number inside a black circle are identified by the same number within brackets, [ ]. Some functional blocks, such as Lo-Pass Filter and Emphasis were described in Chapter 16, and are not repeated here. Another term for VCR Hi-Fi Stereo recording is *AFM*, for *audio frequency modulation*.

Do not be concerned if you don't fully understand everything in these descriptions; just attempt to comprehend the overall manner in which audio is reproduced.

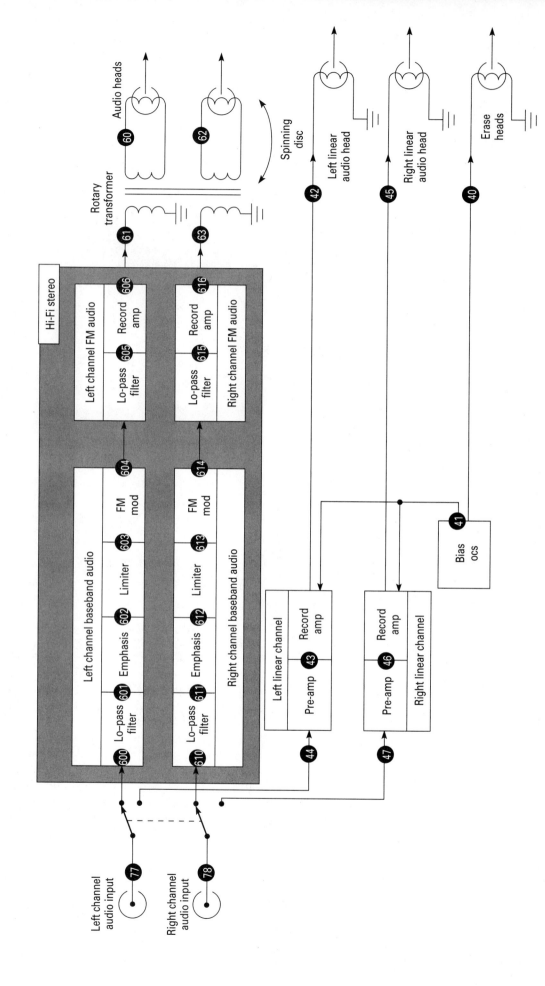

**FIGURE 17.3** Audio record circuits. Courtesy of *Sencore Electronics Inc.*

Left and right audio input [77 and 78] are from the line-input RCA jacks or from demodulated audio on the TV channel selected by the VCR tuner. The diagram *does* show a DPDT switch, which selects whether to send audio to the Hi-Fi Stereo or linear audio sections. In reality, if the user has selected Hi-Fi with a front panel switch during record, audio is also sent to the linear head(s) for playback compatibility on non-Hi-Fi machines.

The limiter stages prior to the FM modulators are similar to an automatic level control, but just limit the peak strength or amplitude of very loud passages.

Although the block diagram looks like left and right channel FM audio signals are sent to separate rotary audio heads, *both signals* are really sent to *both heads* at the same time. Otherwise, each channel would have no signal recorded on tape for 180° of drum revolution. Remember though, that the left channel has an FM carrier of 1.3MHz and the right a carrier of 1.7MHz, so each channel retains its own identity when recorded.

The bias oscillator [41] in the linear audio section is the same high-frequency current source for the full erase head and linear audio erase head. In fact, it is usually called the bias/erase oscillator. A small amount of this high-frequency signal (≈65kHz to 125kHz) is applied to the linear audio head(s) as a bias signal.

Although perhaps not exactly technically correct, you might want to *think* of recording bias as overcoming static friction of the magnetic tape particles so that the audio signal can more easily align them. The bias signal allows audio to be recorded in the linear portion of a magnetic tape's flux density/field intensity curve. No separate bias signal is needed for the rotary heads, either audio or video, because the signals themselves perform this function.

If the bias signal to the linear audio head(s) is missing or too low, the playback will sound very distorted, with a rather "spitty" quality, especially during low-level passages. The very loud audio sections will tend to sound almost right without bias. If the bias/erase oscillator is completely nonfunctioning, previously recorded material, both audio *and* video, will not be erased during subsequent recordings.

Some Hi-Fi Stereo VCRs have only monophonic linear audio, where left and right channel audio signals are combined and sent to a single linear audio Rec/PB head.

Figure 17.4 is a block diagram of both linear stereo and Hi-Fi Stereo audio *playback* circuits. The headswitching pulse [58] is the 30Hz signal derived from the pulse generator (PG) that indicates drum position. As you know, this signal switches output between Videoheads A and B during playback at a 30Hz rate. This signal is then delayed a small amount of time to select output from either audio frequency modulation (AFM) Head A or B. It works exactly the same way as switching video, but because the video and audio heads are offset on the drum, they both can't switch at exactly the same time; (refer back to Figure 17.2).

The BPF in blocks [501] and [511] stands for band pass filter. Each filter allows either the 1.3MHz or 1.7MHz FM signal to pass through, while rejecting the other.

The limiter sections [502 and 512] here work differently than the ones in the record circuits. They prevent any amplitude variations in the FM signal from reaching the FM detectors. Any amplitude variations at this point would be unwanted noise, such as that produced by lightning, because all amplitude variations in the original audio were changed to strictly frequency variations by the frequency modulator stages.

This is why FM radio reception is not bothered by static nearly as much as AM radio. Static and lightning discharges create changes in amplitude. That's how an AM radio receiver works, by detecting changes in amplitude of the carrier signal transmitted by the broadcast station. Therefore it also detects, or demodulates, noise spikes and static.

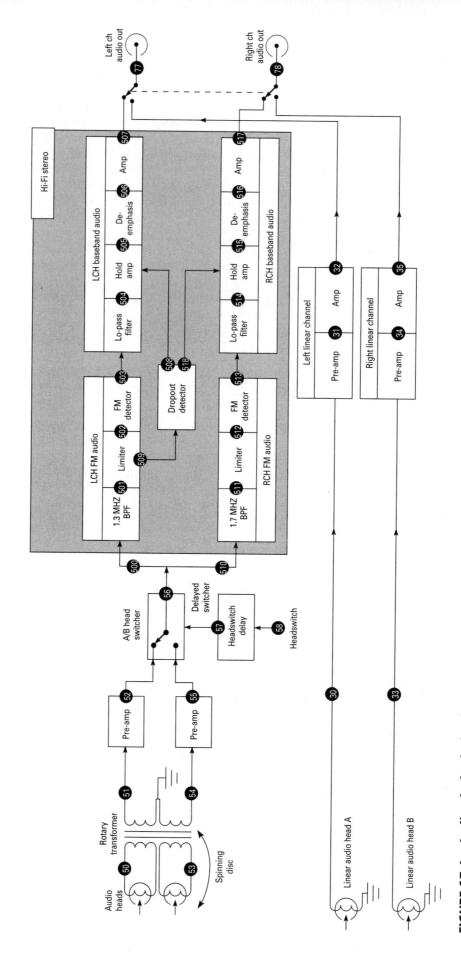

**FIGURE 17.4 Audio playback circuits.** *Courtesy of Sencore Electronics Inc.*

FM works on frequency variations. Instead of audio changing the *amplitude* of the carrier at the transmitter, it changes the *frequency* of the carrier. An FM detector in a radio demodulates the frequency deviations to recover the original audio information. By placing a limiter stage before the FM detector, any amplitude changes, such as noise, are eliminated, and so do not go any further. Quiet reception is the result. Without the limiter stage, amplitude variations would go on to the FM detector, and could be partially detected and interfere, causing noise in the output.

The dropout detector senses a loss in the FM audio carrier, for example during a small surface defect on the magnetic tape. (The hold amps retain the signal level during dropout to aid recovery when the signal returns.)

Although not directly shown on this diagram, the main function of the dropout detector is to switch the audio output [77 and 78] from the Hi-Fi Stereo section to the linear audio section. This keeps audio output of the VCR relatively constant, even though the Hi-Fi Stereo section has a dropout. When a good FM audio carrier is again present, the dropout circuit switches audio output back to the Hi-Fi Stereo section.

Of course, if the tape was not recorded in Hi-Fi Stereo, there will not be any FM signal picked up by the rotary audio heads, and so the VCR will output linear audio no matter what the position of the front panel Hi-Fi/Normal audio selector switch. If this switch is in the Hi-Fi position during record, audio will be recorded by both the rotary audio heads and stationary linear head(s).

A "sputtering" sound in the audio may indicate AFM dropout and that the VCR is continually switching back and forth between the two audio circuits. You can check the output of the dropout detector with a multimeter to see if this is the case. Erratic fluctuations of the meter indicate that the dropout detector is continually switching On and Off because of some problem in the FM audio section. It could be a dirty rotary audio head, poor-quality tape, bad rotary transformer, or some other problem. Consult the manufacturer's service manual for where to probe the output of the dropout detector.

As with video and linear audio, check playback with a quality tape commercially recorded in Hi-Fi Stereo before diagnosing Hi-Fi record/playback problems in a VCR. If playback is fine, then you know the VCR has record problems. If there is a problem with Hi-Fi Stereo playback, troubleshoot and correct this before diagnosing record circuits.

Stereo VCRs can also receive MTS (multi-channel TV sound) broadcasts from cable or antenna. A front panel indicator comes On when a stereo broadcast is received. Some stereo VCRs can also demodulate the SAP (second audio program) subcarrier transmitted by some TV stations. A front panel switch lets the user select either Stereo or SAP.

The SAP signal may be a different language or any other audio signal. Some TV stations place National Weather Service broadcasts on the SAP channel. The presence of a SAP signal is usually indicated on the front panel.

## VHS HIGH QUALITY (HQ)

High Quality (HQ) is a system that enhances the picture quality of standard format VHS while maintaining full compatibility. Tapes recorded in HQ can be played back on non-HQ VCRs, and vice versa, with absolutely no problem or performance degradation.

HQ incorporates two electronic improvements to the video signal: the white clip (WC) level is increased by 20 percent and both luminance and chroma signals are processed by detail enhancement and noise reduction circuitry.

The white clip circuit in a VCR limits the amplitude of positive-going luminance signals, which otherwise might overmodulate and distort the recorded information. Newer circuitry and better magnetic tape formulations have enabled HQ VCRs to raise the point at which the signal is clipped or limited. This provides better high-frequency video response, which results in sharper edges on vertical objects in the picture and overall picture quality improvement.

Detail enhancement circuitry improves both the luminance and chroma signals. High-frequency components of the video signal are selectively reinforced. This increases luminance and chroma output while reducing noise throughout the picture. The result is more detail in objects that are faint in comparison to their background in the picture, and reduced color streaking and patches.

So HQ is mainly some electronic equalization of existing video signals. Actual frequency ranges, FM deviation frequency, and format of the video signals are not changed. *Resolution,* or the number of reproduced horizontal lines making up a picture, is not increased with HQ, but the overall picture looks cleaner, because of noise reduction, and sharper, thanks to detail enhancement.

HQ circuitry is designed to provide the best picture when played back on a TV. This may not necessarily be the best signal for dubbing onto another VCR. Some VCRs have an edit switch that changes the characteristics of signal processing when the machine is the source for video dubbing. Refer to the owner's manual for information on setting any edit or video noise reduction switches when dubbing. Generally, noise reduction should be turned Off to transfer the best quality signal to the recording VCR.

## SUPER VHS

Super VHS VCRs provide significant improvement in picture quality and resolution over standard VHS, even over VHS HQ. This is brought about by using an improved oxide, high-density coating on the videotape; narrower head gaps; and by changes in the format of the recorded luminance signal.

Although S-VHS and standard VHS formats are *not compatible,* all S-VHS machines *are* able to record and play back tapes in standard VHS format. Tapes recorded in S-VHS format, however, cannot be played back on a standard VHS or VHS HQ machine.

To record in S-VHS format requires a special S-VHS cassette with the improved oxide coating. These cassettes are clearly marked "S-VHS" and have an identification hole on the underside, as illustrated in Figure 17.5. A switch on the reel table of a Super VHS machine senses the presence of the hole and allows a recording to be made in S-VHS format. Because of this, a standard cassette cannot be recorded in S-VHS format.

**FIGURE 17.5** Underside of Super VHS cassette. An S-VHS VCR senses the presence of the ID hole, and allows S-VHS format recording.

Most S-VHS machines also have a front panel switch that selects Normal VHS (N-VHS) or S-VHS. If a standard cassette is loaded, recording and playback will be in standard VHS format no matter what position the switch is in. With an S-VHS cassette loaded, recording is in the format selected by the switch. Yes, an S-VHS cassette may be recorded in standard VHS format. An S-VHS machine automatically senses the format of the previously recorded signal during playback and switches to either Standard- or Super-VHS, even though an S-VHS cassette is loaded. A front panel indicator normally lights up to indicate S-VHS operation.

The biggest improvement in picture quality and resolution is due to changes in the luminance signal recorded in S-VHS format, compared with standard VHS. In standard VHS, the lower sideband frequencies of the luminance signal overlap slightly with the upper frequencies of the chroma signal, as shown in Figure 17.6. Super VHS raises the overall frequency of the luminance signal, which reduces luminance and chrominance mixing, producing a sharper, more detailed picture.

Standard VHS has a maximum 1MHz deviation of the frequency modulated luminance signal, from 3.4MHz to 4.4MHz. The lowest frequency represents the tip of the sync signal (*sync tip*), which turns Off, or blanks, the beam during retrace. The highest frequency represents the brightest level, or the *peak white level* of a black and white picture. In Super VHS, the frequency deviation between sync tip and peak white level is increased to 1.6MHz, from 5.4MHz to 7MHz, illustrated in Figure 17.6. The additional 600kHz of video bandwidth over standard VHS greatly increases the resolution of reproduced video.

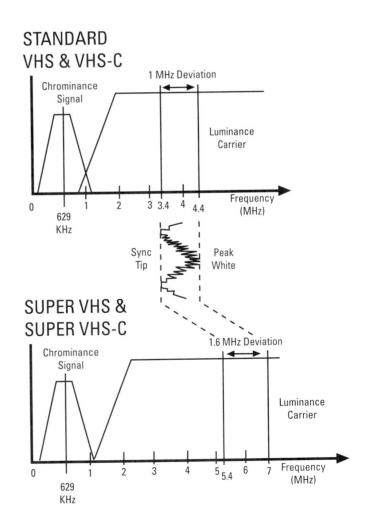

**FIGURE 17.6  VHS Super formats raise the luminance frequency to improve resolution from the existing format. Courtesy of Sencore®.**

A Super VHS VCR has all the same RF and baseband line level inputs and outputs as a standard format VCR, and can be connected to a TV set, monitor, and other A/V equipment in an identical manner. However, an S-VHS VCR also has another enhancement feature that can provide an even higher quality picture: separate Y and C outputs.

Recall that in a standard VHS VCR, the demodulated, or recovered luminance (Y) and chrominance (C) signals are recombined to produce the composite video signal sent to the line-level Video Out RCA jack. The same composite output is also available on a Super VHS unit, but there are also *separate* Y and C outputs, available at a four-pin miniature DIN connector, called an S-Video connector. (Pins 1 and 2 are ground, Pin 3 is the Y signal, Pin 4 is the C signal. DIN stands for Deutsche Industrie Normale, a standards bureau in Germany.)

The separate luminance and chrominance signals can then be connected to a TV/monitor or other piece of video gear that has an S-Video Line-In connector. An S-Video connecting cable with a DIN plug at each end attaches the two pieces of equipment.

Keeping Y and C signals separate in the TV/monitor means that circuitry and filters normally used to separate a composite signal into its Y and C components is bypassed. In general, the fewer stages and the less circuitry a signal has to go through, the less noise and distortion there is introduced onto the signal, and high frequencies containing picture detail suffer less degradation. The result is a more detailed, cleaner picture.

An S-VHS VCR also has an S-Video Line-In connector, as well as the standard RCA jack for composite video line input. Just as with a TV or monitor, VCR circuitry normally required to separate composite video into its Y and C components is bypassed, improving signal-to-noise ratios of both signals with S-Video input. A front panel switch, or on-screen mode selection made from the remote, selects whether input is from the Video In RCA jack or from the S-Video Line-In DIN connector.

Other than the additional S-VHS cassette sensing switch, the tape transport and operations are the same as for a comparable standard VHS machine. One item to remember is that videoheads on a Super-VHS machine have a tendency to clog more easily than heads on a standard VHS format unit, because head gap is about two-thirds narrower. Only high-quality cassettes should be used on an S-VHS machine. This is really true for *any* VCR, but especially applicable to S-VHS units.

## FLYING ERASE HEAD

The stationary full erase head in a VCR is fine for straightforward recording, but it is not suitable for making tight edits, like dubbing over a section of video. This is because the erase head is about 2-1/2 inches away from the videoheads. When you start dubbing over existing video, about 2-1/2 inches of tape will have new video recorded over the top of original video, because this amount of tape is already past the erase head when you go into record mode.

During playback this usually results in an unstable picture with *rainbowing* (diagonal rainbow-like streaks crawling down the screen) until "clean" tape that has been fully erased comes by the videoheads. What is happening is that some of the original video signal interacts or "beats" with the newly laid down video. This distorts the sync and color burst signals, as well as the Y and C signals, creating the rainbow-like video noise.

The solution to this problem is to have a separate head on the video drum that can erase previously recorded video immediately before videoheads record new material. And that is exactly what a *flying erase* (FE) *head* is, an additional rotary head. There is normally a separate oscillator for the flying erase head, which is turned On during edit operations. A front panel edit switch might be labeled something like "Insert Video" or "Insert Audio & Video." This activates the FE head.

Consult the user's manual for proper editing procedures on machines employing a flying erase head.

# OTHER VCR FEATURES

It seems like there is an almost endless stream of features and enhancements as new VCR models come out. The following is not intended to cover each and every one of them, nor to describe any in great detail. Rather, this information is included here so you will know a little about them, at least what the overall purpose or function of each is.

### Still Frame, Slow Motion, Jog and Shuttle

There are a variety of options on VCRs for producing still frames, slow motion, and variable speed forward and backward playback, sometimes called *jog* or *shuttle*. These modes all require precise control of the capstan motor by the system control microprocessor and capstan servo circuits. At first you might think that slow motion could be accomplished by simply slowing down the capstan motor, this is just not so. Rotary heads could not properly position themselves over the video tracks.

For slow motion, the capstan motor runs and stops, runs and stops. It runs; brakes—by electrically reversing the motor for a short period; then stops; then accelerates up to normal speed. Then the sequence repeats. When the motor is stopped, the servo system has positioned tape so that a video track lines up with the rotary heads. Slow motion, therefore, is really a series of start-stop sequences.

On a VCR with four videoheads, where one pair is for SP and the other is for LP and EP speeds, the LP/EP heads may be used for slow/still motion, regardless of the mode in which the tape is recorded.

Some VCRs have an additional special effects or *still field* rotary videohead. This head helps produce very clean still frames and slow motion, with little or no picture tearing, which is fairly normal for VCRs without the additional head. Here is how the fifth head, also called a *freeze frame head*, is used.

With tape stopped, only one of the regular videoheads, either A or B, can read a video track, because only one track on tape is aligned by the capstan servo with the spinning heads. This means there is video output during only 180° of head rotation. Recall that Heads A and B have different azimuths, and therefore can't play back each other's video tracks. The still frame head is positioned *nearly* 180° opposite one of the regular videoheads, for example Head B—and it has the *same azimuth* as Head B. (The fifth head cannot be exactly 180° opposite Head B, because Head A already occupies that place on the drum.)

During *still frame* playback, the servo system stops the capstan with tape positioned so that a Head B video track on tape, containing one video field, is aligned with the rotary heads. Then Head B reads the video track for 180° of drum rotation and the still field head reads the *same* video track for approximately the other 180° of drum rotation. The result is that the same video field is read twice, making a complete video frame sent to the TV screen, rather than having no video output during 180° of drum rotation.

Head A output is not used at all during still frame mode using this technique. It couldn't read the head B videotrack because of the 12° difference in azimuth between head A and the head B videotrack.

For *slow motion* playback, the capstan moves the tape a short distance and then stops for a period of time. Then this sequence repeats. Heads A and B read the tape as they normally do while tape is in motion; head B and the still field head read one Head B video field when tape stops. Some VCRs are capable of reverse slow motion. Essentially it is the same operation, but the capstan moves in the reverse direction. All audio is muted when a VCR is operated at other than normal playback speeds.

Because the fifth head is used only for slow motion and still frame, if it is clogged or associated circuitry is defective, slow motion and still frame operation may produce a terrible image, even though the VCR plays normally.

Some VCRs have a trim pot to make minor adjustments to still frame operation. Often the pot is accessible through a small hole in the front panel, which may or may not be labeled. If not, and you can see the adjustment slot of a PCB potentiometer behind the small front panel hole, you can be fairly confident that this is the "still frame" adjustment pot.

Because of different methods used to produce slow motion and still frame, refer to the user's manual and service literature to determine how to obtain the best quality picture in these modes and to learn how this is accomplished in a particular model VCR.

## Digital Video Systems

Digital circuitry is showing up more and more in video sections of VCRs, TVs, and video monitors. There is such a variety that only a brief description of some of the uses for digital video processing can be undertaken here. Most of these circuits are contained on *very large scale integration* (VLSI) chips, and when things fail, it is usually a matter of replacing an entire circuit board or one of these complex ICs.

Essentially, digital manipulation of video is a three-step process. First, the *analog* video signal is converted into *digital* form with an analog-to-digital (A/D) converter. Here each pixel making up a video line, field, or frame is coded in binary, where multiple digital bits in their On or Off state represent information in the luminance and chrominance analog signals. Secondly, the digital signal can be changed bit by bit, to alter the brightness, color, or position of the pixel on the screen. Then a digital-to-analog (D/A) converter changes the stream of data bits back to analog luminance and chroma signals.

Two general terms describe enhanced TV and VCR video accomplished with *digital signal processing* (DSP): ACTV, for *advanced compatible television* and IDTV, for *improved-definition television*. Most of this circuitry is in TVs, rather than VCRs, and is designed to improve picture quality. Nevertheless, some of the principles are being incorporated in VCRs for improved video reproduction. There are no set standards, and different manufacturers have unique methods for increasing video detail and clarity.

Some systems store a video line in digital memory and then read the line out to the screen between regular horizontal frame lines. This means that each video frame is actually scanned twice. Double scanning at twice the normal scan rate of a conventional TV improves screen brightness, increases apparent resolution, and reduces line flicker. Other systems make up or *interpolate* extra lines of video by averaging the pixel content of two or three actual previous lines. These additional lines are then scanned between regular horizontal scan lines, increasing apparent resolution.

Digital circuitry is also designed to reduce video noise and smearing when rapidly moving objects are present in the picture, like headlights on a speeding car.

Besides improving picture quality, digital video can create special effects, such as picture-in-picture (PIP), where a smaller sized image appears superimposed over the main image, usually positioned by the user to one of the four corners of the screen. This feature on a VCR is handy when editing, with output from another VCR. For example, the main picture might be the pre-recorded material on the recording VCR, while the PIP is of the other (playback) VCR.

A VCR with PIP might also play back a tape as the main image, while a VCR tuner channel produces a PIP of broadcast or cable channels. Main and inserted pictures can be swapped by the user; for example, during a commercial break in a recorded TV show, a live broadcast football game could become the main picture. Many TVs also have PIP capability. Often, the input from a VCR is one picture while the channel selected by the TV's tuner produces the other picture. This allows watching two programs simultaneously, with the VCR's tuner or tape playback providing one of them.

Digital circuitry, consisting of a *character generator* IC, produces on-screen menus, prompts, and other operating information. Most newer VCRs output a solid blue raster when there is no RF envelope from either a tape being played back or from the VCR's tuner. Background color may be user selectable from a menu. Then, when you press Play on the VCR, typically the word **PLAY** appears on the blue raster until tape loads and the tape outputs a good video signal.

This system can be an annoyance when trying to resolve a VCR video problem. Unless there is sufficient RF output from the video heads during playback, the blue field appears, so you don't see the salt-and-pepper raster characteristic of playing a tape with no recorded material. (Some people refer to this raster as "bug races.") If you are trying to resolve a problem, the automatic blue raster will appear with marginal RF output from tape, and you won't be able to see what's really happening as you, for example, adjust transport components.

Look in the service literature and try to locate a pin that you can jumper to ground to kill or mute the background and on-screen display (*OSD*). The pin might be called something like "background mute" or "OSD Blank." It may be shown on the schematic in the vicinity of the OSD or Char. Gen. chip, or in written service procedures.

The OSD or Char. Gen. chip has both Y and C outputs that typically go to one or more switching ICs. If there is insufficient signal from tape in playback or from a received channel, the IC switches the output of the OSD chip to the VCR output.

## Even More VCR Features

On-screen programming has certainly made setting up a VCR for unattended programming quite easy, compared with the multiple steps necessary for earlier machines. There are a couple of VCR programming enhancements that make this even easier and more convenient. In addition to easing unattended recording programming, a few other VCR features are available that add value for some users.

## VCR Plus+

VCR Plus+ is a multiple-digit code that can be entered into the remote control unit for programming a VCR. The local edition of *TV Guide* and local newspapers print a VCR Plus+ code number along with TV program listings. To record the program on a VCR Plus+ Code-capable machine, all that is needed is to go to VCR Plus+ mode and then enter the digits printed in the TV listing. The timer microprocessor converts the code number entered into the day, start time, stop time, and channel number for the recording.

Because different channel numbers in a geographic area often carry the same program, channel numbers must be initially set up for VCR Plus+ operation. For example, an NBC station might be Channel 11 broadcast, Channel 7 on one cable system, and Channel 22 on another cable system. A channel cross-reference in the front of *TV Guide* and in newspaper TV listings indicates which cable channel number equates to the channel number printed in the listing and embedded in the VCR Plus+ code. Then, even though the VCR Plus+ code "says" Channel 11, the VCR will go to Channel 7 to record the program if that is the cable channel that equates to Channel 11 in the listing. Follow the VCR operator's manual to setup VCR Plus+ channels.

This system can be made even simpler by including a printed bar code—similar to the universal product code (UPC) on grocery store items—adjacent to each program listing. The bar code would contain the VCR Plus+ digits. A VCR remote control with built-in wand scanner could then simply read the bar code for the desired program to be recorded.

## Cable Mouse

No, mice aren't just for computers anymore. Some VCRs come with a cable mouse, a small unit about 1/4 the size of a pack of cigarettes, with a cable that plugs into the back of the VCR. Essentially, the cable mouse is an IR transmitter or emitter that selects the channel on a separate cable converter box. It takes the place of the cable box remote control unit for unattended recording.

A cable mouse is positioned so that it can "see" the cable box. The VCR then sends out the correct IR code from the cable mouse to turn the cable box On and select the channel for unattended recording.

Some VCRs don't have a separate cable mouse, but have one or more IR LEDs behind the front panel that send out the data stream to the cable box. In this arrangement, IR light must be able to reach the cable box, either directly or after reflecting off a ceiling or wall.

Refer to the VCR user's manual for instructions on how to use a cable mouse or cable box control. Not all VCR Plus+ machines include cable box control.

## Self Cleaning

*Self-cleaning* is an additional feature found on some VCRs. Arrangements vary, but essentially a chamois roller or arm gently comes up against the video drum to clean the drum and rotary heads. This is obviously no substitute for a thorough manual cleaning of the entire tape path, but it's a help.

A cleaning cycle occurs automatically sometime when tape is *not* loaded, like just before tape loads or after tape unloads. Some models go through a cleaning cycle each time tape loads and unloads. Again, the operator's manual and service literature has information specific to a particular model.

## Closed Captioning

Closed captioning may be a feature on newer VCRs. Manufacturers may react to Federal Communications Commission (FCC) concerns that all full-size TVs need built-in closed caption capabilities, but whether a ruling will carry over to VCRs remains to be seen. Closed captioning is a line or two of text that runs along the bottom portion of the screen, designed primarily for the hearing impaired. It is similar to English subtitles on a foreign film.

At present, a separate decoder box is required to recover the closed captions embedded in a TV broadcast, unless this feature is incorporated into the TV. Many TVs now include closed caption decoders. As closed caption decoding becomes more readily accessible, even non-hearing-impaired viewers will benefit. For instance, captioning would allow late-night viewing in bed with the sound turned Off, so as not to disturb someone else's sleep.

### And More

This is certainly not an exhaustive list of VCR features and enhancements. Some of the higher end machines have connections for synchronizing edits with two (or more) VCRs. For someone editing camcorder tapes, this is a very worthwhile feature. Repeat play between any two points on tape is another feature, especially useful for trade shows and promotions. This operation normally works in conjunction with the tape index counter. For example, a tape will repeat from the beginning to some point where you reset the index counter to 0000.

Some VCRs have real-time counters that indicate actual hours and minutes since the beginning of tape and the time remaining. Some VCRs can even automatically switch to a lower speed under programmed recording if there is insufficient tape left for the higher speed initially selected by the user.

One increasingly popular feature is the ability to set an electronic *"bookmark"* on tape, like at the beginning of each program. During playback, the VCR can then search at high speed until it finds a bookmark, then start playing tape. This feature eliminates fiddling to find the start of a half-hour program somewhere in the middle of a four-hour recording. Another convenient feature is *blank search,* which finds the end of recorded material on a tape automatically, so you can record another program starting at that point.

For "zapping" commercials, a few VCRs have a skip search or short search that skips over 30 seconds of material, at whichever speed the recording was made, with each depression of the Skip button on the remote. One manufacturer, JVC, has a Family Message Center feature on some VCR models that lets someone leave a message for family members or friends on the TV! No doubt tomorrow's machines will have even more features and enhancements, such as digital program identification and related information, similar to that now coded on compact discs.

## SUMMARY

Several enhancements have contributed to improved VHS audio and video reproduction quality. Hi-Fi Stereo VCRs are capable of nearly CD quality audio. Audio frequency response, dynamic range, wow and flutter, and signal-to-noise ratio are vastly better than with linear audio tracks.

Hi-Fi Stereo employs two additional rotary heads on the video drum. During record, left and right channel audio signals are each frequency-modulated to produce two separate signals; the left channel FM carrier is at 1.3MHz, the right at 1.7MHz. This scheme is called AFM, for audio frequency modulation. These two AFM signals lie between the upper frequencies of the down-converted chroma signal and the lower frequencies of the luminance signal recorded on videotape.

AFM signals are first recorded deep in the magnetic oxide layer. Video is then shallow recorded over the AFM tracks. Both AFM signals are applied to both rotary Hi-Fi heads during record. During playback, a delayed videohead switching pulse switches between the two audio heads. This prevents noise pickup from the head that is spinning in air rather than contacting tape.

Hi-Fi audio demands high-quality tape. Videotape having a thin oxide coating will not do well because of the deep recording required for AFM. Thin oxide tapes cause video signals to erase the previously laid down audio tracks.

VHS HQ (high quality) consists of electronic enhancement of signals recorded in standard VHS format. Improved picture quality is achieved by increasing the white clip level by 20 percent and by enhancing frequency response and signal-to-noise ratios of both luminance and chroma signals. VHS HQ is totally compatible with standard VHS.

Super VHS uses magnetic tape with improved oxide and a deeper coating, which is recorded in a different format than standard VHS. An S-VHS cassette has an identification hole on its underside, near the end of the lateral groove at the take-up reel side, which is sensed by an ID switch in an S-VHS VCR.

Sensing the presence of this ID hole after a cassette loads, an S-VHS unit can be set to record either Normal VHS or Super VHS, as determined by a front panel N-VHS/S-VHS switch. An S-VHS cassette therefore can be recorded in either format. However, a normal cassette can only be recorded in the standard VHS format.

An S-VHS VCR can record and play both standard and Super-VHS formats. A standard format VCR cannot play back a tape that has been recorded in S-VHS format.

In the S-VHS recording format the luminance signal is shifted to a higher FM carrier frequency *and* the maximum frequency deviation between sync tip (black screen) and white clip (white screen) is increased by 600kHz. The improved frequency response increases picture resolution, while the shift in luminance carrier frequency keeps luminance signals further separated from the chroma signal (or right channel AFM in Hi-Fi machines).

An S-VHS VCR also keeps the Y and C signals separated when a tape is played, with each signal available at an S-Video connector. When connected to a TV or monitor also having an S-Video connector, circuitry that would normally separate the composite video signal into its Y and C components can be bypassed. This increases the signal-to-noise ratio of both signals, and prevents possible harmful interaction between them.

An S-VHS VCR also has composite video line-in and line-out RCA jacks. Most S-VHS improvements can be realized without using the separate S-Video connections, but TVs and monitors having an S-Video connector provide the best output from a Super-VHS VCR when an S-video interconnect cable is used.

S-VHS video heads have a tendency to clog more easily than heads on standard format machines because of the significantly narrower head gap. Only high-quality cassettes should be used on any VCR, but this is even more important in the case of S-VHS machines; here, premium quality tapes are best.

Digital signal processing is employed in many VCRs to improve picture quality and for special effects, like picture-in-picture (PIP) and freeze frame. First, an analog-to-digital (A/D) converter samples the Y and C analog signals and creates binary or digital representations of pixels making up the image. Next, color and other attributes can be manipulated in the digital domain to improve picture detail, reduce noise, and even relocate pixels within the raster. Finally, a digital-to-analog (D/A) converter recreates analog Y and C signals to be sent to the CRT.

Enhanced features are available on many VCR models, although full-featured VCRs can cost three or four times that of a basic two-head unit. VCR Plus+ programming, cable box control, synchronized editing, electronic bookmarking, blank search, and self-cleaning are just a few of the many extras on some of today's VCRs. Some VCRs are quite sophisticated, capable of many different modes of operation. It is best to read the owner's manual to fully understand how various VCR features operate. Sometimes a VCR is doing exactly what it is supposed to do, but a consumer thinks there's a problem, because it works differently than a previous VCR, or setup instructions have been misinterpreted, or it's being operated improperly.

## SELF-CHECK QUESTIONS

1. Describe the sound characteristics that make Hi-Fi Stereo VHS recording superior to linear audio recording.
2. What is the basic factor that allows Hi-Fi Stereo to be superior to linear audio?
3. Describe what is meant by deep recording.
4. What determines which rotary audio head outputs the two FM signals during playback?
5. What action do most VCRs take during playback when there is a dropout in the AFM signal?
6. What is the result if the bias signal to linear audio heads is missing or of low amplitude?
7. What consideration should be made when copying or dubbing video from a VHS HQ VCR to another VCR?
8. Can videotapes recorded in Super VHS format be played back on a standard VHS VCR? If not, why not?
9. Describe the essential difference in the output signals available at the composite video Out RCA jack and the S-video connector in a Super VHS VCR.
10. What might you need to do when servicing newer VCRs that output a blue raster when there is no playback signal from videotape?

# CHAPTER 18 USING MANUFACTURERS' SERVICE MANUALS

Although many VCR repairs can be accomplished without service literature, most electronic, and some mechanical, repairs are difficult without it. In fact, *electronic* troubleshooting is almost impossible without at least schematic wiring diagrams. Having service data on hand can make even simple mechanical adjustments and parts replacement easier. You will know *for certain* how to set the supply spindle back tension, for example, rather than taking an educated guess. This chapter describes the contents of typical VCR service literature and how to use the various parts of a service manual during maintenance, diagnostic, and repair procedures.

There are many different brands and models of VCRs on the market, but the brand name doesn't necessarily identify the manufacturer. Several companies make more than one brand name, and a given brand name often has diverse manufacturers for different models. This chapter begins by listing some of the numerous brand names cross-referenced to manufacturers.

## CHAPTER OBJECTIVES

**Upon completing this chapter, you should be able to:**
1. Describe the contents of a typical VCR service manual and how to use the various sections.
2. Describe how to follow circuit wiring on bundled schematic cables.

## SO MANY MODELS!

At first glance, it might seem as if there is an overwhelming number of VCR models. The fact is, though, there is nowhere near the number of manufacturers as there are brand names on the market. Most companies making VCRs produce machines with several different brand names. Sometimes two models with different brand names will be identical except for minor cosmetic differences on the front panel.

It is similar to automobiles. For example, there is little difference between some models of Chevrolet, Oldsmobile, Buick, and Pontiac, besides exterior and interior trim, grilles, and tail light arrangements. Many parts can be interchanged among them. Even General Motors and Toyota automobiles are increasingly hard to tell apart, as some GM models are now sold in Japan as Toyotas. Also, a few components, such as headlights, radiator hoses, and tires, are the same for a wide variety of automobiles from many different manufacturers. The same situation exists with VCRs.

All VCRs of a particular brand name are not necessarily made by the same manufacturer. A brand-name VCR may be made by company A one year and by company B the next. Often, you will see the manufacturer's name printed on one or more of the printed circuit boards inside a VCR.

Most manufacturers sell models under their own name, as well as making machines for other companies. Many brand names are made by multiple manufacturers, depending on year, model, or both. It is also interesting to note that even a manufacturer may have a model with its own brand name actually constructed by another company.

The following lists most of the major VCR manufacturers and some of the brand names they currently make or have made in the past. This is *not an exhaustive list,* but covers the majority of VHS VCRs.

## MANUFACTURERS AND BRAND NAMES

- Daewoo:
  Dayton
- Emerson:
  Emerson
  H.H. Scott
- Funai:
  Dynatech
  Funai
  Lloyds
  Multitech
  Sound Design
  Symphonic
  Teac
  Vector Research
- Goldstar:
  Emerson
  Goldstar
  JC Penney
  Memorex
  Totevision
- Hitachi:
  General Electric
  Hitachi
  JC Penney
  Minolta
  Pentax
  RCA
  Sears
- JVC (Victor Company of Japan):
  JVC
  Kenwood
  Teac
  Toshiba
  Zenith
- Matsushita (Panasonic):
  Canon
  Curtis Mathes
  General Electric
  JC Penney
  Kodak
  Magnavox
  Montgomery Ward
  Panasonic
  Philco

  Quasar
  RCA
  Sylvania
  Technics
  Teknica
- Mitsubishi:
  Akai
  Emerson
  Harman Kardon
  Mitsubishi
  Video Concepts
- NEC (Nippon Electric Co.):
  Harman Kardon
  JC Penney
  Kenwood
  Lloyds
  Multitech
  NEC
  Symphonic
  Vector Research
- Phillips
- Samsung:
  Samsung
  Midland
  RCA
  Supra
  Unitech
- Sanyo/Fisher:
  Fisher
  Sanyo
  Sears
- Sharp:
  KMC (K-Mart)
  Montgomery Ward
  Sharp
- Shintom
  Broksonic
  Circuit City
  KLH
  Logik
  Singer
  Toshiba
- Sony
- Phillips

You'll discover that there are far fewer actual differences in the many VCR models manufactured than the sheer number of brand names and models would suggest. The more you work on VCRs, the more frequently you will notice similarities among different brand names, especially in the transport area. This is similar to audio cassette recorders, decks, and car cassette players. There are hundreds of different brand names, but just a few companies manufacture the actual tape transport mechanism.

## OBTAINING SERVICE MANUALS AND PARTS

When ordering a service manual or replacement parts, first go after the *brand name* and model number of the unit on which you are working. Some electronics suppliers that specialize in VCR repair parts have cross-reference lists in their catalogs, as well as telephone assistance, to help you locate the correct replacement parts. Here you might discover that a particular Panasonic replacement video head is the same for a certain Quasar model VCR.

When calling to order parts, be sure to have the model number, serial number, and chassis number, if any, available, in addition to the machine's brand name. Sometimes different serial number ranges for the same model require differences in some of the parts used.

One especially good source of technical service manuals for a wide variety of consumer electronics, including VCRs, is Howard W. Sams & Company. Their yearly index, available for less than $6.00, lists technical service data they have by make and model number. Included in the Sams index is an up-to-date list of manufacturers' and importers' addresses and telephone numbers; this alone is probably worth the modest price. Need to get a service manual for a Sharp VCR? The most recent address and phone number for Sharp is in Sams's yearly index. This reference is also worth having for the many electronic suppliers' ads. Some suppliers specialize in VCR replacement idler assemblies, belts, tires, video heads, and tools. Authorized "stocking core" distributors of Sams products, listed by state and city, are also in the index. For the low price, the yearly Sams index is well worth it!

Howard W. Sams & Company
2647 Waterfront Parkway East Drive
Indianapolis, IN 46214-2041
Customer Service: 1-800 428-7267 (8am–5pm EST)
FAX: 1-800 552-3910
In Central Indiana: 317/298-5566 FAX: 317/299-0952

Sams is also a good source for out-of-date and hard-to-find service information. They have a photocopy service for many manuals that are no longer in print.

Many parts suppliers and electronic companies have toll-free 800 telephone numbers. To find out if a company has one, dial 800 information at 1-800-555-1212.

Magazines, like *Popular Electronics* and *Electronics Now* are loaded with advertisements from electronic parts distributors. Some offer free catalogs.

### Service Manuals

For *most* VCR repairs, you will not need the specific service manual for the model on which you are working. The majority of simple cleaning jobs and part replacements can be done without service literature. But for an electronic problem, a tough mechanical removal and replacement, or precise mechanical or electrical adjustments, technical service data *is* required.

Even though you may be starting out just fixing your own or a friend's VCR, it is probably worth getting the service manual, especially for your own machine. The $20 or so will be well spent. You'll get to see firsthand what is included in typical service literature, and can take your time relating the mechanical adjustment procedures, electronic schematics, and parts illustrations to the real unit in front of you.

One thing you *might* want to consider is experimenting with a VCR for which you have the service manual. You could deliberately maladjust a few simple transport components, like P-2 guide height, and then observe how the VCR behaves, and what sort of picture it produces. First, check that the service literature *does* have procedures for correctly adjusting whatever it is you intend to maladjust, and that you have the tools and test equipment to make the corrections. Some factory adjustments require a special jig or alignment gauge to set things right if they've been disturbed.

It is a good idea to carefully count how many turns you rotate an adjustment screw or nut when maladjusting, so you can return everything to its original position. Don't maladjust more than one thing at a time, and don't overdo! Sometimes just a single revolution of a mechanical adjustment can produce large effects.

Another mode of experimenting is to practice reading power supply output voltages with a multimeter, using the schematic so you know where to measure. Then see how a few syscon microprocessor input pins from switches and sensors change logic levels as the VCR goes through various operations. Trace a complete circuit on a schematic—for example, from the take-up reel sensor that outputs pulses when the spindle rotates to a pin on the system control IC.

Be extra careful not to short adjacent pins on a chip with a test lead clip. Usually you can clip onto a nearby component lead that is connected to an IC pin, like a resistor, diode, or PCB jumper, rather than to the tightly spaced lead on the IC itself. And don't attempt this without miniature, spring-loaded hook-type clips. It is far too easy to accidentally short between component leads with alligator clips or normal probe tips. Shorting between circuits *could* destroy a component or vaporize a printed circuit board wiring trace!

## BASIC SERVICE MANUAL CONTENTS

The following is what you will find in a typical VCR service manual:
1. Unit specifications
   — AC power requirements, dimensions, weight, recording formats, number of heads, inputs and outputs, audio and video reproduction specifications. Refer to Figure 18.1.
2. Complete electronic schematic diagrams
   — Usually has one interconnect wiring diagram that shows cabling and connectors between multiple circuit boards and transport components, such as cassette load and mode motors. Refer to Figure 18.2.
   — Several large sheets with schematics for each printed circuit board. Refer to Figure 18.3.
   — Pictorial of printed circuit boards, showing layout of both component and wiring sides. Refer to Figure 18.4.

Notice in Figure 18.2 that there are several places with a small oval around a wire, with the oval connected to a connector ground pin. There are several examples of this on the Syscon/Servo/Video PWB (printed wiring board) top row connectors. This means that the wire is *shielded* along its length by a braid or metallic foil wrap surrounding the conductor. The shield may or may not be connected to ground at the other end of the conductor.

| | | |
|---|---|---|
| Power Requirements: | | 120V AC, 60Hz |
| Power Consumption: | | 21W (Approximate) |
| Dimensions: | | 4"(H) X 16.9"(W) X 11.6"(D) (Approximate) |
| Weight: | | 12.3 lbs. (Approximate) |
| System Format | | Standard VHS |
| Television Signal System: | | EIA Standard (525 Lines, 60 Fields) NTSC Color Signal |
| Video Recording System: | | Rotary two-head helical scan |
| | Luminance: | FM Azimuth |
| | Chroma: | Converted Subcarrier Phase Shift |
| Audio: | | 1 Track |
| Tape Speed: | SP: | 1-5/16 Inches per second (33.35mm/s) |
| | LP: | 21/32 Inches per second (16.67mm/s) |
| | SLP(EP): | 7/16 Inches per second (11.12mm/s) |
| Maximum Record/Playback Time: | T160 tape | 8 hours |
| FF./Rewind Time: | T120 tape | Less than 6 minutes |
| Heads: | Video: | 2 rotary |
| | Audio/Control: | 1 stationary, 1 track audio |
| | Erase: | 1 full |
| Inputs: | Video: | Video In, 75 ohm unbalanced, 1V p-p |
| | Audio: | Audio In, 50k unbalanced, -10dB |
| | TV Tuner: | VHF Input, 75 ohm unbalanced, TV Channels 2 thru 13, Cable Channels UHF Input, 300 ohm balanced, TV Channels 14 thru 69 |
| Outputs: | Video: | Video Out, 75 ohm unbalanced, 1V p-p |
| | Audio: | Audio Out, 600 ohm unbalanced, -8dB |
| | RF Modulator: | Channel 3 or 4, 75 ohm unbalanced, 72dBu (Open) |
| Video Resolution | Horizontal: | More than 230 lines |
| Audio Response: | SP: | 100Hz to 8kHz |
| | LP: | 100Hz to 6kHz |
| | SLP(EP): | 100Hz to 5kHz |
| Signal to Noise Ratio: | Video: | SP, Better than 43dB LP, SLP (EP), Better than 41dB |
| | Audio: | SP, Better than 42dB LP, SLP (EP), Better than 40dB |
| Operating Temperatures: | | 41°F to 104°F (5°C to 40°C) |
| Operating Humidity: | | 10% to 75% |

**FIGURE 18.1** VCR specifications, Panasonic Model PV-1360. *Courtesy of Howard W. Sams & Co.*

One feature of many complex schematics is *wire bundling*, which reduces the actual number of lines drawn. This makes it easier to follow a line from end to end. Notice the heavy black lines in Figure 18.3 for a wire bundle. The bundle actually represents 15 (or more) individual wires or conductors. Each conductor is assigned a number as it enters and exits the bundle. Two (or more) lines with the same number are the same wire, or are connected wires in the actual circuit.

For example, locate the T REEL PULSE line at the very top left of Figure 18.3. It is assigned line number 2601 at Pin 1 of Connector 2603. To find where the take-up reel pulse line connects, follow along the heavy black lines until you locate one (or more) lines labeled 2601. For this particular line, it connects to Jumper "E" on the circuit board.

Similarly, trace line 2612, V REF, from Connector P2601 Pin 3 to two locations: IC2603 Pin 6 and Jumper "H."

Realize that in an actual VCR or on a printed circuit board there would be multiple wiring traces, individual wires, multi-conductor cables, or combinations thereof for the lines represented in the schematic wire bundle. Remember, a schematic just shows electrical connections, and often has little resemblance to actual physical layout.

Figure 18.4 shows both the wiring and component sides of the character generator PC board in a Sony VHS VCR. Notice the letters A–D along the left edge of the board, and the numbers 1–9 along the bottom edge. These divide the board's real estate into sections so that you can easily locate a component, just like finding a small village on a state road map. To the left of the board is a listing of major components with their coordinates. For example, Transistor Q801 is in Row D, Column 4 on the board.

Continuing with contents of a typical service manual:

3. Electrical alignments and adjustments
   — Instructions on what test equipment to use, where to connect it, and how to adjust some component, like a PCB trim pot, to achieve the proper test equipment indication. Refer to Figure 18.5.
4. Electronic waveforms and voltage charts
   — Images of oscilloscope waveforms that should exist at various test points in the VCR. Used in conjunction with many electrical adjustments. Refer to Figure 18.5.
   — Voltage charts list correct voltage levels at various points in a circuit. Levels may be specified for different modes of operation if applicable.

Figure 18.5 contains two electrical adjustment procedures, both requiring an NTSC color bar generator and an oscilloscope. The TV pattern generator injects standard color bars at the antenna or Line-In terminals of the VCR. For adjustment procedure 1-4-12, trim pots RV104 and RV108 on the YC-92 board are adjusted to obtain specified white and dark clip levels, as observed on an oscilloscope connected to Pin 14 of IC101. In a similar manner, the peak-to-peak voltage of the recording chroma signal is adjusted to 38 millivolts (mV) with pot RV501 on the YC-92/93 board, with the 'scope connected to Pin 6 of IC033.

5. Semiconductor reference
   — Lists manufacturer's part number or device type number for all diodes, transistors, ICs, and the like, often with cross-reference data to alternate, or substitution, replacement components.
6. Mechanical adjustments
   — Information and instructions for aligning or adjusting tape transport components. Refer to Figure 18.6.

Figure 18.6 provides the tension arm adjustment procedures for a Sony VHS VCR. First the *position* of the tension arm is set, using a millimeter scale, so that the movable tension arm guide pole is 4.5mm to the left of stationary guide No. 0 (P-0). Then, using a torque cassette or torque gauge, *tension* is adjusted to between 28 and 34 gram centimeters (gcm). The **Note** says to recheck the position if the tension adjustment is changed. This is fairly common in many mechanical adjustment procedures, where you may have to go back and forth a few times between two (or more) interacting adjustments to meet specifications.

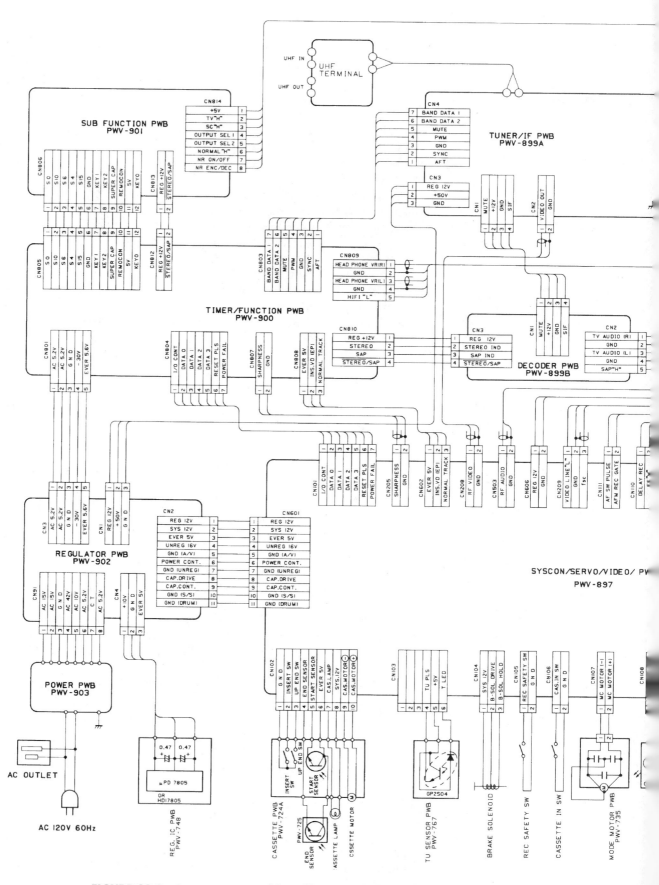

**FIGURE 18.2** Interconnect wiring diagram. *Courtesy of JC Penney Co., Inc.*

## 4-2. CASSETTE SCHEMATIC DIAGRAM

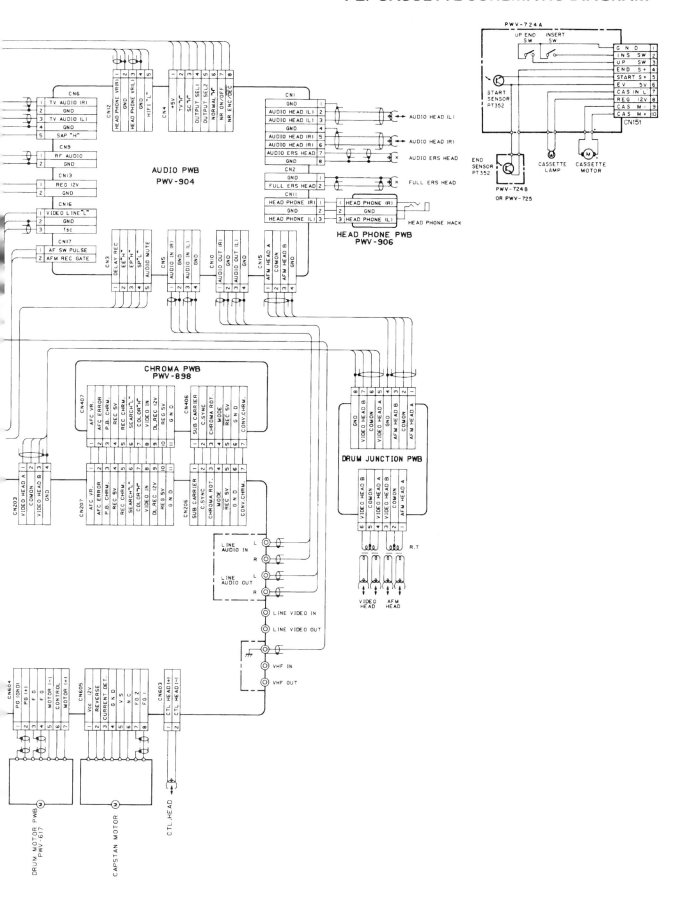

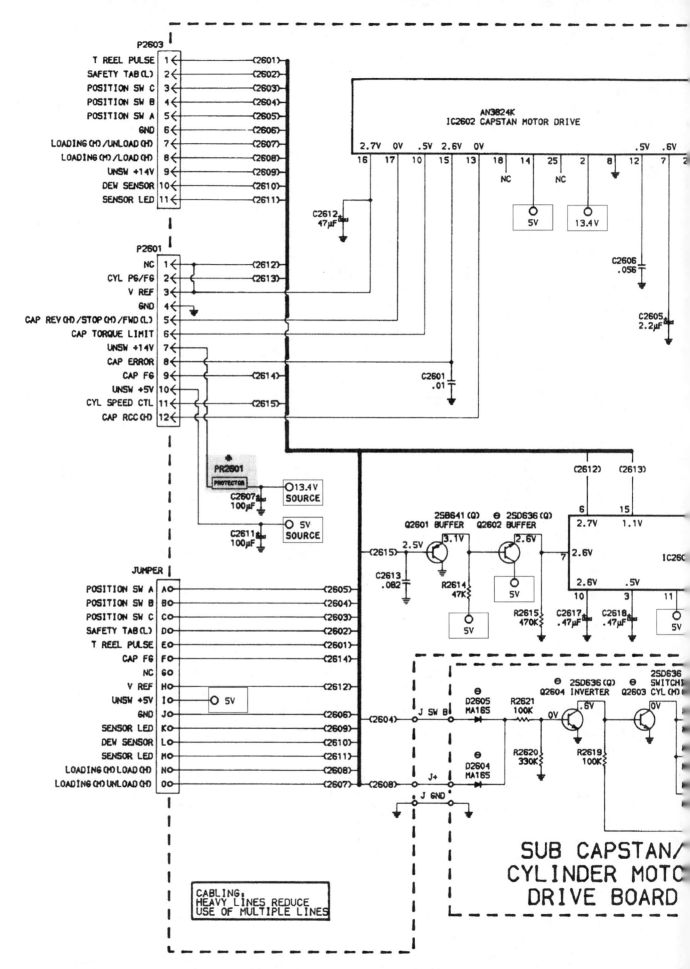

**FIGURE 18.3** Schematic diagram—Capstan/cylinder motor drive board. *Courtesy of Howard W. Sams & Co.*

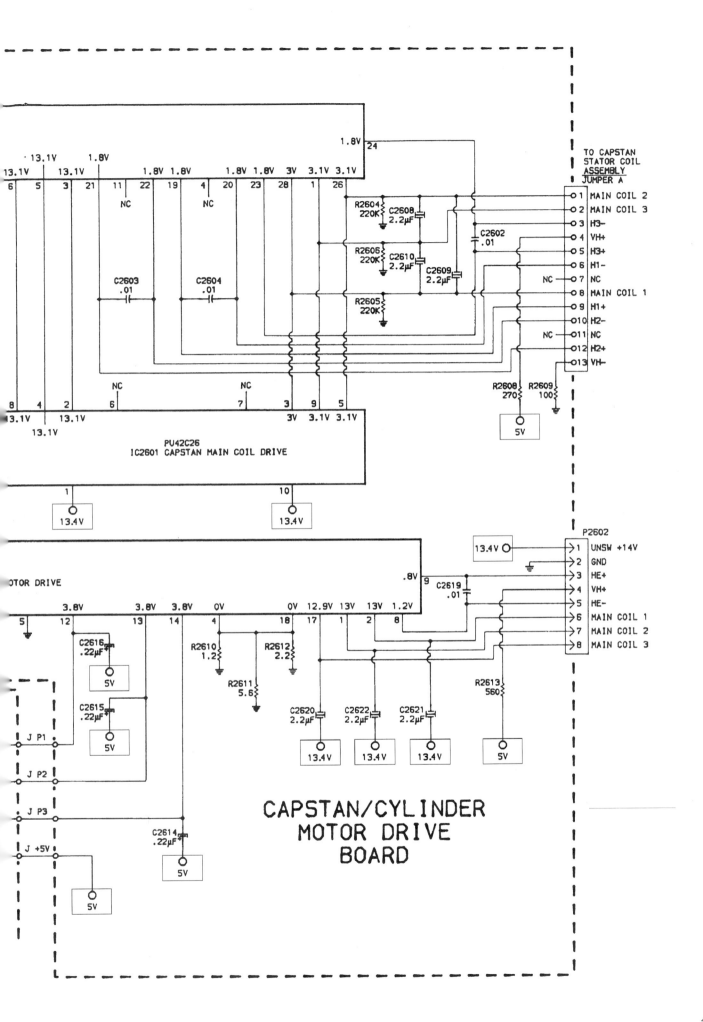

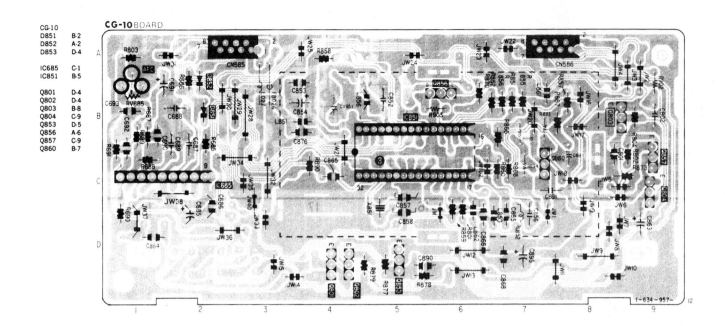

**FIGURE 18.4** PC board pictorial. *Courtesy of Sony.*

### 1-4-12. White Clip/Dark Clip Adjustment (YC-92 Board)

| Mode | E-E(EP) |
|---|---|
| Signal | Color bar |
| Measurement point | IC101 Pin ⑭ |
| Measuring instrument | Oscilloscope |
| Adjustment element | White clip: RV104<br>Dark clip: RV108 |
| Specified value | White clip: 210±10%<br>Dark clip: 70±10% |

[Adjustment Method]
1) Set the EP mode with the REC MODE switch.
2) Adjust with RV104 so that the white clip level is 210±10% to the white (100%) level.
3) Adjust with RV108 so that the dark clip level is 70±10% to the white (100%) level.

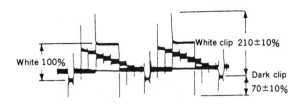

### 1-4-13. Recording Chroma Adjustment (YC-92/93 Board)

| Mode | E-E |
|---|---|
| Signal | Color bar |
| Measurement point | IC033 Pin ⑥ (REC RF C) (YC-92) |
| Measuring instrument | Oscilloscope |
| Adjustment element | RV501 (YC-93) |
| Specified value | 38±5mVp-p |

[Adjustment Method]
1) Set the SP mode with the REC MODE switch.
2) Adjust with RV501 to 38±5mVp-p.

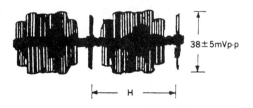

**FIGURE 18.5** Electrical adjustments. *Courtesy of Sony.*

### Tension regulator position/tension adjustment

**Purpose:** Stabilizes contact of the video head and the tape to maintain the tension of the tape so that it feeds at a constant level.

- **Position adjustment**

| Mode | Threading is completed without a cassette loaded. (Refer to section 1-2.) |
|---|---|
| Adjustment locations | Tension band holder |

**[Adjustment method]**
1) Allow the unit to go through the threading procedure without a cassette loaded.
2) Set the VTR unit to playback, then turn the tension band adjuster lever so that the gap between guide No. 0 and tension arm is within 4.5±0.4mm. *(Set the unit to playback without a cassette loaded.)
3) After adjustment, go through the loading procedure once more without a cassette loaded, then check the position of the tension arm.

- **Tension adjustment**

| Mode | Playback |
|---|---|
| Measuring instrument/tool | Torque cassette |
| Adjustment locations | Position for hooking the tension spring |
| Specification | 28 to 34 g·cm |

**[Adjustment method]**
1) Playback the torque cassette.
2) Check that the center value deviation reading on the torque cassette meets with the standards.
3) When the reading is higher than the standards: Move the spring toward direction Ⓐ.
   When the reading is less than the standards: Move the spring toward direction Ⓑ.

**Note:** Move the spring to the tension spring hook position and recheck the tension arm position. If the arm position is misaligned, adjust the position and tension of the tension arm.

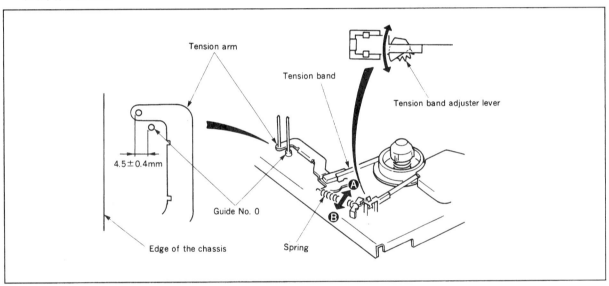

**FIGURE 18.6** Mechanical adjustments. *Courtesy of Sony.*

7. Mechanical views
   — Exploded views of parts in top and bottom of the tape transport, cassette loader assembly, and other areas of the VCR. Refer to Figure 18.7.
8. Parts catalog section
   — List of electronic and mechanical replacement parts. Refer to Figure 18.7 for example of mechanical parts list.

Figure 18.7 is one of two exploded views of components on the top of the tape transport assembly for this particular Sony VCR. Often if you can't locate a part in one view, it will be in another. Also be sure to read any notes in the parts catalog section. Some items may not be readily available. Frequently it is less expensive in the long run to replace a larger assembly than one or more of its component parts.

474 ■ *Practical VCR Repair*

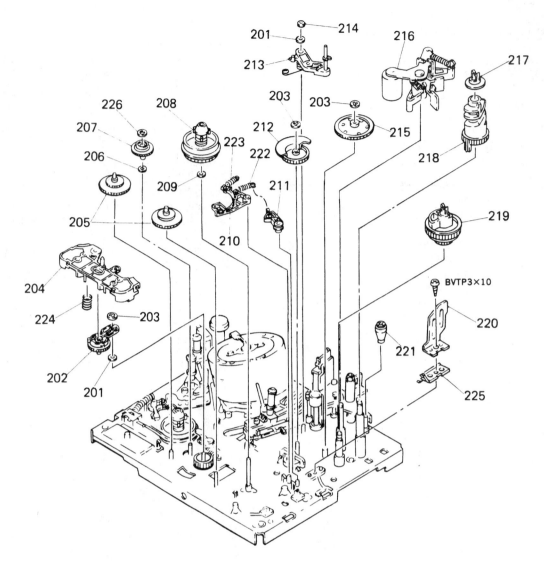

| Ref.No | Part No. | Description |
|---|---|---|
| 201 | 3-701-438-11 | WASHER, 2.5 |
| 202 | X-3727-776-1 | ARM ASSY, PENDULUM |
| 203 | 3-669-595-00 | WASHER (2), STOPPER |
| 204 | 3-736-172-02 | RELEASE, LOCK, REEL |
| 205 | X-3727-795-1 | GEAR ASSY, RELAY |
| 206 | 3-736-074-01 | RETAINER (SMALL), THRUST |
| 207 | 3-736-037-01 | GEAR, REW |
| 208 | X-3727-798-1 | TABLE ASSY, REEL |
| 209 | 3-738-212-21 | RETAINER, THRUST, REEL TABLE |
| 210 | X-3733-335-1 | BRAKE ASSY (AT), T SOFT |
| 211 | 3-736-105-01 | ARM, REV BRAKE |
| 212 | 3-736-143-01 | GEAR, RVS CAM |
| 213 | X-3729-911-1 | ARM ASSY, RVS |
| 214 | 3-736-740-01 | NUT (M2X0.25), NYLON |
| 215 | 3-736-116-01 | GEAR, COMMUNICATION |
| 216 | X-3727-770-1 | PINCH ROLLER BLOCK ASSY |
| 217 | 3-736-111-01 | STOPPER |
| 218 | 3-736-136-01 | CAM, ELEVATOR |
| 219 | 3-736-135-01 | GEAR, PRESS CAM |
| 220 | 3-736-109-01 | PLATE, OPEN, LID |
| 221 | 3-738-250-01 | SCREW, AC ADJUSTMENT |
| 222 | 3-736-025-01 | SPRING (REV BRAKE), TENSION |
| 223 | 3-736-024-01 | SPRING, TENSION |
| 224 | 3-736-020-01 | SPRING, COMPRESSION |
| 225 | 3-744-227-01 | SPRING (ATOM), FL GROUND |
| 226 | 3-736-069-01 | RETAINER, SPRING |

**FIGURE 18.7 Exploded mechanical view and parts list.** *Courtesy of Sony.*

## SUPPLEMENTARY SERVICE MANUAL CONTENTS

The preceding items are pretty much the *minimum* amount of information in most service manuals. Some manuals also include one or more of the following:

9. Theory of operation
   — Conceptual explanations of how various VCR technologies work, such as how video tracks are recorded, similar to the material in Chapter 16.
   — Can vary from broad, high-level theory to very detailed.

10. Description of unit operations
    — Descriptions on how various mechanical and electrical systems operate in the particular model VCR.
    — Can vary in amount of detail. Refer to Figure 18.8.

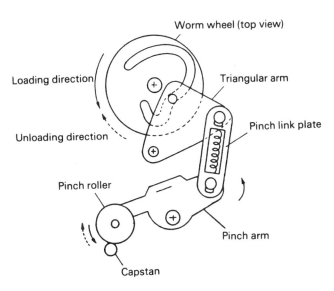

**FIGURE 18.8  Pinch roller operation. The pinch roller moves in response to the worm wheel cam through various parts.** *Courtesy of JC Penney Co., Inc.*

Figure 18.8 illustrates how the pinch roller is loaded against the capstan by a rotating cam on the top of the transport. The tape-load/mode/cam motor turns the cam. Pivot points on the triangular arm and on the pinch arm are indicated by ⊕.

11. Block diagrams or simplified schematics
    — Also called *second-level diagrams;* illustrate how electronic circuits are interconnected and function. Much easier to determine overall circuit operation and basic wiring than with full schematics.
    — Often accompanied by theory and descriptions of operation; similar to the material in Chapters 16 and 17.
    — Some service literature contains separate functional diagrams of integrated circuits, to show equivalent electronic circuits and operations inside each chip.
12. Mechanical timing and operation charts
    — Show positions of cam mechanism, or other transport components, for different operating modes, like Play, Pause, Stop, and the like. Refer to Figure 18.9.
    — Are helpful in understanding mechanical operations.

The cam gear disk in Figure 18.9 turns under control of the tape-load/cam/mode motor. Two IR LED/phototransistor pairs, Sensor 1 and Sensor 2, identify cam assembly position when light shines through apertures in the disk. Each sensor pair consists of an IR LED on one side of the disk and a phototransistor on the other side. When the phototransistor is dark, logical Low is sent to system control; logical High is sent when the phototransistor is illuminated.

Although no units of time are shown in this example, this type of diagram is still commonly called a *timing chart*. Timing also refers to degrees of rotation for meshing gears and related components. For example, ignition timing of an internal combustion gasoline engine refers to the crankshaft position at which sparkplug number 1 fires, expressed in degrees, not in seconds.

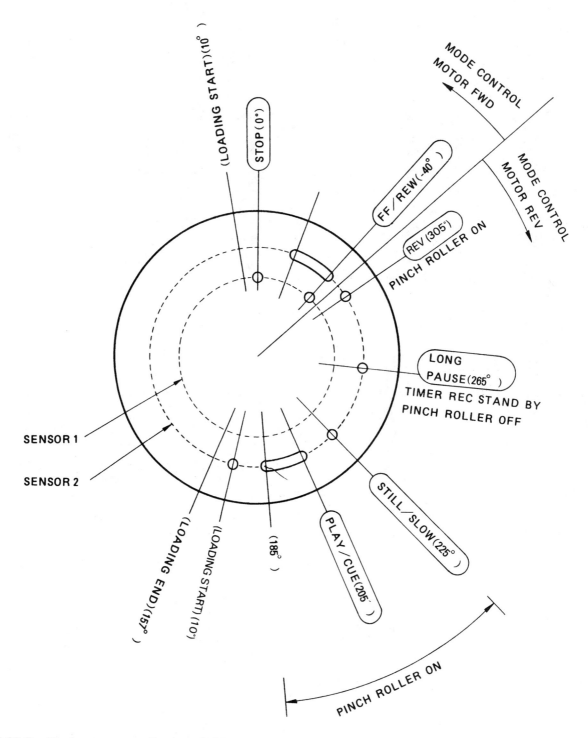

**FIGURE 18.9** **Mode cam gear.** *Courtesy of JC Penney Co., Inc.*

13. Electrical timing charts
    — Show timing relationships, and sometimes electrical waveforms, for electromechanical systems, such as drum and capstan servo systems. Refer to Figure 18.10.

It is not necessary to fully understand the electrical timing chart in Figure 18.10. However, if you suspect capstan servo problems, the chart would be a great help in troubleshooting with an oscilloscope. At the very top is video input during record mode; each horizontal segment represents a video field.

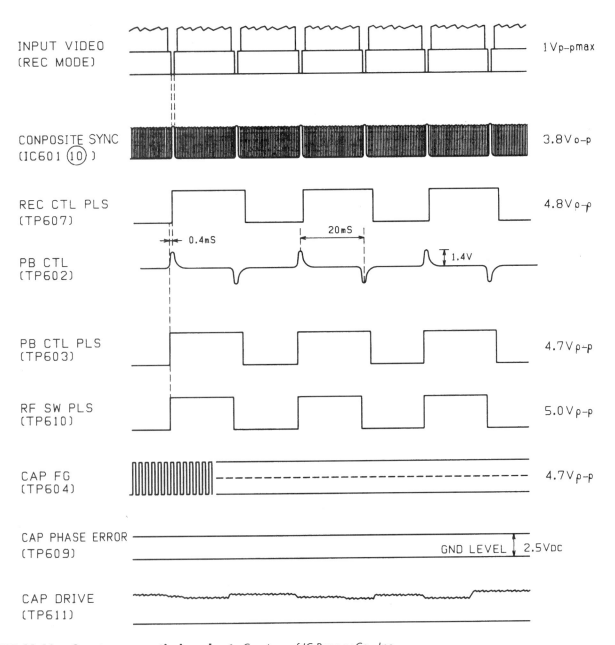

**FIGURE 18.10 Capstan servo timing chart.** *Courtesy of JC Penney Co., Inc.*

For example, you can check the control track logic (CTL) pulse at test point 607 (TP607), the video head switching pulse (RF SW PLS) at TP610, and the capstan frequency generator pulse (CAP FG) at TP604.

14. Disassembly and removal/replacement procedures
    — Detailed instructions about how to take off cabinet covers, swing out circuit boards, and remove subassemblies.
    — Steps to follow to take out and reinstall a part. Sometimes lists electrical or mechanical adjustments, or both, that should be checked *after* a part is replaced. Refer to Figure 18.11.

Note that when removing the upper portion of a video drum, you may need a head puller if it doesn't easily lift off. *Never* pry up on the upper cylinder with a screwdriver blade between cylinder halves. A video head puller is similar to an automotive pulley puller, and normally costs somewhere between $5 and $25 from VCR parts suppliers.

## 1-4. REMOVING THE UPPER DRUM

① Refer to Fig. 1-4 to remove the screw ❶ (P3×5) and then remove the axle ground.
② Completely remove the revolving upper drum PC board and the solder at sixteen points as shown by the arrows.
③ Refer to Fig. 1-5 to remove the two screws ❸ (PSW3×8) and remove the revolving upper drum in the direction shown by arrow Ⓐ. If the drum proves difficult to remove, try tilting slightly backward and left and right.

**Note:** When the drum cannot be removed, check again to ingure that all solder has been removed.

## 1-5. ATTACHING THE UPPER DRUM

① When inserting the revolving drum in the lower drum, be very careful not to leave fingerprints or other soiling on the contact surface.
② Refer to Fig. 1-5 and make sure SP1 on the revolving upper drum PC board is aligned in the same direction as S1 on the rotor transformer PC board (lower drum side). Attach so that the holes on the upper and lower drums are aligned with each other.
③ If it proves difficult to attach the drum, attach while moving the drum slightly backward and foward and left and right.

**Caution:** Be careful not to damage the head at this time. Also make sure that the drum is securely inserted and not rattling.

④ Refer to Fig. 1-5 to tighten the two screws ❸ (PSW3×8).

**Caution:** Lightly tighten one of the screws and then the other screw and check to see that there is no rattling before tightening the screws completely.

⑤ Solder the sixteen locations on the PC board from the revolving drum.
⑥ Tighten screw ❶ (P3×5) so that the protruding portion on the end of axle ground ❷ is in contact with the center of the drum axle.

**Caution:** When attaching the axle ground ❷ be careful not to apply pressure to the spring section of the axle ground.

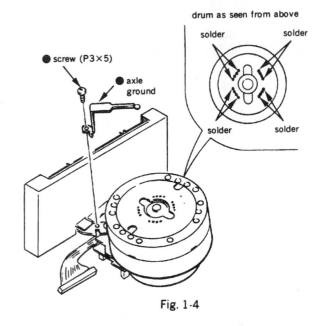

Fig. 1-4

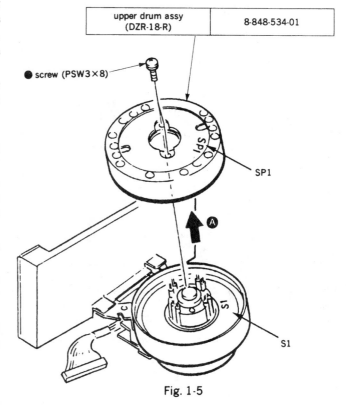

Fig. 1-5

**FIGURE 18.11 Replacing the upper drum.** *Courtesy of Sony.*

This covers most of what you will find in a VCR service manual. Some have just the basics, whereas others contain many helpful tips, second-level diagrams, and detailed descriptions of the unit's operation. Many manuals include preventive maintenance procedures for cleaning, lubrication, and routine adjustments and part replacements. Part numbers for special adjustment jigs, gauges, tools, and alignment tapes are also listed in most service manuals. Compare prices with similar items offered by electronic suppliers of VCR parts and tools; you might realize significant savings over purchasing from the VCR manufacturer.

## SUMMARY

Many different brands of VCRs are made by a fairly small number of manufacturers, sometimes with only cosmetic differences among brands and models. This means that service literature and replacement parts for one brand are often compatible with a different make or model. Order service literature according to the brand name and model number.

It may seem like the $20 or higher price for a typical service manual is a bit steep, but especially when you are just beginning to service VCRs, the cost is well worth it. A service manual can save time, help you learn more about how VCRs work, and perhaps prevent you from breaking a part that costs more than the service literature. For electronic work, it is nearly impossible to proceed without at least the unit's schematic diagrams.

In addition to schematics, service literature contains electrical and mechanical adjustment procedures; oscilloscope waveforms; disassembly, removal, and replacement steps; exploded mechanical views; and replacement parts lists. Some manuals also include theory of operations, detailed explanations of various VCR functions, and second-level or functional block diagrams.

## SELF-CHECK QUESTIONS

1. (True/False) You can be sure that two different model VCRs having the same brand name will have been made by the same manufacturer. Discuss why your answer is right.
2. Refer to the interconnect wiring diagram in Figure 18.2 of this chapter. Where would you probe with a voltmeter to check for +5-volt power supply input to the AUDIO PWB (printed wiring board)? Assume the black or common meter lead is connected to a ground point on the VCR.
3. Refer to the schematic diagram in Figure 18.3 of this chapter. Where does the cylinder speed control (CYL SPEED CTL) line go after it enters P2601, Pin 11?
4. Refer to Figure 18.4 of this chapter. Where on the character generator circuit board would you find diode D851?
5. Refer to Figure 18.6 of this chapter. How would you adjust the distance between the back tension arm guide pole and guide pole P-0 to 4.5 millimeters on this particular VCR?
6. Refer to Figure 18.7 of this chapter. What part number would you order to obtain a replacement pinch roller for this VCR?
7. Refer to Figure 18.9 of this chapter. With this particular VCR in Fast Forward or Rewind mode, what should be the condition of mode Sensor 1 and mode Sensor 2, assuming these sensors are mounted at the zero degree point of the cam gear?
8. Refer to Figure 18.10 of this chapter. Where would you connect a digital logic probe to see whether capstan frequency generator (FG) pulses were being produced when the capstan was spinning?
9. Refer to Figure 18.11 of this chapter. How should the upper drum be oriented when attaching it to the lower, stationary portion of the video drum?

# CHAPTER 19 COMMON AUDIO AND VIDEO PROBLEMS

This chapter is more in the nature of reference material, rather than instructional. Along the way there are a few descriptions of diagnostic and repair procedures that reinforce or supplement the information presented in the preceding chapters, but the chapter is not meant to be studied as are the other chapters in this textbook.

It is not the intent to cover every possible audio and video symptom that might be encountered on a VCR; that would be an enormous undertaking. There are so many varieties of circuits and ways in which components can (and do) malfunction that it would be virtually impossible to list everything that might go wrong.

Also, problems that require advanced electronic test equipment, such as an oscilloscope and NTSC pattern generator, are not covered. Rather, the common symptoms listed here are those most likely to be diagnosed *without* relying on such test gear. However, even these malfunctions can certainly be diagnosed much more accurately with more sophisticated tools.

## CHAPTER OBJECTIVE

The purpose of this chapter is to describe some of the more common VCR audio and video (A/V) failure symptoms, along with some *probable* causes for each. Included are several tips and techniques to help diagnose and repair A/V problems. There are no specific study objectives for this chapter.

## GENERAL DIAGNOSTIC AND REPAIR TIPS

It's been said before in this text, but the vast majority of VCR audio and video reproduction problems are cured by cleaning, lubricating, adjusting, and replacing parts in the tape path and transport. Follow the procedures described elsewhere in this textbook, especially in Chapters 6 and 7. Do not make electronic adjustments, such as tweaking trim pots on circuit boards, until the VCR is mechanically healthy. In general, before looking for or suspecting electronic problems, consider the following steps:

- Thoroughly clean rotary heads and videodrum.
  - Stubborn head clogs may require several cleanings, including spraying cleaner directly into head gaps.
  - Don't overdo to the extent that cleaner gets into drum motor bearings, where it could wash away lubricants and possibly cause the motor to seize up.

- Clean stationary Audio/Control head assembly.
  — Pay particular attention to the control track head gap that reads and writes control track logic (CTL) pulses near the bottom edge of the videotape.
- Clean capstan, pinch roller, all tape guides, and anything else in tape path.
- Lubricate capstan bearings.
  — Check that capstan and pinch roller rotate freely.
- Check that videodrum rotates freely, with no bearing noise or gritty feel.
  — Ensure that static discharge strap is making good electrical contact with top or bottom of drum shaft, depending on model.
- Carefully inspect tape as it goes through transport.
  — Especially notice how it wraps around drum and rides on rabbet, or ledge, on lower portion.
  — There should be no crinkling, skewing, or fluttering anywhere along tape path.
- Check that movable tape guides (P-2 and P-3) are fully seated in their V-stops when tape is loaded.
- Check and adjust tape holdback tension at supply spindle.
- Check take-up reel torque.
  — Clean rubber idler tire; check slip clutch operation.
  — Replace idler and clutch assembly.
- Replace pinch roller.
  — Never hesitate to replace this highly critical, low-cost part.
- Replace belts.

You may be surprised at the large number of VCR failure symptoms that are caused by problems in these portions of the tape transport. Make sure everything is fine in these areas first before digging into the electronics!

A tiny speck of tape oxide, depending on where it is lodged along the tape path, can cause audio or video problems, especially if it's on the video cylinder or A/C head assembly. Check the five anti-stiction grooves running around the rotary head assembly for any particle that might be stuck in one. Inspect the entire tape path carefully with a bright light, and even a magnifying glass, for any foreign particle that may be hampering A/V reproduction.

Always check for tape damage, such as curled, rippled, or crinkled edges, by running a good scratch cassette through an ailing VCR first. If you find no tape damage, then use a high-quality, commercially recorded tape for most diagnostic and alignment procedures, as described in Chapter 7.

Don't forget that many VCRs, including virtually all newer models, mute *both video and audio* if there is not sufficient RF signal from the video heads. VCRs mute playback until CTL pulses are read from tape and the automatic speed control circuitry adjusts capstan drive to either SP, LP, or EP/SLP speed. For some problems, you may have to defeat mute circuits to observe a video symptom (for example, if the raster remains a solid background color, such as blue, when a pre-recorded tape is being played). Consult the service manual for information on disabling the video mute function, such as grounding a test pin, which is often near the on-screen display (OSD) circuit or IC.

When checking VCR audio and video reproduction on a TV tuned to Channel 3 or 4, as opposed to monitoring baseband Line-Out audio and video, don't forget to adjust the fine tuning control on the TV to obtain best picture and sound.

Check VCR operations with video and audio Line-In and Line-Out jacks when there is a problem either recording broadcast programs, or when playing a tape to the Channel 3 or 4 input of a TV. This can help diagnose many VCR A/V problems. For example, if a tape plays back and produces a good picture from the composite Video-Out RCA jack, but the RF modulated signal on Ch 3 is noisy, then the videoheads are *not* the problem. Most likely the RF modulator/antenna switcher is defective.

If you don't have a TV or monitor with a composite Video-In jack, you can use a second known good VCR (VCR-2) to check the composite output of the VCR being diagnosed (VCR-1). Simply connect Video Out of VCR-1 to Video In of VCR-2 with a shielded cable having an RCA plug at each end. Then connect VHF out of VCR-2 to the TV set. It is not necessary to have a cassette loaded in VCR-2. Composite video out from VCR-1 will be modulated by VCR-2 so the image can be seen on TV Channel 3 or 4.

The same technique can be used to check line level audio outputs of VCR-1. Audio Out jacks on VCR-1 can also be connected to Tape, CD, or Aux. inputs on a stereo receiver or audio preamplifier control unit.

A second functioning VCR is also a good tool for diagnosing *record* problems on a failing unit. It can supply quality line level audio and video signals. Connect Video Out from VCR-2 to VCR-1 Video In. Do the same for audio.

Diagnose and repair playback problems with a quality, commercially prerecorded tape *before* attempting to resolve recording failure symptoms.

Correct linear audio problems *before* diagnosing video trouble. For example, repairing the cause of wavering linear audio will often clear up video problems. In general, correct any video failure symptom *before* diagnosing Hi-Fi audio problems.

For multiple troubles, bizarre problems, intermittent malfunctions, and when tracking down suspected electronic failures, many technicians routinely check all power supply (PS) output voltages as their first step. A DC voltage that is too low, or too high, or has a high percentage of AC ripple can cause almost any kind of problem you care to think about. It is good practice to give the power supply a clean bill of health before delving into other electronic circuits, unless there's an obvious defect, like a smoked resistor on a circuit board.

Depending on the particular model VCR, all DC voltages may be produced on a single PS board, *or* some PS outputs may be converted to lower voltages by voltage regulators on other circuit boards. You'll need to look at the schematic to find where to check all PS voltages.

Take time to visually inspect the entire piece of equipment. Look for pinched or broken wires; loose connectors; foreign objects; missing hardware; broken, bent, or loose parts that ought not to be loose; burned or charred components; poor solder connections; and short circuits.

Reseat all multi-pin connectors in circuits related to a problem. For example, if the problem is noisy video, unplug and replug all connectors in video circuits, starting with one or more at the video drum rotary transformer. Refer to interconnection wiring diagram to identify connectors in video signal path. Check connector pins and contacts for corrosion.

Probably the single most frequent cause of electronic trouble is a poor electrical connection. This includes high resistance switch and relay contacts, bad battery contacts, loose plug-in connections, inadequate socket contact tension, faulty crimp of terminals on cable ends, cracked printed circuit wiring, poor potentiometer wiper contact, cold solder joints, and even missing solder, where a connection that should be soldered never was. This doesn't mean a bad connection is always easy to find, but once you are in the right circuit, all it takes is the trusty ohmmeter on R x 1 to verify a good connection. Of course, in some cases, like a cold solder joint, the defect is visible without test equipment.

Although rare in a VCR, a wire can open from being overstretched or flexed too often, without necessarily breaking the insulation. It may look all right, but an end-to-end continuity check will disclose a break.

And then there are shorts. Although not as common as faulty connections, they account for a significant number of electronic defects. Shorts can develop from pinched, punctured, and chafed cables; sloppy soldering; loose hardware; loose solder residue; incorrect assembly (like leaving out an insulator); broken connector bodies; and bent metal covers, shields, brackets, and so on.

A short between closely spaced component leads soldered to a circuit board, as on many ICs, can sometimes be eliminated by scraping between adjacent leads with a scribe, jeweler's screwdriver, or soldering aid.

Problems that occur after the VCR has been powered On for a while, or that show up after several minutes of operation, may be caused by a component breaking down with heat.

For example, an electrolytic capacitor may open as the surrounding temperature increases, or a transistor may become leaky after a few minutes of operation as it heats up. Use spray component cooler to isolate the failing part. Frequently a short blast of coolant on or near the defective part restores operation for awhile.

If you narrow down an electronic failure to a particular circuit, refer to the schematic and check *DC* voltages against what is printed on the diagram. You may find that the voltage input to an IC, usually marked $V_{CC}$, is missing or at the wrong value. Tracing the circuit back to the power supply should locate the cause.

AC and peak-to-peak signal levels may also be indicated on a schematic, but with the exception of the power supply, a multimeter is unable to correctly read the majority of these voltages; an oscilloscope is required.

Following are common VCR audio and video troubles, with some possible causes and a few diagnostic tips. Symptoms and causes are not necessarily listed in any order of probability of occurrence. Realize that these are not absolute fixes for each symptom, but rather likely things to suspect and check.

## LINEAR AUDIO SYMPTOMS

- No audio in playback (or faint output)
  — Check position and electrical continuity of front panel audio switches, usually slide switches.
  — If OK from audio Line-Out jack but not on Ch 3/4, suspect RF modulator. Trace audio path from Audio Out RCA jack back to input of modulator and check for open connection.
  — If no output at line jack, check connections and electrical continuity from audio board to A/C head. Check for possible short between shield and center conductor of cable to audio head.

    Unplug A/C head connector and play pre-recorded tape. Probe contacts on head *cable* with metal scribe or small metallic screwdriver, so your fingers make contact with metal of tool as tool tip touches cable connections. There should be one contact (two if linear stereo) where you will hear a hum, perhaps with a buzz, if audio circuitry after the head is functioning OK. If so, the A/C head may be defective, although this is rather rare. Check solder connections on small PCB mounted to A/C head.

    What happens is that your body acts like an antenna and picks up 60Hz power line energy, plus all sorts of other signals in the air, and injects the combined signal into the audio head preamplifier, in place of the signal from the head. The hum is primarily at 60Hz.

    Do not touch ground with your other hand while probing the audio lead, as this will ground out any signal picked up by your body.
  — If no Line-Out audio, audio mute circuit may be active. Some early VCRs have an audio mute relay. Check electrical continuity of relay contacts.
  — If a linear stereo VCR, and only one channel is dead, you can use playback signal from good channel to inject audio into same stage of dead channel. A schematic is needed!

Connect nonpolarized capacitor (≈2 ~ 10µF @ 16WV, value not critical) between points in circuit of working channel to *same point* in dead channel. Start at Line-Out jack and work your way back toward linear audio head. When injected audio is no longer heard from dead channel, you are at failing stage or component in dead channel.

For example, assume Right channel plays fine, but Left channel is dead. There is a 4.7µF coupling capacitor between the collector of one transistor and the base of the next transistor in the preamp for each channel.

Connect the test cap between the base of the transistor in good channel and the base of the equivalent transistor in the dead channel. You hear audio from the dead channel! That says dead channel is OK from that transistor base onward through rest of circuit.

Next, place test cap between the collectors of the transistor in the previous stage of each channel. Aha!—no output from dead channel. This means the 4.7µF signal coupling cap between transistor stages in the dead channel is probably open.

Note: This basic technique will also work for Hi-Fi Stereo *after* FM demodulator stages. Different audio frequency modulation (AFM) signals are in each channel prior to demodulator, and so injecting the other channel AFM signal won't work as it does with baseband linear audio signals.

- Wow, flutter, or wavering in audio
  — Clean capstan and pinch roller.
  — Check that capstan and pinch roller turn freely, with no binds or indications of bearing problems.
  — Check that capstan flywheel brake, on some models, is fully disengaged.
  — Lubricate capstan bearing.
  — Check that pinch roller and capstan shaft are parallel.
  — Inspect pinch roller for any deformity.
  — Replace pinch roller.
  — Slip clutch, idler, or any take-up spindle drive component causing unsteady or out-of-specification take-up torque. Replace clutch and idler.
  — Incorrect tape holdback tension at supply spindle. Check and adjust.
  — Hard or soft brake pad grabbing supply or take-up spindle.
  — Brake being applied when it shouldn't be, possibly caused by defective brake solenoid coil or driver transistor. On units having a brake solenoid, the solenoid is normally energized to *release* spindle brakes.
  — Anything in tape path or transport mechanism causing irregular tape tension or uneven tape motion.
- Audio plays at wrong speed (SP, LP, EP/SLP), or switches erratically between speeds.
  — Check playback with known good tapes recorded at all three speeds. See if VCR automatically locks onto and stays locked to each speed; observe that only correct front panel indicator (SP, LP, EP/SLP) remains on for each speed, with no tendency to jump back and forth between speeds.

  If speed not locked, suspect trouble reading CTL pulses from tape. Could be problem with control head, capstan FG signal, or other parts of capstan servo circuit.

  Check alignment of A/C head assembly (Chapter 7). Observe lower edge of tape and control head gap for smooth, solid contact. Check control head and capstan motor connections and continuity to servo circuits.
- Hum or buzz in audio
  — Shield of shielded cable between A/C head and audio board not connected to chassis ground, or open shield on other shielded audio cable.

- Defective power supply electrolytic filter capacitor.

   Temporarily connect a known good cap of similar value across each filter cap in PS. Working voltage of test cap must be equal to or greater than original. Observe correct capacitor polarity! Hum will usually disappear if original cap is open or has decreased in value.
- Defective PS diode, could be open or leaky. Unplug VCR, check diodes with multimeter.

   For suspected PS defects, first measure DC output voltages. Then connect a 0.1µF cap at 100WV or greater *in series* with meter lead. Measure supply DC outputs, *but with meter set to read* AC *voltage.* Voltage read is ripple content on DC line. In general, this should be under 5 percent. That is, on a 5VDC line, AC ripple should be less than 0.25VAC (250 millivolts). Some circuits require even purer DC, but the 5 percent figure is a rough guideline. Refer to Chapter 12.

- No audio in record (or faint sound), playback OK.
    - If OK from audio Line-In jack, but not from VCR tuner, check for poor normally closed (N.C.) switch contact on Audio In RCA jack. When RCA plug is inserted, switch contact in jack opens, disconnecting audio received by tuner. If this contact fails to make when there is no RCA plug inserted into the Audio In jack, then broadcast audio won't go to audio record electronics.

       Contact can become dirty or corroded with disuse. Sometimes just plugging an RCA plug in and out of the jack several times will restore operation. Clean N.C. contacts in jack with contact cleaner. (The same thing happens with video signals at the Video In RCA jack, by the way.)
- Recorded audio highly distorted, playback OK
    - Suspect low or missing bias signal, usually from combined record bias/erase oscillator.

       Bias frequency is applied to A/C head along with audio during record mode. If it is missing, audio will be highly distorted, having a clipped or "spitty" quality.

       Check schematic to see if bias oscillator supplies current to full erase head, as it does in most VCRs. If so, check that full erase head erases video on a tape.

       Load a previously recorded tape. Plug one end of RCA cable into Video In jack, leave other end free. Record a section of tape. If section plays back as blank tape, with original video gone, full erase head is functioning. This means that combined bias and erase oscillator is working fine. Check components, connections between oscillator and linear audio head(s).

## VIDEO FAILURE SYMPTOMS

- Image visible and stable, but "snow" or black-and-white pattern overlaying entire raster
    - This is usual result when only one head is producing an image. Only Field 1 or only Field 2 of each video frame is being reproduced.

       Suspect clogged videohead; defective videohead, rotary transformer, head preamp, or related connections.

       Could be problem with head switching IC or 30Hz headswitch signal.

       Refer to schematic that shows multi-pin *video* connector on drum for the particular VCR on which you are working. Now locate this connector on actual videodrum; leave plugged in. While playing a tape, probe connector contacts for Videoheads A and B with a metal scribe or small screwdriver while touching the metal portion of the tool with your fingers. Do not touch chassis ground with other hand. This is similar to injecting hum into a linear audio playback circuit, previously described in this chapter.

On a four-head VCR—that is, one with four *video* heads—probe the head connectors for the two heads being used, depending on the playback speed.

If video playback circuitry is basically working from the rotary transformer output and beyond, there should be at least one contact *for each head* that creates a definite change in the on-screen image when you probe it. You might see black and white bands across the raster or a change in overall brightness.

Injecting electrical noise picked up by your body in this way should cause a noticeable change in the raster as you touch the output connection *for each active videohead.* If not, this indicates a probable electronic failure, such as a dead video preamp or bad head switching circuit at the head preamp outputs. If you *do* see a change in the raster as you touch each videohead's output pin, then either the videohead itself or rotary transformer is likely defective.

Note that on a four-head VCR, you should *not* see any change in raster when injecting noise into the leads from the two videoheads not active at the particular playback speed. If you do, this indicates the head selection IC, or relay in some early VCRs, is malfunctioning. Four-head VCRs often *do* use all four heads during high-speed forward and reverse search to produce a clearer picture.

- Thin horizontal line wandering through image
  — Suspect poor PG signal from videodrum or drum servo failure.
    Check connections and electrical continuity from PG transducer inside videodrum back to servo circuitry.
- One or multiple fairly stationary horizontal lines across image on screen
  — Suspect incorrect tape tension or improper alignment of tape around videodrum.
  — Check that drum entrance guide P-2 and exit guide P-3 are seated in their respective V-mounts at completion of tape load.
  — Check position of bottom edge of tape in relation to rabbet or ledge on stationary part of drum. May require height adjustment of guide P-2 or P-3, or both.
    Minor adjustment of entrance guide P-2 will often eliminate lines along the top portion of the picture; adjusting exit guide P-3 takes care of lines or dashes toward the bottom of the picture. Make sure that the bottom tape edge is positioned properly along the drum ledge and past the two guides.
- Image slowly cycles between good and poor. Color may come and go. Snow or horizontal lines may appear, disappear, and reappear at a regular interval.
  — Suspect capstan servo problem.
    Check capstan for binds, bad bearing, or other problems. Replace belt on belt-driven capstan. Check control head connections and alignment.
  — Could also be caused by missing/poor PG signal from drum. Check connections and cable from drum PG transducer back to servo circuits.
- Image tears, bends, or is distorted near top of screen. This is called *flagging*. Image may also jump sporadically. Image defect could be constant or might come and go.
  — Suspect incorrect tape tension, irregular tape tension, bind in supply or take-up reel spindle, and take-up spindle drive components.

Check supply spindle back tension and take-up spindle torque. Adjust tension, replace clutch/idler assembly.
— May also be exhibited when playing some copy-protected tapes, especially on TV sets manufactured before about 1980. One anti-copy technique changes sync pulses recorded on the tape, making it difficult for older TVs to properly lock onto the video signal.
— Problem can also be due to particular TV, especially if older model.
- Erratic flashes in picture
   — Suspect drum grounding strap not making contact with motor shaft.
   Without the ground strap, a static charge builds up on drum as tape passes over drum surface. When the charge becomes large enough, it suddenly discharges through the drum bearing or elsewhere on the drum, creating, in essence, a small lightning bolt. These voltage discharge spikes create random screen flashes.
- No discernible image at all, just horizontal lines, usually slanting downward from left to right; linear audio OK.
   — Suspect drum servo defect, drum not spinning at correct speed, drum not positioning heads over recorded tracks.
   Audio OK indicates that CTL pulses most likely are being picked up correctly, especially if audio OK at different tape speeds (SP, LP, EP). Could be loss of FG or PG signal from videodrum, or incorrect signals. Check connections and electrical continuity from these transducers back to video servo circuits. Check DC voltages at input to drum motor driver IC and servo ICs.
   — Could be lack of 30Hz video head switching pulses, derived from drum PG pulses.
   — Check composite video at Video Out jack. If OK, could be weak signal from RF modulator.
   — Check in E-E mode (TV/VCR switch to VCR) when *not* playing tape, with VCR tuned to a broadcast or cable channel. If same symptom from RF modulated output on Ch 3/4 but okay from Video Out jack, then RF modulator most likely is defective.
- Color picture cycles to black-and-white, back and forth
   — Suspect deformed pinch roller.
   — Weak color burst signal causes automatic color killer (ACK) circuit to shut down chroma processing, leaving just black-and-white. Could be poor quality signal originally recorded, including weak color burst.
   ACK threshold in VCR or TV is set slightly high; color killer activates too readily. Adjust ACK to retain color with weaker color burst signal; look for ACK trimmer pot in VCR or TV to adjust for lower threshold. If colors go crazy (green milk and purple people), then ACK threshold is set too low and color burst is not synchronizing 3.58MHz oscillator in VCR or TV.
- "Rainbow" lines, usually vertical lines or bands slanted downward to the right, over otherwise generally good screen image. Sometimes called *barberpole effect,* because diagonal color stripes in vertical bands look similar to revolving barber's pole.
   — Often caused by previously recorded video not being completely erased when a subsequent recording is made. Rainbowing is completely normal for first several seconds of tape on VCR without a flying erase head.
   Check full erase head continuity and connections. If linear audio records okay, bias/erase oscillator is working. Check components between oscillator and full erase head.
   — Can also be caused by outside radio frequency interference (RFI). See next symptom.

- *Herringbone effect,* short zig-zag lines in horizontal bands throughout picture. Looks similar to herringbone suit pattern overlaying image. Overall screen brightness may flicker. (Problem may occur only when recording from VCR tuner, and only on certain channels.)
  — Suspect RF interference from commercial FM transmitter.
  
  Disconnect all cables to VCR except Video Out to monitor and play a tape. If symptom is gone, check removed cables for open shield, replace 300-ohm twin-lead with coax (RG-6) (and balun transformer(s), as required), replace interconnect and antenna cables with premium quality cables.
  
  For example, replace RG-59 coax with RG-6. Replace push-on type F connectors with screw-on connectors. All RF connections must be clean and tight. Refer to Chapter 1 Appendix.
  
  Check VCR internal RF connectors at tuner and RF modulator/antenna switcher for good contact.
  
  Add FM trap, available at electronic parts suppliers, between VHF antenna or cable and VCR input F connector. Trap should be as close as practical to VCR VHF input jack.
  
  Ensure all VCR internal ground wires and shields are in place. Ensure *all* top and bottom cover mounting screws are installed.
  
  Nearby CB, amateur radio, or other communications transmitters can also cause interference. Diagonal lines overlaying entire picture accompanied by changing brightness is a typical pattern. May also hear splatter in audio, or even fairly distinct communications. This type of interference is not present all the time, as these transmitters are only On intermittently, whereas a commercial FM station may be on the air 24 hours a day.
  
  Rotating VCR *may* reduce interference. Ensure all cabling is premium quality with sound electrical connections.
  
  For *some* RFI, a separate ground wire to the VCR chassis may help. If the VCR already has a grounded, three-prong plug, try isolating the ground with an AC adapter, leaving the ground connection pigtail disconnected from the outlet cover screw. If the associated TV or monitor has a grounding plug, try lifting the ground here in this manner also. If both TV/monitor and VCR are grounded, it is possible for a ground loop to exacerbate RFI; lifting one of the grounds may help. If this doesn't improve the situation, re-establish the original configuration.

## HI-FI STEREO FAILURE SYMPTOMS

Realize that for most VCRs and commercially pre-recorded tapes, audio recorded in Hi-Fi Stereo, or AFM, is also recorded on linear tracks. Some Hi-Fi Stereo VCRs, however, have only a monaural linear track. When playing back a tape, a front panel switch selects between Hi-Fi and linear audio. (Switch may be labeled Normal—Hi-Fi.)

In the absence of an AFM signal, a VCR will automatically revert to linear audio, regardless of the switch position. In fact, linear audio is normally switched in during dropouts of the AFM signal during playback.

Correct any video problems before diagnosing Hi-Fi audio difficulties.

Check front panel switches for proper setting and electrical continuity.

- Popping sound or static in Hi-Fi sound; frequently reverts to linear audio while playing a videotape recorded in Hi-Fi.
  — Check for dirty, defective AFM rotary head, rotary transformer, or preamp.

— AFM head switching pulse timing requires slight adjustment.
   AFM switching signal is delayed video head switching pulse; trim pot adjusts amount of delay. If video head switching is not working correctly, image will often be overlaid with snow or there will be only a partial image; likely, just the top or bottom half of the screen will be visible. Videohead switching pulses must first be correct before adjusting AFM head switch delay.
— May also be caused by drum ground strap not contacting motor shaft. See *"Erratic flashes in picture"* earlier in this chapter.

## SUMMARY

This chapter provided a brief review of procedures to diagnose and repair the more common VCR failures, without resorting to advanced test equipment. It is generally acknowledged by many electronic service technicians that 80 percent of VCR problems, or more, are mechanical in nature; they can be resolved by cleaning, lubricating, and adjusting tape transport components, and by replacing a few mechanical transport parts that deteriorate. Some electronic failures can be diagnosed using just a multimeter.

Frequently encountered audio and video failure symptoms are described, along with items to suspect as their causes. Tips and techniques for determining the cause of some problems are also included.

In general, linear audio failures should be corrected before diagnosing video problems. Likewise, it is usually wise to fix video defects before tackling audio frequency modulation (AFM) problems (Hi-Fi sound).

Checking Record and Playback operations with A/V Line-In and -Out jacks can help determine whether there is a failure in a VCR tuner and related circuitry or in the RF modulator/antenna switcher.

A second, functioning VCR is a good tool for providing known A/V baseband input signals to the unit under test. It can also accept A/V outputs from an ailing VCR and convert them to VHF Channel 3 or 4. This is extremely helpful when diagnosing an ailing VCR if a TV/monitor with baseband input jacks is unavailable.

Outside radio frequency interference (RFI) can cause rainbowing, a herringbone pattern over the raster, and an image that changes brightness in step with the interfering signal. Providing premium-quality shielded cables and tight, low-resistance RF connections at the VCR and TV is the first step to eliminating RFI.

## SELF-CHECK QUESTIONS

1. When working on an ailing VCR, in general, what should be done *before* looking for an electronic problem or adjusting a trim pot on a printed circuit board?
2. What should you do if you get just a solid blue raster when playing a known good commercially recorded tape on a VCR so you can observe any video that it might be producing?
3. What should you do when RF output on VHF Channel 3 or 4 is poor when playing a tape or monitoring a channel tuned by the VCR's tuner with the VCR in E-E mode?

4. What procedure is highly recommended when an ailing VCR has multiple problems, intermittent failures, or exhibits bizarre behavior in the absence of any obvious visible cause(s)?
5. Symptom: Non-Hi-Fi VCR plays back commercially recorded audio and video all right; video recorded off-the-air from VCR's tuner plays back fine, but there is no audio. Same symptoms on RF and A/V line outputs. What would you do to diagnose this problem?
6. Symptom: When playing back a quality prerecorded tape, there is an image that is fairly stable, but it is entirely overlaid with a salt-and-pepper pattern, similar to what is seen when playing a blank section of videotape on a VCR that has no video mute capability ("snow"). Audio is OK. Same symptoms on RF and A/V line outputs. What is a likely cause of this problem? How might you go about repairing the VCR?
7. Symptom: Picture cycles between good and bad at a regular interval, perhaps a second or so of good video followed by loss of stability, followed by a good picture again. Color may come and go, also. Slight wavering or wow in linear audio. Same symptoms on RF and A/V line outputs. What is a likely cause of this problem? How might you go about repairing the VCR?
8. Symptom: Video and linear audio OK in playback; Hi-Fi audio has popping sound, frequently reverts to linear audio when playing a quality, commercially recorded videotape containing Hi-Fidelity audio. What is a likely cause of this problem? How might you go about repairing the VCR?

# CHAPTER 20 SERVICE CONSIDERATIONS: THE BUSINESS SIDE OF VCR REPAIRS

Knowing how to diagnose and repair mechanical systems and electronic circuitry is not the whole story when it comes to actually servicing VCRs. There are several other points that also must be considered, especially in a commercial repair business. How do you interact with customers? What do you do when taking in a VCR for repair? What should (*or must!*) be done before releasing a repaired unit to a customer? How do you reduce the likelihood that the unit will be back again shortly, with the customer complaining, "You didn't fix it right!"?

The main reason most technicians service VCRs is to make money, either directly for themselves or indirectly by making a profit for their employer, who in turn pays them a wage. With this in mind, there are a few business decisions that should be considered when a VCR is in the shop. After all, if repairing a VCR doesn't turn a profit, it would be better if the unit never got as far as the service bench in the first place. This chapter discusses some aspects of these issues, and more.

## CHAPTER OBJECTIVES

Upon completing this chapter, you should be able to:
1. Describe some recommended dialogue a technician might have with customers, and some things that should be done when a VCR comes in for repair.
2. Describe some of the business decisions that ought to be considered when a VCR is on the service bench.
3. Describe how to reduce the likelihood that a unit will be returned under service warranty.
4. Describe how to perform an AC current leakage test on a repaired VCR before releasing it to the customer.

## YOU AND YOUR CUSTOMERS

This section assumes that you are running your own VCR repair facility, or are working in a small shop where technicians frequently come face-to-face with customers when a unit is brought in for repair. In large repair facilities, there is usually a separate service writer who interfaces with customers and writes up repair tickets.

When customers lug their ailing VCRs up to the counter, they will typically form an overall impression of you, and the shop's abilities, within the first minute of meeting you! That first impression *is* important. You should look, speak, and act professionally. This means good grooming, neat, clean clothes, and a friendly, caring attitude. In short, make customers feel comfortable and important, and help them form a positive attitude about the shop.

A shaggy appearance, a dirty or worn-out shop coat, or a cold manner can easily strip away any confidence the customer has in the business. To paraphrase Tom Peters, the renowned business consultant and quality advocate, "If the passenger gets on the plane, pulls down the tray table and sees a coffee stain, in his mind this implies they don't maintain the engines!" Customers who see a cluttered shop area and unkempt personnel, and receive inarticulate answers, at least subconsciously, think it will be a miracle if their VCR ever gets fixed in this place without their getting ripped off.

Remember, customers are not overly thrilled about being there in the first place. The VCR has died. They may rely on it for entertainment at the end of a long week, or perhaps they use it professionally. Now they're out of luck! In any event, they know it's going to cost money, they've probably got better things to do right then, and they may even feel uncomfortable about being around "techies." Try to keep all these things in mind when you first greet a customer. Be empathetic, which means trying to see things from their perspective.

Now that you're neatly dressed, have greeted the customer with a smile, and courteously asked, "How may I help you?", what's next? Assuming the client says something like, "This won't play tapes," *do* attempt to get a little more information. Ask what the customer means by, "it won't play tapes." Write down on the repair ticket important information you get from the customer. Everything the customer says may not be relevant, but listen for what *is* and jot down the vital facts.

This can help identify the problem more quickly once the VCR is on the bench. Otherwise you might: (a) spend a lot of time trying to get the VCR to fail, (b) conclude nothing is wrong, and (c) return it to the customer—only to find out when she brings it back a day later that it is only *unattended,* programmed recording that doesn't work.

Not all customers will give you much more than "It won't work." If they seem reluctant to answer any further questions, maybe that's as good as you're going to get. Don't press them! But if at all possible, ask questions to try to find out a little more about how the VCR fails. Here are a few suggestions; you'll get the idea:

- Does it fail all the time or just once in a while?
- Does it act up with rental movies or just when playing tapes from your camcorder?
- When you say it doesn't record, can you tell me a little more?
    — Does it fail to record when you set it for unattended recording?
    — Does the tape move at all; is all the tape still on the left reel?
    — Do you get any sound on playback?
    — Describe what the picture looks like. Are there lots of black-and-white lines at the top or bottom of the picture? Loss of color? Vertical rolling? "Snow" all over?
- How does it eat tapes? Is there tape dangling out of the machine when you eject a cassette?
- How long does it usually play before shutting itself Off?
- Do you have a cable box connected to the VCR?

Other questions may also be appropriate, but the main thing is to attempt to learn as much as possible about how and when the unit fails for the customer, or at least what the customer *thinks* is wrong with it. Very few customers are going to be technically knowledgeable, so don't expect precise, technical answers. Be careful not to insult customers by correcting something they say that is not exactly technically correct.

For instance, a customer may say something like, "It won't play back a tape at the end." This *could* mean the VCR *is not rewinding!* To the customer, *playing a tape* is watching the rental movie, and playing *back* a tape is when the tape winds back to the beginning when the movie ends! Sometimes you'll get some pretty bizarre *and inaccurate* explanations of how a VCR behaves (or misbehaves!), and how it is connected to a TV or cable box; you'll get used to it. Just be patient, be a good listener, and communicate so customers are happy and certain that you can make every little thing all right.

Limit your conversation to just two or three questions. Most customers want to be on their way, and really don't want to be interrogated with lots of questions about their VCRs. The bottom line is to make the customer satisfied and confident.

## About Repair Estimates

No doubt an entire book could be written about the pros and cons of charging a fee for estimating VCR repair costs. There are many factors to consider, such as the amount of competition in the area and what *they* are doing regarding estimates, the newness of the shop, and even the present economy of the area. However, consider charging a nominal diagnostic or estimate fee that can be applied toward a repair. Approximately $25 is most likely appropriate.

If you offer free estimates, you may be working a lot of free hours. At a minimum, you'll need to hook up the VCR and spend a few minutes putting it through its paces to see how it behaves, in addition to doing the paperwork associated with receiving the unit and documenting what you find. In many cases, you'll have to remove one or more covers and perhaps poke around a bit to arrive at a repair estimate. If the customer decides to *not* have it repaired—or even worse, takes it to your competition—you haven't gained anything. In fact, you've lost money.

If a customer balks at paying a minimum repair charge, estimate fee, or diagnostic charge, politely explain that it often takes much longer to *determine* the cause of a problem than to actually fix it. As a business person, it is just economically unsatisfactory to spend a significant amount of time finding out what needs fixing without charging a reasonable fee for this labor.

If you *do*, for whatever reason, decide to give free estimates, don't tell the customer the details of what was found to be wrong or what corrective action is needed. Just report the maximum amount it will cost for the repair. If you tell the customer that all the unit needs is new belts and a cleaning, they may decide to tackle this themselves, or take it to a friend to perform the repair. You just gave away the shop!

Repair estimates are just that: *estimates*. However, from the customer's standpoint, an estimate is usually what they expect to pay to get the unit back, and not a dollar more. It is important, therefore, to estimate as closely as possible, including your profit margin, to what you believe will be the total cost of repair. Estimate too low and you lose money. Estimate too high and the customers may decline the repair, or take their business elsewhere in the future, if not this time.

At times your estimates will be wrong. What you thought was a bad video head may turn out to be just a loose tape guide that required minor adjustment: actual repair cost *less* than estimated. Or the capstan motor may need to be replaced in a unit that you figured just had a worn pinch roller: actual repair cost *over* estimate.

You should pass on savings to customers when repairs cost less than expected. This is ethical, makes them feel good, and will probably earn the shop repeat business.

When possible, an excellent approach may be to explain how likely it is that the VCR's symptoms will be fixed by simple repairs, such as replacing a few belts, a rubber wheel, or cleaning the heads and tape path. But also explain that, in some cases, you may discover the need to replace a much more expensive part, such as a motor. This puts the customer on notice that the repair may cost much more than the minimum. Offer to call the customer for authorization to proceed with an expensive repair. Notify the customer in advance what her cost will be if she declines the more expensive repair.

In many cases you won't easily be able to collect *additional money* from customers if repairs cost more than expected, but you can still get some mileage out of this. Explain that you charged less for the repair than it should have cost, thank the customer for her business, and tell her to "Come on back!" Chances are customer loyalty and positive word-of-mouth advertising will more than make up your added costs for the unexpectedly high repair cost.

### Declining a Repair Opportunity

There may be times when you can do yourself and your customer a favor by *not* taking a unit in for repair. If there is evidence the unit has been severely damaged by being dropped, or shows signs of water damage, repairing it could easily become a nightmare. Chances are there are multiple problems, and most likely it's just not worth attempting repair, at least from an economic standpoint.

Another time when it might be prudent to refuse a repair job is if someone else has already been at work on the unit. If a customer brings you a VCR with the covers already unscrewed, or perhaps a unit along with a bag of loose parts that have already been removed, be wary! This might be a unit that a well-meaning neighbor or brother-in-law tried to fix and got in over his head; now they're bringing you the "basket case." There could be missing parts as well as secondary problems added during the attempted home remedy. Trying to get such a unit back on its feet could take more time and parts than it's worth.

You may not realize that repairs have been attempted until you've got the covers off to do an estimate. If you see evidence of poor solder repairs to a circuit board, burned wire insulation from a carelessly handled soldering iron, disconnected wires, or obviously missing parts, that's a big waving red flag; be very cautious about accepting the unit for repair.

In these cases, explain to the customer your reasons for declining to repair the VCR. Emphasize that you are really looking out for *their* best interest, that servicing the unit in its present state is probably not economical, and that the reliability of the unit even after repairs are made might be uncertain.

## RECEIVING A UNIT FOR REPAIR

Suppose you are about to accept a VCR for repair. Is there anything further you need to do, other than record the customer's name, address, daytime phone number, and description of the problem on the repair ticket? You bet! It is important to include a few other items, *especially the customer's signature authorizing the repair.*

First, be sure to establish any minimal, diagnostic, or estimate charge on the repair ticket. Some customers want to be notified by phone after an estimate is made, before going through with the actual repair. Note this fact on the ticket, such as: "call w/est." Otherwise, parts and labor may be put into a machine that the customer doesn't want fixed for the price, and the shop is out that amount. There is no guarantee that a customer will even pick up their unit and pay the diagnostic fee if the repair estimate is above what they want to spend. If you never see them again, at least the shop gains a supply of spare parts, which can help diagnose another VCR of the same or similar model, but this is not the way to turn a profit.

Try to establish a *maximum* repair cost that the customer is willing to pay, and note this amount also. Then the customer can authorize further repair over the phone when the estimate is completed, assuming the estimate is less than or equal to the maximum repair bill the customer has authorized. It is best all the way around—for customer *and* service business—to get the customer's signature authorizing a maximum repair amount in dollars. It can get unpleasant and complicated when a customer comes to pick up a VCR, gets a bill for $158.67, and then starts complaining, "But I didn't know it could cost *that* much!" So it's best to obtain the customer's signature authorizing the repair and agreeing to pay the maximum amount that the repair *could* run, if at all possible.

Of course, be sure to record the make, model number, *and serial number* of any unit received for repair, along with the description of the problem. This avoids mix-ups should two machines with the same model number be in the shop at the same time. It happens!

Make a quick overall power-Off inspection of the VCR, and note anything significant on the repair ticket before the customer signs it. The following are some of the major items you should consider notating:

- Note if a cassette is still in the machine.
    - Perhaps the cassette is jammed in the loader or the VCR has "eaten" the tape, so a loop of tape was stuck in the machine when the cassette was ejected.
    - If possible, note the title or any identifying characteristic of the tape, such as "*Aladdin*" or "TDK Hi-Fi T-120."
    - Note on the repair ticket *and* verbally explain to the customer that the repair shop cannot be responsible for the condition of the cassette when it is jammed or the VCR has eaten a tape.
- Note any physical damage, broken or missing knobs, covers, or other parts, and the general condition of the VCR.
    - Many VCRs have flip-up covers over tuning controls on the top, or a flip-down cover over secondary front panel controls. These covers easily become broken or detached.
    - Operate front panel slide switches and rotary controls. Note any that act like the control or switch may be physically broken. As you'll read in a short while, it is important to return each switch and control to the position to which it was set when the customer put the unit down on the counter, when releasing the unit later.
    - Note any scratches, digs, gouges, or cracks in the VCR cabinet. See if the front panel display lens is cracked or broken and look for any apparently broken plastic latches that hold on covers or panels. Any evidence the unit was dropped or suffered water damage?
    - Note any abnormal condition of the line cord and rear panel jacks.
- Note if the remote control unit or other accessories, like cable mouse or interconnect cables, are with the machine. Otherwise, state "no rmt/accs" on repair ticket.
    - What is the physical appearance of the remote and accessories? Are there batteries in the remote?
    - It's a good idea to label any accessory with a peel-off label having the repair ticket number if there isn't a serial number on the item. (Be absolutely certain to use the type of stick-on label that peels off easily!)

It may sound like a lot of work, but all this can generally be done in about two or three minutes. After making notes on the repair ticket, explain relevant ones to the customers and answer any questions they may have before they add their signatures.

But there's more! Even after the customer signs and has left the shop, there is one more significant action you should take before putting the VCR on the "awaiting repairs" shelf: *document switch positions!* Here's why.

Suppose all the unit needed was the pinch roller replaced plus the usual transport and head cleaning. A few days later the customer picks up the unit. Then, a few hours later, he is on the phone complaining that the VCR won't work at all! You ask him to bring it back in.

He does, you check it out, and everything works fine. As you try to figure out what's going on, you ask him what channel he sets the TV to when playing a tape on the VCR. He replies, "Channel 4." You check the back panel and, sure enough, the Ch-3/Ch-4 switch is set to Channel 3. You ask the customer if he moved the channel switch on the back of the VCR and you hear, "I don't know anything about a switch on the back!" as the reply.

You show him the switch, move it back to Ch 4, and apologize for the inconvenience. Ideally, you would have thought of this as a possibility when the customer phoned, and could have directed him to check the output channel switch. But the point is, if the switch position had been documented when the VCR came in for repair, and it was checked to be in the same position when the customer picked the unit up, everything would have worked fine.

Many customers are not knowledgeable about VCRs, nor do they know what to do when things don't work just right. They may have read the user's guide and figured out how to set up the VCR when they first brought it home two years ago, but now that manual is lost. They've forgotten about any rear panel switches, and so immediately figure that *you* didn't fix the VCR correctly.

Many VCRs have switches that a customer may *never* use, such as input select, audio channel output select, and dubbing and edit switches. Returning a VCR after repair with switch positions changed can cause problems. For example, if the input slide switch is left on "Line" instead of "TV" when the customer picks up the VCR, she may not know to change the switch to make a recording from a broadcast channel. On the other hand, someone may only use the VCR to dub tapes from a camcorder, never recording a program off the air. She may think something's wrong if the unit's input switch is set to "TV" instead of "Line," where it normally is in her setup.

Documenting the position of these switches when a unit comes in can eliminate potential problems. Just make sure switches are set to their original positions before the customer takes it away. And should a customer call on the phone, have them check the obvious connections and switch settings before returning for what may really be operator trouble.

## UNIT REPLACEMENTS

Although the majority of VCR repairs are accomplished by performing proper cleaning, maintenance, and replacement of worn mechanical transport components, such as belts, idler assembly, and pinch roller, there will be VCRs that have electronic problems. Some of these you will be able to repair by replacing an individual component, such as an open power supply diode or shorted motor drive IC. However, there will also be electronic problems that are not so easily diagnosed or repaired.

In many cases, it may take less time and be the least expensive alternative in the long run to replace an entire electronic unit in a VCR, rather than trying to troubleshoot and repair a circuit. Most manufacturers recommend that the tuner section, Ch 3/4 RF modulator, and VHF antenna switcher units be replaced rather than repaired. Most newer VCRs incorporate the RF modulator and antenna switcher in the same unit. If one of these units has failed, look up the part number in the service literature and replace the entire assembly.

These RF units can usually be identified by sheetmetal shielding around the entire assembly or portions of a PCB. Component lead length and positioning is very critical at the high frequencies handled by these units. Repairing one could actually degrade overall performance (one reason why the entire unit should be replaced). Improper repair could also cause the VCR to emit RF radiation, which could interfere with other electronic equipment in the neighborhood.

These same RF units are prone to damage by electrical surges on antenna leads and cable systems caused by nearby lightning. If there is any evidence of charring around RF connectors or elsewhere on these units, replace the entire assembly. It is common for a voltage spike on the antenna lead to take out both the tuner and RF modulator/antenna switcher.

At other times it may also be better to replace an entire circuit board rather than attempting repair. For example, if there are charred wiring patterns on a power supply printed circuit board, this could indicate that the unit received a hefty surge on the power line, again possibly caused by a nearby lightning strike. There could easily be half a dozen or more components destroyed on the board. By the time you locate and replace all the damaged parts, the overall cost might be more than that of a whole new power supply board.

Of course, you first have to find out if the power supply or other circuit board is available, and at what price. Often, entire boards *are* available, but not readily. You may have to call various suppliers to locate a complete circuit board. The board may have to be ordered by a U.S. supplier from overseas, in which case there will be a wait and perhaps an additional shipping charge.

There are really no hard-and-fast rules when it comes to deciding whether to attempt to repair a board or replace it entirely. How confident are you that a new board will correct the problem? You'll need the service manual to at least identify the likely failing board, depending on the function that's failing. With so many interconnections between boards on a VCR, just because it *looks* like the syscon microprocessor is failing doesn't mean the problem isn't really something wrong with the IR detector on the front panel PCB!

Also, a new, complete circuit board probably won't be adjusted if it contains trim pots. These will need adjustment after the board is replaced. Do you have the test equipment and ability to do this?

In contrast, you may have a VCR where the fluorescent display has been shattered or is not working, and you're pretty sure the display itself is at fault. You might elect to solder in a new display, or first find out how much the complete board (including timer/operation microprocessor, front panel switches, and controls) costs. Considering the time it takes to unsolder the old display and solder in the new one, it may be more prudent to put in a whole new board.

## REALISTIC EXPECTATIONS

As you gain electronic knowledge and experience troubleshooting and repairing VCRs, you should feel more confident of your ability to tackle tougher problems. If you are fairly new to electronics, it is totally unrealistic to expect that you can resolve all VCR problems.

First, for some units it will be hard to tell for sure whether you have a mechanical or electronic problem on your hands. Without advanced electronic test equipment, even the most knowledgeable technician would be hard pressed to tell whether a noisy picture is caused by a bad video head, defective video preamp, a faulty head switching IC, or some servo problem. Always suspect a mechanical problem, including clogged heads, as the most *probable* cause for audio and video reproduction and tape handling problems.

But if cleaning, tape path inspection and adjustment, and replacement of worn mechanical parts don't fix the problem, then what? For many picture quality problems, you might want to replace the rotary head assembly, especially for a two-head machine. The more heads, the more expensive this part is. Replacement video head assemblies are available from original manufacturers as well as several other sources, such as electronics suppliers that carry a variety of VCR replacement parts. You can purchase a two-head assembly for around $20 to $40.

If this fixes the problem, fine! If not, perhaps you can use the new video-head assembly on another model. (Suppliers will not accept returns on parts that have been installed in a machine.) This try-and-see approach can get very expensive, however, for upper cylinders having four, five, or more heads. A replacement rotary head assembly alone could cost closer to $100 and beyond.

So what to do when the best attempts at reviving a VCR fail? One possibility is to have *someone else* take care of or help you with units you can't diagnose or can't repair. This might be a person whom you pay part-time on an as needed basis. You probably won't have much success vending out VCRs you can't fix to another service shop, but this is not impossible. In most cases you would be their competition, and there's not much profit in the tough problems anyway. They can take an awful lot of diagnostic and repair time. A service business sometimes *really can't* charge a customer for *all* the time it takes to actually troubleshoot an electronic failure and replace a component, but in the long run the shop makes up for it on the many quick, easy fixes. If all you're going to do is bring another shop the difficult units, it probably won't be long before this ceases to be an economically viable endeavor for all concerned.

But there *are* possibilities, depending on location and what other electronic facilities are in the area. One thing you should know, however: there are ethical, and perhaps legal, considerations if you hire someone part-time who works for a competitor of yours during regular business hours. This person's employer may have spent a considerable sum for training technicians, and now the technician is using the results of that training moonlighting for you. The technician may have signed a contract of employment which prohibits working for any business that could be regarded as competition while employed by the firm. This could even extend for a certain number of months or years after employment is terminated, under certain conditions. You don't want to get involved in anything of this nature, even though you might not be liable. It's not a good idea to hire someone else's employee to perform the same skills, unless the other employer agrees to the arrangement!

As you work on more VCRs, and learn advanced electronics service techniques, you might want to tackle more complex repairs by purchasing an oscilloscope, NTSC pattern generator, and even VCR-specific test gear. Be prepared to spend roughly $2,000 or more for quality equipment that will do the job. The point is, as with almost any human undertaking, we crawl before we walk before we run. Concentrate on doing the *best* job of which you are capable as you repair the 80 percent of VCRs that have mechanical and simple electronic problems. You'll probably be surprised that, before long, you can fix more than 90 percent of what comes your way, with no more test gear than a good multimeter, logic probe, service manuals, *and experience!*

As stated earlier, uncomplicated VCR transport cleaning and adjustment, along with routine replacement of parts, have the potential to bring the most profit. It is difficult to justify expensive test equipment unless you are doing a significant volume of business. So be content to do what you can, and be proud of doing it well. As you'll soon see, you don't want to fix them all anyway!

## BEYOND ECONOMICAL REPAIR

Here's a fact of life in today's highly technological world: it often costs more to repair a piece of equipment than to buy a new one. Nearly all consumer electronics goods are manufactured on highly automated assembly lines. Robotics coupled with computer-aided design and manufacturing (CAD/CAM) have drastically lowered the price for mass-produced TVs, VCRs, PCs, stereo gear, and cellular phones. At the same time, giant strides in the number of circuits contained on large scale integration (LSI) and very large scale integration (VLSI) circuit chips have brought sophisticated digital signal processing to products that would have been prohibitively expensive with the components available only a few years earlier. Just as an example of the strides made in digital integrated circuits, the 8088 processor chip in the original IBM PC, brought to market in 1981, has 29,000 transistors on a single substrate. In 1992 Intel® developed its "Pentium" processor chip, which incorporates a whopping *3.1 million* transistors! This is nearly three times the number of transistors in the i486 processor of just one year earlier.

The result is that today's consumer electronics gear has many more features and enhancements, and at a lower price, than what was previously available. But when it comes to diagnosing and repairing electronics, there is really no robotic, mass servicing that can bring the price down, as in manufacturing. Service work is still pretty much one technician per unit. In some cases, the greater complexity of the circuitry, even if contained in fewer components, can make diagnosing a problem more difficult.

Consider a two-head VCR that is several years old, for example. A typical repair could easily cost more than $150 to replace the video heads and worn rubber parts, clean the transport, and perform one or two mechanical adjustments. A brand-new comparable unit, probably with more features than the original, can be purchased for about $200 at many discount stores and electronic supermarkets. The customer may very well decide that it just isn't worth putting $150 or more into the older unit.

There are other situations when repair isn't economical. A unit that has suffered lightning damage *could* have the power supply plus a few circuit boards fried. Parts and labor costs to repair it are probably more than a new unit would cost. A low-end VCR that has an intermittent electronic problem could demand more time to identify the failing component than it's worth. Replacing a single expensive component, like the entire video drum assembly, including motor, may raise the total repair cost *beyond economical repair* (BER).

Sometimes a customer will decide that a unit is BER based on your repair estimate. Note that the diagnostic fee to prepare an estimate should still be paid. At other times, *you* might decide that a unit is BER; for example if it seems to have multiple troubles or evidence of severe lightning damage. Before even taking the time to estimate what it would cost to repair the unit, given the worse case scenario, you just feel it's a "bag of snakes."

As mentioned earlier, units that have been dropped; that have multiple broken, bent, or missing parts; those that have suffered water damage; those partially disassembled; or units that have been "cobbled" by someone else trying to fix them should sound a loud alarm warning you to stay clear. There's a good chance the unit is BER. Your customers should feel good about your telling them this fact up front, rather than trying to salvage a lost cause.

Of course, the price of a comparable replacement unit must be considered. A customer may have spent more than $1,000 for a Hi-Fi Stereo, Super-VHS VCR in the late 1980s. The present cost for a new unit with similar features might be close to $700. This customer may not balk at all about an estimated $235 repair bill.

## PREVENTING SERVICE RETURNS

No one likes to pick up an automobile, TV, VCR, or computer from the repair shop only to have to return it almost immediately. It may be that the original problem was not entirely fixed, or the unit is starting to act up again in the same way, or it could be a new problem, seemingly unrelated to the original.

The fact is, for *whatever* reason customers need to return something, they are sure to be dissatisfied in some measure with the shop that performed the original repair. This may be entirely unjustifiable, but that's often the way it is. The customer thinks, "The shop had it last, it's broken again, they didn't fix it good enough (or worse, they broke it,) period!"

Remember, customers are always right—right or not—because you have no business without them. Sometimes a customer is entirely wrong, and even unreasonable, but do the best to turn the situation around; make this an opportunity to create a satisfied customer. Remember, negative word-of-mouth is extremely powerful advertising, one you can do without.

There will be times when a repaired item is returned with an entirely different problem than it had just a few days earlier, but in the customer's mind the scenario is, "It wasn't fixed right in the first place," or "They did something to it to cause this new problem." Customer confidence in the repair shop usually slips a few notches. It is important, therefore, to do as much as practical to prevent service returns, or *echoes*. They are a nuisance for shop and customer alike, and can dig into profits; many service returns are fixed free of charge.

It is usually very difficult to convince a customer that a new problem is unrelated to the service work just performed. This is especially true for electronic gear, compared with an automobile. Most consumers wouldn't relate a radiator that starts leaking with the replacement of the master brake cylinder by a shop a week earlier. But what about a VCR that now has sluggish rewind when it was in the shop two weeks ago because it wouldn't eject a cassette? The customer thinks that the shop probably botched up something in the machine. She'll bring it back and demand that it be fixed under service warranty. Often, the shop performs a good will repair in these cases.

The single most significant way to prevent service returns is *attention to detail—at every step,* from the time the unit comes into the shop until it leaves. One item already mentioned can help prevent returns: document switch positions upon receipt and set them the same when the unit goes back out. Following are additional areas where being careful and paying attention to the little things *can* help prevent echoes.

- Fix the *entire* problem
  — Especially with mechanical tape transport components, several different items can contribute to a single observable problem.

    For example, a clogged video head, insufficient supply reel back tension, and an out-of-round pinch roller can all contribute to a distorted and noisy playback image. Suppose a technician just cleans the video heads and tape path, and this seems to clear up the picture just fine. The unit then goes to the completed repairs shelf.

    About a month after the customer picks up the unit, he is back complaining that "the picture sort of comes and goes and the sound warbles." This time around, the technician adjusts the tape holdback tension and replaces the pinch roller: problem fixed.

    Had the technician taken a little extra time, and given the transport area a general checkout when the unit was in the first time, he probably would have discovered that tape holdback tension was too light and that the pinch roller was worn. Taking care of these items *then* would have added little to the repair cost, but would have ensured a much more satisfied customer.

Now the unit has to be logged back in, fixed once more (perhaps under warranty for the first repair), and returned to the customer, who is not all that thrilled about being without the VCR for a second time in a month.

- Fix *all* the problems
  - Something like this has probably happened to you. You take your car in for a new clutch at 95,000 miles. The service writer is filling out the repair ticket, and then you say something like, "Oh, while you're at it, would you fix my backup lights? They're not coming on when I go into reverse."

    Next day you pick up the car, and the clutch works great, but apparently nothing has been done about the failing backup lights.

    This is why it is so essential that you do two things when a customer brings a unit in for repair: *listen and confirm.*
  - *Listen* to everything the customer has to say. Don't get so involved in filling out the ticket that you fail to hear a comment like, "Oh, by the way, while it's in the shop, would you check out the 3 button on the remote. I have to really mash it for it to take."
  - *Confirm* the reported problem(s), and other notes, on repair tickets with customers before they sign them. That way you know nothing has been missed, or that the customer thinks you are going to fix two problems, but you've only written one down.
  - The bench technician is also responsible for paying attention to *all* problems written on the ticket.

- Inspect as much as practical
  - When the unit is on the bench, inspect the entire machine for worn parts or potential sources of problems.

    Cleaning the capstan and replacing the gummy pinch roller has taken care of the "eating tapes" problem, but what about the other rubber parts? Chances are if the pinch roller is badly deteriorated, all rubber belts and the idler wheel tire are also about shot. They may be working fine right now, but will likely cause problems in a few months.

    Look for potential problems while the unit is opened up on the bench. Most customers would much prefer to have everything taken care of at once, rather than having to lug the machine back two months later. Repeat business is fine, but *that* kind of echo can dim the customer's image of a shop that performs quality, thorough work.

    Of major importance is anything in or on the unit that constitutes a potential safety hazard. Inspect the line cord for nicks, cuts, and abrasions that might expose one or more conductors.

    Inspect the AC plug. If it is a three-conductor plug, the ground prong should be present, not clipped off. If it is a two-conductor plug, check that the plug is polarized, with one prong wider than the other. People have been known to file down the wide prong so the plug will fit old fashioned, nonpolarized extension cords. Both of these situations can compromise the electrical safety of the unit.

    Investigate evidence that any part of the unit has overheated, such as warped plastic or charring near the power transformer. If so, check that someone else has not either bypassed a fuse or replaced a fuse with one having a higher rating than specified.

    Any potential safety hazard must be brought to the attention of the customer. It is probably best to refuse to service any unit with a safety hazard that the customer declines to have corrected.

- Check major machine functions
  - Perhaps the machine is in because there is no output at all. A defective RF modulator/antenna switcher assembly is replaced, and now it plays back fine.

Check other major operations while the unit is in the shop. Do fast forward and rewind work alright? Does the tracking control need to be set to one extreme when playing a commercially recorded tape? Is there a squealing sound sometimes when a cassette loads? If the customer left the remote, does it work properly? Taking care of these items, or at least telling the customer you've found a problem, goes a long way toward creating a satisfied customer.

There may be some part of a VCR that doesn't work, but the customer *doesn't* want to spend the money to have it fixed. For example, perhaps one or more segments of the fluorescent time display are not lighting up, or displayed TV channel numbers don't correspond with the actual channel being received. You phone the customer to report the problem, and you hear a response like, "Yeah, that hasn't worked for over a year now, but I know that Channel 5 is really Channel 10. If it's gonna cost me more than about ten dollars to fix it, forget it." Well, it probably *will* cost more than that to fix, but at least your customer knows you've been thorough in checking out the unit.

- Reassemble completely and carefully
  — This may seem obvious, but for some technicians this is a problem. They're in too much of a hurry to get on with the next job, go to lunch, or call it a day. Check for the following before putting the final covers back on; it could prevent an echo.
  — No pinched wires or cables
  — Wires and cables clear of moving mechanical parts
  — Wires and cables not touching hot components, like heatsinks and high-wattage resistors
  — All cable connectors firmly seated
  - All ground wires re-installed
  — All circuit boards fully latched in place
  — All shields and internal covers replaced
  — All removed screws and hardware, such as C-clips, replaced; screws and nuts firmly tightened
  — No extra parts left on the workbench
  — Adjustment screws or nuts that were turned, dabbed with lock paint
  — No loose hardware inside machine
  — No loose wire strands or solder flakes or pellets inside machine
  — No fibers from a cotton swab inside machine
  — *No tools left inside machine!*
  — No oil or grease where there should be none.

    It all seems so obvious, and yet lack of attention to detail about these very items is the cause of countless service echoes.

    Putting the final covers back on must be done with patience and care. On some units, it is easy to pinch a cable or pull a cable partway out of a connector when replacing a cover. Plastic cover latches are also prone to snap off unless reassembly is proper and gentle. Be especially careful that nothing can short out to metal top or bottom covers.

    *All* cover screws must be replaced. Leaving off just a single screw could mean the metal top cover is no longer grounded properly. This might result in the VCR picking up interference from a nearby FM transmitter, or could cause the VCR to interfere with a nearby radio or TV.

- Final checkout
  — Except for performing an AC current leakage test, (described shortly in this chapter), this is the last thing done before the unit goes to the completed repairs shelf.
  — After all covers have been replaced, check that all major VCR functions work. Does a cassette load and unload properly without catching on the compartment opening? Does the flip-down front panel flip down?

- Check that all pushbuttons, switches, and rotary and slide controls operate freely, with no binds. A plastic button extension on the inside of a front panel can break when the panel is replaced, and so the button won't press the microswitch. It would be pretty embarrassing to have the customer come back and say that nothing happens when the Counter Reset or some other button is pressed.
- Return all switches and settings to the positions they were in when the unit was brought in, as noted on the repair ticket. Pay particular attention to Ch 3/4 and TV/Line recording switches.
- Bundle with the VCR any videocassette, remote control unit, or other accessory that was with the unit when it came in.
- A cassette that was in the machine, and which is itself damaged, such as tape twisted or eaten, should be clearly marked with a label that reads something like this:

> CAUTION!: This tape is damaged. Using it in a VCR or camcorder could damage the machine.

No matter what you do, you can never prevent 100 percent of the units from returning after being repaired. There *will* be an occasional, unrelated failure that pops up a short while after the customer gets the unit back home. After all, coincidences *do* happen. But with a bit of care and attention to detail every step of the way—from the moment the customer brings the VCR in until the repaired unit goes back out—the echoes should be infrequent and faint.

## PERFORMING AN AC CURRENT LEAKAGE TEST

Before releasing a repaired unit to a customer, there is one more very important item to take care of. You must test the amount of AC leakage current on exposed metal surfaces. This test should be performed each and every time a line-operated appliance is repaired. In some states, electrical and electronic repair facilities are required by law to perform the test on any plug-in item that has been repaired; the unit must be certified as having passed the test. This may require that a sticker certifying that repaired equipment has passed the AC current leakage test be affixed to the unit.

The purpose of conducting the AC current leakage test is to make certain that no hazardous voltages are present on any exposed metal of the repaired unit. It is possible, for example, that during service a wire got pinched under a cover mounting screw or a component was installed incorrectly. This could place dangerous voltages on the metal chassis and covers. Someone touching the unit while also contacting a grounded piece of equipment could receive a potentially lethal shock!

Even very small amounts of AC leakage current can cause very serious results. The shock itself may not be enough to produce bodily injury, but the involuntary muscle contraction resulting from a small electrical shock can be disastrous. You can imagine what might happen if a customer received a shock while reconnecting a VCR and fell into a glass-topped coffee table.

The AC current leakage test should be performed after the unit is *completely* reassembled, with all covers, shields, knobs, and the like replaced. The unit under test should *not* be connected to any other piece of equipment. For example, *a VCR must not be connected to a TV or monitor when the AC current leakage test is performed.*

For a unit to pass the test, AC leakage from any exposed metal part on the unit to earth ground, and from all exposed metal parts to any exposed metal part that itself is connected to the chassis ground, must be no greater than 0.5mA (500 microamperes). This means that if a person touches between exposed metal on the equipment and any other exposed metal on the unit, or between any exposed metal on the equipment and something at earth ground potential, the *maximum* current available is only 0.5mA.

The test basically consists of connecting one lead of an *AC milliammeter* to earth ground, and then touching all exposed metal on the equipment under test with the other test lead. At no time should the measuring instrument read more than 0.5mA. The test should also be conducted between exposed metal parts on the unit; for example, between the outer shells of F connectors, RCA jacks, and UHF screw-type antenna terminals, and the metal bottom cover of a VCR.

AC current leakage testing is performed with the equipment being tested plugged into a standard 120VAC outlet, supplying rated power. *Do not plug the unit into an isolation transformer for this test!* That would give a false indication, because the purpose of the transformer is to isolate the unit from the power line during service activities in the first place. An AC leakage test checks the degree of isolation of any exposed metal on the unit itself from the power line.

One additional requirement is that the test be performed under two conditions: with the unit turned On, and with it turned Off. It may come as a surprise to you, but it is entirely possible for a device to have acceptable leakage current when turned On, and yet have extremely dangerous amounts of leakage current available when turned Off. How this can happen is shown in Figure 20.1.

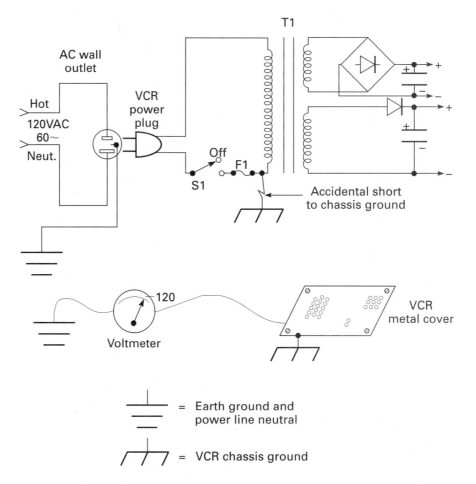

**FIGURE 20.1 Electrical safety hazard. With S1 Off, dangerous voltage is present on metal cover.**

Here is a piece of electronic equipment *without a polarized AC plug*, or maybe someone has filed down the wide, neutral prong on a polarized plug. It could even be a line cord with a polarized plug that someone has replaced incorrectly, interchanging hot and neutral conductors inside the unit.

The narrow slot of a correctly-wired standard 120-volt AC wall receptacle is the hot line. The wide slot is the neutral side of the power line, which connects to earth ground at the main circuit breaker panel. A unit's On/Off switch *should* always be on the hot line, not the Neutral.

Notice in Figure 20.1 that there is an accidental short between Fuse F1 and chassis ground. This short circuit could be created by a wire that got pinched when the bottom cover was reinstalled, or perhaps a technician used an overly long screw to replace a part inside the unit; the tip of the screw is now touching a terminal on the fuse holder.

Trace the circuit with the power switch On, and then with it Off.

- With Switch S1 On:
  — Neutral/earth ground side of 120-volt power line is connected to Fuse F1, and thus to the bottom of power transformer T1's primary winding.
    The accidental short between chassis ground and Fuse F1 connects the metal cover of the unit to the neutral/earth ground side of the 120V power line.
  — If a person touched the metal cover of the unit and ground at the same time, she would *not* receive any shock, because the cover is already at earth ground potential.
- With Switch S1 Off:
  — The hot side of the AC line is applied to Fuse F1 through T1's primary winding.
  — The accidental short therefore places 120 volts from power line hot, via T1 primary, onto the metal cover.
  — If a person touched the metal cover of the unit and ground at the same time, he could easily receive a *severe* shock! The person's body would complete the circuit path from the bottom of T1's primary winding to the neutral/earth ground side of the 120V power line. Current that flows through S1 when it is closed, so T1 receives 120V power, now tries to flow *through the person* touching the metal cover and ground at the same time. In effect, the person's body acts like the power switch in its On position, and T1 primary current attempts to go through the body.

The AC current leakage test *can* be performed with a multimeter by measuring the voltage drop across an external resistor and capacitor. (Remember, many multimeters *do not have* an AC milliammeter function.) Testing *can* be done with a quality analog multimeter, but it *must have* an accurate low-voltage AC scale, on which 0.75VAC can be reliably measured. A battery-operated digital meter with a 2-volt AC range should also be suitable.

If a multimeter is used to perform the AC current leakage test, it must be battery-operated. Never use a meter that is AC powered to perform the test, unless it is specifically designed for taking AC current leakage measurements. Figure 20.2 shows the test setup using a low-voltage AC range on a multimeter or AC voltmeter to perform the leakage current test.

Notice that a 1,500-ohm resistor and a 0.15-microfarad capacitor are connected in parallel between an earth ground point and exposed metal parts on the unit being tested. The multimeter or AC voltmeter measures the voltage drop across the RC combination. A reading above 0.75VAC indicates excessive AC leakage current. The cause for the leakage must be determined and corrected, and the test conducted again, with no reading greater than 0.75 volts, before the unit can be released to the customer.

**FIGURE 20.2** Using an AC voltmeter to check AC leakage. Courtesy of Sony Electronics.

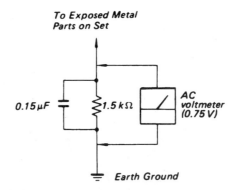

A one-watt carbon resistor and a ceramic capacitor with a 400-volt or greater rating are adequate for the test components. A metallic cold water pipe, or the U-shaped grounding contact, or the cover mounting screw of a properly wired AC grounded outlet can be used as ground. *Do not use the wide, neutral slot of an AC outlet as an earth ground reference.*

Verify that you have a good earth ground. An incandescent light, like a standard 60W bulb, should light at normal brilliance with no flickering when connected between a good earth ground and the *narrow* (hot) slot of an AC outlet.

Be careful if you conduct this test! You are dealing with 120 volts.

- Always attach the earth ground point to the *outside shell* of the light bulb socket.
- The *round base* of the lamp socket should be attached to the narrow blade connection of a standard polarized AC plug.
- Always make the earth ground connection first. Then and only then, plug the polarized plug into a standard AC outlet.
- Always unplug the plug as soon as a good earth ground has been confirmed.
- Disconnect the ground wire to the lamp last.

Note that PVC or other nonmetallic pipe is used in some sections of the country for cold water lines. Even though there is metal pipe at some points in the system, like at a sink or lavatory, this is no assurance that there is a good earth ground. If the bulb is dimmer than when it is in a regular plug-in lamp, or if it flickers, this indicates a poor ground reference for conducting the AC current leakage test.

Never use a natural or propane gas line, nor any conduit or pipe other than a cold water pipe, as an electrical ground reference for the AC current leakage test.

Also, testing the quality of earth ground with a light bulb as described here will cause a ground fault interrupter (GFI) outlet or circuit breaker to trip. This is normal. In fact, it proves the GFI device is performing as it should.

Once you have the RC combination assembled and have established a good earth ground reference, conduct the test by touching all exposed metal parts of the unit under test and observing the meter reading. Remember to conduct the test with power both On and Off when the unit is plugged into a standard 120-volt AC outlet. Of course, a VCR is really powered On at all times, or at least some of it, but the test should still be conducted under both conditions. On any piece of equipment having a primary Power On/Off switch, the test should be conducted with the switch in each position.

Just in case there is a direct short to chassis ground, similar to that shown in Figure 20.1, it is always a good idea to perform a preliminary continuity check with an ohmmeter *before* plugging the VCR into an AC outlet for the leakage test. Here's how:

1. Unplug the unit to be tested.

2. Connect a jumper wire, like a short test lead with an alligator clip at each end, across the two prongs of the AC line cord plug.
3. Select R x 1000 or higher ohmmeter range.
4. Connect one ohmmeter lead to the two shorted AC plug prongs.
5. Touch the other ohmmeter lead to all exposed metal parts on the VCR, including screw-type and coaxial antenna terminals, and the like.
6. The reading should be at least 500,000 ohms, and will generally be over 1 megohm, if no abnormal condition exits. Be sure the ohmmeter range you select can measure at least 500KΩ. You may get no reading at all, that is, ∞Ω, which is all right.
   — If greater than 500KΩ, proceed with the AC current leakage test.
6. If the reading is significantly lower than 500KΩ, track down the cause of the problem with an ohmmeter *before plugging the VCR into an AC outlet!*
   Note: A reading substantially less than 500KΩ is normal for some units if you measure to the center conductor of an RCA jack or F connector.

Should you have access to a battery-operated AC milliammeter, conduct the test *without* the RC combination. A reading over 500 microamps (0.5mA) indicates excessive AC leakage.

The best method for conducting an AC leakage test is to use a piece of equipment designed for the purpose. Simpson Electric Company's *"Model 228 Current Leakage Tester"* is designed specifically to meet the stringent leakage current specifications (C101.1) adopted by the American National Standards Institute in 1991. Sencore's *PR570 "Powerite II®"*, which also contains a variable isolation transformer and amp/watt meter, is also an excellent instrument for conducting electrical safety leakage tests.

Be sure to follow the manufacturers' instructions on how to use these and any other AC leakage testers. Also, VCR service manuals have procedures for conducting safety checkouts when repairs have been completed. This information normally is located directly after the table of contents.

## What if the VCR Fails the AC Current Leakage Test?

Assuming the test is conducted correctly, but an ohmmeter measures less than 1/2 megohm from exposed metal to AC plug prongs, or AC leakage current is above specifications, here are a few things to check inside the unit that may be causing or contributing to the problem:

- Pinched wire contacting metal part
- Frayed wire or cable touching component
- Wire or cable with melted insulation touching hot component
- Incorrect hardware
  — Screw that is too long short circuiting to electrical component
  — Metal fastener used in place of insulating plastic fastener
- Components reassembled without required insulating materials, such as fish paper between wiring side of PCB and bottom cover
- Loose hardware, solder pellets, flakes, or wire strands shorting between components and exposed metal
- Solder bridge between circuit board wiring patterns or between wire trace and PCB mounting screw or chassis
- Short or leakage between primary and secondary windings of power supply transformer
- Short or leakage between winding(s) of power supply transformer and transformer frame
- Defective component in switching mode power supply
  — Transformer
  — Carbon arc on PCB

- Exposed metal part, like a cover, not making proper connection to chassis ground
- Missing or disconnected ground wires, ground straps, or metal shields
- Leakage path on any line voltage insulator, such as a PCB containing 120V main line fuse and line filter components (choke coils and capacitors)
  — A lightning-induced power surge can actually turn an insulator into a resistor or conductor. Look for evidence of arcing or a carbon path on any component connected to AC input voltage. Carbon path looks like burnt toast.
- Leaky AC line RF filter capacitor, if present
  — A small value capacitor, like 130pF, connected between neutral side of AC line and chassis.
- Defective line filter component
  — Any resistor, coil, or capacitor connected to 120V input that might be leaky or somehow shorted to exposed metal parts.

When tracking down the source of excessive AC leakage, concentrate first on all components that are directly connected to the AC line. There may be more than one problem; leakage sources can be cumulative.

## SUMMARY

This chapter discussed several important aspects of overall VCR service activity. Listening to the customer describe a problem, asking some questions for clarification, and then writing down on the repair ticket how a VCR fails can save service time. This also helps ensure that *all* problems are fixed.

Note any missing or broken parts, cabinet damage, and included accessories on the repair form. Mark down switch positions, such as the Ch-3/Ch-4 switch. Return switches to where they were when the VCR came in, before releasing a repaired unit; this can eliminate operator problems when the VCR gets back home. Include a maximum or estimated cost for repair, and confirm items on the repair authorization forms with customers before they sign them.

About 80 percent or more of VCR problems are resolved with routine cleaning, lubrication, adjustment, and replacement of tape transport components. Many electronic problems can be diagnosed with just a multimeter and digital logic probe, and can be repaired by replacing a discrete component, like a diode, transistor, or IC. When a malfunction is found in the tuner, RF modulator, or VHF antenna switcher sections, most manufacturers recommend replacing the entire assembly. It may also be less expensive to replace a complete printed circuit board containing all components than to spend time diagnosing and repairing it.

*Beyond economic repair* (BER) describes a VCR that would cost more to fix than the purchase price of a new comparable machine. A unit that has multiple problems, a potentially hard-to-diagnose intermittent problem, or one that requires extensive parts replacement may be BER, considering the costs of diagnostic time, repair time, and parts. A VCR with broken or bent parts and cracked circuit boards from being dropped, or one that has suffered extensive lightning or water damage, is usually BER, especially if it is an older, low-end model.

One goal of any repair business *should be* to reduce, as much as possible, service returns or echoes. These are machines that customers bring back within a short period of time after they've been repaired. An echo might have the same *or a different* problem than that for which it was originally brought in. Attention to detail at every phase of service activity, from the moment the customer brings the machine in until the repaired unit leaves the shop, will reduce echoes.

A competent technician inspects the entire machine and checks all major operating modes for potential problems, in addition to those written on the repair ticket. Replacing a worn belt, although it may not be the cause of the reported problem, can prevent the unit from showing up again shortly. This gains customer satisfaction and loyalty. All parts must be properly replaced and the machine fully checked out after covers have been reinstalled before releasing it. Many returns can be prevented with just these simple but crucial steps.

Before releasing any repaired line-voltage-operated equipment to a customer, an AC current leakage test should be performed. Some states require that this test be certified with a sticker placed on the unit. The leakage test is to ensure that no more than 500 microamperes (0.5mA) of current would flow through a person touching any exposed metal on the unit and earth ground at the same time. The test is conducted by measuring current with an AC milliammeter between earth ground and all exposed metal with the unit plugged into a standard AC outlet. The unit must not be connected to anything else during the test. Measurements shall be taken with the equipment turned both On and Off while it is plugged into a standard 120VAC receptacle.

The AC current leakage test *can* be performed with a multimeter having an accurate low-voltage scale, such as 1.5V or 2V. A 1.5K resistor is wired in parallel with a 0.15 microfarad capacitor; this RC combination is then connected between earth ground and all exposed metal on the unit being tested. The multimeter measures the voltage drop across the RC combination. A reading of more than 0.75VAC indicates there is excessive leakage. Components in line voltage circuits, like a power transformer primary winding that is partially shorted to the transformer frame, are prime suspects in creating AC leakage current.

## SELF-CHECK QUESTIONS

1. Describe some of the things you can do when dealing with customers so they develop a positive attitude about you and the repair facility in which you work?
2. Discuss the pros and cons of charging customers a repair estimate or diagnostic fee on a VCR.
3. List two or three instances when it might be best for the service facility, as well as the customer, to refuse to accept a unit for repair.
4. Discuss some suggested practices when receiving a unit for repair and filling out a repair ticket.
5. List a few VCR units that are frequently more economical to replace than repair or which some manufacturers recommend replacing rather than repairing.
6. What does the term *BER* mean? Give an example of when a VCR may be BER.
7. Discuss a few recommended practices that can prevent service "echoes."
8. What is the maximum amount of AC leakage current allowed between a metal cover of a 120-volt appliance, like a VCR, and earth ground? Describe how to make the AC leakage test with a VOM or DMM that does not have an AC current function.

## CONCLUSION

VCRs are sure to be with us for several years yet, and may still be popular at the turn of the century. No doubt in the early 21$^{st}$ Century something better will replace the VHS cassette and VCR for consumer A/V reproduction, just as the compact disc replaced vinyl LP records for audio. Will it be recordable laser disks? Or will compact "Media Paks" with no moving parts—massive amounts of memory that can hold more than two hours of high-definition video and high-fidelity multi-track audio in a package about the size of a pack of cigarettes—be the norm? Will interactive media and virtual reality be incorporated in most popular entertainment? Probably twenty years from now, what we now have will look as archaic as the flickering black-and-white "Talkies" in the early Twentieth Century.

In the meantime, basic videotape recording and transport principles presented in this book are unlikely to change a great deal for consumer products. There will be refinements, improvements, and wider acceptance of 8mm and VHS-C for camcorder use. So as long as magnetic tape is the medium, you are well positioned to service yesterday's, today's, and tomorrow's VCRs. There is great opportunity in the field of VCR maintenance and repair for years to come.

Although this concludes the textbook, I sincerely hope it is just the beginning for many students who find electronics fascinating and want to learn more about how VCRs and other electronic gear work and how to diagnose and repair them when they fail. Congratulations for sticking with your studies! I wish you all the best of success in your electronics future.

# CHAPTER 1 APPENDIX

This appendix contains more detailed explanations and supplemental information for material presented in Chapter 1 of this book.

## FRONT PANEL OPERATOR CONTROLS

Following are further descriptions of VCR front panel operator controls, functions, and displays. Refer to Figure 1.3 in Chapter 1.

### Power Button

Depressing the Power button turns the VCR On or Off. Many VCRs indicate power On with a lighted rectangle about 3/8-inch square within the display/indicator panel. On some VCRs, a separate light emitting diode (LED) on the front panel indicates power On. Day of the week and time of day are displayed whether the power switch is On or Off.

Even with the VCR powered Off, several portions of the machine are still energized as long as the unit is plugged into a 120-volt AC outlet. These are the timer/operations microprocessor, system control (syscon) microprocessor, and infrared (IR) wireless remote control detector circuit. Programmed recording requires the timer/operations microprocessor to operate with the VCR Off; the IR detector must operate with power Off so that the VCR can be turned On from the handheld remote.

Some VCRs automatically power On when a cassette is inserted into the cassette compartment. This requires the system control microprocessor to be powered up all the time to detect when a cassette is inserted.

A VCR's AC line cord must be unplugged from the wall receptacle for the unit to be *completely* turned Off. Be sure to unplug a VCR before removing any covers to prevent possible electrical shock and VCR damage.

### Cassette-In Indicator

Once a cassette is loaded onto the tape transport within the VCR, a *"Cassette-in"* indicator illuminates on the display/indicator panel. It looks like the one shown in Figure A1.1:

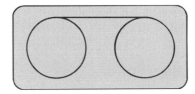

**FIGURE A1.1** Cassette-in, or cassette-loaded, indicator.

### Timer Button

Unless you've spent some time contemplating this button on your own VCR, you might be surprised to realize that the Timer button itself is not used to program a VCR for unattended recording of a TV program. However, *after* the VCR timer is programmed—using other front panel buttons—pressing the Timer button shuts the VCR Off, where it waits for the day and time to automatically turn On and record. With a video cassette loaded, a timer recording indicator lights on the display/indicator panel. The timer recording indicator looks like an icon of a conventional, analog clock face as shown in Figure A1.2.

When the Timer button is pressed but no cassette is loaded, or if the loaded cassette's plastic

**FIGURE A1.2** Timer recording indicator.

record safety tab is broken off, the timer recording indicator blinks On and Off at about a one-second rate. This tells the user to insert a recordable cassette. When the cassette record safety tab is missing, a VCR will not go into Record mode, whether initiated manually or by the pre-programmed timer. This is explained in more detail in Chapter 4.

If unattended recording has been programmed and the Timer button is pressed, pressing the Power button *will not* power up the VCR. With the clock icon displayed, it is necessary to press the Timer button to turn the VCR On for manual operation. There are many tales of new VCR owners bringing their machines back to the retailer complaining that the units wouldn't turn On when the Power button was pressed. Usually this was because the VCR had been set for unattended recording, and the owner didn't know to press the Timer button to power the VCR On. When the clock icon is displayed, the VCR is under the control of the timer/operations microprocessor and cannot be operated manually!

### TV/VCR Select Button

This control determines the output signal from the VCR going to the TV set. When VCR mode is selected, you can monitor recordings being made by the VCR on the attached TV set. This position is also used to play back a tape. VCR mode is indicated by an illuminated **VCR** on the display/indicator panel.

With TV mode selected, or with the VCR powered Off, VHF antenna or cable signals going into the VCR are directly connected to the VCR VHF Out terminal by the VCR antenna switcher. TV mode allows you to record one TV channel selected by the VCR's tuner while simultaneously watching a different channel tuned in by the TV. When taping one program while watching another, you can switch back and forth between programs to be seen on the TV by pressing this button to select VCR or TV mode. In VCR mode, the program selected by the VCR tuner, and perhaps being recorded, is displayed on TV Channel 3 or 4; in TV mode, VHF antenna or cable signals can be selected by the TV tuner.

NOTE: On many VCRs, UHF signals are not switched by the TV/VCR mode button; UHF In signals are always available at the UHF Out terminals on the VCR.

VCRs automatically switch to VCR mode when starting to record or play a tape, so practically the only time you need to select TV mode is when viewing one channel while recording another. When turned Off, a VCR goes into TV mode, so all antenna or cable signals coming into the VCR VHF In F connector go directly to the VCR's VHF Out F connector, and then on to the TV.

Although most manufacturers label this button **TV/VCR**, some use just **VCR** or **TV** or even **VTR**, which stands for video tape recorder. No matter what the button is labeled, the basic function is the same.

**Special Term:** E-E, which stands for electronic-to-electronic, is a term that describes operation in VCR mode when *not* playing a tape. Video and audio signals at the output of the VCR going to the TV are those received by the tuner section of the VCR, or from A/V line input jacks used with a video camera or other source. This lets you monitor signals before actually placing the VCR into Record mode, as well as during recording.

### Rewind/Reverse Scan Button

This button has two functions:

- Rewinds tape back onto the supply reel on the cassette's left side after the VCR is first put into Stop mode. No video from the tape is seen on the TV.
- Moves tape rapidly in reverse during Play mode to review or scan previously seen pictures. Video quality is usually not as good as in normal forward Play, with some tearing and horizontal noise lines present. Audio is muted during Reverse Scan.

This button may also be labeled **REW/REV SEARCH** or **REW/REVIEW** along with a double-headed arrow pointing to the left. In most VCRs, reverse search or review mode works only while the button is held in; releasing the button resumes normal forward Play. Typically, a blinking left-pointing arrow head on the indicator/display panel comes on while in Reverse Scan.

### Play Button

Pressing Play with a pre-recorded cassette loaded causes tape to load and then be played back. On many VCRs, a right-pointing arrow head or the word **PLAY** illuminates on the display/indicator panel. **PLAY** also displays on a blue background on the TV screen for a second or so before the program image from tape displays.

### Fast-Forward/Forward Scan Button

Similar to Rewind/Reverse Scan, this button serves two functions:

- Fast-forwards tape onto the take-up reel on the cassette's right side after the VCR is in Stop mode. No video from the tape is seen on the TV.
- Moves tape rapidly forward during Play mode, to scan ahead. Video quality is usually somewhat

deteriorated from normal Play. Audio is muted during Forward Scan.

Sometimes this button is labeled **FF/FWD SEARCH** or **FF/CUE** along with a double-headed arrow pointing to the right. Usually Forward Search or Cue mode operates only while the button is held down. When released, normal Play resumes. A blinking right-pointing arrow head on the indicator/display panel shows that the VCR is in Forward Scan.

Throughout this text, the terms *Reverse Scan* and *Forward Scan* are used, respectively, for moving tape rapidly backward and forward during Play, with fast-motion video on the TV screen.

## Record Button

With many VCRs, you must hold the REC button in and then press the Play button to put the VCR into Record mode. This is to prevent recording should you accidentally hit the record button. Some models, though, go directly into Record when this button is pushed.

Most VCRs show **REC** on the display/indicator panel when in Record mode.

## Eject Button

Pressing Eject causes the front loader mechanism to lift the cassette up off the tape transport and push it partway out the compartment door. You can then grasp the rear of the cassette and remove it from the VCR. On top loading VCRs, the cassette basket rises up through the top cover.

Some VCR models will eject a cassette even with power Off. Of course, the VCR still must be plugged into AC power.

On most VCRs, it is not necessary to press Stop first when playing a tape, before pressing Eject. When you press Eject, the system control microprocessor stops tape motion, unloads the tape, and then ejects the cassette.

## Stop Button

Pressing this button stops all tape motion and drops the machine out of its present mode, such as Record or Play. Tape is unloaded on many machines; on some newer models, tape remains loaded or half loaded during Stop mode.

## Pause/Still Button

The Pause/Still button stops forward tape motion in both Play and Record modes. Pressing the button again resumes the previously paused mode. In Play, Pause produces a still video frame on screen. As with Forward and Reverse Scan modes, streaking and noise bars often appear in the picture during pauses.

When Play is paused, Forward and Reverse Scan buttons rapidly move tape forward or backward while either button is held down. Releasing a scan button returns to paused play at that point on tape. Forward and reverse scan are inoperative in Record mode.

In some VCRs, the system control microprocessor automatically puts the VCR into Stop mode (tape unloads) if Pause is continuously active for five minutes. Other models resume Play after five minutes of Pause, but go into Stop if in Record mode. All this is to reduce damage to the stationary section of videotape wrapped around the still-spinning video drum, which would happen if Pause continued indefinitely.

## Frame Advance Button

Used in conjunction with the Pause/Still button during Play, each depression of this button moves the tape ahead to display the next video still frame. Frame advance is inoperative during Record and other modes.

## Reset Button

Whenever videotape is moving forward or backward in the VCR, a four-digit electronic tape counter on the display/indicator panel increments or decrements as the take-up reel turns clockwise for forward tape motion or counter-clockwise in reverse. Pressing the Reset button at any time, during any tape mode, resets the tape counter to "0000."

This is handy for locating a particular spot on a tape. Suppose on day one you record a half-hour program starting at the beginning of the tape, and on day two you press Reset before recording a second program. You can now locate the beginning of the second program by rewinding the tape until the counter displays "0000."

Contrary to what some people think, the tape counter has no relationship to the number of inches or feet of tape passing through the VCR; instead, it relates to the number of take-up reel revolutions. At the beginning of a tape, the take-up reel turns faster than toward the end of the tape, when the take-up reel diameter is much larger. Tape counters on two different machines may not indicate the same for an equal number of take-up reel revolutions.

On some VCRs, the Reset button may also be used to clear the program during timer programming.

On earlier VCRs, a mechanical tape counter, similar to the odometer in an automobile, is driven by a belt from the take-up spindle on the tape transport, beneath the cassette take-up reel. For these, a mechanical pushbutton at the side of the counter resets it to "0000."

Many later model VCRs have a real-time tape counter that indicates elapsed play or record time in actual hours, minutes, and seconds. A real-time counter works by using control track pulses on the tape during playback. Therefore, the counter will not change for sections of unrecorded tape during playback.

### Speed Select Switch

This is usually a three-position slide switch, marked SP LP EP (SLP) for the three tape speeds. Some VCRs may offer only two speeds (for example, SP and EP). Selecting one of these positions determines tape speed during Record only. For playback, the switch has no effect. VCRs sense at what speed the tape was originally recorded from control track logic (CTL) pulses on the tape, and automatically switch to the appropriate playback speed. You will read more about these control pulses in later chapters.

### Tracking Control

This is a rotary control, similar to a volume control. A center detent position marks its normal position. During playback, adjusting this control to either side of the center detent may be required to produce a stable picture and eliminate horizontal noise bars. Turning the tracking control makes minor timing adjustments as videotape passes the spinning video heads. This compensates for small differences in videotape timing recorded on another machine.

For tapes recorded on the same VCR, correct tracking is usually obtained with the control at or near its center detent position.

The tracking control has no effect during Record mode.

Some VCRs have two pushbuttons marked "+" and "-" or marked with arrows for adjusting playback tracking. Pushing both buttons together sets the electronic tracking adjustment to the center of its range. Still other, newer models have auto tracking. Electronic circuitry adjusts the position of the spinning video heads in relation to the tape to obtain the best video signal during playback.

### OTR Button

Many VCRs have a One Touch Recording button that allows you to go instantly into Record mode for up to two or three hours, in half-hour increments. On some models, one touch recording is up to five hours. The VCR must be turned On, the appropriate channel selected on the tuner, and a cassette loaded that has the record safety tab intact. At the end of the selected OTR time, the VCR goes to Stop mode and powers itself Off.

Suppose you just start to watch your favorite program. Five minutes into it, a friend drops by. You could quickly load a blank cassette, and then depress OTR three times to record for an hour. You could then leave with your friend, your program would record, and the VCR would shut down after recording for one hour.

Each depression of the OTR button adds 30 minutes recording time. Press it once and "0:00" appears on the clock display. A second press displays "0:30"; the third displays "1:00." As the recording progresses, the display decrements one minute at a time until "0:00," at which time the VCR goes to Stop mode and powers down.

Should you wish to manually stop recording before the one hour has elapsed, press OTR until "0:00" displays. The VCR goes into Stop mode after a few seconds.

*Segment Record* is another term for one touch recording on some VCR models.

### Program/Clock/Timer Buttons

With these buttons the user sets the correct day of the week and time of day after a VCR is first plugged into a 120-volt outlet. They are usually used to also program the timer/operations microprocessor for unattended recording.

Clock and timer programming methods vary, depending on the VCR model. Following is a description of the way in which day, time, and unattended recording programming is accomplished by some VCRs using three pushbuttons, labeled *Select*, *Set*, and *On/Off*.

### Select Button

This button selects what it is you are going to program with the Set button. Push Select once and the day of the week on the display/indicator panel starts to blink (for example, **SUN**). Now press the Set button until the correct day of the week is displayed (for instance, **WED**).

Press Select a second time, and the hour portion of the display blinks. Again, press Set until the correct hour of the day displays (for example, **PM 5:xx**—the x's could be any number of minutes from :00 through :59).

Pressing Select a third time causes the minutes part of the display to blink. Press Set until the correct minutes display, like **PM 5:27**. Pressing Select a final time ends timer/clock programming.

## Set Button

As described for the Select button, Set advances the day of the week, hour, and minutes being programmed. Release the button when the correct day or time displays.

Most VCRs are capable of unattended recording of two or more programs. Once the On/Off button has been pushed once to set up the turn-On time, and Select has been pushed once so the day of the week blinks, Set is used to determine which unattended recording program is being entered.

Program 1 is indicated by **I** on the display. Keeping Set depressed scrolls through all seven days of the week for Program One, and then **II** displays, indicating Program Two.

Most VCRs can also be programmed for recording during the same hours for all seven days of the week. Pressing Set until all days of the week display and blink simultaneously starts seven-day programming. Select and Set are then used to program On and Off times, as described earlier.

On some VCRs, Select and Set also determine which VCR tuner preset channel will be recorded. After the On time minutes are set, an additional push of the Select button brings up a **CH** prompt on the display/indicator panel. Pushing Set then increments through VCR channels for the desired pretuned TV channel.

## On/Off Button

Pressing this button once prepares the timer for programming the turn-On time for unattended recording, as determined by the Select and Set buttons. The word **ON** appears on the display indicator panel. Once the day of the week and turn-On time have been programmed, a second push of the button allows the Off time to be programmed; the word **OFF** is displayed.

As stated previously, methods for setting the time-of-day clock and programming unattended recording vary widely among different VCR models. Some have separate buttons labeled PGM, DAY, HOUR, and MINUTE—somewhat easier to program than toggling between Select and Set buttons. Newer VCRs have on-screen menus for programming unattended recording. Some new VCRs have VCR Plus+, where unattended recording is programmed by simply entering a four- to eight-digit code printed in the local television listing.

Referring to the user's manual is one sure way of determining what the many controls on a VCR do, and how to set the clock and program for unattended recording. Without such an operator's guide, a little experimentation will usually reveal the secrets for operating a particular VCR. In general, the newer breed of VCRs are easier to program, with on-screen prompts.

Enhanced-feature VCRs, such as those with Hi-Fi Stereo, Super VHS, and VCR Plus+ are discussed in Chapter 17.

## Infrared Remote Detector

This is the "eye" that receives coded infrared light pulses from a handheld, wireless remote control. It consists of an infrared (IR) lens, IR photodiode, preamplifier, waveshaping circuits, and detector mounted in an IR detector module. All these components are usually contained in a small metal enclosure soldered to a front panel printed circuit board.

Infrared light is invisible to the human eye. We describe two ways to detect infrared light beams elsewhere in this book. Chapter 11 has construction plans for an electronic IR detector using just a few low-cost parts.

Some older VCRs are equipped with a *wired* remote control, which plugs into a small jack, either on the front or rear panel of the machine.

## Channel Select Buttons and Indicators

On many VCRs, you pre-set 12 (or more) TV channels. Each VCR preset can be either a VHF, UHF, or (on cable-ready VCRs) a cable channel. The desired preset channel is then selected by pushing one of the VCR preset buttons. A light-emitting diode (LED) adjacent to each button illuminates to indicate which VCR preset channel is active.

In place of 12 (or more) individual channel select buttons, some VCRs have only two buttons, labeled Ch Up and Ch Down. Pushing one or the other moves to higher or lower preset VCR channels, at about a one-second rate. When the desired VCR channel number lights, the Channel Up or Down button is released.

On some VCRs with this channel selection scheme, the user can insert small plastic numbers indicating the *actual* preset TV channel number in a snap-in holder in front of the VCR channel indicator LEDs. Other VCRs have nonchangeable numbers that light up, usually 2 through 13. But this *can* cause confusion! Suppose a VCR is the type where the user cannot change the little plastic numbers in front of the channel select LEDs. The first channel select button may be preset to VHF Channel 5 and the second button to Channel 8. Yet when these buttons are pushed, they light as **2** and **3**, respectively. Just be aware, with VCRs employing this type of channel selection, that the lit channel number you see on the panel may have no relationship at all to the channel actually being received. These channel indicators are only relative (and would probably have been better labeled A, B, C, D, etc., to avoid channel number confusion; thankfully, some VCRs *do* label their channels in this manner).

## TUNING VCR CHANNEL PRESETS

Following is a typical procedure for presetting a channel on a VCR with individual tuning controls for each preset. Refer to Figure 1.5 in Chapter 1.

- With TV antenna(s) connected to the VCR, turn on the TV connected to the output of the VCR so you can see the results of tuning the VCR. (VCR connections are described in Chapter 1 and later in this Appendix.)
- Play a pre-recorded tape and make whatever fine tunings and other adjustments on the TV that produce the best picture. Then stop the tape.
- Set the VCR/TV switch on the front panel of the VCR to the VCR position. Usually an indicator light labeled VCR will come On, or **VCR** illuminates on the indicator/display panel. You are now in E-to-E mode, where TV signals tuned by the VCR are sent to the TV set.
- Set the automatic fine tuning (AFT) switch on the VCR to Off. This allows more precise tuning of the tuner, without the AFT circuit trying to compensate for a slightly off-tuned TV channel. With some tuners, leaving the switch On may facilitate tuning a weak station.
- Press the VCR channel select button for the VCR channel you wish to preset.
- Set the small band selector switch associated with the VCR channel just selected to the range in which the TV channel to be tuned lies:
  — $V_L$ for VHF Channels 2–6
  — $V_H$ for VHF Channels 7–13
  — U for UHF Channels 14–83

Although the VCR may not specifically be labeled as cable-ready, you can tune in cable channels that fall within the frequency range of the VHF Low, VHF High, and UHF bands. Consult Tables A1.1 to A1.3 to see which cable channels fall within these three TV bands. Some cable channels outside of these three bands may also be capable of being tuned on some VCRs. Consult the owner's manual, or experiment.

Some cable-ready VCRs with this style of channel preset have a fourth position on the small individual range selector switches, labeled **CATV** or **Cable**. Again, consult the owner's manual to determine which cable channels can be tuned with the band switches in various positions.

- While looking at the TV screen, slowly rotate the tuning wheel for the VCR channel being preset to obtain the best picture on the desired TV channel.

This can cause some confusion. For example, suppose VHF Channels 7, 10 and 12 are broadcast in your area. Because these are all in the VHF High band, it is sometimes difficult to tell whether you've tuned in Channel 10 or 12 as you rotate the tuning wheel.

One way is to start with the tuning wheel, and the associated mechanical indicator, at the low end of its range. Then, as the wheel is turned, the first channel to come in will be 7, followed by 10 and then 12. Comparing the program you see with a TV listing will also identify which channel you've tuned. You could also tune in the channel with the TV's tuner, and compare the program with that from the VCR's tuner.

- Once the best quality picture is obtained with the tuning wheels for each VCR preset, turn the AFT switch back On.

This type of VCR tuner is electronic, and does not rely on the many contacts of mechanical turret tuners. Electronic tuners are much more reliable.

## REAR PANEL FEATURES

Additional information about some of the items on the rear panel of a typical VCR is given here. Refer to Figure 1.7 in Chapter 1.

### UHF Input and Output Terminals

A 300-ohm flat, twin-lead cable connects a UHF antenna to the two screw terminals marked UHF IN. Another 300-ohm twin-lead cable connects the UHF OUT terminals on the VCR to the UHF Antenna terminals on the TV set. For a combined VHF/UHF antenna, a splitter is needed at the VCR to separate VHF and UHF signals.

A UHF antenna is normally the only device ever connected to VCR UHF IN terminals.

Both UHF IN and UHF OUT terminals are connected with a small RF splitting transformer just inside the back panel of the VCR; there is no antenna switching as with VHF. Inside the VCR, a short 300-ohm twin-lead cable runs to the UHF section of the VCR tuner.

Twin-lead is connected to UHF terminals by loosening the screws, slipping the bared ends of wire (or better yet, wire ends with spade lugs) underneath the screwheads, and then turning the screws or thumbscrews clockwise so the wire ends, or spade lugs, make good, firm electrical connections. Make sure there are no loose strands of wire showing that could short out to the other UHF terminal or to the metallic VCR chassis. Twin-lead with spade lugs on the wire ends is preferred for this reason.

## VHF Input Terminal

Both the VHF Input and VHF Output terminals are female 75-ohm, F-type connectors. A male F connector on the end of round 75-ohm coaxial cable attaches to the panel mounted F connector, as shown in Figure A1.3. Seventy-five-ohm coax cable has a solid copper inner conductor, which goes into the center contact of the female F connector. The cable shield, or outer conductor, is attached to the outer shell of the F connector, which screws onto the outside threaded portion of the female F connector.

Panel mount female F connector

Male F connector on coax cable

**FIGURE A1.3  Coaxial 75-Ohm, F-Type Connectors.**

Some 75-ohm cables are equipped with push-on F connectors. These have spring-loaded inner shell surfaces that make electrical contact with the outside threads on a female F connector, without having to be screwed on and off.

VHF input sources to a VCR can be from:
- VHF antenna
- CATV or cable company system
- Cable TV converter/decoder
- TV computer game (such as Nintendo or Sega-Genesis)

VHF input signals go to the VCR tuner via a short length of coax cable inside the unit.

## VHF Output Terminal

VHF Input and Output terminals are *not* connected with just an RF transformer, like the UHF In and UHF Out terminals are. It is very important to realize that the signal at the VHF Out terminal can come from two distinctly different sources, depending on the mode of the front panel TV/VCR switch when the VCR is powered On. (You may wish to refer back to "TV/VCR Select Button" and "Special Term: E-E" sections earlier in this appendix.)

- When the VCR is powered Off, or when it is powered On in **TV** mode, the VHF Out terminal is directly connected to the VHF In terminal by the antenna switcher. Thus, all VHF In signals are available at VHF Out.

For example, the rooftop VHF antenna receives Channels 2, 5, 8, and 11. With the VCR powered Off, these channels are available at the VCR VHF Out connector. With VCR turned On, *and* in TV mode, as determined by the TV/VCR select button, VHF Channels 2, 5, 8, and 11 are also present at the VCR's VHF Out terminal.

With the VCR Off, or On in TV mode, signals at VCR VHF Out are *not affected in any way* by the VCR's tuner or any tape that may be playing or recording. VHF Output signals are *identical* to VHF Input signals.

- With the VCR turned On, *and* in **VCR** mode (usually indicated by **VCR** on the indicator/display panel or an LED labeled VCR), VHF Out will *always* be on either VHF Channel 3 or 4.

Inside the VCR is an electronic circuit called a *radio frequency* (RF) *modulator*. The *RF modulator* is really like a very low-power TV station operating on either Channel 3 or 4 of the VHF Low band. It takes baseband audio and video signals within the VCR and produces an amplitude modulated (AM) video carrier and a frequency modulated (FM) audio carrier, similar to the signals created by the video and audio transmitters at a TV station.

*Output* from the RF modulator goes to the VCR VHF Out F connector. The signal here is the same as if a TV were receiving either Channel 3 or 4 from a local broadcast station.

Audio or video *input* signals, or both, to the VCR RF modulator can come from four sources:

1. From a videotape being played
2. From audio and video being received, as determined by the TV channel to which the VCR tuner is tuned
3. From A/V Line In jacks
4. VCR-generated on-screen displays, such as a menu for programming unattended recording or the word "REWIND" while a tape is being rewound. (NOTE: Many earlier model VCRs do not have any on-screen displays.)

To summarize, the VCR VHF Out signal is:

- The same as VHF In when the VCR is turned Off, or when On and in TV mode.

- On VHF Channel 3 or 4, when in VCR mode. Output signal is videotape being played, TV channel selected by VCR tuner, or A/V input, such as from a camcorder.

VHF output on Channel 3 or 4 of the program selected by the VCR tuner, or from A/V input, is called *electronic-to-electronic*, or *E-E*, operation. That is, in E-E the VCR Ch 3/4 output is the program material received by the VCR tuner or A/V Line In jacks.

The VCR circuitry that switches the VHF Out jack between the VHF In jack and the output of the RF modulator is sometimes called the *antenna switcher*. In most VCRs, the RF modulator and antenna switcher is a single circuit board entirely enclosed within a small metallic case, with the VHF In and Out F connectors part of the RF modulator/antenna switcher assembly itself. The metal case is an RF shield.

Some newer VCRs do not have separate VHF and UHF input and output connectors. Instead, a single input and a single output F-type connector carries both VHF and UHF signals, as well as cable TV channels. The overall operation is the same as just described, except that UHF channels are output only when the VCR is in TV mode.)

## RF Channel Select Switch

This two-position slide switch selects whether the output of the VCR RF modulator is on VHF Channel 3 or 4. The TV to which the VCR VHF Output is connected must be tuned to either Channel 3 or 4 to receive RF output from the VCR when playing a tape or monitoring a recording from the VCR's tuner (or A/V Line In jacks).

If a local TV station is broadcasting on either Channel 3 or 4, the VCR RF channel selector should be placed to the other channel to prevent possible interference between the TV station and the VCR RF modulator. That is, if TV Channel 3 is on the air, the selector switch should be set to Channel 4.

Because most cable TV converters and video games also have an RF modulator that outputs on either Channel 3 or 4, certain hookups may require that the VCR be set to a different channel than one of these components, to avoid interference.

On most VCRs, the Channel 3/4 switch is on the back panel; however, it could be somewhere else, even on the bottom. Once set, there is normally no reason to change this switch for a given installation. On some VCRs, a small screwdriver or other tool is needed to change the Channel 3/4 switch.

Just like the VHF In and Out F connectors, the Channel 3/4 switch is usually part of the RF modulator/antenna switcher assembly itself.

## Video In and Out Jacks

Recall the Chapter 1 discussion of baseband, or composite, video. Jacks labeled Video In and Video Out handle *un*modulated video signals. Both connectors are RCA phono jacks, the same as on most consumer audio equipment, such as Tape In and Tape Out jacks on a receiver or preamplifier. In almost all cases, the plastic insulator surrounding the center contact of the video In/Out jacks is yellow.

Video In and Out jacks connect the VCR to video components having standard composite video connections. Possible VCR composite video input sources include:

- Video camera
- Camcorder—camera or tape playback
- Laserdisc video player
- Another VCR—could be VHS, Beta, or 8mm
- Personal computer display adapter card or separate PC video interface unit with composite video output jack
- Monitor-type TV—signal received by set's tuner
- Separate component TV tuner (no display screen).

On some VCRs, a two-position front panel switch selects video input from either the composite Video In jack, or video received by the VCR tuner; the switch is usually labeled something like TV/Line. In other VCRs, inserting an RCA plug into the Video In RCA jack automatically disconnects video received by the VCR's tuner. A switch contact within the RCA jack accomplishes this switching.

Possible devices to receive VCR composite video output include:

- Camcorder—tape being recorded (acting as VCR)
- Another VCR or video tape recorder (VTR)
- Video monitor or TV/monitor
- Computer video digitizer—converts composite video analog signal to digital, which can be stored and processed on a computer.

Some TV receiver/monitors have a Video Out jack with composite video of whatever is displayed on the screen, as well as a Video In jack. This is handy for "daisy chaining" video monitors, for example, when several screens must display the same video in a large conference room or auditorium.

*Daisy chaining* simply means interconnecting two or more devices so that the output of one becomes the input of the next in line. For example, suppose we wish to play a tape on three large-screen monitors at the same time. We could daisy chain the equipment as follows:

- VCR Video Out to Monitor # 1 Video In
- Monitor # 1 Video Out to Monitor # 2 Video In
- Monitor # 2 Video Out to Monitor # 3 Video In.

High-quality coax cable and plugs should be used to interconnect line level video jacks. Although run-of-the-mill audio interconnects have the same type RCA plugs, in many cases they are poor choices for video. What's OK for audio may be marginal for video. Insufficient shielding can allow external RF signals or noise to get into the cable. High interconductor cable capacitance on some interconnects causes deterioration of high-frequency video signals, resulting in loss of picture detail and sharpness, unstable picture, or both. Be sure to use coax cable with installed RCA plugs designed for video. "Gourmet" audio interconnects, the expensive ones, will work fine for composite video.

The outer shell of the RCA plugs on cables should tightly grip the outside surface of the RCA jacks to make a good electrical connection. Loose RCA plug shells can be tightened with longnose or duckbill pliers by slightly squeezing in on the shell segments. Contact surfaces must be clean and free of corrosion. A thin film of oxidation often forms around RCA plugs and jacks. Cleaning with a nonlubricating contact cleaner and reseating plugs several times in their jacks helps establish solid electrical contact. Gold-plated plugs and jacks prevent surface oxidation and maintain good electrical contact.

## Audio In and Out Jacks

Line level audio input and output connectors are also RCA phono jacks, like those on a stereo receiver. On monophonic VCRs, there is one input jack and one output jack. Stereo VCRs have Left and Right channel input jacks, and L and R output jacks.

In keeping with the standard for consumer audio equipment, Right channel jacks usually have a red plastic insulator surrounding the center contact of the RCA jack, whereas Left channel jacks have a white insulator. Even though jacks are labeled, just remember "Red equals Right" to identify audio channels on stereo A/V equipment.

As mentioned earlier, line level Audio In and Audio Out jacks on VCRs are totally compatible with Tape/Aux In and Out line level RCA jacks on consumer audio gear. Line level jacks on audio equipment may also be referred to as high-level, as opposed to inputs for a microphone or magnetic phono cartridge, which are low-level inputs.

Some VCRs also have a microphone input jack (two for stereo), usually located on the front panel. Mic jacks are normally 1/8-inch mini phone plugs. Plugging in a microphone typically disables any audio input from the line level RCA jacks and from any channel received by the VCR's tuner. An external mixer is required to mix microphone and line level audio sources.

## Dolby Noise Reduction

While discussing audio, one other item found on many VCRs deserves mention: Dolby Noise Reduction (NR). *Dolby NR* is a double-ended noise reduction system that reduces high-frequency tape noise (hiss) when playing back a tape. Double-ended means a signal is modified during both Recording *and* Playback. Dolby NR is a product of Dolby Laboratories.

Just like virtually all audio cassette decks, some VCRs also employ Dolby NR, activated with a front panel On/Off switch, to reduce playback tape hiss, which is most noticeable during quiet or low-level passages of music. In brief, this is how Dolby NR works:

- When Recording, low-volume high-frequency sounds are boosted in volume, but high-level sounds are pretty much recorded as is.
- In playback, Dolby NR circuitry reduces the volume of high-frequency sounds by the same amount as they were originally boosted during recording. The overall tonal balance is thus preserved.
- With the volume of high frequencies automatically turned down during playback, especially during quieter moments in the taped program, tape hiss volume is also reduced, making it less noticeable.

Dolby NR is generally found on stereo and Hi-Fi Stereo VCRs. Most commercially produced VHS movie cassettes are recorded with Dolby noise reduction. Dolby processing applies to standard VHS audio tracks, called *linear audio*, not to Hi-Fi tracks. Linear and Hi-Fi audio are described in Chapters 3 and 19, respectively.

## TV AND CABLE CHANNELS

Getting back to TV channel allocations, each 6 MHz-wide TV channel carries both video and audio information. Video signals are amplitude modulated, just as with AM radio; audio signals are frequency modulated, like FM radio. In fact, TV stations actually have two separate transmitters, one for video and the other for audio.

Just as your AM radio tunes to an RF carrier at 980 kilohertz (kHz) and your FM tuner tunes a 98.5 MHz carrier, TV and VCR tuners tune audio and video carriers within each 6 MHz TV channel. A *carrier* is an RF signal that is changed, or *modulated*, by an audio, video, or some other signal.

In amplitude modulation (AM), the *strength* of the carrier changes in accordance with the applied signal (video) at the transmitter; the carrier frequency remains constant.

In frequency modulation (FM), the strength, or amplitude, of the carrier remains the same, but the carrier *frequency* varies in accordance with the signal applied at the transmitter (audio).

Don't be concerned if you don't know how AM and FM work. You don't need any detailed knowledge to understand the material presented in this textbook.

TV channels occupy two parts of the radio spectrum:

- VHF (very high frequency)
  Channels 2–13
  54–216 MHz
- UHF (ultra-high frequency)
  Channels 14–83
  470–890 MHz

Tables A1.1 and A1.2 list these TV channels, along with the video and audio carrier frequencies. Notice that each channel is 6 MHz wide. TV signals do not occupy the entire VHF and UHF bands. The entire VHF band is from 30 megahertz to 300 MHz, but there are no TV channels below 54 MHz nor above 216 MHz. The entire UHF band covers 300 MHz to 3,000 MHz; TV channels occupy frequencies between 470 MHz and 890 MHz.

### Table A1.1 VHF TV Channels

| CHANNEL NUMBER | BANDWIDTH (MHz) | VIDEO CARRIER | AUDIO CARRIER |
|---|---|---|---|
| 2 | 54–60 | 55.25 | 59.75 |
| 3 | 60–66 | 61.25 | 65.75 |
| 4 | 66–72 | 67.25 | 71.75 |
| 5 | 76–82 | 77.25 | 81.75 |
| 6 | 82–88 | 83.25 | 87.75 |

- FM radio band: 88–108 MHz
- Aircraft navigation, two-way radio: 108–174 MHz

| 7 | 174–180 | 175.25 | 179.75 |
|---|---|---|---|
| 8 | 180–186 | 181.25 | 185.75 |
| 9 | 186–182 | 187.25 | 191.75 |
| 10 | 192–198 | 193.25 | 197.75 |
| 11 | 198–204 | 199.25 | 203.75 |
| 12 | 204–210 | 205.25 | 209.75 |
| 13 | 210–216 | 211.25 | 215.75 |

Notice in Table A1.1 that there is quite a gap in the band of frequencies between VHF Channels 6 and 7. (There is also a much smaller 4 MHz gap between Channels 4 and 5.) TV Channels 2 through 6 are called *VHF Low band*; Channels 7 through 13 are *VHF High band*.

You may have observed that some rooftop TV receiving antenna systems have two separate VHF arrays, besides the UHF array, which is often one or more "bow-ties" mounted within an angled reflector. Separate VHF arrays can each be designed to best capture frequencies within the VHF Low and High bands, rather than a "compromise" single array to cover Channels 2 through 13.

### Table A1.2 UHF TV Channels

| CHANNEL NUMBER | BANDWIDTH (MHz) | VIDEO CARRIER | AUDIO CARRIER |
|---|---|---|---|
| 14 | 470–476 | 471.25 | 475.75 |
| 15 | 476–482 | 477.25 | 481.75 |
| 16 | 482–488 | 483.25 | 487.75 |
| 17 | 488–494 | 489.25 | 493.75 |
| 18 | 494–500 | 495.25 | 499.75 |
| 19 | 500–506 | 501.25 | 505.75 |

Channels 20 through 77 have exact same frequency relationships, with no gaps.

| 78 | 854–860 | 855.25 | 859.75 |
|---|---|---|---|
| 79 | 860–866 | 861.25 | 865.75 |
| 80 | 866–872 | 867.25 | 871.75 |
| 81 | 872–878 | 873.25 | 877.75 |
| 82 | 878–884 | 879.25 | 883.75 |
| 83 | 884–890 | 885.25 | 889.75 |

Referring to Tables A1.1 and A1.2, you can see there is a tremendous 254 MHz gap between TV Channels 13 and 14. Because of this large frequency difference, separate VHF and UHF antennas are required. VHF and UHF arrays are often combined in a single antenna, called a *multi-array antenna*. Some combination antennas also effectively receive FM radio frequencies, 88–108 MHz.

Table A1.3 lists standard cable TV bands and the cable channel numbers contained in each band.

### Table A1.3 Standard CATV/Cable Channels

| CABLE BAND | FREQUENCY RANGE (MHz) | CABLE/CATV CHANNELS |
|---|---|---|
| Low | 54–114 | 2, 3, 4, 1, 5, 6, 95, 96, 97, 98, 99 |
| Mid | 120–168 | 14–22, inclusive |
| High | 174–210 | 7–13, inclusive |
| Super | 216–294 | 23–36, inclusive |
| Hyper | 300–462 | 37–64, inclusive |
| Ultra | 468–798 | 65–94, inclusive 100–125, inclusive |

Table A1.4, at the end of this appendix, lists *alternate names* for cable channel numbers in Table A1.3.

A quick comparison of the frequency ranges and channels in Tables A1.1, A1.2, and A1.3 reveals that there is not necessarily a direct correlation between VHF/UHF broadcast TV channel numbers and cable channel numbers. So don't be confused when broadcast VHF Channel 6 is supplied as cable Channel 19, but is tuned in on VCR Channel 3. Be aware of the possibility of channel number confusion, especially when talking with someone who may not know what you now do.

A cable-company-supplied descrambler may still be required with cable-ready VCRs for pay-per-view and premium channels.

## BASIC ANTENNA-TO-VCR CONNECTIONS

Many possibilities exist for connecting one or more antennas to the RF input terminals on a VCR. Figure A1.4 illustrates an installation with separate antennas for VHF and UHF.

In this setup, the VHF antenna's downlead is a round 75-ohm coax cable with a male F connector, and is directly attached to the VHF IN terminal on the VCR. The downlead from the UHF antenna is a 300-ohm flat twin-lead, which is directly connected to the two UHF IN terminals.

But perhaps the VHF antenna downlead is 300-ohm twin-lead instead of 75-ohm coax, or the UHF antenna has a 75-ohm downlead rather than 300-ohm twin-lead. In either case, a small matching transformer is required to connect the antenna lead to the VCR. In this application, the transformer converts a 75-ohm unbalanced line to a 300-ohm balanced line, or vice versa. Figure A1.5 illustrates two types of matching transformers. Although they both perform exactly the same function, converting a 75-ohm line to 300 ohms or a 300-ohm line to 75 ohms, the two different styles are designed for specific hookups.

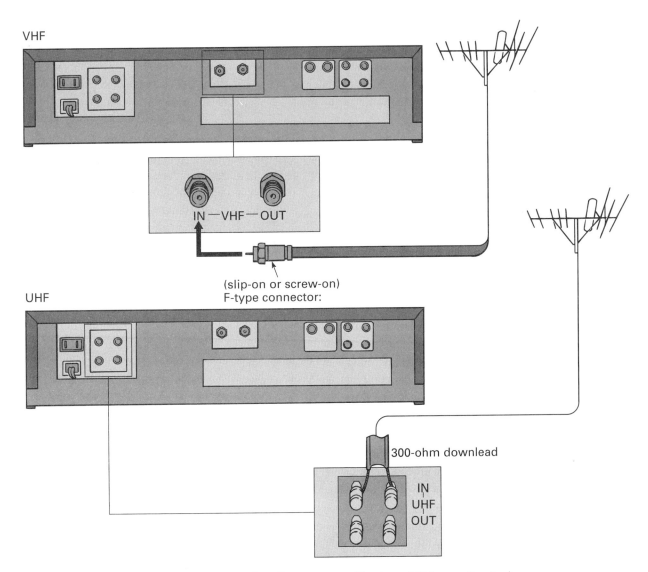

**FIGURE A1.4** **Connecting separate VHF and UHF antennas.** *(Courtesy of JC Penney Co., Inc.)*

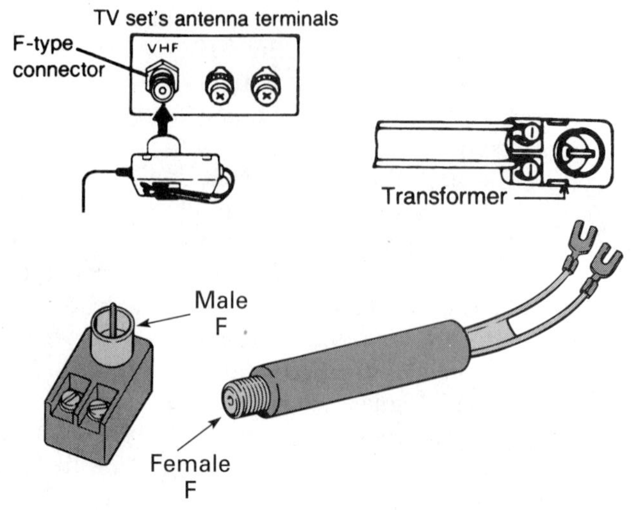

**522** ■ Practical VCR Repair

**FIGURE A1.5  RF transformers for matching 75-ohm and 300-ohm lines.** *(Courtesy of JC Penney Co., Inc.)*

verted. They work the same for VHF and UHF (FM also). Figure A1.6 shows a matching transformer connecting a 300-ohm VHF antenna lead to the 75-ohm VHF F connector on a VCR.

Other terms for an RF matching transformer are *balun* and *impedance matching transformer*. *Balun* is a combination of the words "balanced" and "unbalanced." An electrical line is said to be balanced when neither of its two conductors are connected to electrical ground, but are each equally distant from ground, electrically. Three-hundred-ohm twin-lead is a balanced line, because neither of the two conductors is at ground potential. A regular 2-conductor, 600-ohm telephone line is another example of a balanced line. Both conductors are equally "above" ground, electrically.

An unbalanced line has one of its conductors connected to ground. Household 120-volt AC power is unbalanced, because the neutral side of the line is at ground potential. Coax cable is unbalanced, because the shield conductor is connected to chassis ground. Therefore, *balun* describes a transformer that matches a *bal*anced line to an *un*balanced line.

The term *impedance* refers to the total resistance of a component to the flow of an electric current, rated in ohms. Especially for RF circuits, the cable connecting the signal source (antenna) and the load (VCR) should closely match the characteristic impedance of the source and load for maximum signal transfer, that is, with the least amount of signal loss on the cable. Thus a 75-ohm antenna and a 75-ohm VCR input should be connected with 75-ohm cable so the signal doesn't deteriorate. An impedance matching transformer converts one impedance to another. In the case of RF transformers for VCR

In Figure A1.5, the two transformers on the left convert balanced 300-ohm twin-lead to an unbalanced 75-ohm connection. Bare wire or spade lug ends on twin-lead cable attach to the two screws; the push-on male F connector then plugs into a female F connector. On the right are two 75-ohm to 300-ohm matching transformers. The male connector on a 75-ohm coax cable screws onto the transformer's female F connector, and the 300-ohm pigtails attach to screw terminals on a TV or VCR.

Matching transformers should be used whenever 300-ohm and 75-ohm RF lines need to be con-

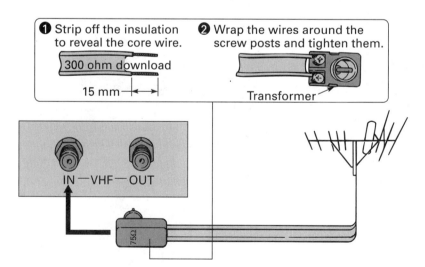

**FIGURE A1.6  Matching transformer connecting 300-ohm VHF antenna to 75-ohm VHF In F connector on VCR.** *(Courtesy of JC Penney Co., Inc.)*

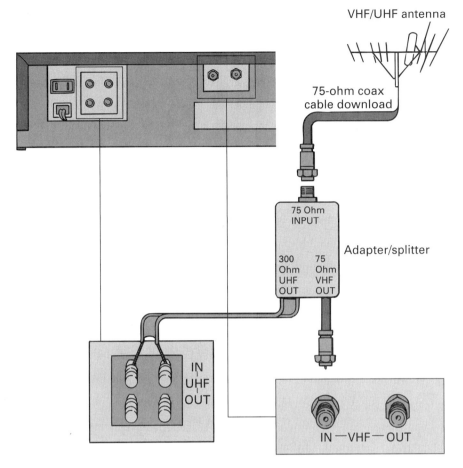

**FIGURE A1.7** **Signal splitter/matching transformer connecting combination 75-ohm VHF/UHF downlead to VCR VHF and UHF inputs.** *(Courtesy of JC Penney Co., Inc.)*

and TV hookups, they convert balanced and unbalanced lines, *as well as changing line impedance.*

Many homes have a combination VHF/UHF antenna with a single 75-ohm coax cable downlead. To connect this to a VCR (or TV) with separate VHF and UHF inputs, a combination signal splitter/matching transformer is needed. Figure A1.7 shows how such a device is connected.

It may be that the combination VHF/UHF antenna has a single 300-ohm downlead, instead of a 75-ohm coax line as shown in the Figure A1.7. In this case, a similar signal splitter/transformer with a 300-ohm input is called for.

Some newer TVs and VCRs have a single F connector for *both* VHF and UHF input. With a combination VHF/UHF antenna, a common 75-ohm downlead can simply be connected to the VHF/UHF IN terminal. Of course, if the common downlead is 300-ohm twin-lead, a matching transformer is required between antenna lead and TV or VCR.

Even though a VHF/UHF signal splitter/transformer may have its connections clearly labeled "IN" and "OUT," there is really nothing directional about these devices. If there are two separate antenna leads for VHF and UHF, a signal splitter can be connected "backwards." In this way, the splitter really functions as a *combiner*, providing a single VHF/UHF output from separate VHF and UHF inputs. In fact, this device is frequently labeled a VCR/TV splitter/combiner.

## BASIC VCR-TO-TV CONNECTIONS

Figure A1.8 shows typical connections between a VCR and TV set. A 75-ohm coax cable connects the VHF OUT F connector on the VCR to the 75-ohm VHF IN F connector on the TV. Twin-lead likewise connects the VCR UHF OUT terminals to the TV UHF IN terminals.

Older TV sets may have two screw terminals for VHF input instead of an F connector. In this case, a balun transformer is required to convert the 75-ohm VCR output to the 300-ohm VHF input. This could be accomplished in one of two ways:

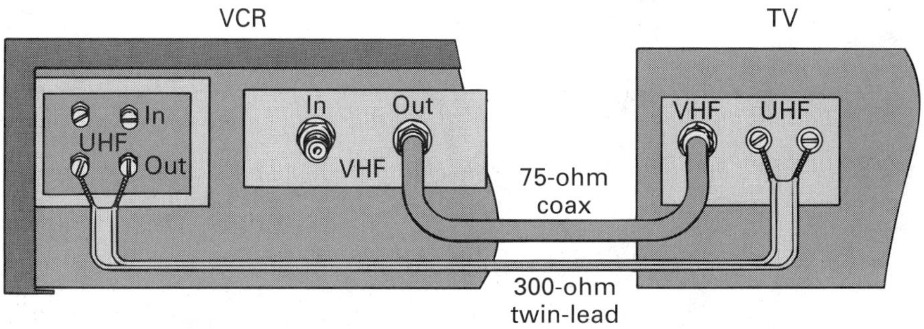

**FIGURE A1.8** VCR to TV VHF and UHF connections.

1. Transformer at VCR OUT F connector, with twin-lead going to TV set
2. Coax cable connecting VCR OUT F connector to transformer at TV set 300-ohm terminals.

Given a choice, 75-ohm coax cable between VCR and TV rather than twin-lead is usually better; its RF shielding keeps out interference that could be picked up on unshielded twin-lead.

Newer TV sets have a single 75-ohm input for VHF and UHF. A VHF/UHF splitter/combiner can be installed at the TV, as previously described, to combine the separate VHF and UHF outputs from the VCR.

With a TV that is actually a TV/video monitor, it is desirable to connect the VCR Video Out and Audio Out line jacks to the A/V Line In jacks on the monitor. When playing a tape or monitoring signals from the VCR's tuner or Line In jacks, the TV/Monitor is placed in audio/video or audio/visual (A/V) mode rather than tuning to VCR output on VHF Channel 3 or 4.

To record one program while watching another, the TV/monitor will still require RF input so its tuner can select a TV channel while the VCR is tuned to a different channel.

## CABLE SYSTEM CONNECTIONS

Figure A1.9 shows the least complicated cable/VCR/TV hookup. In this example, both VCR and TV can directly tune cable TV channels; that is, they are cable-ready. Cable TV systems use 75-ohm coaxial cable, so a direct connection can be made to the VHF IN connector on a VCR. Notice in Figure A1.9 that there is no cable converter box. Both VCR and TV, being cable-ready, can directly tune cable channels without a converter.

Most cable boxes convert cable channels to VHF Channel 2, 3, or 4, as determined by a small switch on the back panel or bottom of the converter. The TV receiver's tuner is set to the output channel of the cable converter, and all channel selections are made at the converter or its handheld remote.

With the hookup in Figure A1.9, one cable channel can be recorded while a different channel is viewed on the TV, with the majority of cable systems. On these systems, all cable channels are present on the cable at the same time; the cable-ready VCR and TV tuners select the desired one.

Some cable companies use an addressing system, for which a converter box is always required. In

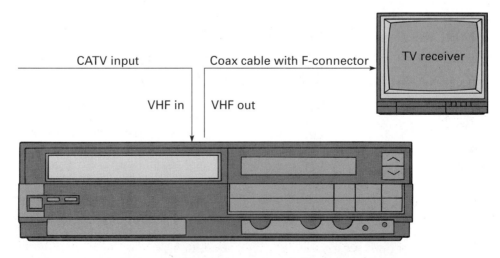

**FIGURE A1.9** Cable without converter box connected to cable-ready VCR and TV.

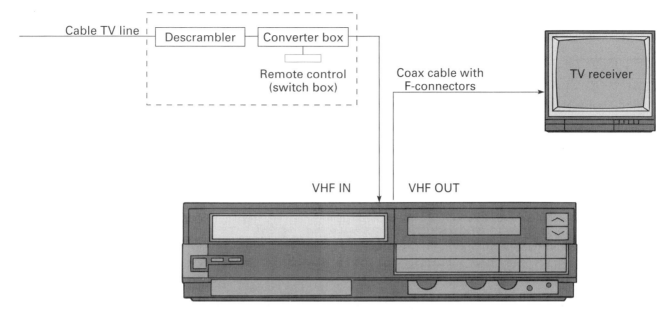

**FIGURE A1.10  Cable converter/descrambler connected to VCR and TV.**

this type of system, *only one channel is sent down the cable to the subscriber at a time*, determined by the selection made on the cable box. To determine the cable system type, ask the cable company if channels can be directly tuned by a cable-ready TV receiver without a converter. If the answer is no, then this is one of the rare systems using addressing.

Assuming cable-ready VCR and TV, and a nonaddressing cable system with all channels present, the hookup in Figure A1.9 should work fine, except for one thing: pay-per-view and premium channels that are scrambled by the cable company cannot be viewed or recorded. Figure A1.10 shows the required configuration for viewing and recording scrambled, as well as nonscrambled, cable channels. Note that in this hookup, only one channel is output by the cable box, so that viewing one program while recording another is not possible. Because a cable converter *is* used, neither VCR nor TV needs to be cable-ready.

With a cable-ready TV, a somewhat more complicated hookup allows the following operations:

- Nonscrambled channels can be viewed on TV while recording a different (non-scrambled or scrambled) channel
- A scrambled or nonscrambled channel can be viewed on TV while recording the same program.

Figure A1.11 illustrates how this can be done. In this configuration, the cable company signal is divided by a 2-way, 75-ohm RF splitter. This type of splitter provides two identical 75-ohm outputs; it does not separate the input signal into separate VHF and UHF bands as the previously described splitters do. One output of the RF splitter feeds a cable converter/descrambler box, which then connects to the VCR. The VCR tuner is set to the output channel of the cable box, and so scrambled and nonscrambled channels can be recorded.

The second output of the two-way splitter goes to the A side of an A/B selector switch; the switch common line connects to a cable-ready TV. VCR output goes to the B side of the A/B switch.

To see how this hookup might work, assume:

1. Cable Channel 27 is a premium, scrambled channel for which the subscriber is paying.
2. Channels 8 and 11 are nonscrambled cable channels.
3. A cable company converter/descrambler outputs on VHF Ch 3.
D. The VCR outputs on VHF Ch 4.

Following are some operational possibilities with this hookup.

- Watch Ch 8 while recording Ch 27
  — Put A/B switch to A
  — Select Ch 8 with TV tuner
  — Select Ch 27 on cable box
  — Select Ch 3 with VCR tuner
- Watch Ch 27 while recording Ch 27
  —Put A/B switch to B
  — Select Ch 4 with TV tuner
  — Select Ch 27 on cable box
  — Select Ch 3 with VCR tuner
- Watch Ch 8 while recording Ch 11
  — Put A/B switch to A
  — Select Ch 8 with TV tuner
  — Select Ch 11 on cable box
  — Select Ch 3 with VCR tuner

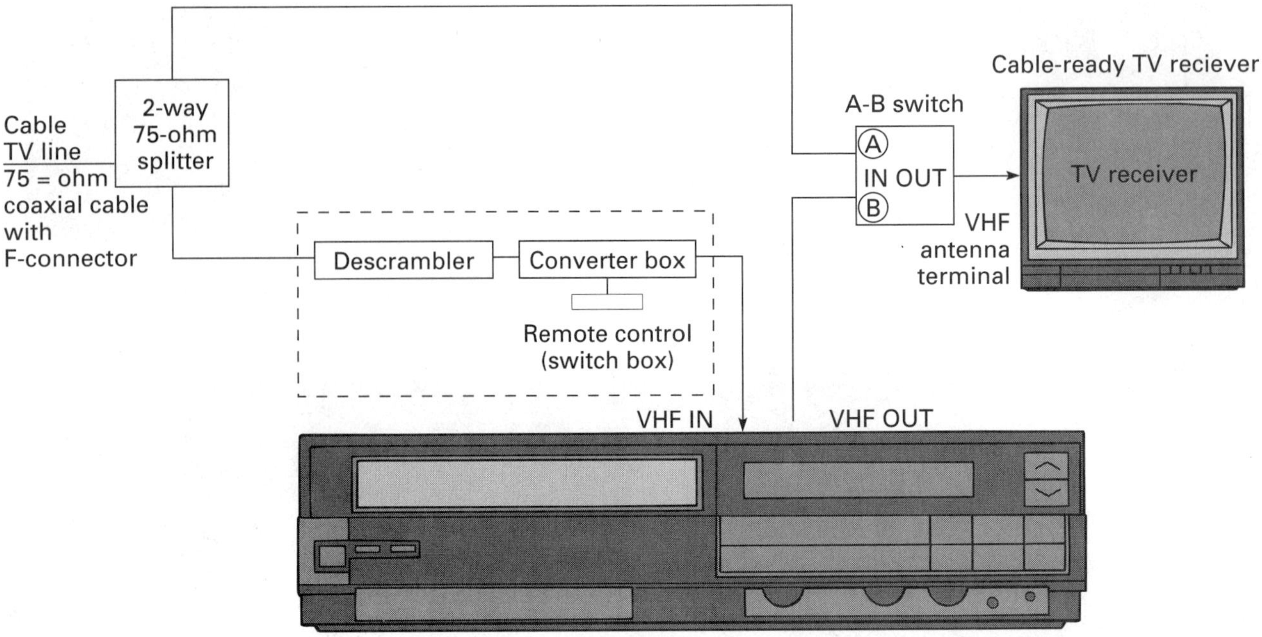

**FIGURE A1.11** Hookup for recording scrambled channels while viewing nonscrambled channels with cable-ready TV.

An almost limitless number of configurations for interconnecting cable systems, antennas, VCRs, TVs, and other A/V gear is possible. We hope the previous examples will give the reader the knowledge to handle just about any situation. Talking with the cable company before hand can often make things go smoother, and eliminate *some* experimentation, when connecting equipment. Depending on the particular cable company, and the person you happen to talk with, you *may* get anything from extremely helpful information all the way to completely inaccurate information. But it's usually worth a try! Ask to speak with someone in the engineering department, not sales.

Following are some general guidelines to follow when hooking things up.

## About RF Cables

All cables are *not* created equal. Several coaxial cables look very similar, but have different characteristics, especially the nominal impedance and quality of the shield. Some common 75-ohm coax cable types are:

- RG-59
- RG-11
- RG-6

Of these, RG-59 is the thinnest and most flexible, whereas RG-6 is thicker with a heavier shield. RG-6 also has less loss per foot at high frequencies than RG-59. Use the best quality cable you can afford.

The same goes for the F connectors. Generally, the old adage "You get what you pay for" is true for cable; the more expensive cables and gold-plated connectors have less signal loss, are less prone to outside RF interference, and resist corrosion at connections. The extra money spent is usually worth it, providing clearer pictures, less ghosting, and decreased susceptibility to outside electrical and RF noise.

RG-59 fitted with RCA plugs is excellent for connecting composite video and A/V line jacks. RG-6 is a better choice for RF frequencies, such as antenna downleads and VCR-to-TV VHF connections, although in some installations RG-59 may work fine, especially for short runs.

Ham and CB radio operators use coax cable that looks nearly identical to RG-59. It is RG-58, or RG-58/U, which is 52-ohm cable. Using 52-ohm cable on a 75-ohm circuit can cause signal reflections on the cable, due to the mismatch in impedance. This results in ghosts and loss of signal strength.

There are many pre-assembled 75-ohm cables on the market for connecting VCRs, TVs, and the like—and there is a difference! One particular home video installation the author recalls had moderate ghosting and occasional noise bars on several cable channels. Replacing just a *single*, inexpensive premade six-foot cable between the cable system wall outlet and the VCR with a quality RG-6 cable with gold-plated F connectors cleared everything up!

As Figure A1.11 readily illustrates, there can be several connections in a video hookup. Each hunk of wire and connection tends to deteriorate signal strength, as well as allowing outside RF interference to get in. This becomes more pronounced with higher TV/cable channel frequencies. Cables should be as short as practical; any excess cable should *not* be coiled. Connections should be clean and tight. Screw-on type F connectors provide a better shield connection than the push-on type do.

## RF Splitters and A/B Switches

Signal splitters are required any time an RF signal must be sent to two devices, as illustrated in Figure A1.11. If a simple parallel, or Y, connection were made, the impedance of the circuit would be upset, leading to ghosting and signal loss. Signal splitters maintain the required 75-ohm circuit impedance, while dividing the signal equally.

Each time a signal passes through a splitter, signal strength is decreased. That is, each output of a two-way splitter has one-half the signal strength of the input signal. For this reason, the number of signal splitters should be kept at a minimum.

To maintain system impedance and prevent signal reflections on RF cables, unused outputs of signal splitters should be terminated with a 75-ohm terminator. For example, suppose a 75-ohm downlead from a rooftop antenna goes to a two-way splitter in the attic. One splitter output goes to a wall jack in the downstairs family room; the other goes to a jack in the master bedroom.

Now suppose there is no TV connected in the bedroom. A 75-ohm terminator should be placed on the wall outlet F connector in the bedroom to properly load the cable. Without the terminator, it's possible for signal reflections on the unterminated line to cause ghosting and other problems on the TV in the family room.

What's been said about RF cable goes for A/B switches as well: cheap is cheap, and can cause problems. Some inexpensive A/B switches have insufficient isolation between their contacts at high radio frequencies. This means that even though the switch is in the B position, some signal from the A input might still leak across to the output. Leakage is more pronounced at higher frequencies.

The same is true of some RF switchers, typically a group of pushbuttons on a unit that can direct RF signals from different sources to different devices. For example, an RF switcher with a 3-by-4 matrix can take three separate RF inputs and connect each to four separate outputs:

### *RF Switcher Inputs*

- Antenna
- VCR 1 VHF OUT
- VCR 2 VHF OUT

### *RF Switcher Outputs*

- TV 1 VHF IN
- TV 2 VHF IN
- VCR 1 VHF IN
- VCR 2 VHF IN

Inexpensive RF switchers can have excessive leakage from inputs that are supposedly Off. There is also some signal loss going through a switcher, which is really a combination signal splitter and switch bank. In general, the simpler the RF hookup, the better the picture and the fewer the problems. Switching line level audio and video (composite) signals is much less prone to problems than switching RF signals.

## SELECTING VCR INPUT SOURCE

VCRs employ two different methods for selecting whether a recording is made from signals tuned in by the VCR's tuner or from the A/V Line In jacks. On some, a front panel switch selects the input source. A typical stereo VCR might have a three-position switch labeled **Tuner–Simul–Line**. Here's what the three switch positions do:

**Tuner:** Video and audio recorded on tape are from the TV/cable program selected by the VCR's tuner. This switch position may be labeled **TV**.

**Simul:** This position is intended for TV/FM simulcasts, where stereo audio for the TV program is carried on an FM radio station at the same time. In the Simul switch position, video comes from the TV program selected by the VCR's tuner, but audio is from the line level RCA input jacks.

An FM tuner or receiver supplies the stereo audio signal. Tape Out jacks on a receiver, or preamplifier with attached FM tuner, are routed to the VCR's Audio In jacks.

**Line:** In this position, audio and video come from the RCA line jacks, such as when copying, or dubbing, a tape from another VCR or camcorder.

Many VCRs do not have an input selector switch. Instead, switching occurs automatically when a cable is plugged into the Line In jacks. Switch contacts inside the A/V RCA jacks automatically disconnect audio or video derived from the VCR's tuner.

For example, suppose you have the TV/VCR mode button set to VCR; the program selected by the VCR's tuner is seen and heard on the TV connected to the VCR. You then plug a cable into the Video In RCA jack on the VCR; the other end of the cable is not connected to anything at this time. Normally closed (N.C.) switch contacts in the Video In RCA jack open as you insert the RCA plug. This disconnects the video signal derived from the VCR's tuner. There will be a blank screen on the TV, but audio will still be that of the TV program.

Plugging a cable into the Audio In line jack works similarly: Sound derived from the VCR's tuner is cut off, and audio now comes into the VCR from the cable. On many stereo VCRs, plugging a single cable into just the Right (Red) Audio In RCA jack places the input signal on *both* Right *and* Left channels. This is handy when recording a mono program on a stereo VCR. It puts the same signal on left and right channels without having to use a separate Y audio adapter to feed the single source to both left and right channels.

Automatic switching when plugging a cable into a jack is probably not new to you. Perhaps you have a transistor radio, boom box, or TV that cuts the speaker(s) Off when you plug in a headset. The same principle of opening N.C. contacts inside the Line In jacks on a VCR cuts off audio or video derived from the tuner.

## LINE JACKS VERSUS RF JACKS

Connecting video equipment with RCA cables between line jacks provides higher quality video (and audio) than when cabling between RF connectors. Suppose you have a combination TV/video monitor, a TV set that also has A/V Line In jacks. Connecting a VCR with RCA cables to the TV/Monitor Line In jacks provides a clearer, more detailed picture and cleaner audio than an RF connection. Here's why.

Suppose you play a tape on a VCR connected to a combination TV/video monitor. A 75-ohm cable connects the VHF Out on the VCR to the TV's VHF Antenna terminal.

In the VCR, video information is read by rotating tape heads. Head signals are processed by VCR video circuits to produce a composite video signal. This composite signal goes *two places*: to the Video Out RCA jack *and* to the RF modulator, which outputs on VHF Channel 3 or 4.

Thus the signal goes out on VHF Channel 3 (or 4). At the TV/monitor, the signal is received by the tuner set to Channel 3 (or 4), amplified, and then demodulated to recover the composite video (and audio) signals. Demodulation is just the opposite of modulation: video and audio baseband signals are separated from their amplitude-modulated and frequency-modulated RF carriers.

Figure A1.12 is a simplified block diagram of the signal path. Note the points where line level audio and composite video signals are in relation to the RF signals.

By running cables between VCR A/V Line Out jacks and A/V Line In jacks on the TV/monitor, signals do *not* have to pass through these circuits:

- RF modulator in VCR
- Tuner section of TV/monitor
- Signal amplifiers in TV/monitor
- Demodulator in TV/monitor

With fewer circuits to go through, signal quality is significantly higher with A/V connections, compared to taking the RF modulation/demodulation route. In general, an electronic signal deteriorates the more circuitry it has to pass through. (Of course in some cases, added electronic filters, noise reduction, or equalizing circuits may *actually improve* signal quality.)

Just remember, video and audio signals pass through less circuitry with Line than with RF connections, providing higher quality picture and sound.

## Dubbing with Two VCRs

By now it should be fairly obvious how to copy, or dub, a tape from one VCR to another: simply connect the output of the Play VCR to the input of the Record VCR. Sounds simple, but a few things can help make a good copy, and there are some other items of which you should be aware.

Copying a videocassette that contains copyrighted material is probably illegal in most cases. It definitely is if you intend to sell the copy or make money in any other manner from the copyrighted tape without permission from the copyright holder.

A copy of an original tape is never as good as the original. Color is usually the first thing to deteriorate, often giving a reddish/orange halo around images. Picture detail will be fuzzier. Higher audio frequencies will not come through as clearly on a copy. The first copy is called a second generation tape, and will usually be acceptable, assuming an excellent quality original. A copy of a copy, or third-generation tape, will be unacceptable in most cases. For the best quality copy, record at the SP speed.

As previously explained, higher quality results are obtained using A/V outputs and inputs instead of RF connections when copying from one VCR to another. Connect the two VCRs between the separate Video and Audio In/Out jacks to achieve the best results, rather than connecting the VHF In/Out connectors.

To dub in this manner, use quality shielded cable to connect the VIDEO OUT jack on the Play VCR to the VIDEO IN jack on the Record VCR. Likewise, connect the AUDIO OUT jack(s) on the Play VCR to the AUDIO IN jack(s) on the Record VCR with shielded cable(s).

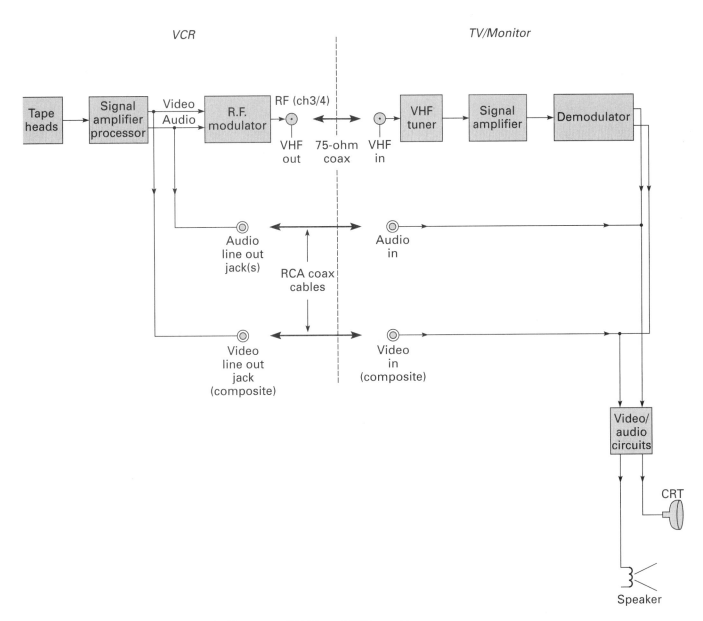

**FIGURE A1.12  Simplified block diagram of VCR and TV/monitor.**

For easy hookup, some VCRs have A/V Line In and Out jacks on the front panel. There may also be a parallel set of RCA jacks on the rear panel. Operation is the same as previously described. You may have to set an input selector switch for recording from line input sources, or just insert a plug into an RCA input jack to automatically switch out signals derived from the VCR's tuner.

A TV/video monitor should be connected to the VCR doing the *recording* so you can observe the signal actually coming in. Adjust the tracking control on the Play VCR for the best picture.

Purchase only name-brand, quality blank tapes on which to record. Discount brand or "no name" bargain tapes are often not a bargain at all. Inferior bonding of the ferric oxide magnetic coating to the mylar tape can cause clogged video heads, signal dropout, and loss of picture detail. They are usually substandard for Hi-Fi audio recording, as well, because of thin oxide layers.

Make a test copy of a short segment of the original. Check that it plays back OK. Start making the actual copy a short ways into the tape, not at the very beginning—let several turns of tape wrap around the cassette take-up reel first. Especially with used machines and cassettes, there is frequently some speed irregularity at the very beginning of a tape, which results in some picture instability.

Once you are ready to make the dub, locate the section to be copied on the Play VCR, then press the Pause button. This machine is now in Play/Pause mode. Place the Record VCR in Record mode, let a few turns of tape wind around the take-up reel, then hit Pause on the Record VCR. This machine is now in Record/Pause mode. To start the dub, press Pause on the Record VCR slightly before hitting Pause on the Play VCR. When stopping, hit Stop on the Record VCR first. This helps create clean starts and stops on the copy.

Some commercially recorded video cassettes employ a copy protection or "copyguard" system to discourage dubbing. One method is to omit or even reverse some or all of the vertical synchronization 60 Hz pulses recorded on the tape. This may make the original tape itself difficult to play back, especially on older TVs. Vertical rolling and instability is an indication of this scheme. Another scheme, called Macrovision, disrupts the automatic gain control (AGC) circuit in a VCR, causing brightness changes in the picture and occasional bright flashes on the screen when playing back the copy.

*Video stabilizers*, accessory devices that connect between the Video Out and Video In jacks of two VCRs, can help overcome some of the ill effects of copy protection, but a full discussion of these devices and copy protection is not within the scope of this textbook. Various video magazines have related articles and advertisements for video stabilizers and other accessories.

## CABLE CHANNEL NUMBERS AND NAMES

Table A1.4 lists CATV and cable TV channel numbers along with alternate channel names that are sometimes used in the industry.

### Table A1.4 Alternate Cable Channel Names

| CATV/CABLE CHANNEL | ALTERNATE CHANNEL NAMES | CATV/CABLE CHANNEL | ALTERNATE CHANNEL NAMES |
|---|---|---|---|
| 1 | A-8, C54, J54, G64, 4+, 5A | 45 | II, W+9 |
| 5 | A-7, C55, J55, G65 | 46 | JJ, W+10 |
| 6 | A-6, C56, J56, G66 | 47 | KK, W+11 |
| 95 | A-5, C57, J57 | 48 | LL, W+12 |
| 96 | A-4, C58, J58 | 49 | MM, W+13 |
| 97 | A-3, C59, J59 | 50 | NN, W+14 |
| 98 | A-2, C60, J60, G60 | 51 | OO, W+15 |
| 99 | A-1, C61, J61, G61 | 52 | PP, W+16 |
| 14 | A | 53 | QQ, W+17 |
| 15 | B | 54 | RR, W+18, C62 |
| 16 | C | 55 | SS, W+19, C63 |
| 17 | D | 56 | TT, W+20, C64 |
| 18 | E | 57 | UU, W+21, C65 |
| 19 | F | 58 | VV, W+22, C66 |
| 20 | G | 59 | WW, W+23, C67 |
| 21 | H | 60 | AAA, W+24, C68 |
| 22 | I | 61 | BBB, W+25, C69 |
| 23 | J | 62 | CCC, W+26, C70 |
| 24 | K | 63 | DDD, W+27, C71 |
| 25 | L | 64 | EEE, W+28 |
| 26 | M | 65 | U14, FFF, W+29 |
| 27 | N | 66 | U15, GGG, W+30 |
| 28 | O | 67 | U16, HHH, W+31 |
| 29 | P | 68 | U17, III, W+32 |
| 30 | Q | 69 | U18, JJJ, W+33 |
| 31 | R | 70 | U19, KKK, W+34 |
| 32 | S | 71 | U20, LLL, W+35 |
| 33 | T | 72 | U21, MMM, W+36 |
| 34 | U | 73 | U22, NNN, W+37 |
| 35 | V | 74 | U23, OOO, W+38 |
| 36 | W | 75 | U24, PPP, W+39 |
| 37 | AA, W+1 | 76 | U25, QQQ, W+40 |
| 38 | BB, W+2 | 77 | U26, RRR, W+41 |
| 39 | CC, W+3 | 78 | U27, SSS, W+42 |
| 40 | DD, W+4 | 79 | U28, TTT, W+43 |
| 41 | EE, W+5 | 80 | U29, UUU, W+44 |
| 42 | FF, W+6 | 81 | U30, VVV, W+45 |
| 43 | GG, W+7 | 82 | U31, WWW, W+46 |
| 44 | HH, W+8 | 83 | U32, AAAA, W+47 |

Table A1.4 Alternate Cable Channel Names

| CATV/CABLE CHANNEL | ALTERNATE CHANNEL NAMES | CATV/CABLE CHANNEL | ALTERNATE CHANNEL NAMES |
|---|---|---|---|
| 84 | U33, BBBB, W+48 | 108 | U52, UUUU, W+67 |
| 85 | U34, CCCC, W+49 | 109 | U53, VVVV, W+68 |
| 86 | U35, DDDD, W+50 | 110 | U54, WWWW, W+69 |
| 87 | U36, EEEE, W+51 | 111 | U55, AAAAA, W+70 |
| 88 | U37, FFFF, W+52 | 112 | U56, BBBBB, W+71 |
| 89 | U38, GGGG, W+53 | 113 | U57, CCCCC, W+72 |
| 90 | U39, HHHH, W+54 | 114 | U58, DDDDD, W+73 |
| 91 | U40, IIII, W+55 | 115 | U59, EEEEE, W+74 |
| 92 | U41, JJJJ, W+56 | 116 | U60, FFFFF, W+75 |
| 93 | U42, KKKK, W+57 | 117 | U61, GGGGG, W+76 |
| 94 | U43, LLLL, W+58 | 118 | U62, HHHHH, W+77 |
| 100 | U44, MMMM, W+59 | 119 | U63, IIIII, W+78 |
| 101 | U45, NNNN, W+60 | 120 | U64, JJJJJ, W+79 |
| 102 | U46, OOOO, W+61 | 121 | U65, KKKKK, W+80 |
| 103 | U47, PPPP, W+62 | 122 | U66, LLLLL, W+81 |
| 104 | U48, QQQQ, W+63 | 123 | U67, MMMMM, W+82 |
| 105 | U49, RRRR, W+64 | 124 | U68, NNNNN, W+83 |
| 106 | U50, SSSS, W+65 | 125 | U69, OOOOO, W+84 |
| 107 | U51, TTTT, W+66 | | |

# CHAPTER 3 APPENDIX

This appendix contains additional information about topics discussed in Chapter 3.

## HOW TAPE LOAD SHUTTLES EXTEND AND RETRACT

Although not strictly a part of the tape path itself, this is an appropriate point at which to describe the mechanism that moves the two shuttle assemblies and P-guides 2 and 3 along their guide tracks in the transport. All this action occurs in the undercarriage, the underside of the tape transport assembly.

When the VCR goes into Play or Record mode, the video drum starts spinning. Once drum speed is detected, tape load operation begins.

There are many methods to move the two shuttle assemblies along their tracks. Figure A3.1 shows a portion of a representative undercarriage, with the shuttle assemblies in their tape unloaded or home position. The tape load mechanism typically consists of gears driven by a 12-volt DC tape load motor. Other names for this motor are *cam motor* and *mode motor*. For tape loading, the motor spins in one direction, and through a gear train and linkage arms, P-guides 2 and 3 move out of the cassette and travel to their respective V-stops. Next, the pinch roller comes up against the capstan and the cam motor stops. Spring tension supplied by the undercarriage tape load mechanism holds guides P-2 and P-3 firmly against their V-mounts after the load motor stops.

Notice in Figure A3.1 that there are *two* ring gears (B), one above the other. Gear A drives only one of these ring gears. The two small reversing gears (C) transfer motion from the first ring gear to the second, causing it to turn in the opposite direction from the first ring gear. A pivoting linkage arm (D) attached to each ring gear moves the shuttle assemblies (E).

To unload tape, a reverse DC voltage is applied to the tape load motor, causing it to spin in the opposite direction. After the pinch roller moves away from the capstan, the gears and linkage arms now move the other way, causing the shuttles to move in their tracks back into the cassette. The video head drum also stops spinning when tape is unloaded.

Typical VCR operation is to unload tape:

- When machine is placed in Stop mode
- After five minutes of continuous Pause

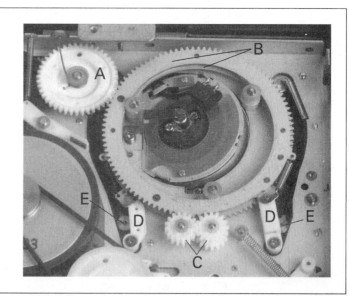

**FIGURE A3.1** One style of undercarriage tape load mechanism. A: Gear powered by cam gear and load motor B: Two tape load ring gears C: Reversing gears D: Linkage arms E: Shuttles

(some units *resume* play after a five-minute pause)
- When not actually playing or recording, such as when timed recording is set
- During fast forward or rewind.
- If video drum is not spinning or is not spinning fast enough
- If capstan is not spinning when it should be
- For several other faults detected by system control, such as the take-up reel not turning during play or record.

Although this happens in a typical VCR, some models keep tape loaded even when the machine is in Stop mode. As you can imagine, this puts additional wear on rotary heads when tape is not actually being played. This is done so the machine can rapidly go into Play from Stop mode, without having to first go through a tape load cycle. For these quick play VCRs, the cassette should be ejected or the VCR should be powered Off when not in use.

You can determine the type of machine by listening as it goes from Stop to Play. You will definitely hear some mechanical sound for a few seconds as transport mechanics load and unload tape. Again, this occurs on models that do *not* keep tape loaded at all times after a cassette is loaded. Quick play VCRs start playing tape almost instantaneously.

There is a wide variety of undercarriage tape load/unload mechanisms. Rotating the load motor shaft by hand with power Off and observing what happens is probably the best way to learn how any particular machine accomplishes these functions. As you manually go through the tape load or unload cycle, be careful not to force anything. Too much force at the end of travel of a gear train, cam, or other component can break parts.

When hand cycling, apply your "finger drive" at the *source* of power to the mechanism, such as at the pulley or worm gear on a cassette load motor or mode motor shaft, and at the flywheel of a direct drive capstan. Trying to turn an intermediate gear or some other component can cause parts breakage. Figure A3.2 illustrates hand cycling the tape load and cam mechanism by turning the mode motor shaft. Go slowly! Observe which gears and cams turn and the movement of linkages and other parts. Note any resistance as you hand cycle: you might be at the end of travel in one direction for the mechanism; there could be a foreign object jamming a gear; or parts could be binding, bent, or broken.

In most VCRs, the tape load (cam, or mode) motor controls other functions as well, such as moving the pinch roller against the capstan shaft and braking the capstan flywheel during Pause. On a top

**FIGURE A3.2** Hand cycling tape load and cam mechanism.

loader, the cam mechanism may also operate to release the cassette basket latch when the Eject button is pressed, allowing the basket to pop up out of the top of the machine. Again, VCRs are designed in *many* different ways.

Other variations may use one or more solenoids—electromagnetic plungers that pull on a linkage when their magnetic coils are energized. You can easily push in on a solenoid plunger to see what happens when the solenoid is electrically energized.

A solenoid may be used to position an idler wheel, or apply spindle brakes, or move some transport linkage or part. Older VCRs with piano-style function keys (Play, Stop, FF, Rewind) typically employ a solenoid to stop the machine. This occurs when Stop is pressed or if the system control microprocessor detects that the take-up reel is not turning when it should be, and for other reasons.

In short, many configurations are possible. High-quality VCRs tend to have less complex mechanical assemblies, using multiple motors instead. Just be aware that VCRs have a variety of transport designs. Undercarriage components are discussed in greater depth, both later in this appendix and in Chapters 5 and 6.

To continue along the video tape path, recall that during tape load, guide posts P-2 and P-3 move into their V-stops and precisely position tape around the front half of the spinning video drum.

## VIDEO HEADS

On a new upper cylinder, the tips of the video heads protrude (extend) about 45 microns beyond the drum surface. This is about 0.0018 inch (0.001 inch = 25.4 microns). To get a rough idea of the amount of head tip protrusion, a dollar bill is roughly 127 microns or 0.005 inch thick. So a new head protrudes about one-third the thickness of a dollar bill beyond the drum surface.

As a VCR is used, rotary head tips wear. At a tip protrusion somewhere around 5 to 10 microns (less than one-fourth to one-eighth the original protrusion), the upper cylinder containing the rotary heads needs to be replaced. In a well-cared-for, quality machine, this could be after 4,000 to 5,000 hours of use.

Measuring head tip protrusion is done with a highly accurate, and somewhat expensive, micrometer. This is usually a gauge with a round scale and pointer that measures small amounts of mechanical displacement. Testers that electronically evaluate rotary head wear are also available from electronics supply houses that carry VCR service tools and replacement parts.

## VIDEO DRUM

You might be wondering about machines that have four, five, or more heads: "What are the other heads for if only two heads are needed for helical scan video?" The answer is that better quality VCRs have two heads for SP speed, and a second pair for LP and EP/SLP speeds. Head gaps can be optimized for different tape speeds, resulting in sharper picture definition than when one head pair is used for all three speeds. VCRs with four video heads can also produce better quality pictures during Pause, Forward and Reverse Scan or Search, and Slow Motion Playback modes.

This is worth knowing; a four-head machine might do fine at SP, but produce a poor picture at LP and EP/SLP tape speeds. This can tell you a lot about what's working correctly in the machine, and areas that may have problems. Because separate pairs of heads are used for the different speeds, a VCR that plays all right at SP but not at LP may have one of the two heads used for LP and EP clogged with magnetic tape residue.

Some VCRs incorporate an additional head for video special effects, such as still frame and slow motion. The extra head helps create a stable, noise-free picture in these modes. Normally, this added head is incorporated only on machines that already have four video heads for normal operation.

To complicate matters just a bit, a VCR that has four rotary heads may use two heads on all tape speeds for video, and two other spinning heads for Hi-Fi Stereo audio. Some high-end models also incorporate yet another spinning head, called a flying erase head. Hi-Fi audio and the flying erase head are covered in Chapter 19.

As you can see, the *minimum* number of spinning heads is two, with the theoretical possibility that the upper cylinder *could* have as many as eight heads:

- 2—SP video
- 2—LP & EP video
- 2—Hi-Fi Stereo
- 1—Special effects or still-frame head
- 1—Flying erase head (a few models have two FE heads)

Most likely there is no machine that actually has all these possible rotary heads.

No matter how many heads a VCR has, they are all contained on the same upper rotating cylinder of the video drum assembly. A particular VCR is generally described by the number of *video* heads it has. Thus a machine advertised as a two-head Hi-Fi VCR will have at least four heads on the upper cylinder: two for video at all speeds, and two for high-fidelity audio. A VCR that is advertised as a four-head machine has four video heads, and if it's a 4-head Hi-Fi unit, it will have six rotary heads.

In all likelihood, the majority of VCRs that you will be servicing will have either two or four rotary heads. Cost goes up as the number of rotary heads increases; the mass VCR market is made up of the less expensive machines.

You may be wondering how the signal to or from the rapidly spinning heads is transferred electronically. There certainly can't be a direct wire connection to the spinning heads! You might be thinking that this could be accomplished with slip rings and brushes, like in an automobile alternator, but a different method is used, called a *rotary transformer*. (Slip rings and brushes would introduce too much electrical noise into the extremely low-level signals coming from the heads.) As shown in Figure A3.3, each head has two associated rotary transformer windings. One winding is in the upper, rotating cylinder along with the heads; the other is in the lower, stationary drum assembly. Alternating current signals are electromagnetically coupled between the two rotary transformer coils during record or play. The rotary transformer itself is not a replaceable item; its failure requires replacement of the entire drum assembly. Rotary transformer defects are rare.

Although in early VCRs, notably Betas, the video drum is driven by a separate motor connected by a belt to the rotating cylinder, virtually all VHS machines have a direct drive motor incorporated into the drum assembly itself. This motor's speed *and position* are tightly controlled by electronic circuitry during Record and Play.

First, it is important that the heads rotate at the correct speed (1,800 rpm) to match the NTSC standard of 30 video frames per second. Therefore, the drum motor drive circuitry must know at what speed the drum is rotating to ensure correct drum speed at 30 revolutions per second.

Second, the rotary position of the upper cylinder, and thus the position of Heads A and B, must be precisely controlled so that each head scans its own videotape tracks. Refer back to Figure 3.13 in Chapter 3. During playback, it is absolutely essential that Head A scan the video tracks originally recorded by Head A. Likewise, Head B must precisely scan the tracks laid down by Head B. This is true even if a tape was originally recorded on another machine, as when playing a commercially produced movie. Head A must track only video tracks recorded by Head A and likewise for Head B.

With these two requirements in mind, there needs to be some way to report drum speed and drum position to the electronic circuit controlling the drum motor. This is accomplished with two separate sensors, or transducers, which are built into the drum assembly. They produce a frequency generator (FG) signal to indicate drum speed and a pulse generator (PG) signal to tell where the drum is within each revolution.

Within the drum assembly, one arrangement is to have two permanent magnets on the rotating upper cylinder and two magnetic pickup coils on the lower cylinder. As the upper cylinder rotates, the coils produce an output pulse whenever a magnet passes over them. These pulses are sent to the electronics for determining drum speed and position of the video heads. Figure A3.4 shows drum transducers.

Many VHS video drums produce FG (frequency generator for speed) and PG (pulse generator for position) signals with *Hall effect devices* instead of pickup coils. A *Hall device* is a solid state component that allows current flow with an applied magnetic field. The end result is the same no matter what the type of drum transducer: feedback signals representing the drum rotational speed (FG) and position (PG) are generated so electronics can control drum speed and rotational position.

Drum transducers are part of the drum motor assembly, and cannot be individually serviced or replaced.

For now, don't be too concerned with details on how the drum is kept at an accurate speed and rotary position. This tight control, called the *drum servo*, is explained in Chapter 14. As long as you

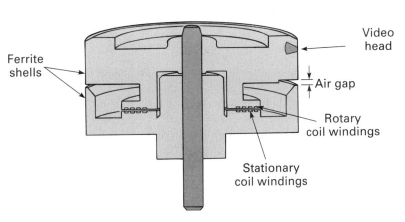

**FIGURE A3.3 Cutaway view of rotary transformer windings.**

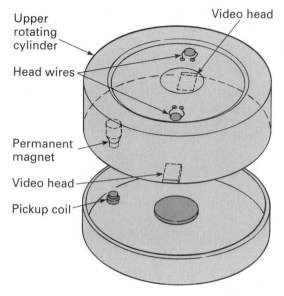

**FIGURE A3.4** Drum transducers for producing FG and PG signals.

have the big picture and understand the need for this tight control of the drum, you're doing fine!

One more item about the video drum assembly needs discussion. An electrostatic charge is produced by the drum spinning against videotape. If this charge is allowed to build up, it can arc across drum ball bearings, damaging their lubrication, and can cause both audio and video noise during record and playback. To bleed off any static charge, a ground strap, attached to chassis ground, contacts the drum motor shaft. An example of an electrostatic discharge (ESD) ground strap on the bottom of a video drum is shown in Figure A3.5.

Depending on the manufacturer of the video drum, the ground strap could be on the top or bottom of the assembly, or even inside the drum, between the upper and lower portions. In the latter case, the drum assembly will have what is called a *rotary cap* over the top of the drum assembly. Remove the mounting screws and separate the two cylinder halves to reach the internal grounding strap.

Some degree of force may be needed to lift the upper cylinder off the lower cylinder, because permanent magnets in the direct drive drum motor may tend to keep the two halves attracted in some

**FIGURE A3.5** Video drum ground strap.

designs. *Never use a tool to pry the drum sections apart!* A video head puller tool is available from several electronic parts suppliers that carry VCR service tools and replacement parts.

Over time, a small dab of lubricant applied where the strap contacts the drum shaft can dry out. The result is often a high-pitched acoustic squeal, either constant or intermittent, when the cylinder is spinning. The remedy is to clean and then relubricate the contact area with a *small* amount of phono lube or similar grease.

## A/C HEAD ASSEMBLY

Hi-Fi Stereo audio, described in Chapter 19, is reproduced with two rotary heads on the video drum, not by the stationary linear audio head. However, during Hi-Fi recording, audio also goes to the linear audio head so that the tape can play back on non-Hi-Fi VCRs.

Some VCRs with linear stereo can rerecord, or edit, one audio channel while retaining existing audio on the other. For example, right channel audio can be changed without affecting existing video or left channel audio. To perform this type of editing, a stereo VCR must have two audio erase heads in the A/C assembly. Erase current is applied to both during normal recording, but to only one when laying down new audio on one track channel while retaining existing audio on the other channel. Hi-Fi audio editing is not possible, because Hi-Fi audio is recorded along with video tracks, not on its own, separate audio track, as is linear audio.

Now some additional information about the control track, which is recorded and played back by the control head portion of the A/C assembly.

Remember that Heads A and B each record 30 diagonal video tracks per second, corresponding with 30 video frames each second. During record, 30-Hz pulses are recorded on the control track by the control head. This signal is called the *control track logic* (CTL) *signal*. Each CTL pulse corresponds with each of the video frames recorded by Heads A and B. The *position* of each 30-Hz control pulse defines *which* head, A or B, is starting to record its video track.

Another way of stating this is that the CTL signal is recorded in synchronization with the drum's precise rotational position, and hence the position of Heads A and B. During playback, the control head reads these synchronization pulses. Electronic servo circuitry then uses this control signal to control the rotational position of the drum so it is the same as when the tape was recorded. Thus, Heads A and B trace the same video tracks originally recorded by Heads A and B, respectively.

Control track pulses serve an additional function. Because pulses are recorded at a 30-Hz rate regardless of tape speed (SP, LP, or EP), an automatic playback speed control circuit monitors the pulse rate during playback, and then changes the capstan speed to match that of the original recording, *independent of the front panel speed control switch*.

As long as you get the general idea of what the control track is for, that's fine at this point. You'll be taking a closer look at servo circuit operation in Chapter 14.

The A/C head assembly is mounted on a bracket that allows adjustment of head assembly *height*, *tilt*, and *azimuth*.

- *Height* describes the vertical head position with respect to the tape, looking at the face of the head assembly. Head height must be adjusted so head gaps are vertically positioned or aligned over the linear audio and control tracks on tape.
- *Tilt* is any forward or backward leaning of the head assembly. The assembly should be straight up and down, so as you look straight down at tape passing over the heads you see the top edge of the tape as a perfectly straight hairline.
- *Azimuth* describes the angle of the vertical head gaps to the tape. Head gaps should be exactly 90 degrees perpendicular to the audio and control tracks, as illustrated in Figure A3.6.

These three adjustments, height, tilt, and azimuth, are the same as head adjustments on conventional audio cassette or reel-to-reel tape recorders.

There is yet another A/C head assembly adjustment. This one is critical to tracking: video Heads A and B passing precisely over their respective diagonal video tracks on tape during playback.

Mechanical tracking adjustment moves the entire A/C head assembly left or right along the tape path. This changes the distance between the control head gap and the video drum, and thus controls *when* the 30-Hz control track logic pulses pass the control head gap during playback, in relation to the position of the video heads. As you look at the face of the A/C head assembly, moving it to the left causes control track pulses to arrive a little bit earlier.

Figure A3.7 illustrates a typical arrangement for moving the A/C head assembly. The head assembly mounting plate is spring loaded against a cone-shaped adjustment nut. As the nut is turned higher or lower on its threaded mounting stud, the head assembly moves slightly to the right or left, respectively.

Some VCRs have no A/C tracking alignment nut; the entire head assembly mounting bracket must be loosened and then slid left or right to accomplish the adjustment. Adjustment the A/C head assembly is described in Chapter 7.

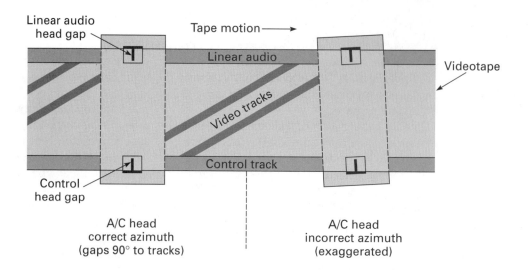

**FIGURE A3.6** Audio/Control head azimuth.

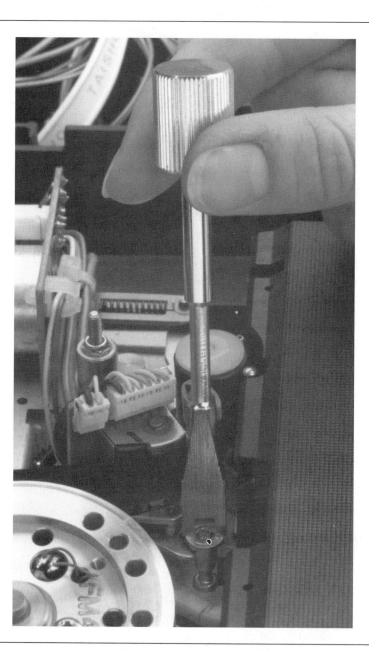

**FIGURE A3.7** Mechanical tracking adjustment of A/C head assembly.

## CAPSTAN AND PINCH ROLLER

To achieve smooth tape movement, the pinch roller must contact the capstan uniformly from top to bottom. That is, as the roller approaches the capstan, both the top and bottom of the roller should touch the capstan shaft at the same time. Tape skew occurs when the capstan and pinch roller are not perfectly parallel with each other. *Tape skew* is any upward or downward slant of the tape anywhere along the tape path, rather than being straight across.

Because many VCRs power more than just the capstan shaft with the capstan motor, it is often helpful to turn the capstan motor by hand to study how things operate (or how they are failing to operate!). Figure A3.8 shows how to hand cycle a tape transport drive by turning the large flywheel at the bottom of a direct drive capstan motor.

Apply rotation with your fingers at the *source* of power to the transport mechanism you're investigating or servicing (in this case the capstan motor flywheel). Viewed from the undercarriage, the capstan flywheel goes counterclockwise for play or any forward tape motion, clockwise for rewind or reverse scan.

Whether the capstan is belt drive or direct drive, its motor is a critical component to proper VCR operation. Just as with the video drum motor, its speed is tightly controlled.

An arrangement similar to that for video drum motor control is used. The capstan motor produces frequency generator (FG) pulses that electronic servo circuits monitor to maintain precise capstan speed during record and playback. FG pulses are produced with a magnetic emitter, Hall effect device, or a rotating permanent magnet and stationary pickup coil.

Hall devices are normally part of direct drive motors. A magnetic emitter often consists of a rotating multi-toothed metallic wheel, similar to a gear. The teeth pass a few thousandths of an inch away from a magnetic coil, or transducer, to produce FG pulses. Fixed magnetic coils near the circumference of a permanent magnet capstan flywheel is another arrangement.

An FG emitter could consist of an infrared light-emitting diode (IR LED) shining onto a phototransistor through numerous holes or slits in an opaque disk, which is attached to the capstan shaft. As this encoder disk spins with the capstan, the phototransistor alternately goes dark and light as the IR LED beam is interrupted. A variation is to place both

**FIGURE A3.8** Hand powering direct drive capstan motor flywheel.

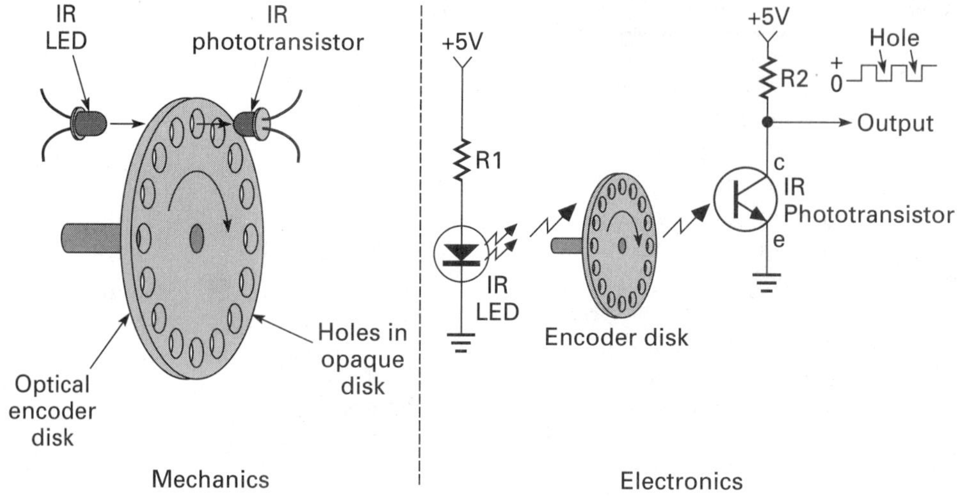

**FIGURE A3.9** How an encoder disk produces electrical pulses.

Mechanics | Electronics

IR LED and phototransistor on the same side of an encoder, or emitter, disk that has regularly spaced reflective strips, as described in Chapter 3.

Figure A3.9 illustrates an optical encoder. The main point is that, whatever the FG transducer type, signals are sent from the capstan motor to the capstan servo circuits which monitor and maintain correct capstan speed.

During recording, capstan speed is controlled by comparing capstan FG pulses with a synchronization signal in the video being recorded. This video could be from a broadcast station, cable system, video camera, or other composite video source, such as Video Out from another VCR. Precise control of tape speed is necessary so that the video tracks laid down by Heads A and B in the spinning upper cylinder are timed correctly with the 30 frames per second of video being "received" and recorded. In a similar manner, during playback, capstan FG pulses are compared with the 30-Hz control track pulses from the A/C head. Electronic correction of capstan speed ensures that tape movement places the correct video tracks under Heads A and B as they scan the tape.

Minor changes in the capstan speed during playback are controlled by the front panel tracking control, which enables adjustment for tapes made on another machine. Essentially the tracking control adjusts for minor differences in the distance between the control head gap and video drum on different VCRs.

Many newer VCRs have auto tracking. The video signal from tape, called the *RF envelope*, is monitored. Corrections are made to control the capstan so that a maximum amplitude signal results. This occurs when Heads A and B are precisely tracking their respective video tracks on tape.From this discussion, and from previous descriptions of video drum motor speed and phase control, you are probably getting the idea that the capstan and video drum motors must work very closely together both in record and playback to ensure correct head positioning with respect to their video tracks on tape. If so, you're right on track! (pun intended)!

Rotational speed and phase of both drum and capstan motors, control track pulses, and video synchronization signals are all closely related. This is true during record *and* playback. VCR servo circuits analyze the frequency generator (FG) and pulse generator (PG) signals from the drum, FG pulses from the capstan, sync signals contained in video, and control pulses from the tape to keep video heads and tape in step with each other. We take a closer look at VCR servo circuits in Chapter 14.

# CHAPTER 8 APPENDIX

## IMPEDANCE: SUPPLEMENTARY INFORMATION

You may be familiar with the term *impedance* in relationship to loudspeakers. Most loudspeakers have a nominal impedance of either 4 or 8 ohms. However, if you were to measure the DC resistance of the voice coil of a typical speaker with an ohmmeter, you would find the resistance to be much lower than 4 or 8 ohms, usually less than 1 ohm. At audio frequencies, however, the loudspeaker appears more like a 4- or 8-ohm resistor across the amplifier output rather than a 1-ohm or lower load.

Here's another example. Suppose you measure the DC resistance between the two prongs on the AC cord for a VCR designed for 120V, 60 Hz. Your ohmmeter reads 9 ohms, which is the DC resistance of the primary winding of the transformer in the power supply. Knowing the voltage (120) and the resistance, you decide to see how much power the VCR takes. From the formula chart you select:

$P = E^2/R$    Substituting known values:
$P = 120 \times 120 \div 9 = $ **1,600 watts**

But wait! You think that's a little bit high, and you're right. The information plate on the back panel of the VCR reads:

| 120V | 60Hz | 29W |

Sure is a big difference between 1,600 watts and 29 watts! What gives?

You go back and check your meter reading: yes, it's 9 ohms. You check your math: 1,600 watts it is. So what's wrong?

What's wrong is that the DC resistance of the VCR power transformer primary winding has no reasonable relationship to its impedance at 60 Hertz.

So what impedance *would* the VCR present to the 120V, 60-Hz power line? Remember, impedance is just AC resistance, so we need to solve for R, but when solving for AC resistance, we use Z (for impedance) to replace R. We know the voltage (120V) and the power (29W) from the VCR rear panel rating plate. From the formula chart select:

$R = E^2/P$    Z is the symbol for impedance, so:
$Z = E^2/P$    Substituting known values:
$Z = 120^2/29 = 120 \times 120 \div 29 = $ **497 ohms**

Even though the DC resistance of the VCR power transformer primary winding is 9 ohms, when the VCR is operating on 60-Hz AC power, the impedance is closer to 500 ohms. Quite a difference!

In AC circuits, values of impedance (Z) can be calculated in series or parallel just like resistance (R) in DC circuits.

What is the nominal impedance presented to the output of an audio amplifier that has five 8-ohm speakers connected to it in parallel? Answer:

$R_T = R/n$    Substituting Z for R:
$Z_T = Z/n$    Substituting values:
$Z_T = 8 \div 5 = $ **1.6Ω**

This is much too low a load impedance for many amplifiers to drive. Most are designed to drive a speaker load no lower than 4 ohms. The difference between the amplifier source impedance and combined speaker load impedance is called an *impedance mismatch*. Damage to the output transistors could easily result from the increased current demand. That's why it's not a good idea to keep hooking up speakers in parallel across your stereo outputs.

In a similar manner, an impedance mismatch occurs if two TV sets are directly connected to the output of a VCR without using a proper RF splitter. The 75Ω impedance VHF output source transfers maximum signal to the load when the load impedance is also 75Ω. Directly connecting two 75Ω loads in parallel to the VCR output would present a load impedance of 37.5 ohms to the 75Ω source:

$Z_T = Z/n = 75 \div 2 = $ **37.5 ohms**

While not damaging to the VCR, this impedance mismatch would reduce signal strength at both

TVs as well as causing RF reflections, or echoes, on the interconnecting cables. This causes "ghosts."

As much as possible, especially for RF signals, source and load impedances should be matched. Interconnecting cable should also have the same impedance as source and load for maximum signal transfer from source to load.

# CHAPTER 10 APPENDIX

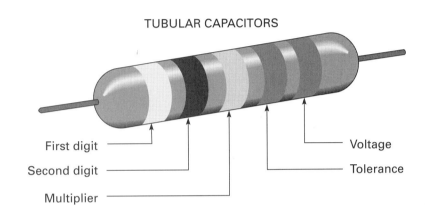

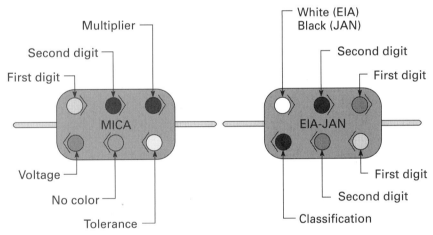

**FIGURE A10.1
Identification of mica
and tubular capacitors**

| Color | Number | Multiplier | Tolerance (%) | Voltage |
|---|---|---|---|---|
| No color | | | 20 | 500 |
| Black | 0 | 1 | | |
| Brown | 1 | 10 | 1 | 100 |
| Red | 2 | 100 | 2 | 200 |
| Orange | 3 | 1000 | 3 | 300 |
| Yellow | 4 | 10,000 | 4 | 400 |
| Green | 5 | 100,000 | 5 (EIA) | 500 |
| Blue | 6 | 1,000,000 | 6 | 600 |
| Violet | 7 | 10,000,000 | 7 | 700 |
| Gray | 8 | 100,000,000 | 8 | 800 |
| White | 9 | 1,000,000,000 | 9 | 900 |
| Gold | | 0.1 | 5 (JAN) | 1000 |
| Silver | | 0.01 | 10 | 2000 |

Figures A10.1 through A10.4 are charts to determine the values of different types of color-coded capacitors.

Mica caps like those pictured in Figure A10.1 are hardly ever seen any more; they had their heyday in the days of tube-type equipment, prior to solid state electronics. Even tube equipment manufactured after about the early 1950s rarely used this style of mica cap. EIA stands for Electronic Industry Association; JAN stands for Joint Army Navy.

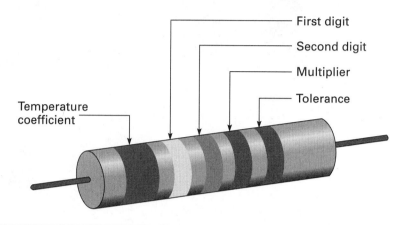

**FIGURE A10.2** Color codes for ceramic capacitors

| Color | Number | Multiplier | Tolerance Over 10 pf | Tolerance 10 pf or less | Temp. Coeff. |
|---|---|---|---|---|---|
| Black | 0 | 1 | 20% | 2.0 pf | 0 |
| Brown | 1 | 10 | 1% |  | N30 |
| Red | 2 | 100 | 2% |  | N80 |
| Orange | 3 | 1000 |  |  | N150 |
| Yellow | 4 |  |  |  | N220 |
| Green | 5 |  |  |  | N330 |
| Blue | 6 |  | 5% | 0.5 pf | N470 |
| Violet | 7 |  |  |  | N750 |
| Gray | 8 | 0.01 |  | 0.25 pf | P30 |
| White | 9 | 0.1 | 10% | 1.0 pf | P500 |

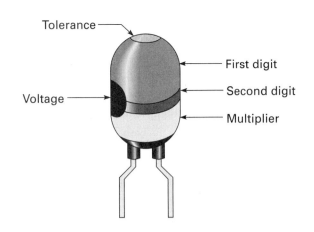

**FIGURE A10.3 Dipped tantalum capacitors**

| Color  | Number | Multiplier | Tolerance (%) | Voltage |
|--------|--------|------------|---------------|---------|
|        |        |            | No dot 20     |         |
| Black  | 0      |            |               | 4       |
| Brown  | 1      |            |               | 6       |
| Red    | 2      |            |               | 10      |
| Orange | 3      |            |               | 15      |
| Yellow | 4      | 10,000     |               | 20      |
| Green  | 5      | 100,000    |               | 25      |
| Blue   | 6      | 1,000,000  |               | 35      |
| Violet | 7      | 10,000,000 |               | 50      |
| Gray   | 8      |            |               |         |
| White  | 9      |            |               | 3       |
| Gold   |        |            | 5             |         |
| Silver |        |            | 10            |         |

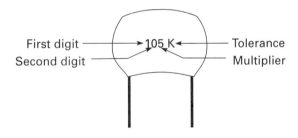

**FIGURE A10.4 Film type capacitors**

| Number | Multiplier | | Tolerance | |
|--------|-----------|---|---|---|
| | | | 10 pf or less | Over 10 pf |
| 0 | 1       | B | 0.1 pf |      |
| 1 | 10      | C | 0.25 pf |     |
| 2 | 100     | D | 0.5 pf |      |
| 3 | 1000    | F | 1.0 pf | 1%   |
| 4 | 10,000  | G | 2.0 pf | 2%   |
| 5 | 100,000 | H |        | 3%   |
| 6 |         | J |        | 5%   |
| 7 |         | K |        | 10%  |
| 8 | 0.01    | M |        | 20%  |
| 9 | 0.1     |   |        |      |

# APPENDIX M: METRIC SYSTEM AND CONVERSIONS

## METRIC SYSTEM AND CONVERSIONS

The following is excerpted with permission from **The Muscle Wires® Project Book**—*A Hands-on Guide to Amazing Robotic Muscles that Shorten When Electrically Powered.* Copyright 1993 by Mondo-Tronics, Inc., San Anselmo, CA 94960.

* * * *

The following chart describes the measurements of the metric system, first developed and implemented in France in 1795. The United States Congress approved the metric system for use in the United States in 1866.

The metric system consists of seven standard units of measure (see Figure AM.1). Other useful measures come from the basic seven, such as the *liter* (one thousand cubic centimeters), and the *newton* (the force needed to accelerate one kilogram one meter per second squared).

The tables in Figures AM.2 and AM.3 have some units of length, mass, weight and temperature. Note that *mass* is the *amount* of substance, and *weight* is the *force* exerted by a mass when pulled by a gravity field.

| The 7 Basic Metric Units | |
|---|---|
| *meter* | length |
| *kilogram* | mass |
| *second* | time |
| *ampere* | electric current |
| Kelvin | temp. above absolute zero |
| mole | amount of substance |
| *candela* | luminous intensity |

**FIGURE AM.1**

|  | Measure | Abbreviation | Amount |
|---|---|---|---|
| **LENGTH** | micrometer (or micron) | μm | 0.000,001 |
|  | millimeter | mm | 0.001 |
|  | centimeter | cm | 0.01 |
|  | decimeter | dm | 0.1 |
|  | meter | m | 1 |
|  | deckameter | dam | 10 |
|  | hectometer | hm | 100 |
|  | kilometer | km | 1,000 |
| **MASS** | milligram | mg | 0.001 |
|  | gram | g | 1 |
|  | kilogram | kg | 1,000 |
|  | metric ton | t | 1,000,000 |

**WEIGHT**  Multiply mass in kilograms by 9.8 m/s$^2$ (the acceleration of gravity at the Earth's surface) to obtain weight in newtons.

**TEMPERATURE**  When you know temperature in Celsius (°C): Add 273.16 to get Kelvin.

When you know temperature in Kelvin (K): Subtract 273.16 to get Celsius.

**FIGURE AM.2**

## Metric – U.S. / U.S. – Metric Conversions

| | If you know | Multiply by | To get |
|---|---|---|---|
| **LENGTH** | micrometer | 0.04 | mil (thousandths of an inch) |
| | millimeter | 0.04 | inches |
| | centimeters | 0.39 | inches |
| | meters | 3.28 | feet |
| | meters | 1.09 | yards |
| | kilometers | 0.62 | miles |
| | mil | 25.40 | micrometer |
| | inches | 25.40 | millimeters |
| | inches | 2.54 | centimeters |
| | feet | 30.48 | centimeters |
| | yards | 0.91 | meters |
| | miles | 1.61 | kilometers |
| **AREA** | square centimeters | 0.16 | square inches |
| | square meters | 10.8 | square feet |
| | square meters | 1.20 | square yards |
| | square kilometers | 0.39 | square miles |
| | square inches | 6.45 | square centimeters |
| | square feet | 0.09 | square meters |
| | square yards | 0.84 | square meters |
| | square miles | 2.60 | square kilometers |
| **MASS** | grams | 0.035 | ounces |
| | kilograms | 2.21 | pound mass |
| | metric tons (1000 kg) | 1.10 | short tons |
| | ounces | 28.35 | grams |
| | pound mass | 0.45 | kilograms |
| | short tons (2000 lbs) | 0.91 | metric tons |
| **FORCE & WEIGHT** | newton | 0.2248 | pound force |
| | pound force | 4.448 | newton |
| **PRESSURE** | megapascal | 145 | pounds per square inch |
| | pounds per square inch | 0.006895 | megapascal |
| **VOLUME** | milliliters | 0.03 | fluid ounces |
| | liters | 0.26 | gallons |
| | fluid ounces | 29.57 | milliliters |
| | gallons | 3.79 | liters |

**TEMPERATURE**

When you know temperature in Celsius (°C):
Multiply by 9, divide by 5, and add 32 to get Fahrenheit.

When you know temperature in Fahrenheit (°F):
Subtract 32, multiply by 5, and divide by 9 to get Celsius.

**FIGURE AM.3**

# GLOSSARY

Words, terms, and acronyms in this glossary are defined or described as they pertain to video cassette recorders and material within this textbook.

**'scope** abbreviated form of *oscilloscope* (see).

**λ** Greek letter lambda; stands for the word *wavelength* (see).

**μ** Greek letter mu; stands for *micro*, metric prefix meaning 1/1,000,000 or $10^{-6}$. Also used in abbreviations, such as μprocessor for microprocessor.

**μF** Microfarad; unit of capacitance equal to 1/1,000,000 of a farad.

**μH** Microhenry; unit of inductance equal to 1/1,000,000 of a Henry.

**Ω** Symbol for the word *ohm*.

**≈** Symbol used preceding a numeric quantity to mean "approximately" or "roughly." For example, most VHS VCRs require ≈30 to 40 watts of power at 120VAC when operating.

**60/40 solder** Solder composed of 60 percent tin and 40 percent lead.

**A** Ampere; unit of electrical current.

**AC** Alternating current; voltage that varies over time following the sine function. Power from a standard household wall outlet is AC.

**A/C** Audio/Control; refers to combined magnetic tape head assembly that records and plays back linear audio and the control track on videotape. This assembly also contains the linear audio erase head. Also called ACE head.

**ACE** Refers to combined magnetic tape head assembly that contains linear audio Rec/PB, control track Rec/PB, and linear audio erase heads. Alternate name for A/C head assembly.

**ACK** Automatic color killer; circuit in VCR and TV that shuts off processing of chroma signal, reverting to black-and-white, in absence of 3.58MHz color burst signal on back porch of horizontal blanking pulse. Weak color burst signal can activate ACK.

**ACTV** Advanced compatible television; enhanced TV and VCR video through the use of *digital signal processing* (DSP).

**A/D** Analog-to-digital converter; circuit that changes an analog signal into a binary data stream.

**AF** Audio frequency. Usually ≈50Hz to 8,000Hz for VCR linear audio, and 20Hz to 20kHz for high-fidelity audio.

**AFM** Audio frequency modulation; describes method of recording Hi-Fi audio on VHS videotape. AFM uses two rotary heads for recording and playing back left and right stereo channels.

**AGC** Automatic gain control; circuit that controls amplitude or intensity of a signal, such as audio or video.

**alignment** To bring two or more parts into proper arrangement or agreement with each other. For example, the A/C head is aligned to the linear audio and control tracks on videotape.

**allen wrench** A six-sided wrench, also called hexagonal or hex wrench.

**AM** Amplitude modulation. Applied signal varies the amplitude of the radio frequency carrier. Broadcast and cable TV *video* are amplitude modulated. (Audio is frequency modulated.)

**ammeter** Meter that measures the amount of electric current flowing in a circuit.

**ampere** Unit of electrical current.

**analog** A signal that is constantly changing in amplitude and frequency (for example the audio signal going to a loudspeaker). Also, a display or indirect readout consisting of a gauge or scale with a movable pointer, such as older style automobile speedometers.

**anode** The positive terminal of an electrical device, such as a diode. In vacuum tubes, *anode* is synonymous with the plate element.

**aperture grill** Same as *shadow mask* (see).

**A/V** Audio/video; describes equipment and baseband signals, such as at audio and video input and output RCA jacks on consumer VCRs and TV/monitors. Also, used for audio/visual.

**AWG** American wire gauge; the cross-sectional size of an electrical conductor. The larger the wire, the smaller the AWG. Common two-conductor 120V lamp cord is 18-gauge wire.

**axial** "On axis" or sharing the same axis. Axial-lead components, such as resistors and capacitors, have a wire lead at the center, or axis, at each end.

**azimuth** Describes the angle of a magnetic tape head gap with respect to the tape track it records or plays back.

**back circuit** An alternate path for current to flow. Especially important when performing in-circuit component tests. A back circuit could give a false indication of a component's condition. For example, a resistor in parallel with a capacitor would make it appear as if the cap had leakage, if the resistor were not accounted for (see lift).

**balanced** An electrical circuit in which both signal conductors are equally distant electrically from ground. Neither signal conductor is at ground potential. 300-ohm twinlead is a balanced signal line. Contrast with *unbalanced* (see).

**balun** Combination of words *balanced* and *unbalanced*. Describes a transformer that converts a balanced line to an unbalanced line or vice versa. A transformer that converts a 300-ohm balanced twin-lead line to a 75-ohm unbalanced coaxial line is a balun.

**bandwidth** The range of frequencies from the lowest to the highest for a particular signal. For example, VHF Channel 4, including frequencies from 66MHz to 72MHz, has a 6MHz bandwidth.

**base** Region of a transistor between the emitter and the collector. One of the three leads of a transistor. A signal applied to the base of a transistor controls the amount of emitter-collector current flow. The base is either N-type or P-type material. See *emitter; collector; NPN; PNP.*

**baseband** Refers to an unmodulated signal. A/V input and output jacks on VCRs and TV/monitors are for baseband signals.

**baseplate** Refers to the main chassis of a transport mechanism, such as a VCR, audio cassette tape, or compact disk transport.

**bass** Low audio frequencies, generally in the range of 20Hz to 500Hz.

**BER** Beyond economical repair. Repair costs that are considered too high, usually in reference to the age of the machine and what a new replacement unit would cost.

**bias** A DC voltage applied to an electronic device, such as a transistor, that establishes its operating range. A diode conducts when it is forward biased. Also refers to a high-frequency signal applied along with audio to a magnetic tape head during record, like the linear audio head in a VCR.

**bias oscillator** Electronic circuit that produces high-frequency signal, usually in the range of 60kHz to 120kHz, that is applied along with audio to a tape head during recording. Also supplies erase current for erase heads.

**Bit** Stands for *binary digit*, meaning a digit that can have two values, in this case 0 or 1. The state of a digital signal. A bit can be either Off (0) or On (1).

**black clip** Describes circuit that limits the amplitude of the luminance signal for black, or lowest intensity. This prevents possible smearing or blurring of the reproduced image caused by overshoot of the signal.

**blank** To turn off the electron beam in a CRT at the end of a trace while the beam is repositioned to scan another line. The beam is blanked during retrace by a blanking pulse.

**break or broken** Means that a relay or switch contact or connection is open and no current can flow.

**bridge** Four similar devices in a balanced circuit, such as four diodes comprising a full-wave bridge rectifier. Also means to place a component across a portion of an electrical circuit. A crack in a circuit board can be bridged with a piece of wire. A suspected faulty electrolytic filter capacitor is bridged temporarily with a known good cap to see if correct circuit operation returns.

**BW** or **B&W** Black-and-white image on a TV screen

**C** Capacitor. Also describes the *chrominance* or *chroma* signal in a TV transmission or composite video signal; C signal contains color information.

**C-clip** A small metal part, similar to a washer that is open on one side, used to secure a gear or pivot bracket to a shaft. Clip seats into a groove around the shaft. Resembles the letter C.

**cable** Two or more conductors within a common sheath or bundle; also a short form for cable TV.

**cable mouse** An IR source that connects to some VCRs, which allows the VCR to change channels on a cable TV converter by sending coded infrared signals to the cable box. Used to select cable TV channels for unattended recording.

**capacitor** Electrical device consisting of two metal plates separated by an insulating material called the *dielectric*. A capacitor blocks DC but allows AC current to flow.

**capstan** Rotating metal shaft in a tape transport that moves tape at a uniform speed along the tape path. A rubber pinch roller presses the magnetic tape against the rotating capstan.

**carrier** Radio frequency signal that is changed, or modulated, by some signal to be transmitted. For example, VHF Channel 2 has a 55.25MHz amplitude-modulated video carrier and a 59.75MHz frequency-modulated audio carrier.

**cassette** Plastic enclosure containing supply and take-up reels and magnetic tape, such as an audio cassette or videotape cassette.

**cassette loading** The process of placing a video cassette onto the reel table. Transport spindles are engaged with the two cassette reels and the cassette door is opened.

**cathode** The negative terminal of an electrical device, such as a diode. In vacuum tubes, the cathode is heated by a filament and emits electrons.

**CATV** Community antenna television; synonymous with cable TV. Refer to tables in the Chapter 1 Appendix.

**CCW** Counterclockwise; rotation that is in the opposite direction to the movement of hands on an analog clock face.

**cell** Any device constructed of two dissimilar metals in an acidic, alkaline, or salt solution and which produces a DC voltage. The word *battery* is often used for a cell, such as an AA-, C- or D-cell battery.

**chamois** Soft leather made from the skin of a goat, deer, or sheep. Chamois-tipped swabs are ideal for cleaning delicate rotary heads in a VCR.

**charge** The amount of electrons on some material. When a capacitor is placed across a battery, electrons are removed from one of the capacitor plates and electrons are added to the other capacitor plate. The capacitor is said to be charged. The plate with added electrons has a negative charge, while the other plate has a positive charge.

**chip** A slice of silicon on which are manufactured multiple transistors, diodes, capacitors, and resistors. Synonymous with integrated circuit (IC).

**choke** A coil or inductor, which inhibits or chokes the flow of alternating current. A choke is used as a filter component in some DC power supplies. A choke can prevent radio frequencies from entering or leaving a circuit.

**chroma** Abbreviated form of word *chrominance*.

**chrominance** The color portion of a video signal; C signal.

**closed** Refers to a switch or relay contact or a connection that allows current to flow. A SPST switch that is turned On is closed.

**clutch** A device for controlling the transfer of rotational energy from one device to another. Many VCRs have a slip clutch in the drive train between a motor and the take-up spindle.

**CMOS** Stands for complementary metallic oxide semiconductor; very low power consumption integrated circuits (ICs), often used to store channel presets and timer programming information in a VCR. CMOS devices are very susceptible to damage by electrostatic discharge (ESD), and therefore need to be handled with care, such as wearing a conductive wrist strap.

**coax** Abbreviated form of the word *coaxial*, meaning to share the same axis. A coax cable's inner conductor and outer shield have the same axis.

**cog belt** A flat belt with regularly spaced ridges or teeth that engage with similar ridges in a cog pulley. A mechanical system using cog belt(s) and pulley(s) keeps rotating devices in time or in synchronization with each other, as the cogs prevent slippage that would occur with other types of belts.

**cog pulley** A pulley that has regularly spaced ridges or teeth that engage a cog belt. See *cog belt*.

**coil** One or more turns of wire wound around some core material, including air, iron, or an insulating material like ceramic. A coil forms an electrical inductor. See *choke; inductor*.

**cold solder joint** A poor soldered connection, usually caused by insufficient heat or movement of the joint before solder has completely solidified. A cold solder joint has a dull gray, grainy appearance rather than the smooth, shiny look of a good solder joint.

**collector** A semiconductor region of a transistor. One of the three leads of a transistor. See *base; emitter; NPN; PNP*.

**color burst** Approximately 8 to 10 cycles of 3.58 MHz chrominance subcarrier transmitted during each horizontal blanking pulse to synchronize the oscillator in a TV or VCR with that at the TV station.

**color-under format** Method used to record chrominance signal in VCRs where the chroma signal is modulated at a lower frequency than the luminance signal. In VHS, the chroma carrier is 629kHZ, for Beta it is 688kHz. The 3.58MHz chroma signal is down-converted before being recorded, and so lies "under" the luminance signal in frequency.

**composite** Baseband video signal containing luminance (Y), chroma (C), and sync signals. Describes signal at Video Out and Video In jacks on A/V equipment.

**compression ring** A small metal part that clamps around a shaft to hold a gear or other part onto the shaft.

**condenser** Another, older word for *capacitor*, usually used only in the automotive industry. The capacitor across the breaker points in older type internal combustion engine ignition systems.

**conductor** A material that allows electrons to flow easily. Metals are good conductors. Common conductors are copper, aluminum, gold, and silver.

**continuity** An electrical connection having negligible resistance between two points. A closed SPST switch has continuity. Individual conductors in a cable can be checked from end-to-end for continuity with an ohmmeter.

**control track** A recorded track near the bottom edge of VHS videotape that has 30Hz pulses used to synchronize video tracks with rotating video heads during playback. A VCR automatically selects playback speed (SP/LP/EP) based on pulses recorded on the control track. See *A/C; CTL.*

**conventional current flow** Electric current flow that is thought to be from positive to negative. Used before theory of electron current flow from negative to positive was substantiated. Some people still use conventional current flow; on electronic schematics this is sometimes easier, because current then can be thought to flow *with,* or in the same direction as, the arrow head on devices such as diodes and transistors. See *electron current flow.*

**convergence** Adjustments made to direct the three color beams in a CRT so that they impinge on their respective colors within the same color triad at all places on the TV raster.

**copyguard** System to discourage dubbing of commercially recorded videotapes. One method omits or reverses some or all of the vertical sync pulses recorded on the tape. This may make the original tape itself difficult to play back, especially on older TVs. Vertical rolling and instability is an indication of this scheme.

**CPS** Cycles per second; older term, replaced by hertz, for unit of frequency.

**CPU** Central processing unit. Used in some service literature to describe the system control microprocessor.

**CRT** Cathode ray tube; synonymous with *TV tube* or *picture tube.* Computer terminals and monitors or displays also employ CRTs.

**crystal** A thin slice of quartz that vibrates at a precise frequency when excited by an alternating electric current. Used as the basis for very stable oscillators, such as the 3.58MHz color oscillator in TVs and VCRs. Also used in some RF tuners. A quartz crystal can also form the timing reference for electronic clocks, watches, and timers. Abbreviated Xtal. or Xtl.

**CTL** control track logic signal, consisting of 30Hz pulses recorded on the control track of VHS videotape by the control head gap in the A/C head assembly. CTL signal is derived from vertical sync pulses in the video signal. The VCR uses CTL pulses in playback to synchronize the rotary heads with tape movement so that video heads A and B each track their respective video tracks on tape. See *A/C; control track.*

**current** The amount of electrons that are flowing in a circuit. Symbolized by the letter "I" in formulas.

**CW** Clockwise; rotation in the same direction as the movement of hands on an analog clock face.

**cylinder** Another term for the video drum in a VCR. Also for the rotating and stationary portions of the video drum, as in the upper rotating cylinder of a VHS VCR.

**D/A** Digital-to-analog converter. Circuit that converts a binary data stream to an analog signal.

**DAC** Digital-to-analog converter. See D/A.

**dB** See *decibel.*

**DC** Direct current. A battery supplies DC. Electronic equipment works on one or more DC voltages produced by the power supply.

**DD** Direct drive; describes a device directly driven by a motor, such as a DD capstan, as opposed to a capstan that is driven via a belt and pulley from a separate motor. All VHS VCRs employ a DD video drum, in which the motor is an integral part of the video drum assembly.

**Decibel** A measurement of the intensity of acoustic sound and signal strength. A doubling of the voltage of a signal is an increase in signal strength of 6 decibels (dB).

**deflection** Describes movement of an electron beam within a CRT by electrostatic plates or magnetic coils in the deflection yoke, or both.

**demodulate** To recover signal information from a modulated carrier. A TV set demodulates the amplitude-modulated video RF signal to produce a baseband video signal. The frequency-modulated audio signal is demodulated to recover baseband audio.

**desoldering braid** Braided copper wire used to draw molten solder away from a solder joint. Also known as *solder wick.*

**dew sensor** A two-lead device in a VCR, the resistance of which varies in direct proportion to changes in humidity. At some value of resistance and higher values, indicating moisture within the VCR cabinet, the system control microcomputer prevents the VCR from loading tape. Loading tape could result in tape sticking to a rotary cylinder that has condensation on its surface. Most VCRs with a dew sensor will not power On if dew is sensed. The word **DEW** appears on the display panel of many VCRs having a dew sensor when moisture is sensed. On some, a Power-On LED may blink, and the VCR won't turn On.

**dielectric** The insulating material between the two plates of a capacitor. Air, vacuum, ceramic, paper, and tantalum are common dielectrics, although vacuum capacitors are not found in consumer electronics.

**digital** Refers to a signal composed of binary digits. Digital signals have only two logic levels, 0 or Low, and 1 or High. Also, equipment that displays a direct readout in numbers, such as a digital clock, rather than with a gauge or scale and a movable pointer.

**DIN** Type of connector. DIN stands for *Deutsche Industrie Normale,* a standards bureau in Germany. DIN-type connectors are round, with several pins or contacts arranged in a semicircle around the inside of the connector shell. The S-Video connector on Super-VHS VCRs is a four-pin miniature DIN connector. Most personal computer keyboards attach to the system unit with one form or another of DIN connector.

**diode** An electronic component that allows current to flow through it in only one direction, from the cathode to the anode. Used to change AC into DC, demodulate an RF signal, and to prevent back circuits in DC voltage circuits.

**DIP** Dual inline package; an IC or similar device having two parallel rows of pins, one on each long side of the component.

**discharge** Electrons leaving the negative plate of a capacitor. Electron flow from a capacitor or battery.

**DMM** Digital multimeter. See *VOM.*

**DOC** Dropout compensator; VCR electronic circuit containing a delay line that replays a previous line of video information during short-term dropouts of the video signal read from videotape.

**Dolby NR** A system for reducing high-frequency noise, especially tape hiss, from linear audio recordings. NR = noise reduction.

**DPDT** Double pole, double throw switch or relay.

**DPST** Double pole, single throw switch or relay.

**driver** Circuit consisting of one or more transistors or an IC and related components to energize a motor, LED, solenoid, relay, or other output device.

**drop** Term used to describe de-energizing a relay or solenoid. Removing the voltage to a relay coil causes the relay to drop. See pick. Can also refer to the voltage across a component, such as the voltage drop across a resistor.

**dropout** Momentary loss of audio or video signal played back from tape, caused by defect in the tape or tape movement through the tape path. For example, a loss of tape oxide at one point in the tape, or tape becoming slack so it doesn't intimately contact magnetic tape heads, can cause dropout.

**DSP** Digital signal processor. Usually some form of microprocessor that alters or changes a digital signal. Picture-in-picture screen image is produced by a digital processor in some VCRs and TVs.

**dummy cassette** A videocassette without any tape, tape reels, or other internal parts, which also has the two windows removed, and end-of-tape light exit openings blocked. Useful for diagnosing some tape transport problems without actually loading a real cassette. See *test jig.*

**DVM** Digital volt meter. See *VOM.*

**E** Symbol for voltage in electrical formulas.

**E-clip** A small metal part, similar to a washer that is open on one side, used to secure a gear or pivot bracket to a shaft. Clip seats into a groove around the shaft. Sort of resembles the letter E.

**E-E** See *E-to-E.*

**E-to-E** Electronic-to-electronic; Ch 3/4 RF output and baseband A/V video output of VCR is signal being tuned by VCR tuner or from line input jacks. TV/VCR button set to VCR. Use this position to monitor channel being recorded by VCR. In TV position, VHF antenna input to VCR goes on to TV.

**electrolytic capacitor** Polarity-sensitive capacitor used for power supply filtering, interstage signal coupling of audio frequencies in transistor equipment, and as supply voltage decoupling capacitors at various points throughout a unit. Values can be from a few microfarads to many thousand microfarads. Connecting an electrolytic capacitor backwards will destroy it. Cap must have working voltage at least equal to circuit voltage.

**electron** One of the three parts of an atom; has a negative electrical charge. (The proton has a positive charge and the neutron has a neutral charge.) Electrons move from atom to atom to produce the flow of an electric current.

**electron current flow** Electrical current flow from negative to positive. Current flows *against* the arrow head of schematic symbols, such as an LED or transistor. More commonly used than conventional current flow, and used exclusively in this textbook. See *conventional current flow.*

**EMF** Electromotive force, the pressure or force behind electrons that push them along a conductor; synonymous with *voltage.* Also, *electromagnetic field,* such as that developed by an electrical coil.

**EMI** Electro-magnetic interference. Electrical noise, such as that caused by lightning.

**emitter** A semiconductor region of a transistor. One of the three leads of a transistor. See *base; collector; NPN; PNP.*

**encoder disk** A round, opaque disk with holes or slits in it through which light can shine. Many VCRs have an encoder disk driven by the mode motor. Two IR LEDs on one side of the encoder disk shine through openings in the disk upon two IR phototransistors mounted on the other side. The outputs of the phototransistors are sent to the system control microprocessor, which determines the position of mechanisms driven by the mode motor.

**equalization** Changing the characteristics of a signal, usually by altering the intensity or strength of different frequencies within the signal's bandwidth.

**ESD** Electrostatic discharge. Sudden discharge of a static electricity charge, usually from a person, which can be damaging to delicate electronic components like integrated circuits. An ESD wrist strap should be worn when working on electronic equipment to prevent ESD damage.

**F connector** 75-ohm coaxial RF connector used extensively in video systems, antenna systems, cable TV, VCRs, and related video equipment.

**feedback** Output from a system that goes back to the input. Feedback can be detrimental or helpful. Positive feedback from loudspeaker to microphone in a public address (PA) system causes a squeal or howling sound. Negative feedback in an amplifier is used to reduce distortion products of the amplifier. A servo system uses feedback: pulses that tell how fast the video drum in a VCR is rotating are fed back to circuitry that regulates the drum speed.

**ferrite** Magnetic core material used in high-frequency transformers, coils, and chokes. A compound of ferric oxide and metallic oxide with high magnetic permeability (magnetic lines of force easily pass through the material). Ferrite cores are usually the color of pencil lead and are quite brittle. Balun transformers are made by winding two coils upon a ferrite donut.

**FET** Field effect transistor. An FET has a much higher input impedance than a bipolar transistor. Used in very low-level circuits, such as AF and RF preamplifiers and RF tuners.

**FG** Frequency generator; signal produced by video drum and capstan to tell the servo circuitry how fast each is spinning. Usually produced with a stationary Hall device or magnetic coil and a spinning permanent magnet.

**field** In NTSC television, 262-1/2 horizontal lines of video information. Two such fields make up a video frame. Each field takes 1/60 of a second to "paint" the face of a CRT. Also an energy domain, such as an electrostatic field or magnetic field.

**filter** An electronic device or circuit that reduces or eliminates certain frequencies while allowing others to pass through. A filter capacitor reduces DC power supply ripple. A comb filter in a VCR or TV separates the luminance (Y) and chrominance (C) signals.

**fish paper** An insulating material, similar to stiff shirt cardboard, usually dark gray in color. May be used to insulate the wiring side of part of a printed circuit board, so that circuits don't short out to nearby metal pieces. May be used around high-voltage areas to help prevent accidental electrical shock when working on equipment.

**flange** A rim or collar on a wheel; brake pads press against the lower flanges of supply and take-up spindles in a VCR when the machine is placed in Stop mode.

**FM** Frequency modulation. Applied signal varies the frequency of the radio frequency carrier. Broadcast and cable TV audio are frequency-modulated.

**forward bias** Voltage applied to a PN junction so that the device conducts. A diode that has a negative voltage applied to its cathode, with respect to the voltage applied to its anode, is forward biased and will conduct electrons.

**frame** In NTSC video, a complete TV image composed of 525 lines, made up of video field 1 and video field 2. There are 30 frames produced each second.

**frequency** The rate of repetition of a signal, usually expressed in hertz; the number of complete AC sine wave cycles in one second.

**front loader** VCR that has the cassette opening in the front panel. A cassette is gently pushed partway into the compartment opening. The cassette load motor then energizes and the cassette is drawn in and down, in an inverted L-shaped path, to be loaded onto the reel table. Also, the mechanism itself that loads a cassette onto the reel table in a front loading VCR.

**full scale** Refers to the maximum reading on any scale of a meter. For example, a VOM function switch is set to its 50V position: the full scale, or maximum voltage that can be read at this setting, is 50 volts.

**full-wave rectification** Converting both half cycles of an AC signal to DC. A bridge rectifier consisting of four diodes forms a full-wave rectifier. A full-wave rectifier requires two diodes when a transformer secondary has a center tap.

**function** Refers to the mode or way in which a multimeter operates. It can function or operate as a voltmeter, ohmmeter, or milliammeter. The function switch selects how the meter works.

**ghost** One or more secondary video images on a TV screen, displaced horizontally from the main image. Caused by the broadcast RF TV signal being reflected off buildings or other structures, which creates an echo signal at the receiving antenna. Most echoes arrive after the main received signal, placing the ghost image slightly to the right of the primary image, but an RF reflection *can* also arrive at the TV antenna prior to the main signal.

**guide pole** See *guide post*.

**guide post** A tape transport component that positions tape at the correct height within the tape path. A guide post can have a rotating sleeve between an upper and lower shoulder, or it can be stationary. Roller guide posts P-2 and P-3 position tape at the correct height around 180 degrees of the video drum.

**H** Symbol and abbreviation for *henry* (see).

**half-wave rectification** Converting only one half cycle of an AC signal to DC. The other half cycle is not used. A single diode forms a half-wave rectifier.

**Hall device** A semiconductor device that responds to an external magnetic field. A Hall-effect device functions like a switch, turning the flow of electrons On and Off in the presence of a moving permanent magnet. Many DD drum and capstan motors have Hall devices for producing FG and PG signals. A permanent magnet that rotates with the motor passes over the Hall device, which produces output pulses.

**hand cycle** To manually operate a mechanism with power turned Off. For example, a VCR tape load mechanism can be hand cycled by manually turning the cam motor shaft with fingertips or the tip of a screwdriver.

**heatsink** A metal device that removes excess heat from a component, such as a power transistor, driver IC, or voltage regulator. Most VCRs have one or more heatsinks, usually in the power supply area, composed of a piece of aluminum with fins. Various metal parts of the transport may also be used as a heatsink. Heatsink also describes a small spring-loaded clamp that can be placed on a component lead, such as a diode, to remove excessive heat when the part is being soldered into a circuit. This protects the part from possible damage. The heatsink should be clamped to the wire lead between the body of the component and the solder joint.

**helical scan** Method of recording video on tape in diagonal tracks, caused by the 5° tilt of the video drum in a VHS VCR relative to the movement of tape. Tracks are laid down in a configuration similar to a helix.

**henry** Unit of electrical inductance. Coils are rated in microhenries (μH) or millihenries (mH).

**hertz** Basic international unit of frequency, equivalent to transitions in each second; usually synonymous with *cycles-per-second* (CPS).

**high** Describes the On state of a digital logic circuit or signal. For example, a TTL level of +4.8V is logical High. Equivalent to logic 1 state.

**hold coil** A low-current coil that keeps a relay or solenoid energized after it has been energized or picked by a high-current pick coil.

**hot chassis** Describes a piece of equipment where the AC line is connected to all or part of the chassis or circuit ground, rather than being connected directly to the primary of a step-down transformer. A VCR with a switching mode power supply (SMPS) is an example of a hot chassis unit. A 120V isolation transformer should be used when working on this type of equipment to prevent electrical shock and damage to test equipment.

**HQ** High Quality; enhances VHS format by improving luminance and chroma noise reduction circuitry, and by increasing the white clip level by 20 percent.

**I** Symbol for current in amperes in electrical formulas.

**IC** Integrated circuit; a slice of silicon on which are manufactured multiple transistors, diodes, capacitors and resistors. Synonymous with *chip*.

**icon** A stylized line drawing representing a function, operation, or control. For example, a display of an analog clock face is an icon indicating that the VCR is programmed and set for unattended recording.

**idler gear** A gear placed between two others that transfers rotary motion from one to the other at the same speed and in the same direction.

**idler wheel** A wheel having a rubber tire that is placed between a driving wheel and driven wheel that transfers rotary motion in the same direction and at the same speed from one to the other. For example, in many VCRs an idler wheel transfers motion from the output of a slip clutch to the take-up spindle. Idler can also describe a spring-loaded pulley or wheel which presses against the side of a belt to give it tension.

**IDTV** Improved-definition TV; TV and VCR video enhanced through the use of *digital signal processing* (DSP).

**impedance** Resistance to the flow of an electric current, rated in ohms. Used for "resistance" of components in AC circuits. Inputs and outputs are generally classified as low impedance (Lo-Z) or high impedance (Hi-Z). For example, an 8Ω loudspeaker is a Lo-Z device. The audio Line-In and -Out jacks on a VCR, usually somewhere near 10KΩ, are considered Hi-Z.

**induced** An electrical current produced by a magnetic coil in the presence of a moving or changing magnetic field. For example, a rotating permanent magnet on a capstan flywheel induces a voltage in a magnetic coil which it passes to produce capstan FG pulses.

**inductor** An electrical coil consisting of one or more turns of wire wound around some core material, including air, iron, ceramic, wood, or any other substance. An inductor inhibits the flow of alternating current while freely allowing direct current flow. See *choke; coil.*

**inertia roller** A spring-loaded roller, similar to a small thread spool, which presses against videotape to smooth out minor irregularities in tape tension. Usually located after the full erase head, but can be before the head or even after drum exit guide P-3. It's possible for a VCR to have two inertia rollers, one on each side of the video drum. Also called *impedance roller.*

**insulator** Material through which an electric current cannot flow; rubber, plastic, ceramic, glass, and paper are examples.

**interlace** Describes the scanning of a complete video image on the face of a CRT, where odd-numbered horizontal lines are first scanned during video field 1 and then even-numbered lines are scanned during field 2 to produce a complete frame. See *field, frame.*

**inverse** Relationship when one quantity goes up the other goes down. For example, in Ohm's Law formula I = E/R, for a constant voltage (E), current (I) and resistance (R) have an inverse relationship. If the resistance goes *up,* the current goes *down* with the applied voltage remaining the same. Conversely, if the resistance *decreases,* then current *increases* for the same value of applied voltage.

**IR** Infrared. Invisible rays just beyond the visible red portion of the spectrum, having longer wavelengths. Most VCR optical or photo sensors and remote control systems use infrared. An IR LED is usually the source, with an IR phototransistor or IR photodiode as the sensing device. IR energy can be detected with an IR sensor card (described in Chapter 3) or two-way tester (described in Chapter 11).

**isolation transformer** A power transformer that produces a secondary output voltage the same as the primary input voltage, but where there is no electrical continuity between the primary and secondary. A 120V isolation transformer should be used when working on equipment with a hot chassis, such as a VCR with a switching mode power supply. See *hot chassis.*

**jack** Another name for a connector, usually a female chassis mount, like an RCA jack or 1/8-inch headphone jack.

**K** Stands for the metric prefix kilo, meaning times 1,000. For example, a 47KΩ resistor is 47,000 ohms. Also used as symbol for a relay on a schematic diagram or circuit board, such as relay K-4.

**kHz** Times 1000 Hertz. For example 629kHz is 629,000 hertz.

**kickback diode** Diode placed across a DC-operated coil, such as a relay coil or solenoid, so that it is reverse biased when the coil is energized. When current through the coil drops, the back EMF from the coil is swamped out by the diode, which now conducts in its forward biased condition. This protects coil driver components from damaging high voltage spikes and reduces electrical noise resulting from back EMF when a magnetic field collapses.

**kilohm** Times 1,000 ohms. See *K.*

**L** Symbol for a coil, choke, or inductor, such as L101. Also, symbol for amount of inductance in henries in electrical formulas.

**LC** Combination of a coil (L) and capacitor (C), either in series or parallel. An LC combination is resonant at a given frequency or over a range of frequencies.

**LCD** Liquid crystal display. Display technology that uses very low current in comparison to LED and fluorescent displays. Does not itself emit light. A liquid crystal display is frequently backlit with an incandescent or fluorescent light source for viewing in the dark.

**leader** Nonmagnetic material at the beginning and end of tape in a video cassette. The leader is transparent in VHS cassettes, and approximately six inches long.

**LED** Light-emitting diode. A diode that produces light when current flows through it, when it is forward biased. Light can be visible, such as red, green, and amber, or infrared.

**lift** To disconnect one or more component leads, especially when performing in-circuit tests of an individual component. For example, it may be necessary to lift one lead of a diode when testing for reverse leakage when a diode is wired into a circuit. Lifting the component lead isolates it from any possible back circuits.

**linear audio** Audio track(s) along the top millimeter of videotape. Linear audio is recorded and played back by the stationary audio/control (A/C) head assembly.

**logic** Refers to digital circuits that have two states: 0 and 1. These two states are also synonymous with Low and High, Off and On, False and True, No and Yes, Down and Up. Logic describes the interconnection of digital circuits.

**logic probe** A tool for determining the logic level of TTL and CMOS circuits. LEDs on the probe indicate whether a circuit is logical Hi or Lo. Most probes can also latch their LEDs On to indicate that a brief Up or Down level pulse has occurred. This feature is handy for babysitting a circuit.

**logic pulser** A tool that can inject TTL or CMOS technology level pulses for testing circuit operation. A logic pulser produces very short logic High pulses at the correct voltage levels; typical pulse width is 10μsec (microseconds). With some logic pulsers, the user can select from two or more pulse repetition rates, for example, 1 pulse per second or 400 pulses per second.

**low** Describes the Off or 0 state of a digital logic circuit or signal. For example, a TTL level of +0.4V is logical Low.

**luminance** The part of the overall video signal that contains black-and-white or brightness information, along with synchronization pulses, such as horizontal sync, vertical sync, (and color burst for color programs). Also called the *Y signal* and *monochrome signal*.

**m** Stands for metric prefix milli, meaning 1/1000 or $10^{-3}$. For example, a current of 500mA is 500 milliampere, which is equal to 500/1,000 of an amp or 1/2 amp.

**M** Can stand for either of two metric prefixes:
— Mega, meaning times one million (times $10^6$). For example, a 2.2MΩ resistor is a 2.2-megohm resistor, or 2,200,000 ohms.
— Micro, meaning 1/1,000,000 (times $10^{-6}$). For example, a 10MF electrolytic capacitor is a 10-microfarad cap.

**mA** Milliamp; 1/1000$^{th}$ of an ampere. See *m*.

**Macrovision** A videotape copy protection scheme. Macrovision disrupts the automatic gain control (AGC) circuit in a VCR, causing brightness changes in the picture and occasional bright flashes on the screen when playing back a copy.

**make or made** Means that a relay or switch contact or connection is closed and current can flow.

**megohm** One million ohms.

**MF** Microfarad; unit of capacitance equal to one millionth of a farad.

**MFD** Same as *MF*; microfarad.

**mH** Millihenry; unit of inductance equal to one thousandth of a henry.

**MHz** Megahertz; equal to one million hertz.

**microfarad** Unit of capacitance equal to one millionth of a farad.

**milliamp** Unit of electrical current equal to one thousandth of an ampere. For example, 250 milliamps equals 1/4 ampere.

**MMF** Micro-microfarad; measure of capacitance equal to one millionth-of-a-millionth of a farad, or one trillionth, or times $10^{-12}$. Same as *picofarad*.

**MMFD** Same as *MMF*; micro-microfarad or picofarad.

**modulate** To change the characteristics of an RF carrier, usually by altering its strength or amplitude (AM) or by altering its frequency (FM) for transmission of information, such as audio or video. Pulse width modulation (PWM) is another form.

**MTS** Multi-channel Television Sound; synonymous with stereo TV sound, but can also include *SAP* (see).

**multimeter** Multiple-function electronic tester, usually capable of measuring resistance, AC and DC voltage and DC current. Synonymous with *VOM*.

**NC** No connection. Schematic notation, such as adjacent to a switch contact or connector pin that is not used in the particular circuit.

**N.C.** Normally closed. Refers to a switch, relay, or pushbutton contact. For example, a relay contact that is made when the relay coil is de-energized is a N.C. contact. Energizing the relay causes the contact to open.

**negative** Refers to the source of electrons in a DC circuit or a component lead that is normally connected to the minus side of a DC circuit. For example, a DC power supply has negative (−) and positive (+) output terminals. An electrolytic capacitor has one negative lead and one positive lead.

**N.O.** Normally open. Refers to a switch, relay, or pushbutton contact. For example, a relay contact that is *not* made when the relay coil is de-energized is a N.O. contact. Energizing the relay causes the contact to close.

**noise** Broadly means any undesired component in a signal. Examples of audio noise include static, clicks, pops, hiss, hum, distortion; video noise includes ghosts, streaks, incorrect color, rainbowing and intensity flashes.

**NPN** Transistor type where the base is composed of P-type material and the emitter and collector are composed of N-type material. Emitter arrow head in schematic symbol points outward. Normal circuit operation is for the collector to be positive with respect to the emitter. A positive signal at the base increases emitter-collector current flow. (An NPN transistor may be thought of as being *similar* to a triode tube, where the emitter acts as cathode, the base acts as control grid, and the collector acts as the plate.) See *PNP; transistor*.

**NR** Noise reduction. Any electronic circuitry, including filters, equalizers, and compensation networks that reduces overall noise components in a signal. See Dolby NR.

**NTSC** National Television System Committee; organization that specifies standards for color TV broadcasts in the United States. Often pronounced "NIT-see."

**ohm** Unit of electrical resistance.

**ohms-per-volt** Measure of a voltmeter's input sensitivity, which describes the loading effect on a circuit when the meter makes a measurement. The higher the $\Omega/V$ rating, the less the meter will load down the circuit under test. A VOM with 20,000 ohms-per-volt DC sensitivity means that, for example, on its 10V DC scale, it will place the equivalent of a 200,000-ohm resistor (20,000 $\times$ 10) across the circuit. Digital multimeters (DMMs) typically have a DC input sensitivity of 10 megohms or greater, while 20K or 30K is more the norm for analog VOMs.

**open** Refers to a switch or relay contact or a connection that does not allow current to flow. A SPST switch that is turned Off is open. An open circuit is one in which the current path has been disrupted and no current flows.

**optical** Refers to components that work with light. Synonymous with *photo*. For example, VCR end sensors are optical detectors, usually consisting of IR phototransistors that detect light from an IR LED when transparent leader is present, or when no cassette is loaded.

**oscilloscope** Electronic test equipment that displays an electrical signal or waveform on the face of a CRT. An oscilloscope displays amplitude on the Y-axis (vertical) in reference to time along the X-axis (horizontal). A dual-trace oscilloscope can display two separate signals simultaneously for comparison purposes. Often abbreviated as *'scope*.

**OSD** On-screen display; selection menu or prompts produced by digital character generator in a video device. The IC chip that produces on-screen characters.

**p-p** See *peak-to-peak*.

**PAL** Phase alternation line; video standard used in other parts of the world; incompatible with NTSC. Used primarily in Britain; provides 625 horizontal lines of picture resolution. NTSC is 525 lines.

**parallel** Describes a circuit in which two or more components are connected so that any voltage appearing across one of them also appears across the others. The two headlights of a car are connected in parallel.

**pawl** A mechanical arm that rides along the teeth of a ratchet wheel that allows the wheel to turn in one direction only. Two pawls inside a video cassette keep the supply and take-up reels from turning in a direction that would spill tape whenever the cassette is *not* loaded onto the VCR reel table.

**PB** Play back.

**PCB** Printed circuit board.

**peak** Maximum amplitude of a signal or waveform.

**peak-to-peak** The voltage between the most negative and the most positive peaks of an analog signal or waveform.

**PEL** Picture element; the smallest part of a reproduced image. For example, a single dot triad on the face of a color CRT forms a PEL. Synonymous with *pixel*.

**pF** See *picofarad*.

**PG** Pulse generator signal; pulse developed by both the capstan and video drum to tell the servo circuitry rotational speed. Often developed with a small permanent magnet that rotates past a magnetic pickup coil or Hall-effect IC. See *FG*.

**photo** Refers to components that work with light. Synonymous with *optical*. For example, VCR end sensors are photodetectors, usually consisting of IR phototransistors that detect light from an IR LED when transparent leader is present, or when no cassette is loaded.

**pick** Term to describe energizing a relay or solenoid, which attracts a movable armature to the core. Applying voltage to a relay coil causes the relay to pick. See *drop*.

**picofarad** Measure of capacitance equal to one millionth-of-a-millionth of a farad, or one trillionth, or times $10^{-12}$. Same as micro-microfarad.

**pinch roller** Rubber roller that pinches tape against a rotating capstan shaft to pull tape through a magnetic tape path.

**PIP** Picture-in-picture. Capability of TV or other video device to display two pictures at once, where a second (or more) video program occupies a small rectangular area within the main image raster. Some TVs with PIP use the TV tuner for one picture and output from a VCR for the other picture. The TV's remote can select which of the two is the main picture, and which is the PIP insert.

**PIV** Peak inverse voltage; the maximum reverse-polarity voltage that an electronic device is designed to handle. For example, a 1N4002 rectifier diode has a PIV of 100 volts, and so should not be used in circuits approaching or exceeding this voltage across the diode when it is reverse biased. Synonymous with *PRV* (peak reverse voltage).

**pixel** Picture element; the smallest part of a reproduced image. For example, a single dot triad on the face of a color CRT forms a PEL. Synonymous with the acronym *pel*.

**PN junction** Area where P-type and N-type semiconductor material meet. A diode is composed of a

PN junction; P-type material is the anode and N-type material is the cathode. Electrons flow from N to P across the junction when the diode junction is forward biased, that is, when the cathode is negative with respect to the anode.

**PNP** Transistor type where the base is composed of N-type material and the emitter and collector are composed of P-type material. Emitter arrow head in schematic symbol points inward. Normal circuit operation is for the collector to be negative with respect to the emitter. A negative signal at the base increases emitter-collector current flow. See *NPN; transistor.*

**polarity** Describes the direction of current flow; describes relationship of positive and negative voltages in a circuit. DC voltage to the cassette load motor reverses polarity to cause it to spin in the opposite direction.

**positive** Refers to depletion of electrons in a DC circuit or a component lead that is normally connected to the plus side of a DC circuit. Electrons flow from negative to positive. For example, a DC power supply has positive (+) and negative (−) output terminals. An electrolytic capacitor has one positive lead and one negative lead.

**pot.** Abbreviation for *potentiometer.*

**potentiometer** A variable resistor having three terminals, one at each end of the resistance and a movable contact called a wiper that can tap any place along the resistor between the low and high ends. Manual VCR tracking controls are frequently rotary or slide potentiometers. Abbreviated pot.

**power** Electrical unit for the amount of work performed by an electric circuit or device, measured in watts and symbolized in formulas by P. In electronics, power is often associated with the amount of heat a device, such as a resistor, can dissipate. Power is equal to the current through a device multiplied by the voltage across the device: $P = IE$.

**PRV** Peak reverse voltage; same as PIV (see).

**PS** Power supply.

**PWB** Printed wiring board; same as *PCB.*

**Q** Frequently used as the symbol for a transistor in a schematic diagram, parts list, or marked on a circuit board; for example, Q305. (Q also stands for the quality of a coil, the amount of coil resistance in comparison with the inductive reactance of the coil. Q also stands for quadrature signal, which is a combination of red, blue, and luminance (Y) signals, used in color reproduction within a TV set.)

**R** Symbol for resistance in electrical formulas.

**rabbet** Small ledge around the lower, stationary portion of the video drum. The bottom edge of videotape should just rest on the rabbet.

**radial** Refers to a component whose leads extend from only one end, as opposed to axial-lead components, which have one lead coming out of each end. Radial leads are normally at the center and at the radius of a component, such as an electrolytic filter capacitor.

**range** Describes electrical values that can be measured with a given setting of a multimeter's function switch. For example, a 5-volt DC power supply output can be measured on the 10-volt or 20-volt range of a multimeter, but not on a 2-volt range. Equates with full scale reading of meter.

**raster** The area of a CRT that is scanned and lighted by the electron beam(s).

**raw** Describes unfiltered DC, which has large ripple content.

**RC** Resistor-capacitor combination. R and C can be in series or in parallel. Usually used to describe resistor and a capacitor that together form the basis for a timing or filter circuit.

**rcvr.** Abbreviation for *receiver.* May also be used for *recover,* as to regain a particular signal characteristic.

**RGB** Red, green, blue; also used to describe the chrominance signal, although the transmitted C-signal itself does not directly contain green information. Also refers to color monitors or other video and computer devices having separate signals for red, green, blue, and intensity.

**reactance** The opposition to the flow of an AC signal by an inductance (coil) or capacitance (capacitor); inductive reactance and capacitive reactance.

**Rec.** Abbreviation for *receive* or *receiver.*

**Rec/PB** Record/Playback. Frequently used to describe some component that is used for both modes, such as the A/C head.

**rectifier** Device for changing AC to DC. One diode forms a half-wave rectifier; four diodes make up a full-wave bridge rectifier. Sometimes synonymous with *diode.*

**rectify** To change an AC voltage into a DC voltage.

**reel** Spool upon which tape is wound. A video cassette has a supply reel and a take-up reel.

**reel table** Top surface of the tape transport, especially the portion with supply and take-up spindles that engage tape reels in a cassette.

**relay** Electromagnetic switch. An electrical coil attracts a movable armature when it is energized. The armature moves electrical contacts.

**resistance** Opposition to the flow of electrons in a DC circuit. Unit of resistance is the ohm.

**resistor** Device that opposes electrical current flow. Common resistors are carbon composition, carbon film, metallic film, metal oxide, and wire wound. Resistors are rated by their ohmic value, wattage, and tolerance.

**resonate** A circuit in which the current or voltage is at a maximum at a particular frequency, or over a range of frequencies. The natural frequency of an electrical circuit. An LC combination forms a resonant circuit. For L and C in parallel, maximum voltage will appear across the components at resonance. For L and C in series, maximum current will flow through them at resonance.

**retrace** Term used to describe electron beam movement or deflection in a CRT when the beam is turned Off. There are two different retraces:
— Horizontal: at the completion of a horizontal scan line, the beam is deflected from the right side back to the left side of the screen.
— Vertical: at the completion of a video field, the beam is deflected from the bottom to the top of the screen.

The electron beam is blanked or turned Off during retrace.

**reverse bias** Voltage applied to a PN junction so that the device does not conduct. A diode that has a positive voltage applied to its cathode, with respect to the voltage applied to its anode, is reverse biased and will not conduct electrons.

**RF** Radio frequency. Also refers to video tracks and signals.

**RFI** Radio frequency interference.

**rheostat** A variable resistor having two terminals. Similar to a potentiometer, but having a terminal at only one end of the resistance material and a terminal for the wiper. Often *incorrectly* used to describe a lamp dimmer.

**ripple** AC components in a DC power supply output; undesirable. Ripple is 120Hz for a full-wave rectifier and 60Hz for a half- wave rectifier, assuming a line frequency of 60Hz. It is the job of filter components, such as electrolytic filter capacitors, to remove ripple or reduce it to acceptable levels. Raw DC has a large amount of ripple. See raw.

**RMS** Root mean square. The *effective* power of an AC voltage. AC voltmeters read the RMS value. An AC source having a peak value of 100 volts has an RMS value of 0.707 times 100 or 70.7 volts. This AC voltage will produce the same amount of heat in a resistor as a DC source of 70.7 volts.

**roller guide** A tape guide with a rotating sleeve between upper and lower flanges, or collars, against which magnetic tape presses. See *guide post*.

**rosin** Flux material inside the core of electronic solder that helps prevent oxidation of metals during soldering.

**rotary heads** Video and VHS Hi-Fi audio heads that are in the spinning portion of a VCR video drum.

**rotary transformer** Transformer that transfers signals between a rotary head and the stationary portion of the video drum.

**rpm** Revolutions per minute.

**S** Often used in VCR service literature for "Supply"; for example, the "S end sensor" is the supply reel end-of-tape sensor.

**S-video connector** Miniature four-pin DIN connector that has separate connections for luminance (Y) and chroma (C) signals.

**S-VHS** Super VHS: enhanced VHS recording format, produces higher resolution video than standard VHS format. Videotape recorded in S-VHS format cannot be played back on a standard VHS VCR. A Super VHS VCR is able to record and play standard as well as Super VHS format tapes. S-VHS VCRs also have S-video connectors for even greater quality video when connected to a TV/monitor equipped for S-video input.

**safety tab** Small plastic tab on the rear surface of a cassette. When present, the tab allows recording; when broken out, the VCR will not go into Record mode, protecting material already on the tape. Also called *record safety tab*.

**Sams** Howard W. Sams & Company. Source of *VCRfacts*™, technical service data for many VCR brands.

**SAP** Second audio program or special audio program; an alternate sound signal, such as a different language dialogue, used in some stereo TV broadcasts.

**scanner** Another name for the video drum in a VCR.

**schematic** Representation of the interconnection of electronic components in a piece of equipment. Also called wiring diagram.

**SECAM** *Sequential couleur á memoire;* television standard used in some parts of the world, notably France; provides 819 horizontal lines in the picture. Incompatible with NTSC, which is 525 horizontal lines per video frame.

**semiconductor** Generally germanium or silicon that is the fundamental component in diodes, transistors, and integrated circuits. Generally synonymous with solid state.

**series** Describes two or more components that are connected end-to-end. Current is the same through each component connected in series.

**servo** A system that uses feedback for self-regulation. For example, the drum and capstan servo circuits in a VCR monitor the rotation of both the drum and capstan, and compare this with a reference signal and CTL pulses from tape. The servo then adjusts capstan and drum rotation so that video heads align properly with recorded tracks on tape. Abbreviated form for servomechanism.

**shadow mask** A component behind the face of a color CRT which helps aim the three electron beams at their respective color phosphors (red, green, blue) on the inside surface of the screen. The shadow mask is a screen with many holes or slits in it, through which the electron beams can pass. Synonymous with *aperture grill*.

**shield plate** A thin metal plate that acts as a shield to prevent outside extraneous signals from affecting a low-level signal. For example, many VCRs have a shield plate covering the video drum, which blocks stray RF energy from affecting the very small playback signals developed by the video heads.

**short** An undesirable low-resistance electrical path around a circuit. Same as short circuit. For example, a poorly soldered PCB connection may cause solder to flow between adjacent wire traces on the board, shorting out a circuit. A short circuit has negligible resistance, causing maximum current to flow.

**shunt** An electrical device that is placed in parallel with another component. For example, an ammeter has one or more low-resistance shunts in parallel with the meter movement. Most current flows through the shunt resistor, with a small amount going through the meter itself. Also describes a very simple type of voltage regulator, generally consisting of just a Zener diode and a resistor in series, placed across a DC voltage. The voltage across the Zener diode is regulated by the action of the diode.

**shuttle assembly** Movable bracket or mounting plate containing a roller guide and a slant pole that extracts tape from a video cassette and positions tape around the video drum. There are two shuttle assemblies; one is the drum entrance shuttle with guide P-2, the other is the exit shuttle with guide P-3.

**sine** The waveform of an AC voltage.

**sinusoidal** A voltage or waveform that is a sine wave or follows the trigonometric sine of the angle of the AC generator.

**skew** Any undesirable slant or twist to tape along the tape path, such as tape angling upward or downward instead of passing straight between capstan and pinch roller. Also, an error in tape tension, causing a change in the position, size, or shape of video tracks on tape between record and playback.

**Slant pole** Nonroller guide post that is at an angle from vertical to position tape around the tilted video drum in a VCR. Part of the drum entrance and exit shuttle assemblies.

**slant track** Refers to diagonal video tracks on tape. See *helical scan*.

**SMD** Surface mount device; describes integrated circuits, resistors, and other electronic components that are soldered to small pads on the surface of a PCB instead of having a pin-in-hole connection.

**SMPS** Switching mode power supply; type of supply where line voltage is directly rectified to DC, which then operates a high frequency power oscillator. The oscillator feeds the primary of a transformer. Secondary windings are used to develop several low-voltage DC outputs. Can be built smaller, lighter, and cheaper than a conventional PS, where line voltage goes to the primary of a power transformer first. Use an isolation transformer to power equipment with a SMPS during service activities to prevent accidental shock and damage to test equipment. Also called *switching power supply* and *transistor switching supply*.

**SMT** Surface mount technology; packaging system where components are soldered to the surface of printed circuit boards, rather than pin-in-hole. See *SMD*.

**soft brake** A brake that provides some back tension to a tape spindle during fast forward and rewind, rather than completely braking the spindle. Prevents tape from reeling out too fast and spilling.

**solder wick** Same as *desoldering braid* (see), used for removing solder. (*Soder Wick*, without the L, is a registered brand name for one company's desoldering braid product.)

**solenoid** An electromagnetically operated plunger. An electromagnet pulls an iron plunger into its core when energized. Plunger is also called *armature*. Used to release spindle brakes in some VCRs and to unlatch cassette basket in some top loaders.

**SPDT** Single pole, double throw switch or relay.

**spindle** Reel table component that engages with cassette supply and take-up reels.

**splines** Vertical ridges in the hubs of cassette reels that engage projections on transport spindles.

**SPST** Single pole, single throw switch or relay.

**stiction** Combination of the words *sticking* and *friction;* describes tendency for video tape to adhere to the spinning video drum. Moisture on the drum or a highly polished, worn drum that prevents a slight air bearing from being established between tape and drum increase the possibility of stiction.

**SW** Nomenclature for switch on a schematic and parts list; for example, SW-105. Also means switched, as used to denote a VCR power supply output that is switched On and Off with the front panel Power button.

**switched** Describes a power supply output that turns On and Off with the front panel Power control, in contrast to unswitched power supply outputs.

**sync tip** The most negative portion of the composite video waveform, which is the horizontal sync pulse.

**sync.** Abbreviation for *synchronization* (see).

**synchronization** Two (or more) systems that are in step with each other. When the horizontal and vertical scanning circuits in a TV set are in step with the scanning within the TV camera producing the picture, the TV and camera are said to be synchronized.

**syscon** Abbreviation for *system control microprocessor* within a VCR. May also be referred to as CPU, for central processing unit. Syscon is the microcontroller responsible for VCR functions and operations; often works with a timer/operations microcontroller.

**T** Symbol for transformer on a schematic; for example T101. Can also mean "time" on a timing chart: for example, T1 may indicate the first part of an operation, while T3 describes the third event to take place.

**tape guide** A pin, post, or pole that positions tape correctly in the tape path. Primarily establishes tape height. See *guide post*.

**tape loading** Process of extracting a loop of tape from a video cassette and threading it through the tape path, positioning it around the video drum. Accomplished by roller guides P-2 and P-3 and their associated slant poles on movable shuttle assemblies.

**Tentelometer®** Gauge for precisely measuring tape tension, manufactured by Tentel Corporation.

**Test jig** Usually refers to a testing tool which can be loaded onto the reel table to simulate a real videocassette. It contains no videotape, tape reels, or other internal parts of a regular cassette. End-of-tape light exits are blocked to simulate tape presence at supply and take-up reels. The top is open for access to reel table spindles. A videocassette test jig is an aid to troubleshooting transport problems, such as insufficient spindle torque.

Test jig also refers to a variety of special tools or gauges for setting height or position of tape path components, such as a manufacturer's jig for setting the height of supply guide P-1 and take-up guide P-4.

**THM** Through-hole mount; componets whose leads go through holes in a printed circuit board. Contrast with surface mount devices (SMD).

**timing chart** A printed chart included with some VCR service literature that shows timing relationships of mechanical components or electrical signals for various modes of operation.

**timing hole** A hole through a gear, gear segment, moving part, and machine chassis through which a timing pin is passed to ensure that parts are aligned, or timed, correctly with each other. Timing holes and a timing pin are used when reassembling parts in correct relationship with each other.

**timing mark** A mark such as a dot, dimple, or arrow head on a gear, gear segment, or other part used to align two or more related parts. For example, a mechanical VCR mode switch is adjusted so that the timing mark on the switch body aligns with the timing mark on the movable switch lever.

**timing pin** Thin rod, such as a finishing nail or similar tool, that is passed through aligned timing holes in gears, baseplate, etc. when reassembling moving mechanical components that have a timing relationship with each other.

**tinning** Preparing the tip of a soldering iron by applying solder to a new, hot tip and allowing it to cook for a while so that the molten tin/lead alloy fills the pores of the tip, coating it thoroughly.

**top loader** A VCR that has the cassette opening in the top cover. A cassette is inserted into a basket or tray that rises from the top cover. The basket is then pushed down by hand to load a cassette onto the reel table. Also refers to the mechanism itself that loads a cassette in a top loading VCR.

**torque** The amount of rotational or braking power of a reel spindle. For example, during play, the slip clutch in many VCRs controls the amount of torque applied by the take-up spindle. Torque can be measured with special torque gauges designed for VCR servicing.

**TR** Sometimes used in schematics and parts list to label transistors, such as TR-216. (Can also stand for transmitter.)

**trace** Scanning of the electron beam(s) in a CRT to produce horizontal lines of video on the screen. Contrast with retrace. Also can refer to a single wiring circuit on a printed circuit board: a wire trace. Also can refer to rotary heads correctly tracking their respective tracks on tape. Also means to follow from one point to another in a circuit, as to trace the Power-On signal from the syscon IC to the power supply on a schematic diagram. Also can mean a small amount, as a trace amount of ripple on a power supply output.

**tracking** Video heads A and B aligning with the diagonal video tracks on tape originally recorded by heads A and B. Maximum RF output from the heads is achieved when tracking is correct.

**transfer** Refers to switch and relay contacts. Flipping a DPDT toggle switch causes its contacts to transfer: contacts that were closed, open; and contacts that were open, close.

**transformer** Electrical device for changing the voltage or the impedance of an AC signal. Consists of at least two windings, a primary and a secondary, wound around a common core.

**transistor** A three-lead semiconductor device for controlling the flow of electrons. The three leads are the emitter, base, and collector. A small change in applied base voltage can effect a large change in current flow between the emitter and collector. Thus, a transistor is an amplifier. See *base; emitter; collector; NPN; PNP*.

**transport** The entire mechanical assembly for moving tape in a VCR, including tape guides, magnetic tape heads, motors, and related mechanisms.

**trim** See trimmer.

**trimmer** A variable resistor (potentiometer) or small-value variable capacitor used to make minor adjustments to an electronic circuit. Usually mounted on circuit boards with other components; not user controls. For example, a trim pot establishes the amount of bias current sent to the linear audio head(s) during recording.

**TTL** Transistor-transistor logic. Technology for digital circuits operating from +5VDC. Voltages near +5V are logical High; levels near 0V are logical Low.

**TU** Often used in VCR service literature for Take-up, for example, the "TU end sensor" is the take-up reel end-of-tape sensor.

**UHF** Ultra high frequency. For TV, frequencies between 470MHz (Channel 14) and 890MHz (Channel 83). Overall UHF spectrum is 300 MHz to 3,000MHz. Refer to tables in the Chapter 1 Appendix.

**UL** Underwriters Laboratory; tests and certifies equipment in United States as meeting certain safety specifications.

**unbalanced** Describes a circuit in which one of the two signal conductors is at ground potential. RCA and F connectors are unbalanced, as the shell of the connector is at ground. Coax cables are used for unbalanced signals. Contrast with balanced signal connections. See *balanced*.

**Undercarriage** The entire underside of the transport assembly, containing most of the tape load mechanism, and the majority of the pulleys, belts, and gears that operate the transport.

**UNSW** Abbreviation for *unswitched*.

**unswitched** Refers to power supply outputs that are active as long as the line cord of a VCR is plugged into a hot outlet. Contrast with switched outputs that turn On and Off with the front panel Power button.

**V** Symbol for voltage, as in +5VDC. Also used in place of E in electrical formulas. That is, V = IR is the same as E = IR.

**V-mounts** Precisely positioned, stationary metal stop blocks with V-shaped notches into which roller guides P-2 and P-3 seat at the completion of tape loading. The V-mounts determine how far the entrance guide P-2 and exit guide P-3 extend. Also called V-stops.

**V-stops** Same as *V-mounts*.

**V-V** Video-to-video; playback picture from a pre-recorded videotape.

**Varactor** A diode that changes the amount of capacitance between its P and N material with the amount of applied reverse bias voltage. Used extensively in electronic tuners. Essentially a voltage-controlled variable capacitor.

**Varicap** Same as *varactor*.

**VCR** Video cassette recorder.

**VCR Plus+** System used to program a VCR for unattended recording, where the user enters a four- to eight-digit code with the remote control unit. The VCR Plus+ code, which appears along with TV program listings in the local newspaper and local editions of *TV Guide,* tells the VCR's timer/operations microcomputer when to turn On, what channel to tune to, and the period of time to remain in Record mode.

**VCRfacts** Technical service data, available for many VCR models from Howard W. Sams & Company.

**VCP** Video cassette player; similar to VCR but without tuner and record capability.

**VHF** Very high frequency. For TV, frequencies between 54MHz (Channel 2) and 216MHz (Channel 13). Overall VHF spectrum is 30MHz to 300 MHz. Refer to tables in the Chapter 1 Appendix.

**VHS** Video home system. Describes popular VCR helical scan format using 1/2-inch-wide magnetic tape housed in a plastic cassette.

**videocassette** Plastic housing containing magnetic videotape wound on supply and take-up reels.

**videocassette test jig** See *test jig*.

**video drum** Part of VCR transport that contains rotating video heads. On a VHS machine, the drum spins at 1,800 rpm and is tilted slightly more than 5° with respect to the videotape path to achieve helical scan recording. On a VHS VCR, the upper portion of the drum containing magnetic tape heads rotates while the bottom part remains stationary. The video drum contains rotary transformers, which transfer signals between the rotating heads and stationary portion.

**VLSI** Very large scale integration; a complex integrated circuit chip containing thousands, or even *millions*, of transistors. A VLSI chip can have over 100 leads to connect it into a circuit.

**voltage regulator** An electronic circuit that controls the output of a power supply so that the voltage remains relatively constant with changing loads and input voltages. Most VCRs use a single integrated circuit, usually with three leads ($V_{IN}$, $V_{OUT}$, ground), that looks similar to a medium-power transistor, for each regulated output. The TO-220 and TO-126 type cases are common packages for voltage regulators. Voltage regulator ICs are normally mounted to a heatsink.

**voltage** The unit of electrical pressure, force, or potential. Synonymous with *electromotive force* (EMF). Symbolized in formulas by both V and E. Unit of measurement is the volt.

**VOM** Volt-ohm-milliammeter. A piece of test equipment that can measure AC and DC voltage, resistance, and DC current. Used fairly synonymously with multimeter and multi-tester. Generally refers to an analog meter, whereas *DMM* (digital multimeter) or *DVM* (digital volt meter) usually describe a digital readout multimeter.

**VR** Abbreviations for variable resistor and for voltage regulator on schematics, parts lists, and circuit boards; for example VR-2, VR305.

**VTP** Video tape player. Usually a video tape machine, such as VHS, that can play back pre-recorded tapes but has no tuner or recording capabilities.

**VTR** Video tape recorder. Usually used to describe a reel-to-reel-type machine used in commercial video studios and TV stations, but sometimes used to describe a video cassette recorder.

**VTVM** Vacuum tube voltmeter. Meter with a vacuum-tube amplifier for measuring AC and DC voltage, and also resistance, that has a very high input resistance, normally around 11 megohms. This high ohms-per-volt sensitivity places minimal load on the circuit being tested. FET multimeters are a solid state version of the VTVM, and also have a high input sensitivity. They have largely replaced the older VTVM in service work.

**VU** Volume unit. A VU meter on professional and semi-professional A/V equipment measures average audio intensity in dB.

**W** Abbreviation for watt.

**watt** Unit of electrical power.

**wavelength** The distance that a full wave of energy at some particular frequency occupies. For magnetic tape, the distance on tape required to record a full AC cycle, which is dependent on tape-to-head-gap speed and the frequency of the signal to be recorded. Greek letter $\lambda$ symbolizes wavelength.

**white clip** Describes circuit that limits the amplitude of the luminance signal for white, or highest intensity. This prevents possible smearing or blurring of the reproduced image caused by overshoot of the signal.

**wink** A slight amount of free play between gear teeth so that they don't press tightly against each other. Can apply to other moving mechanical parts that interact with each other.

**wiper** The movable contact in a potentiometer or rheostat that taps off different amounts of resistance as the variable resistor is operated.

**WV** Working voltage. A capacitor rating that specifies the highest voltage that should appear across the device for reliable and safe operation.

**WVDC** Working voltage DC. Essentially the same as WV.

**Xfer.** Transfer; a relay or switch contact that is operated to its alternate state. A normally closed (N.C.) contact becomes open when the contact transfers.

**Xfr.** See Xfer.

**Xfrmr.** Abbreviation for *transformer*.

**Xistor.** Abbreviation for *transistor*.

**Xmt.** Abbreviation for *transmit* or *transmitter*.

**Xmtr.** Abbreviation for *transmitter*.

**Xtal.** Abbreviation for crystal, such as a quartz crystal used in an oscillator.

**Xtl.** Same as Xtal.

**Y** Luminance portion of a color video signal. Conveys black-and-white intensity information; also includes horizontal and vertical sync signals and the 3.58MHz color burst for a color program.

**yoke** Magnetic coils situated around the neck of a CRT that deflect the electron beam(s) within the tube to form a raster on the screen. Also called *deflection yoke*; contains vertical and horizontal deflection coils.

**Z** Symbol for impedance in ohms in AC electrical formulas. Used in place of R in AC formulas.

**Zener diode** A diode that has a specific breakdown voltage at which it conducts in the reverse biased direction. Used as voltage references and voltage regulators. For example, a 6-volt Zener diode will not conduct in the reverse biased direction until the applied voltage reaches 6 volts. At this point—and for any higher applied voltage—the diode conducts heavily.

# INDEX

AC convenience outlet, 17
AC current leakage test, 503–8
ACE head, 64
Acetone, 141, 150
A/C head, 184–190
    and azimuth adjustment, 187–89, 197
    and height adjustment, 186–87, 197
    and tilt adjustment, 186, 197
    and tracking adjustment, 189–90, 197
AC line cord, 17
Advanced compatible television (ACTV), 456
AFM (audio frequency modulation), 459
Aligning tapes and making adjustments, 173–99
Alternating current (AC), 228–34
Ammeter, 210
Analog circuits, 234–35
Analog-to-digital (A/D) converter, 236–37, 456
Analog VOM, 242, 245, 261–63
Audio/control head assembly, 64–65
Audio In, 18
Audio Out, 19
Audio PCB, 46
Audio symptoms, 482–85
Automatic/Direct entry (digital), 13–14
Azimuth recording, 436–37

Back tension guide pole, 54–56, 175
Bandwidth, 11
Baseband signals, 15–16
    as audio, 15–16
Basic electronics, 200–240
Basket
    See Cassette tray
Belts, idlers, and pulleys, 159–64
Beta I, 4
Beta II, 4
Beta III, 4
Beyond economical repair (BER), 499
Blank search, 459
Bookmark (electronic), 459
Brake bands, 160–61
    and shoe band, 54
Brand names and manufacturers of VCRs, 462–64
Business aspects of VCR repairs, 491–509

Cable mouse, 458
Cables, 273
Cam motor, 46, 52
Cam position sensor, 122–24
Cams and gears, 161–64
Cam switch, 77–80, 122–24, 194–96
Capacitors, 306–15, D-2–4
Capstan and pinch roller, 65–67, 129–30, 156–59
    and capstan cleaning, 156
    and capstan oiling, 156–57
Capstan/cylinder motor drive board schematic diagram, 470–71
Capstan motor and flywheel, 46
Capstan servo timing chart, 477
Cassette basket assembly
    See Cassette loading, cassette loader assembly
Cassette-down switch
    See Cassette-in switch
Cassette-in switch, 71–72
Cassette loading, 8–9, 51, 84–85
    and cassette-insert switch, 8
    and cassette loader assembly, 85
    and end loading, 8, 109–16
    and load motor, 45
    problems with, 107–17
    and top loaders, 8, 107–9
Cassettes, 8
    inside parts of, 89–92
    reassembly, 96–97
    and recording times, 3–5
    reel locks of, 87–88
    repair of broken splice or leader, 92–96
    taking apart, 88–89
    and transferring videotape, 96
    underside of, 86
    See also Videocassettes
Casette schematic diagram, 469
Cassette tray, 9
C-clips, compression rings, o-rings, and slit washers, 161–64
Ch. 3/Ch. 4, 17
Channel memory button, 15
Channel presets, 13
Channel up/Channel down buttons, 14
Circuit, 202–4

Circuit boards, releasing, 38–44
Cleaner/Degreaser, 139–40
Cleaning, lubrication, and inspection, 144–49, 480–81
Cleaning swabs, 141–43
Closed captioning, 458
Cog belts, 159
Coils, 315–17, 320
Components and PCBS, 45–47
Composite video, 16
Compressed air, 141
Compression rings, 161
Connections, VCR basic, 19
Control track logic (CTL), 477
Converted subcarrier direct recording method, 435
Cords
   See Cables
Cover removal, 28–34
   and bottom covers, 32–34
   and front loaders, 32
   and screw removal, 29–30
   and top loaders, 30–31
CRT (cathode ray tube), 420, 421–29
Current, 206–9
   measurement of, 248–50
Customers, 491–93
   and beyond economical repair (BER), 499
   declining repair opportunity from, 494
   and preventing service returns, 500–503
   realistic expectations of, 497–98
   receiving a unit for repair from, 494–96
   and repair estimates, 493–94
   and unit replacements, 496–97
Cylinder
   See Video drum

DEW, 22, 80–82
Diagnostic tips, 480–82
Digital circuits, 234, 235–37, 405–10
Digital logic levels, 400–401
Digital logic probe, 410–12
Digital multimeters (DMM), 242, 256, 261–63
Digital signal processing (DSP), 456
Digital-to-analog (D/A) converter, 237, 456
Digital video systems, 456–57
Diodes, 288–97
Direct current (DC), 228–34
Dropout, 53
Drum entrance and exit roller guides P-2 and P-3, 58–61
Drum entrance roller guide, 58
Drum exit roller guide, 58
Drum, upper
   replacement of, 478

Eject, 7
Electrical adjustments, 472

Electrical connections
   cleaning of, 166–68
Electrical contact cleaner, 140
Electrical power, 216–24
Electrical shock, 24–27
   and secondary reactions, 26
Electrical units of measure, 206–9
Electron flow, 201–6
Electronic components, 271–324
Electronics, basic, 200–240
*Electronics Now,* 464
Electrostatic discharge (ESD), 27–28
Encoder disk, 74
EP, 4
Estimates for repairs, 493–94
Exit roller guide, 58–59
Expectations, realistic, 497–98
Eye injury, 27

FF/FWD search, 6, 124–27
Fields, 62
Fish paper, 39
Flying erase head, 454–55
Frame, 62
Frame advance, 7
Front loader, 45, 109–16
Front panel removal, 35–38
Full erase head, 56–57
Fuses, 274–75

Gear-type (GT) belts
   See Cog belts

Headphone jack, 38
Hertz, 11
Hi-fi stereo VCRS, 445–51
   failure symptoms, 488–89
Holdback tension, 53
Hookups and connections, 17–19
HQ (VHS high quality), 451–52

Idler, 53
Impedance, 231, C-2–3
Impedance roller
   See Inertia roller
Improved-definition television (IDTV), 456
Individual LED indicators, 38
Inertia roller, 57–58
Integrated circuits
   microprocessor, 401–5
   and pin identification, 398–400
Integrated display indicator panel, 38
IR photodetector, 38

Jog, 455–56

Lightweight grease, 141

Lightweight oil, 141
Line cord, 25
Loader and transport malfunctions, 104–32
Load motor, 52
LP, 4

Maintenance and common repairs, 133–72
Manufactuers' service manuals, use of, 462–79
    contents of, 465–74
    obtaining, 464–65
    supplementary, 474–78
Mechanical adjustments, 473–74
Microprocessors, 116–17, 398, 405–10
Microswitches, 37, 116
Mode cam gear and sensor, 46, 476
Models of VCRs, 462–64
Mode motor, 46, 52
Mode switch
    See Cam switch
Motors
    multi-phase AC, 384–86
    two-terminal DC, 379–84
Multimeter
    knowledge of, 264–68
    maintenance of, 263–64
    selecting, 260–63
    use of, 241–70
Multi-pin connectors, 38
Mute function, 481

Negative terminal, 202
Noise, 34
Normally closed (N.C.) switches, 72, 277
Normally open (N.O.) switches, 72, 277
Nylon lock nut, 179

Ohm's Law, 209–24, 237–38
Operational abilities, 2–5
Operations and controls, 1–20
Operator controls, 5–7
Optical sensors, 386–89
O-rings, 163
OTR, 7

P-0 guide
    See Back tension guide pole
P-1
    See Supply guide P-1
P-2
    See Drum entrance roller guide
P-3
    See Exit roller guide
P-4, 190
P-07, 14
P-guides, 10
Parallel circuits, 224–28

Parts
    list, 474
    replacement, 464–65
Pause/Still, 7
PC board pictorial, 472
PCBS
    See Components and PCBS
Picture-in-picture (PIP), 456–57, 460
Pinch roller, 140, 191–92
    cleaning of, 157–58
    removal and replacement of, 158–59, 475
    See also Capstan and pinch roller
Playback circuitry block diagram, 440–42
Polarity, 232–34
Poles
    drum entrance, 59
    drum exit, 59
    thrust, 59
*Popular Electronics,* 464
Positive terminal, 202
Potentiometers, 285–86
Power, 5–6
Power cord, 25
Power formulas, 218–24
Power supply, 46, 345–78
    always, unswitched, and ever output valtages, 355–60
    conventional power supply schematic, 365–69
    full-wave rectification, 347–51
    overview, 346–47
    rectifier and filter failures, 351–64
    schematic, 354–57
    switched or power-on outputs, 360–64
    switching mode, 369–75
Power transformer, 46
Pozidriv, 137

Rabbet, 129
Reactance, 231
Real-time counters, 459
REC, 7
Receiving a unit for repair, 494–96
Record circuitry block diagram, 437–40
Recording on VHS videotape, 430–43
Record interlock switch, 72–73
Record safety tab, 85
Rectifiers, 291, 347–64
Reel, 52, 54
Reel sensors
    supply, 73–75
    take-up reel, 73–75, 84
    tape end, 75–77
Reel table, 8, 50, 52
Relays, 278–81
Remote control, 389–95
Removing covers, 21–49

Repeat play between any two points on tape, 459
Replacement
 of units, 496–99
 *See also* Parts
Reset, 7
Resistance, 206–9, 211
 measurement of, 250–56
Resistors, 221, 281–88
 carbon composition, 203, 208, 283
 color code, 281–84
 fixed, 281
 variable, 285–86
 wire wound, 284–85
REW/REV search, 6, 124–27
RF tuner section, 46
Rheostats, 287
Roller guide posts, 180–84
Root mean square (RMS) voltage, 230–31
Rotary heads, 149–54
 assembly replacement and, 152–54
 cleaning of, 149–50
 replacing of, 150–52
Rubber cleaner/revitalizer, 140–41

Safety, 24–27, 243–44, 251, 315
 when soldering, 328, 329
 switching mode power supply, 372–73
Sams, Howard W. & Company, 465
Scan button, 14–15
Scanner
 *See* Video drum
Schematic symbols, 247–48
Scratch cassette, 71, 175–76
Self-cleaning, 458
Semiconductor, 204
Sensors
 cleaning of, 168–70
Series circuits, 224–28
Service manuals (manufacturers'), use of, 462–79
Service returns, prevention of, 500–503
Servo circuits, 412–15, 416–17
Shield plates, 34
Shuttering, 58
Shuttle, 455–56
Shuttle assembly, 58
Sinewave, 230
Skip, 459
Slide switches, 37
Slit washers, 163–64
Slow motion, 455–56
Solder, 325–44
 building a two-way tester, 336–41
 insulating connections, 333
 successful techniques of, 327–33
 and surface mount devices, 332
 tools and supplies, 326
 unsoldering, 333–36
 using the two-way tester, 341–43
Solenoid, 108
SP, 4
Speed, 7
Spindles, 8, 52–54, 192–94
 height, 194
 reel, 160–61
 take-up, 68–70, 84
Stiction, 81
Still frame (slow motion, jog and shuttle), 455–56
Stop, 7
Super VHS (S-VHS), 444, 452–54, 460
Supply guide P-1, 55, 176, 178–80
Supply spindle and tape holdback tension band, 52–54
Switches, 37, 116, 275–78
 cleaning of, 165–66
Switches, connectors, and sensors, 164–70
Synchronization of edits with more than one VCR, 459
Systematic approach to diagnosing problems, 105–7
System control microprocessor, 116–17
System control, servo, and video board, 45
 and system control microprocessor, 45

Take-up side guide pole, 190
Tape, 8
Tape guides, 10
 cleaning of, 148–49
Tape heads
 cleaner, 139
 demagnetizer, 143
 stationary 154–56
Tape loading, 9–10, 50–52, 85
 problems with, 117–24
Tape load motor, 46, 52
Tape load ring gears, 46
Tape motion problems, 124–30
 edge curling, 128
 fast forward/rewind, 124–27
 fluttering, 128
 height variations, 127–28
 observing tape around video drum, 128–29
 observing tape at capstan and pinch roller, 129–30
 playback, 127–29
 tension variations, 128
Tape path
 demagnetizing of, 147–48
Tape tension, 175–78
Tape transport, 8, 46, 52
 servicing of, 71
 switches and sensors of, 71–80
Tension roller
 *See* Inertia roller

Time counters, 459
Timer, 6
Timer/Operations microcomputer, 116–17
Tools and supplies, 133–44
   and hand tools, 134–38
   and supplies, 138–44
Tuner/Control cleaner, 140
Tracking, 7
Tracking control potentiometer, 38
Transformers, 317–20
Transistors, 297–305
TV pictures, creation of
   black and white, 420–24
   color, 424–27
TV tuner cleaner, 140
TV/VCR control, 6
Two-head helical scan format, 62
Timing belts
   *See* cog belts
Transport malfunctions, 104–32
Tuners, 12–15

UHF In/Out, 18
Undercarriage, 46

VCR Plus+, 457–58, 460
VHF In, 18
VHF Out, 18
VHF, UHF, and cable TV signals, 10–12
VHS cassettes, 144

VHS format, 437
VHS high quality (HQ), 451–52, 459–60
Videocassettes
   damage to outside, 97–8
   examination and repair, 84–103
   outside parts and features, 84–88
   tape damage, 98–100
   test jig, 71, 100–103
   *See also* Cassettes
Video drum, 46, 61–63, 128–29
Videodrum motor, 46
Video failure symptoms, 485–88
Video fields and frames, 422–24
Video head azimuth, 435–36
Video in, 18
Video out, 18
V-Mounts, 184
Voltage, 206–8, 230–31
   drop, 210
   forward/reverse, 293
   measurement of, 244–48
   root mean square (RMS), 230–31
Voltage regulator heatsink, 46
Volt-ohm milliammeter (VOM), 242

Watt, 217
Wire, 272–74
Wiring diagram, 468–71

Zener diode, 291–93

# 4. SCHEMATIC/CIRCUIT BOARD DIAGRAMS
## 4-1. FRAME WIRING

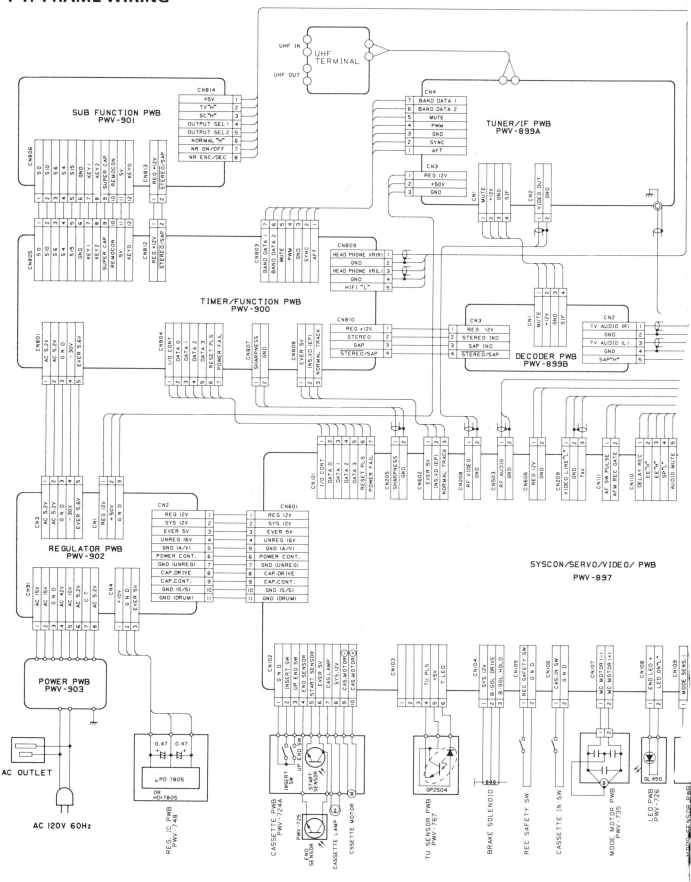

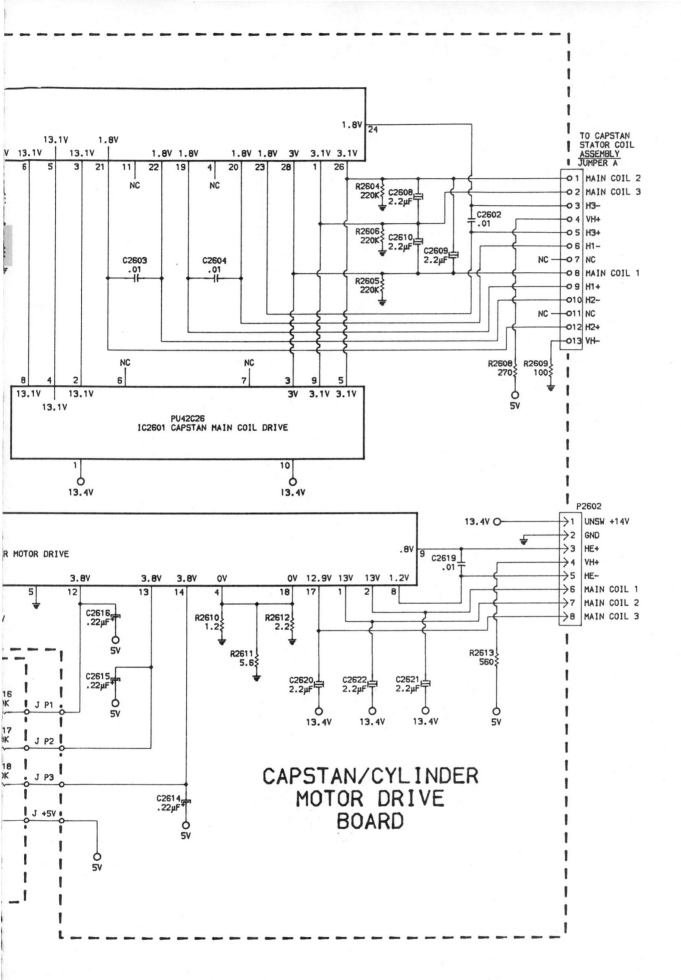

CAPSTAN/CYLINDER MOTOR DRIVE, SUBCAPSTAN/CYLINDER MOTOR DRIVE SCHEMATICS

## 2-6. TIMER/FUNCTION BLOCK DIAGRAM

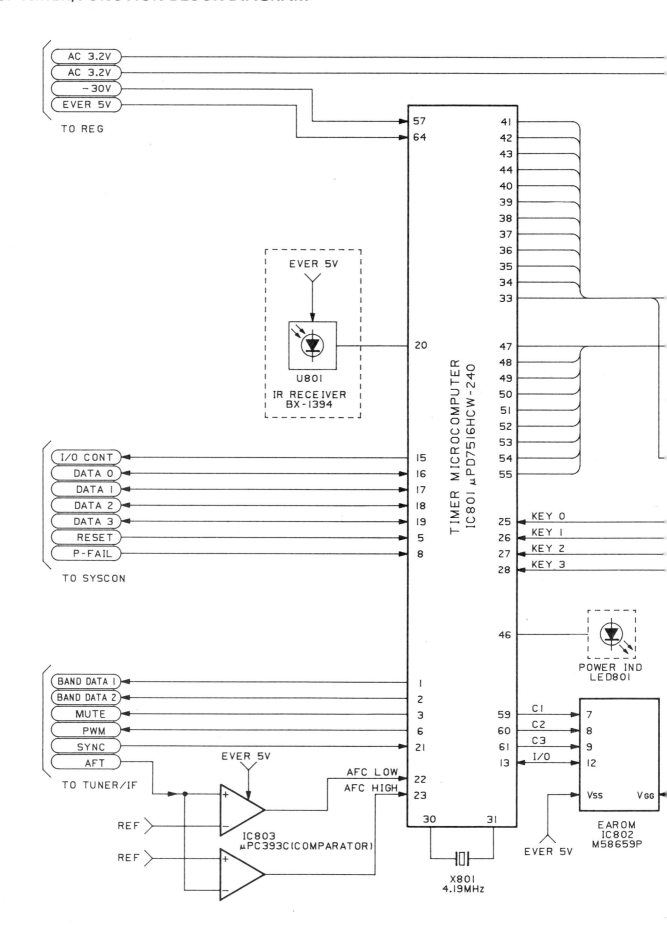

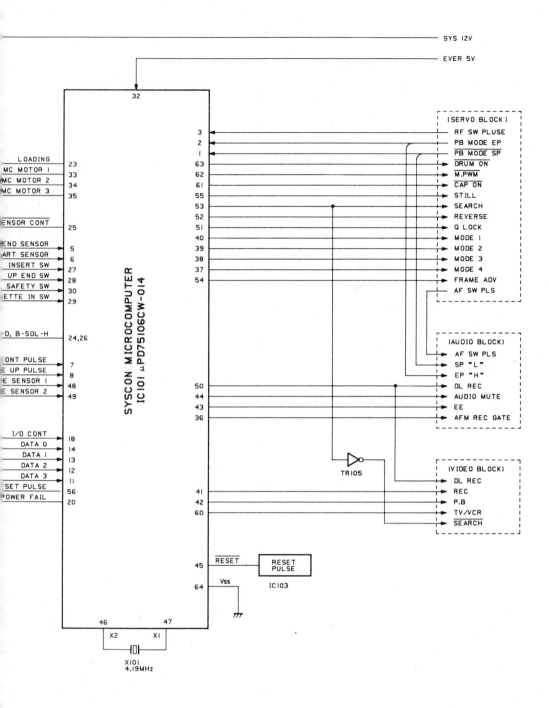

## 2. BLOCK DIAGRAMS
### 2-1. SYSTEM CONTROL BLOCK DIAGRAM

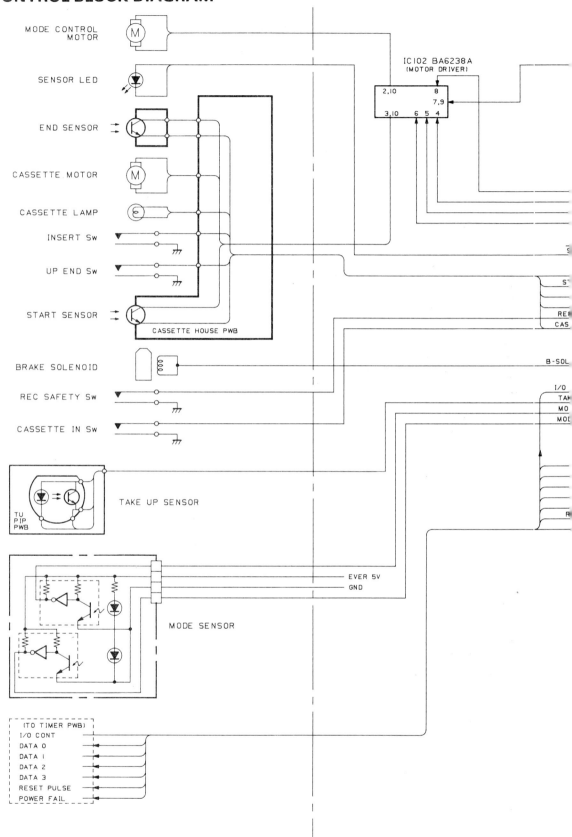

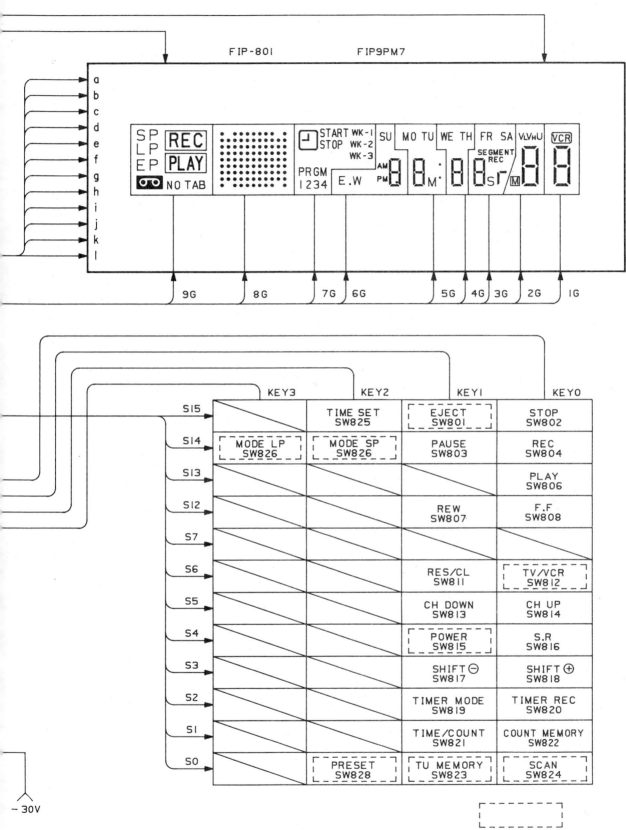

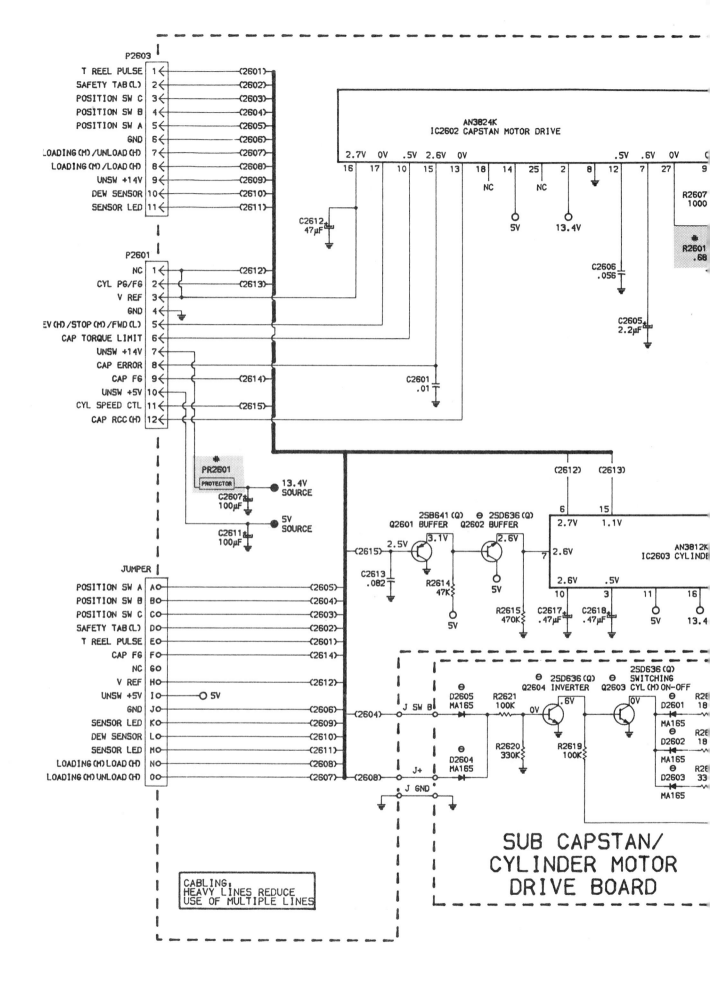

## 4-2. CASSETTE SCHEMATIC DIAGRAM

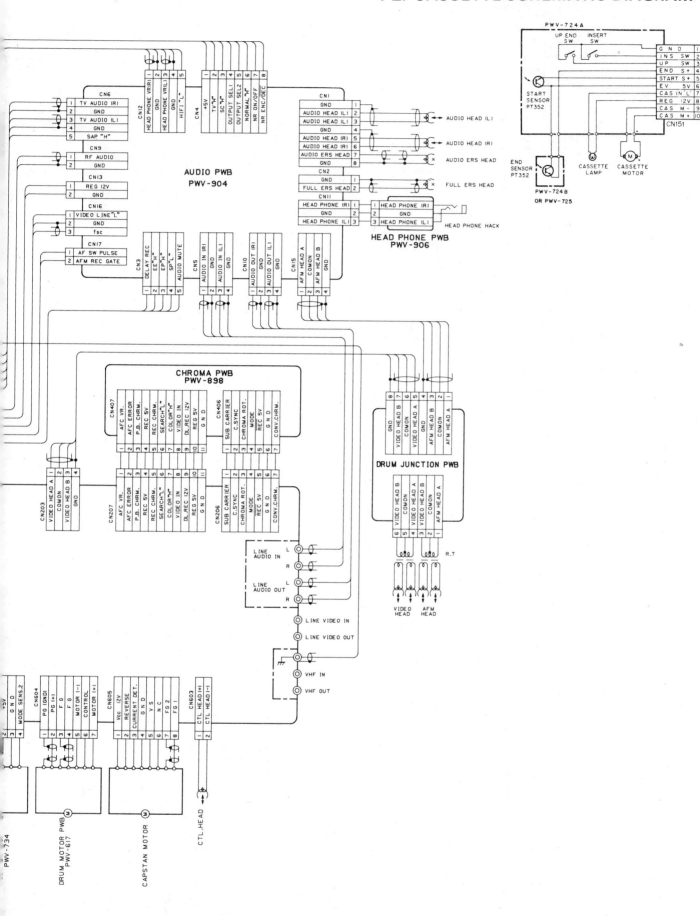

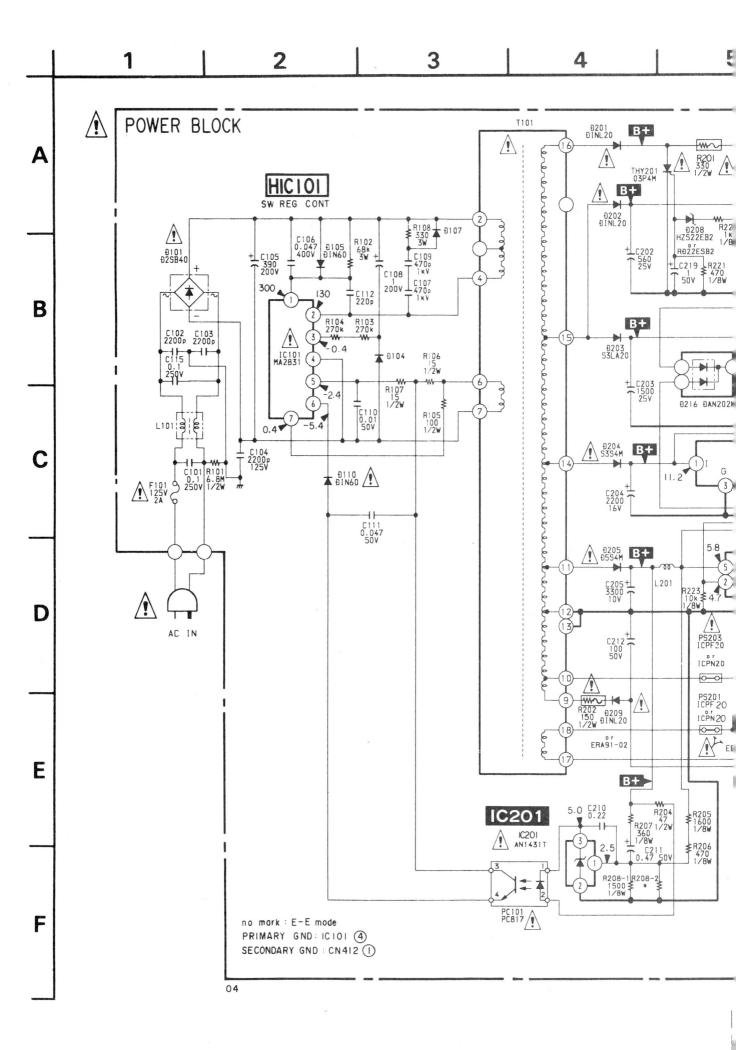

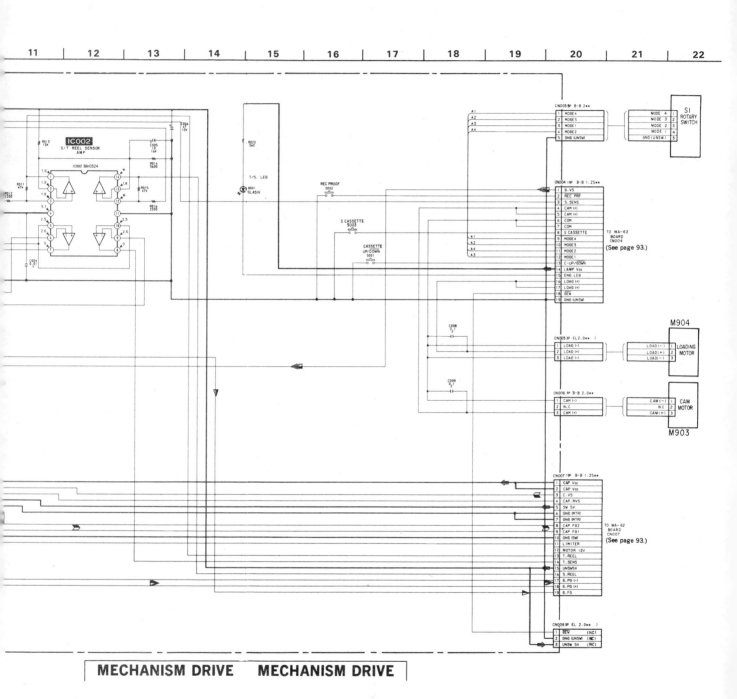

MECHANISM DRIVE    MECHANISM DRIVE

## 4-12. SYSCON SCHEMATIC DIAGRAM

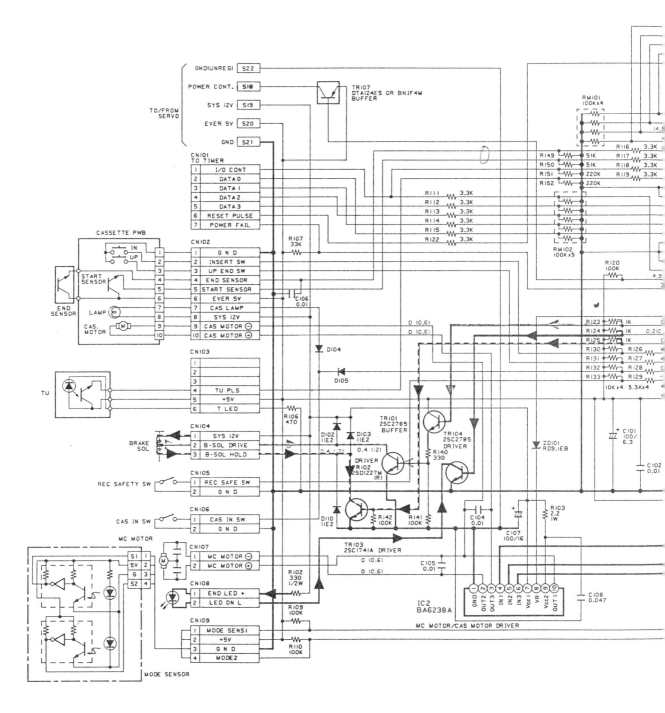

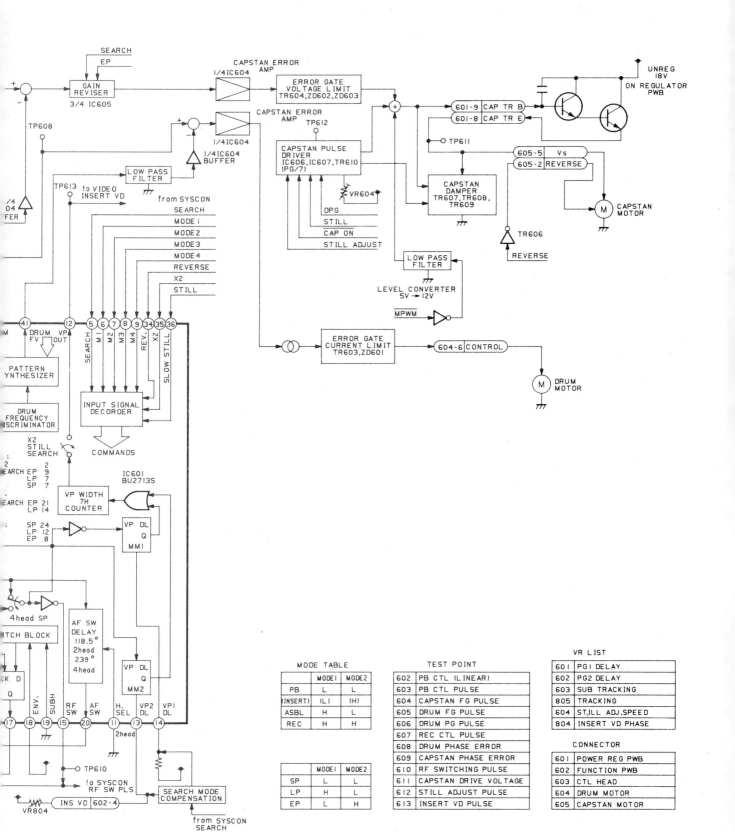

## 2-2. SERVO CONTROL BLOCK DIAGRAM

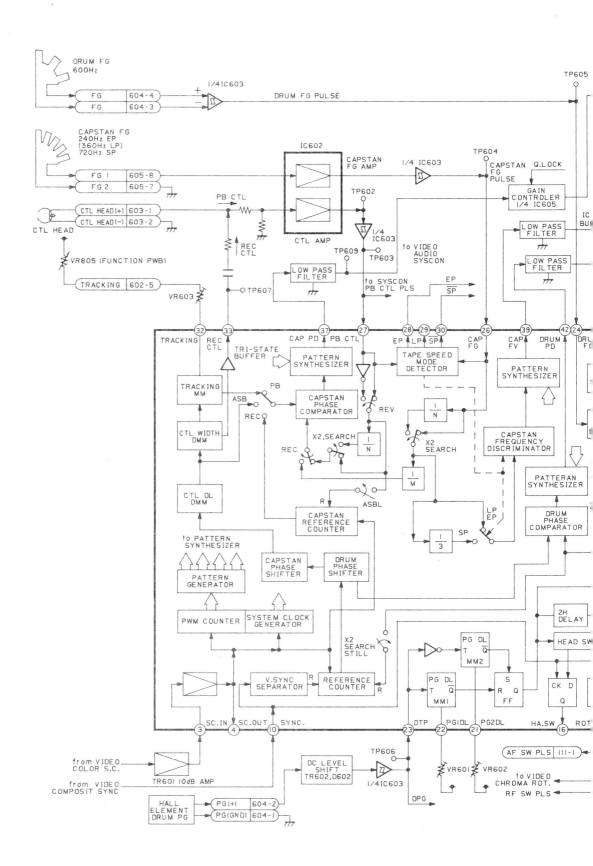

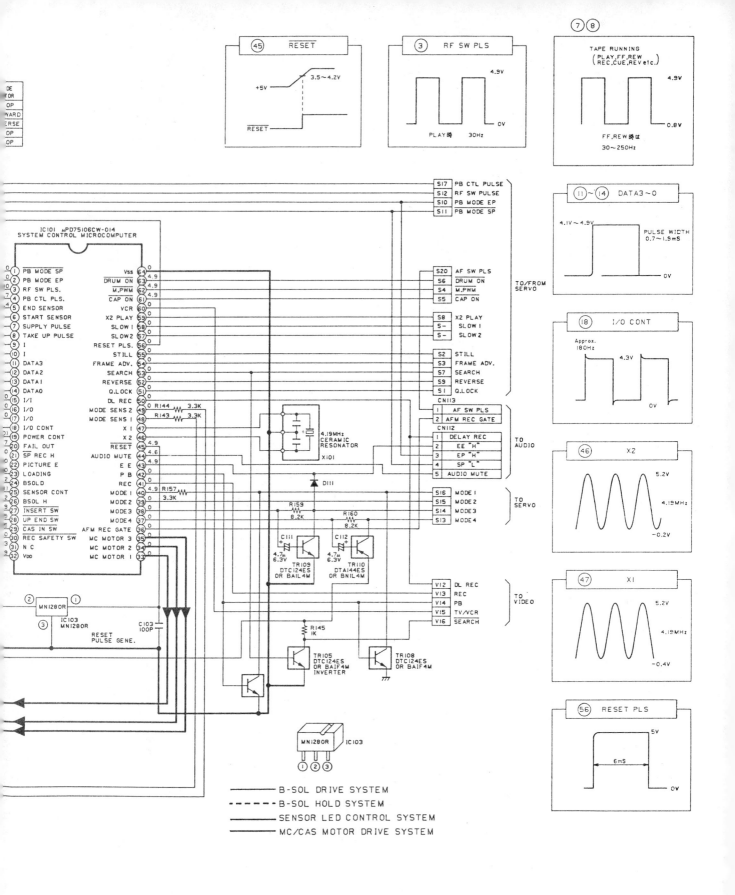

## MD-49 (MECHANISM DRIVE) SCHEMATIC DIAGRAM
—Ref. No. MD-49 BOARD: 7000 series—

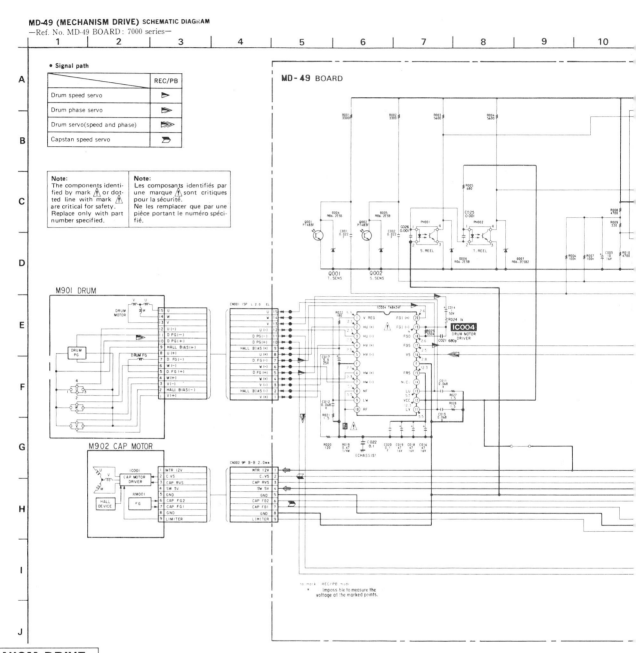

**MECHANISM DRIVE**

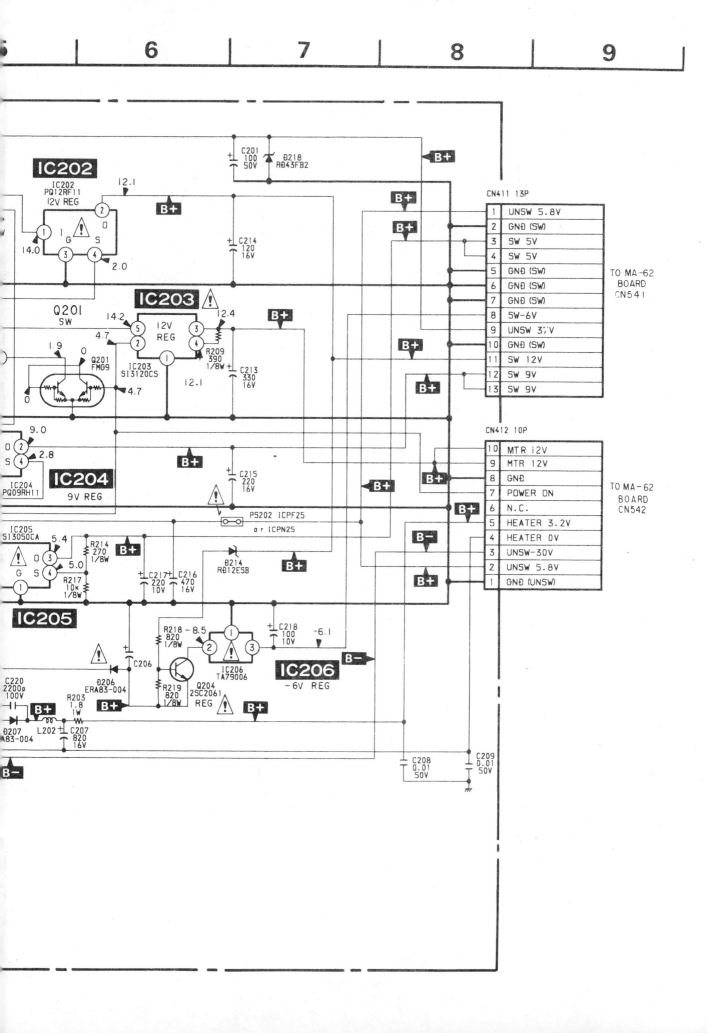